Advanced Nanomaterials for Water Engineering, Treatment, and Hydraulics

Tawfik A. Saleh
King Fahd University of Petroleum and Minerals, Saudi Arabia

A volume in the Advances in Environmental Engineering and Green Technologies (AEEGT) Book Series

IGI GLOBAL
DISSEMINATOR OF KNOWLEDGE
www.igi-global.com

Published in the United States of America by
IGI Global
Engineering Science Reference (an imprint of IGI Global)
701 E. Chocolate Avenue
Hershey PA 17033
Tel: 717-533-8845
Fax: 717-533-8661
E-mail: cust@igi-global.com
Web site: http://www.igi-global.com

Library of Congress Cataloging-in-Publication Data

Names: Saleh, Tawfik A., editor.
Title: Advanced nanomaterials for water engineering, treatment, and hydraulics / Tawfik A. Saleh, editor.
Description: Hershy, PA : Engineering Science Reference, 2017. | Includes bibliographical references.
Identifiers: LCCN 2016054446| ISBN 9781522521365 (hardcover) | ISBN 9781522521372 (ebook)
Subjects: LCSH: Water--Purification--Materials. | Nanostructured materials--Industrial applications.
Classification: LCC TD433 .A378 2017 | DDC 628.1/40284--dc23 LC record available at https://lccn.loc.gov/2016054446

This book is published in the IGI Global book series Advances in Environmental Engineering and Green Technologies (AEEGT) (ISSN: 2326-9162; eISSN: 2326-9170)

British Cataloguing in Publication Data
A Cataloguing in Publication record for this book is available from the British Library.

Advances in Environmental Engineering and Green Technologies (AEEGT) Book Series

ISSN:2326-9162
EISSN:2326-9170

Mission

Growing awareness and an increased focus on environmental issues such as climate change, energy use, and loss of non-renewable resources have brought about a greater need for research that provides potential solutions to these problems. Research in environmental science and engineering continues to play a vital role in uncovering new opportunities for a "green" future.

The **Advances in Environmental Engineering and Green Technologies (AEEGT)** book series is a mouthpiece for research in all aspects of environmental science, earth science, and green initiatives. This series supports the ongoing research in this field through publishing books that discuss topics within environmental engineering or that deal with the interdisciplinary field of green technologies.

Coverage

- Biofilters and Biofiltration
- Sustainable Communities
- Renewable Energy
- Cleantech
- Electric Vehicles
- Waste Management
- Air Quality
- Industrial Waste Management and Minimization
- Green Technology
- Policies Involving Green Technologies and Environmental Engineering

IGI Global is currently accepting manuscripts for publication within this series. To submit a proposal for a volume in this series, please contact our Acquisition Editors at Acquisitions@igi-global.com or visit: http://www.igi-global.com/publish/.

The Advances in Environmental Engineering and Green Technologies (AEEGT) Book Series (ISSN 2326-9162) is published by IGI Global, 701 E. Chocolate Avenue, Hershey, PA 17033-1240, USA, www.igi-global.com. This series is composed of titles available for purchase individually; each title is edited to be contextually exclusive from any other title within the series. For pricing and ordering information please visit http://www.igi-global.com/book-series/advances-environmental-engineering-green-technologies/73679. Postmaster: Send all address changes to above address.

Titles in this Series

For a list of additional titles in this series, please visit: www.igi-global.com

Applied Environmental Materials Science for Sustainability
Takaomi Kobayashi (Nagaoka University of Technology, Japan)
Information Science Reference • ©2017 • 416pp • H/C (ISBN: 9781522519713) • US $205.00

Environmental Issues Surrounding Human Overpopulation
Rajeev Pratap Singh (Banaras Hindu University, India) Anita Singh (University of Allahabad, India) and Vaibhav Srivastava (Banaras Hindu University, India)
Information Science Reference • ©2017 • 325pp • H/C (ISBN: 9781522516835) • US $200.00

Reconsidering the Impact of Climate Change on Global Water Supply, Use, and Management
Prakash Rao (Symbiosis International University, India) and Yogesh Patil (Symbiosis International University, India)
Information Science Reference • ©2017 • 430pp • H/C (ISBN: 9781522510468) • US $215.00

Environmental Sustainability and Climate Change Adaptation Strategies
Wayne Ganpat (The University of the West Indies, Trinidad and Tobago) and Wendy-Ann Isaac (The University of the West Indies, Trinidad and Tobago)
Information Science Reference • ©2017 • 406pp • H/C (ISBN: 9781522516071) • US $200.00

Technologies for the Treatment and Recovery of Nutrients from Industrial Wastewater
Ángeles Val del Río (University of Santiago de Compostela, Spain) José Luis Campos Gómez (University Adolfo Ibáñez, Chile) and Anuska Mosquera Corral (University of Santiago de Compostela, Spain)
Information Science Reference • ©2017 • 391pp • H/C (ISBN: 9781522510376) • US $200.00

Agricultural Development and Food Security in Developing Nations
Wayne G. Ganpat (The University of the West Indies, Trinidad and Tobago) Ronald Dyer (Grenoble School of Management, France) and Wendy-Ann P. Isaac (The University of the West Indies, Trinidad and Tobago)
Information Science Reference • ©2017 • 370pp • H/C (ISBN: 9781522509424) • US $200.00

Oceanographic and Marine Cross-Domain Data Management for Sustainable Development
Paolo Diviacco (Istituto Nazionale di Oceanografia e di Geofisica Sperimentale (OGS), Italy) Adam Leadbetter (Marine Institute, Ireland) and Helen Glaves (British Geological Survey, UK)
Information Science Reference • ©2017 • 425pp • H/C (ISBN: 9781522507000) • US $205.00

IGI GLOBAL
DISSEMINATOR OF KNOWLEDGE

www.igi-global.com

701 E. Chocolate Ave., Hershey, PA 17033
Order online at www.igi-global.com or call 717-533-8845 x100
To place a standing order for titles released in this series,
contact: cust@igi-global.com
Mon-Fri 8:00 am - 5:00 pm (est) or fax 24 hours a day 717-533-8661

Table of Contents

Detailed Table of Contents

There has been a growing interest in the investigation of the possibility of nanomaterials utilization as efficient materials for water purification. The efficiency of the techniques used for water treatment depends mostly on the efficiency of the material involved in the process. The objective of this chapter is to introduce an overview of the nanomaterials classification and general schemes of their synthesis. In addition, the chapter highlights the techniques used for the characterization of the nanomaterials and nanocomposites. In addition, technologies used for water purification where the nanomaterials play a role are discussed.

Adsorption techniques are widely used for the removal of various classes of pollutants from water due to their mild and facile operating conditions. The operations involved in the adsorption techniques are environmentally friendly, economical, highly selective on pollutants, highly efficient and easily operative. However, the adsorption

The development of cost-effective, efficient and stable materials helps to provide the affordable solutions to get safe and fresh water to increasing population with health guidelines of emerging contaminants. Nanomaterials (NMs)-based techniques involve the design, synthesis, manipulation, characterization and exploitation of materials for adsorption and separation of target species from the contaminated and waste water. NMs show better adsorption capacity and catalytic for number chemical species and microbes because of their small size and large surface area that favors the purification and treatment of waste or contaminated environmental water. Here, we present the chemical properties, adsorption/removal mechanism and applications of advanced NMs such as magnetic nanoparticles (MNPs), carbon nanotubes (CNTs), graphene and graphene oxide (GO), titanium oxide (TiO2), silica (SiO2), silver (Ag), gold (Au) NPs and zeolites in effective and efficient removal of toxic metal ions, organic and inorganic chemical substances and disease-causing microbes from contaminated and wastewater.

The focus is to highlight the catalytic technologies to convert lignocellulosic biomass into the activated carbon (AC) which can be used in photocatalysis applications. The drawback of carbon production raised by energy assumption and product selectivity has encouraged the development of sustainable carbon synthesis process, where the catalytic approach is considered. This treatment via either homogenous or heterogeneous catalytic approach relative in mild condition provided a bulk, mesoporous and nanostructure AC materials. Those characteristics of AC materials are basic requirements for the efficient photocatalytic system. Due to the excellent oxidizing behavior and stability, semiconductor materials have been widely used in the photocatalytic system. However, they lead to some drawbacks in terms of the separation steps and loss of the photocatalyst. So, attention has been paid to supported semiconductor catalysts which carbon materials were explored. AC reported as a potential support in photocatalytic systems.

The increasing awareness of the environment has raised so much concerns in the way we live and the manner to which we dispose our waste. The rapid growth in polymer production has resulted in increasing concerns about the consumption of nonrenewable resources and the environmental impacts associated with its production and disposal. Polymer waste is one of the major components in municipal solid waste and is increasingly becoming a huge burden in industrialized nations. The rise in the use of plastic, coupled with increasing concerns about its disposal, has led to a renewed interest in its recycling and recovery. The technologies involved in the recovery and recycling of polymers are rapidly growing, however, there is no specific pattern of treating polymer waste. The technology depends on the type of material used in the production and consumption pattern. This chapter, therefore, discusses the patterns of polymer consumption, the environmental concerns, and different modes of recycling polymer waste.

Chitosan is a naturally occurring biopolymer originating from several microbial species as well as crustacean species, such as shrimp and lobster. Chitosan has excellent physical and chemical properties that allow its use in various environmental applications especially in water treatment. It is a biodegradable polymer, and it is inexpensive providing an environmentally friendly and economic option for water and wastewater treatment. Chitosan offers a myriad of applications through chemical coagulation and flocculation, antimicrobial properties, adsorption capabilities, and nanofiltration and can provide a sustainable route for water and wastewater treatment. This book chapter elaborates the recent developments in chitosan applications in water and wastewater treatment.

Hexavalent chromium Cr(VI) is toxic to living systems and must be removed from wastewater. Chitosan is a cationic, biocompatible, biodegradable, biopolymer obtained from marine wastes. The performance of chitosan particles (CH) and chitosan nanoparticles (CHN) to remove Cr(VI) from aqueous solutions is discussed in the present chapter. CHN were obtained by reticulation with tripolyphosphate (TPP), and physico-chemically characterized. The performance of CHN decreased at higher pH due to the cross-linking process with TPP. Langmuir isotherm described the equilibrium adsorption values and pseudo-second order rate provided the best fitting to the kinetic data. Chemical analysis to determine the oxidation state of the adsorbed Cr, showed that Cr(VI) was adsorbed on CH particles without further reduction; in contrast Cr(VI) removed from the solution was reduced and bound to the CHN as Cr(III). Chitosan crosslinking was essential to adsorb Cr(VI) at $pH<3$ due to the dissolution of CH in acid media.

Development of new technologies, progressive urbanization, increasing consumerism and industrial boom in developing countries has led to elevated pollution of the environment. The spectrum of pollutants produced and released to the environment has increased in the last few decades including the agricultural, industrial, pharmaceutical and plastic industries. In the developed and underdeveloped countries where environmental pollution goes on increasing day by day, the concern of mankind to the threat of humanity increases which comes from anthropogenic degradation of the environment. An analytical chemist is always in search of cheaper, quicker, more sensitive, more reliable, precise methods of analysis. To achieve such a goal many properties of the materials are studied. Nanotechnology meets many of the conditions mentioned above and is very economic. So, analytical nanotechnology is an important tool for preconcentration and separation of pollutants at low levels.

Preface

Nanotechnology has been developed as a consequence of significant advances in the materials science community. In the excitement surrounding these materials and technologies, however, their potential has been, frequently overhyped. The book explores these kinds of materials and forward-looking potential applications for water engineering. The book chapters are organized in a good way and make extensive use of illustrations.

The first chapter introduces an overview of the water technologies and nanomaterials for water purification. From chapter two to chapter five, the aim is to provide the readers with information about the advanced nanomaterials and their methods of preparation with more focus on green methods such as microwave-assisted methods for the synthesis of nanomaterials and fundamentals and sources of magnetic nanoparticles. Biomaterials are good sources for the preparation of nanomaterials and their use as a support as well. Thus, chapters six and seven introduce scientific insights into modified and non-modified biomaterials and the production of activated carbon for pollutants removal. Next chapter will introduce cost-effective sources and ideas for utilizing waste polymers either as adsorbents or as sources for the production of adsorbents.

Sources such as chitosan have been proven to be potential efficient adsorbents which will be discussed in chapter ten. The next chapter will focus more providing examples on the use of chitosan micro/nanoparticles to remove hexavalent chromium from waters. The last chapter will highlight the applications of nanomaterials for water treatment with types of future research suggestion in the field.

We welcome suggestions from readers towards improvements that can be incorporated in future editions of this book.

We hope that you enjoy the book!

Tawfik A. Saleh
King Fahd University of Petroleum and Minerals, Saudi Arabia

Acknowledgment

I would like to acknowledge the support provided by King Fahd University of Petroleum & Minerals. I would like to express my gratitude to the many people who saw me through this book; to all those who provided support, read, wrote, offered comments, allowed me to quote their remarks and assisted in the editing, proofreading and design. Above all, I want to thank my parents, relatives, wife, children and the rest of my family, who supported and encouraged me in spite of all the time it took me away from them. Last and not least: I beg forgiveness of all those who have been with me over the course of the months and whose names I have failed to mention.

Many thanks.

Tawfik A. Saleh
King Fahd University of Petroleum and Minerals, Saudi Arabia

Chapter 1
An Overview of Nanomaterials for Water Technology

Tawfik A. Saleh
King Fahd University of Petroleum and Minerals, Saudi Arabia

ABSTRACT

There has been a growing interest in the investigation of the possibility of nanomaterials utilization as efficient materials for water purification. The efficiency of the techniques used for water treatment depends mostly on the efficiency of the material involved in the process. The objective of this chapter is to introduce an overview of the nanomaterials classification and general schemes of their synthesis. In addition, the chapter highlights the techniques used for the characterization of the nanomaterials and nanocomposites. In addition, technologies used for water purification where the nanomaterials play a role are discussed.

1. INTRODUCTION

The water point and non-point polluting sources include industries, domestic and agricultural activities and other environmental and global changes. The need of water quality improvement and preservation is growing continuously. Thus, attempts to develop various water treatment and recycling technologies. The efficiency of most of these technologies depends on materials design. In the following two sections, the water technologies for water treatment and nanotechnology for material design will be introduced.

DOI: 10.4018/978-1-5225-2136-5.ch001

2. WATER TECHNOLOGIES

There are several water treatment technologies used for

1. Pollutant source reduction,
2. Wastewater treatment and
3. Water recycling.

The water technologies include primary, secondary, and tertiary unit operations. Primary treatment deals with physical and chemical purification processes while the secondary treatment deals with the biological treatment. Tertiary treatment converts wastewater into a good quality water. Figure 1 illustrates the wastewater treatment units. These techniques involve the utilization of efficient materials. Among these, as examples are separation, filtration, extraction, oxidation, photodegradation, ion exchange, reverse osmosis, and adsorption.

Figure 1. Classification of Water recycling technologies

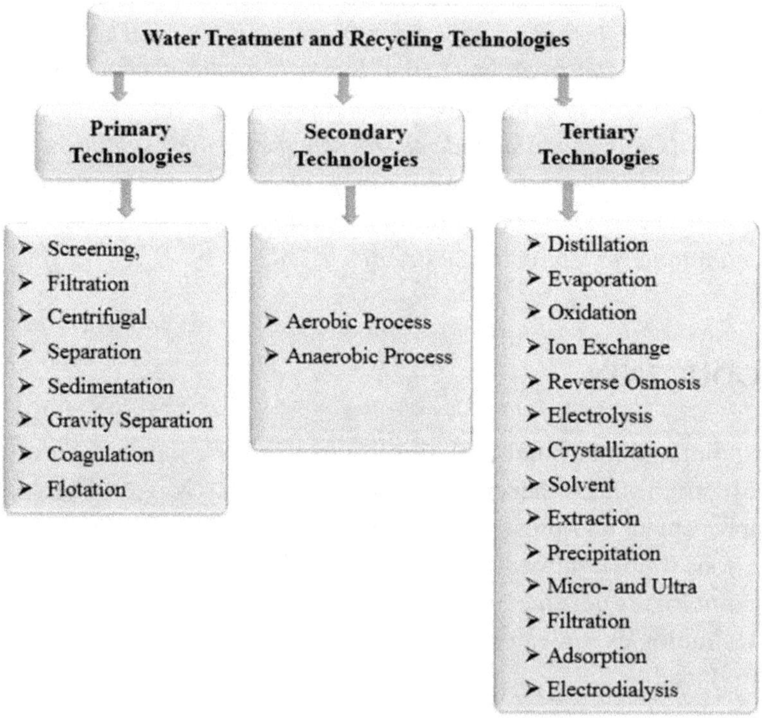

Figure 2. A general scheme of the most available water treatment methods

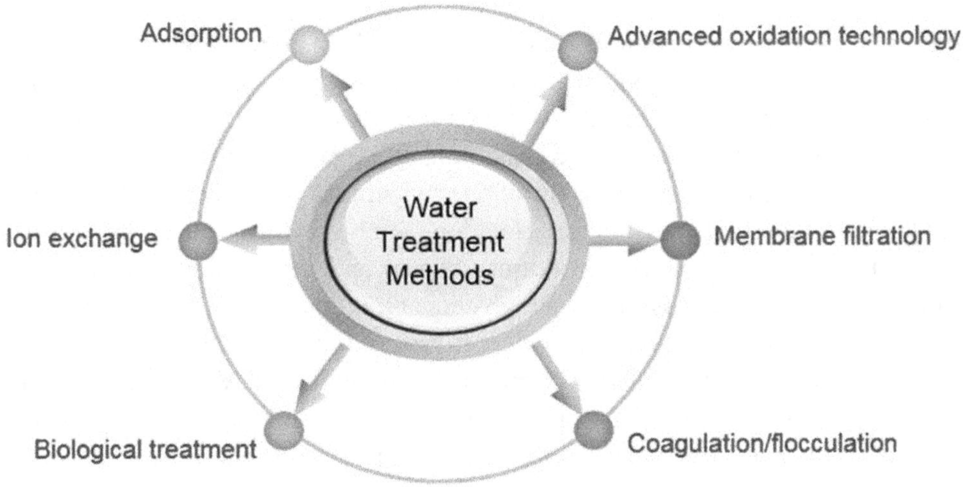

2.1. Selection of Water Treatment Technologies

The type of the required wastewater treatment determines the selection of the treatment technologies. The water treatment methods can be summarized as shown in Figure 2.

The highly polluted water having color, solid wastes are first treated with primary and secondary processes followed by the tertiary water treatment technologies. If the biological oxygen demand is negligible then the secondary process is not required. If water is colorless without any solid and polluted due to inorganic, organic and biological pollutants, then only tertiary water treatment is required. In general, groundwater is polluted by toxic metal ions and anions and only tertiary water technologies are required for its treatment. Surface water is being contaminated due to inorganic, organic and biological pollutants require secondary and tertiary treatment methods. Wastewater is highly polluted and it may be colored with solid wastes containing inorganic, organic, biological pollutants, which requires a good hyphenation of primary, secondary and tertiary treatment technologies. The choice of the tertiary water treatment technologies depends on the types of the pollutants present in water and the optimum selection can be done by considering the required quality of water, type of pollutants and their sources.

2.2. Wastewater Management

Wastewater is generated in industry, water purification and recycling technologies. Therefore, the waste management is very important to the environmental and health point of views because, generally, waste products are toxic. The waste generated may enter into water resources and contaminant it again. A large amount of solid waste is generated in screening, filtration, centrifugal separation, sedimentation and gravity separation, coagulation, flotation, aerobic process, anaerobic process, evaporation and precipitation processes. Micro- and ultra-filtration and reverse osmosis produce water containing a high concentration of pollutants. On the other hand, some gasses are produced in electrolysis and anaerobic processes that are dangerous to the environment. In adsorption, the production of exhaustive adsorbents is quite high. Many methods have been developed and used for the management of waste products generated during water treatment processes. Most important waste management methods are recycling of waste products into useful products such as fertilizers, fillers, building material etc. Some toxic waste products used to be burnt. Suggestions have also been given to dig the waste products underground in airtight iron or plastic containers. Nowadays, fly ash, red mud, sand etc. are being used as the useful and economic adsorbents for the removal of pollutants from water. The large quantities of these exhausted adsorbents produced have been used as fillers and the building materials successfully.

3. NANOTECHNOLOGIES

3.1. Overview of Nanomaterials

The "nano" is used as the prefix for the billionth from the range 9 to 10. Particles which have two or more dimensions in the size range as 1 to 100 nm are defined as nanoparticles. Nanoparticles have unique chemical, mechanical and physical properties as compared to their solid bulk materials because of their high surface area and electronic properties. Nanoparticles have been utilized in many applications, for example, water purification, oil separation, electrochemistry, photochemical, and biomedicine. Nanomaterials can be in:

- Zero dimension (0D); for example Q-dot, Powder, carbon nanosphere
- One dimension (1D); for example wire/nanotube, carbon nanofiber, carbon nanotube, titania nanorods, core-shell
- Two dimensions (2D); for example, thin films, multilayers, graphene, flower-like nanosheets,
- Three dimensions (3D); crumpled graphene, hybrid nanostructures

Carbon-based materials and nanomaterials are among the most common materials used in water purification. Figure 3 depicts different examples of carbon structures. Graphene, as an example of this class, is a single carbon layer of the graphite structure. It is a two-dimensional carbon allotrope made up of a conjugated system of sp^2 carbons arranged in a honeycomb structure. It is similar in nature to polycyclic aromatic hydrocarbons of quasi-infinite size. Figure 4 presents a typical TEM image of a graphene nanosheet as an example of 2D nanostructured materials, while a SEM image as an example of the nanocomposite of magnetic iron oxide-loaded carbon nanotubes is shown in Figure 5. Some carbons structures can be used to form 3D nanostructured materials. For examples, carbon nanotubes are used as a spacer between graphene nanosheets, Figure 6.

3.2. Classification of Nanomaterials

The shape and structures of the nanomaterials play a major role in their efficiency. Figure 7 introduces examples of various shapes of nanomaterials (Hummers and

Figure 3. different examples of carbon structures

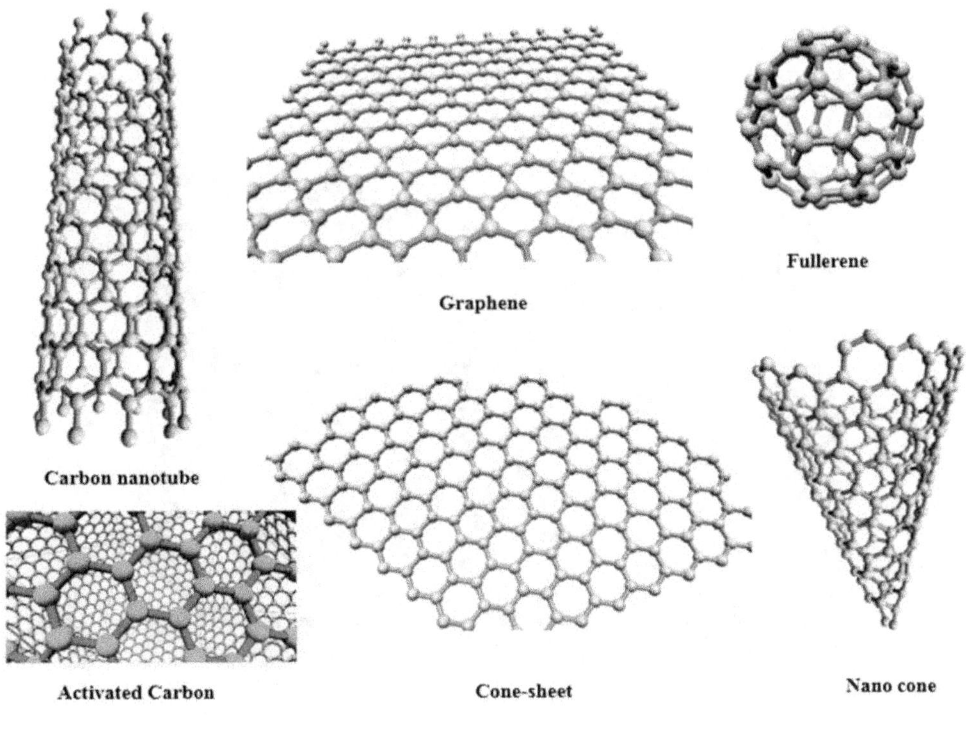

Figure 4. Typical TEM image of a graphene nanosheet

Figure 5. SEM image of a nanocomposite of magnetic iron oxide-loaded carbon nanotubes

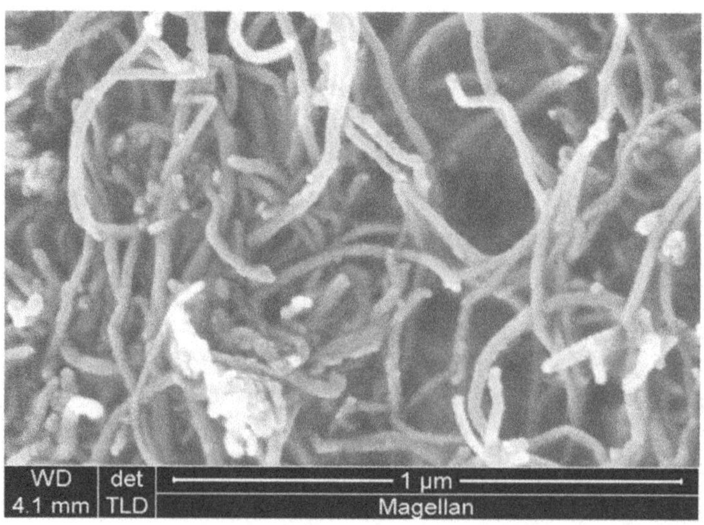

Figure 6. Carbon nanotubes used as a spacer between graphene nanosheets

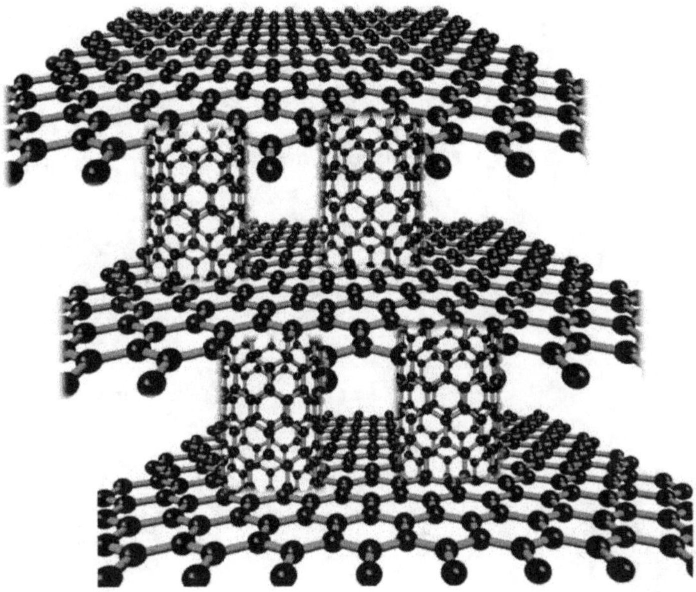

Offeman 1958; Saleh et al. 2016a,b). Nanomaterials can be categorized based on their chemical composition as:

- Carbon-based nanomaterials, which consist of carbon with different structures, including nanotubes, fullerenes, graphene, and a nanosphere.
- Metal based materials where the main components of these materials are metals. Examples are silver, gold and copper nanoparticles.
- Metal oxide nanomaterials are materials consists of metal and oxygen, like titania, silica, and alumina.
- Dendrimers, which are materials with highly branched macromolecules with dimensions in the nanoscale. The surface of dendrimers have numerous chains which can be modified toward specific functions. Another type is called nanocomposites, which consists of more than one type of nanomaterials.
- Nanocomposites that are described as multiphase materials where at least one of the phases has at least one dimension in nanoscale. They are special classes of materials formed by combining two phases of different materials and they tend to retain the properties of each material used in their formation. The properties of nanocomposites are a function of the properties of the constituent phases, their relative amounts, and the geometry of the dispersed phase, the shape of the nanoparticles, size, distribution and orientation. A matrix, filler, and resin, and a reinforcement, like fibers, are required for the formation of a composite.

Figure 7. Illustration of various shapes of nanomaterials

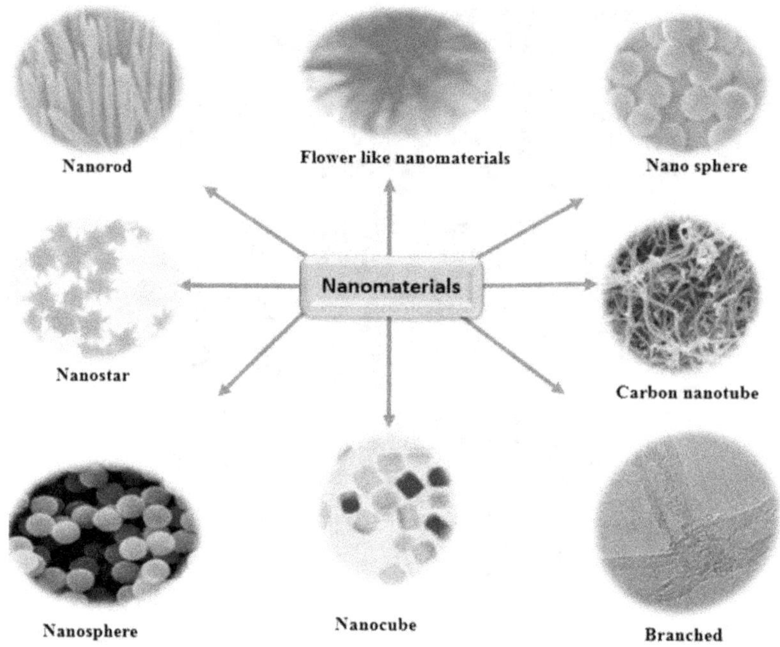

3.3. Synthesis of Nanomaterials

The synthesis routes of nanomaterials can be classified into two different approaches; the bottom-up approach and the top-down approach. In the bottom-up approach, the nanoparticles are first obtained at the atomic level and later integrated into the desired material. Examples include the formation of nanoparticles from the colloidal dispersion and the formation of powders from the sol-gel method followed by integration. The top-down approach starts with a bulk material at the macroscopic level followed by trimming of the material to the desired nanoparticles. Examples include etching and ball milling. These methods can be classified into chemical, physical and biological methods, as shown in Figure 8. Nanomaterials can be prepared by various methods including intercalation, the sol-gel method, the molecular composite formation method, high energy sonication, the nanofiller direct dispersion method, hydrothermal synthesis, the polymerized complex method, surfactant assisted processing, solution-evaporation methods, electrochemical synthesis of polymers, chemical vapor deposition, microwave synthesis, and the ball milling process. More details about these methods can be found in the references (Huang et al. 2012; Iijima 1991; Saleh and Gupta 2016)

Figure 8. General classification of the methods for the synthesis of nanomaterials

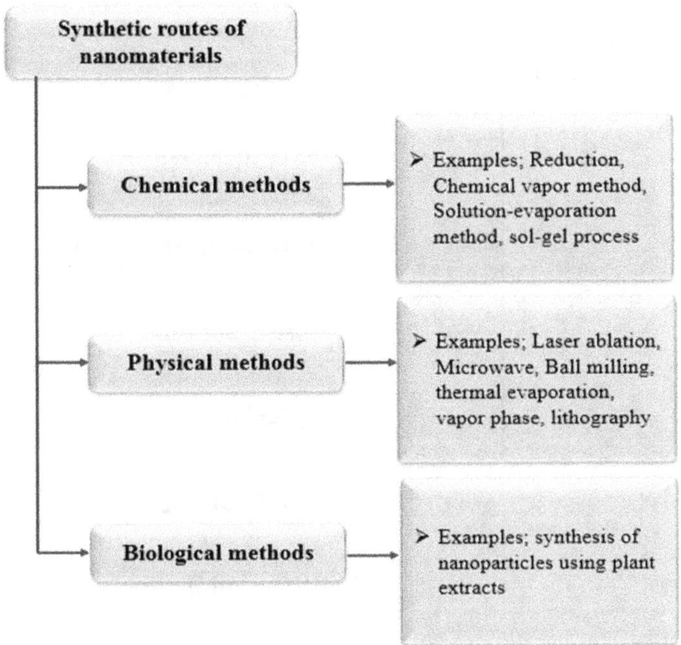

3.4. Main Techniques Used for the Preparation Carbons

As shown in Figure 9, there are various techniques have been developed for preparing carbon nanomaterials with different structures, including fullerene, bucky-balls, nano-cone, carbon nanotube (CNT), capped carbon nanotube, tube bundle, graphene sheet and cone-sheet. The buckyball is the colloquial term used for buckminster fullerene, the first molecule discovered in the fullerene family of carbon allotropes. The term fullerene is used to indicate spherical fullerenes, which are also called buckyballs; they resemble the balls used in football (soccer). Cylindrical ones are called carbon nanotubes or buckytubes (Ito et al. 2014; Saleh 2016; Saleh and Gupta 2012; Saleh et al. 2017; Saleh and Gupta 2014).

3.5. Characterization of Nanomaterials

The characterization of nanomaterials can be conducted using different techniques including X-Ray Diffraction (XRD), field emission scanning electron microscope (FESEM), high-resolution transmission electron microscope (HRTEM), Energy-dispersive X-ray spectroscopy (EDX), X-ray photoelectron spectroscopy (XPS).and infrared transform absorption spectrometer (FTIR); to explain the morphology of

Figure 9. Main techniques used in preparation of nanostructured carbons

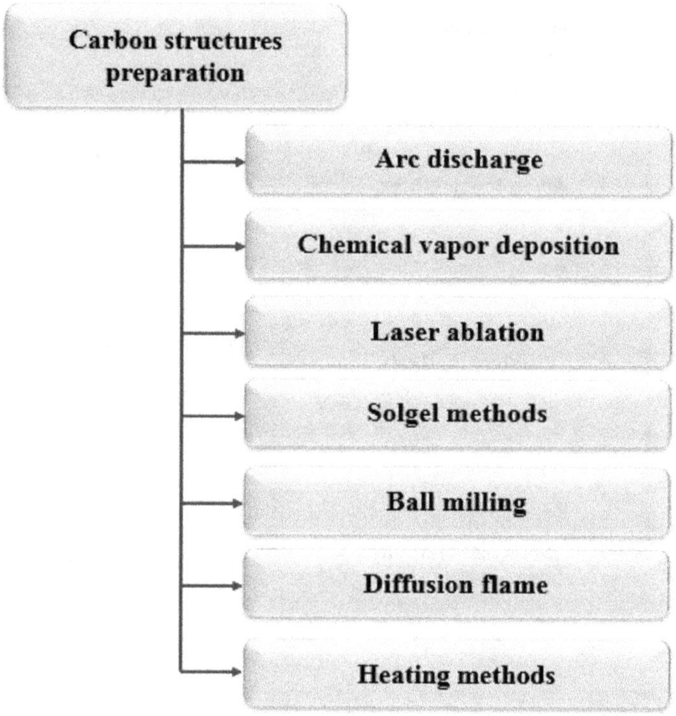

the composites, crystalline phases, and average size. The morphology of the composites can be characterized by SEM, AFM, and TEM. The presence of functional groups on the surface of nanotubes can be carried out by FTIR, Raman, XPS and EDX. The thermal stability cab is studied by thermogravimeter (TGA) (Kosynkin et al. 2009; Li et al. 2011; Park and Ruoff 2009).

The surface area, porosity, pore size and pore distributions can be evaluated by physisorption and chemisorption analyzers. Temperature programmed desorption (TPD) analyses determine the number, type, and strength of active sites available on the surface of a nanomaterial by measuring the amount of gas desorbed at various temperatures. If multiple active materials are present, the gas desorbs at different temperatures. Temperature programmed reduction (TPR) determines the number of reducible metal species present in the nanomaterial and the temperature at which reduction occurs. This can be performed by measuring the amount of analysis gas (like hydrogen) that reacts with the catalyst at different temperatures. Temperature programmed oxidation (TPO) provides the extent to which the nanomaterials can reoxidize.

4. CONCLUSION

In this chapter, a brief overview of the nanomaterials, their major synthesis routes and shapes were introduced. There are several methods for the synthesis of nanomaterials including chemical, mechanical and physical methods. Many kinds of nanoparticles, such as metals, metal oxides, carbon structures (graphene, nanotubes, and nanosphere) and nanocomposites could be used in water treatment.

Several examples and more details about the principles, fundamentals of nanomaterials will be discussed in this book in the following chapters. In addition, the applications of nanoscale composite materials as adsorbents in different modes, including batch and fixed bed will be introduced in many chapters with some cases. The applications of the composites in catalysis and photocatalytic degradation are also discussed.

ACKNOWLEDGMENT

The author acknowledges the support of the chemistry department, and King Fahd University of Petroleum and Minerals, (KFUPM) Dhahran, Saudi Arabia.

REFERENCES

Huang, X., Qi, X., Boey, F., & Zhang, H. (2012). Graphene-based composites. *Chemical Society Reviews*, *41*(2), 666–686. doi:10.1039/C1CS15078B PMID:21796314

Hummers, W. Jr, & Offeman, R. (1958). Preparation of graphitic oxide. *Journal of the American Chemical Society*, *80*(6), 1339. doi:10.1021/ja01539a017

Iijima, S. (1991). Helical microtubules of graphitic carbon. *Nature*, *354*(6348), 56–58. doi:10.1038/354056a0

Ito, A., Sato, T., & Goto, T. (2014). Transparent anatase and rutile TiO2 films grown by laser chemical vapor deposition. *Thin Solid Films*, *551*, 37–41. doi:10.1016/j.tsf.2013.11.089

Kosynkin, D. V., Higginbotham, A. L., Sinitskii, A., Lomeda, J. R., Dimiev, A., Price, B. K., & Tour, J. M. (2009). Mechanically Assisted Exfoliation and Functionalization of Thermally Converted Graphene Sheets. *Nature*, *458*, 872–876. doi:10.1038/nature07872 PMID:19370030

Krishnan, D., Raidongia, R., Shao, J., & Huang, J. (2014). Graphene Oxide Assisted Hydrothermal Carbonization of Carbon Hydrates. *ACS Nano*, *8*(1), 449–457. doi:10.1021/nn404805p PMID:24298909

Li, X., Qin, Y., Picraux, S. T., & Guo, Z.-X. (2011). Noncovalent assembly of carbon nanotube-inorganic hybrids. *Journal of Materials Chemistry*, *21*(21), 7527–7547. doi:10.1039/c1jm10516g

Park, S., & Ruoff, R. S. (2009). Chemical methods for the production of graphenes. *Nature Nanotechnology*, *4*(4), 217–224. doi:10.1038/nnano.2009.58 PMID:19350030

Saleh, T. A., & Gupta, V. (2012). Synthesis and characterization of alumina nanoparticles polyamide membrane with enhanced flux rejection performance. *Separation and Purification Technology*, *89*, 245–251. doi:10.1016/j.seppur.2012.01.039

Saleh, T. A., & Gupta, V. K. (2014). Processing methods, characteristics and adsorption behavior of tire derived carbons: A review. *Advances in Colloid and Interface Science*, *211*, 93–101. doi:10.1016/j.cis.2014.06.006 PMID:25001042

Saleh, T. A., & Gupta, V. K. (2016). Nanomaterial and Polymer Membranes. *Synthesis, Characterization, and Applications.*

Saleh, T. A., & Mohammed, A. (2016). *Graphene Science Handbook Fabrication Methods*. CRC Press, Taylor & Francis Inc. doi: 10.1201/b19606-39

Saleh, T. A., Muhammad, A. M., Tawabini, B., & Ali, S. A. (2016). Aminomethylphosphonate Chelating Ligand and Octadecyl Alkyl Chain in a Resin for Simultaneous Removal of Co (II) Ions and Organic Contaminants. *Journal of Chemical & Engineering Data*, *61*(9), 3377–3385. doi:10.1021/acs.jced.6b00475

Saleh, T. A., Sarı, A., & Tuzen, M. (2017). Effective adsorption of antimony (III) from aqueous solutions by polyamide-graphene composite as a novel adsorbent. *Chemical Engineering Journal*, *307*, 230–238. doi:10.1016/j.cej.2016.08.070

Chapter 2
Scientific Insights Into Modified and Non-Modified Biomaterials for Sorption of Heavy Metals From Water

Tawfik A. Saleh
King Fahd University of Petroleum and Minerals, Saudi Arabia

Salawu Omobayo Adio
King Fahd University of Petroleum and Minerals, Saudi Arabia

Prakash Parthasarathy
Independent Researcher, India

Gaddafi I. Danmaliki
King Fahd University of Petroleum and Minerals, Saudi Arabia

ABSTRACT

Adsorption techniques are widely used for the removal of various classes of pollutants from water due to their mild and facile operating conditions. The operations involved in the adsorption techniques are environmentally friendly, economical, highly selective on pollutants, highly efficient and easily operative. However, the adsorption of heavy metals from water using biomaterials (biosorption) is a relatively new and interesting technique which holds a great potential. Its effectiveness in lowering the heavy metals concentration to sub-ppb levels is appealing and has attracted increasing attention. The technique is believed to replace the existing technologies in the near future. This chapter discusses the prospects of biomaterials in the removal of heavy metals from waste water, the modification techniques that can enhance biosorption efficiency, and the factors influencing the biosorption processes.

DOI: 10.4018/978-1-5225-2136-5.ch002

1. INTRODUCTION

The increase in population, advances in technology and excessive use of resources have disturbed the ultimate existence of human beings. Water is the platform from which lives evolved and on which life exists. Notwithstanding, the amount of fresh water necessary to sustain the geometric rise in population is limited. More than one billion people live in a shortage of drinking water, and the figure is expected to reach four billion by 2025 (UNEP 2008). The available fresh water that is unevenly distributed within nations is faced with the problems of climate change, pollution, and inefficient management strategies. No denial, the world is seriously facing water crisis.

The presence of heavy metals (cadmium, mercury, lead, arsenic, chromium, nickel, zinc, and selenium) in the aquatic environment due to industrial, agricultural and domestic activities has confounded the existing problems of water scarcity with pollution. Heavy metals are widely known for their toxicity, bioaccumulation, and their propensity to pose serious threats to the ecosystem (Saleh and Gupta 2012). The conventional methods of removing heavy metals from effluents include flotation, coagulation, chemical precipitation, reverse osmosis, ion exchange, electro-dialysis, ultrafiltration, nano--filtration, *etc*. (Gupta *et al*., 2012). However, the conventional methods have several disadvantages such as lack of specificity to the metal ions, relatively expensive, and generate excessive sludge as a by-product (Lakherwal 2014).

The use of adsorbents is an efficient way to isolate heavy metals and other pollutants from industrial, agricultural, and domestic effluents (Saleh *et al*., 2016; Gupta *et al*., 2013). It has become the most preferred method of heavy metals removal because it is inexpensive, versatile, simple, and highly effective. Despite its advantages over the conventional methods, the pursuit for more efficient, cheap and environmentally-friendly adsorbents has been the focus of researchers in recent years (Ngah *et al*., 2008; Hua *et al*., 2012). Biosorption is an emerging technique in this regard, possessing a tremendous scope, especially in the developing countries where potable water has limited accessibility. Biosorption is the technique of utilizing living and non-living biological materials specifically for the purpose of removing contaminants such as heavy metals from aqueous solutions and effluents (Salam *et al*., 2011; Feng *et al*., 2011; Demirbas 2008). This chapter, therefore, discusses the application of modified and non-modified biomaterials for efficient heavy metals removal.

The first section of the chapter reviews the pollution and toxicity of heavy metals. The second section of the chapter discusses the adsorbents in use for heavy metals isolation, including activated carbon, nano-materials, biomaterials, and low-cost adsorbents. The application of non-modified and modified biomaterials for the removal of heavy metals is also reviewed in this section. The last section, which is the heart of this chapter discusses the modification strategies associated with biosorbents for efficient adsorption, possible biosorption mechanisms and the factors affecting the biosorption process.

2. POLLUTION AND TOXICITY OF HEAVY METALS

Heavy metals pollution is caused by natural and artificial sources. The volcanic eruptions, erosion of metal deposits and unavoidable atmospheric pollution are the major points of entry of heavy metals naturally into the environment. Nevertheless, nature has an inherent capacity to curtail the impact of natural pollution. Artificial sources of heavy metal pollution are mainly from agricultural activities, tannery operations, mining operations, metal plating activities, production activities of commercial products (*e.g*. batteries, paints, pigments, glass, ceramics), leachates from landfills and contaminated groundwater, *etc*. (Förstner and Wittmann 2012). Since the survival and comfort of human beings are directly linked to the above activities, the continuous release of heavy metals into the environment is inevitable.

Cadmium, mercury, lead, arsenic and chromium (even at low concentrations) are regarded as the most toxic heavy metals due to their significant impact on the environment (Tchounwou *et al*., 2012). They are systemic toxicants classified by the international agency for research on cancer (IARC) and the US EPA as "known or probable" carcinogens with the potential of inducing organ damage. These heavy metals are very much denser than water and the water laden with these heavy metals turns very toxic. The microbial population, which grows on this toxic water also turns toxic. When the water bodies flow into the seas, the polluted microbial population accumulates in the sea bed through which the heavy metals largely enter into the overall food chain of many organisms.

It has been proven that heavy metals affect cellular components and organelles such as the mitochondria, nuclei, cell membrane, lysosomes and enzymes involved in detoxification, metabolism and damage them (Wang and Shi 2001). They interact with cell components such as nuclear proteins and DNA and introduce significant changes to cell cycle leading to carcinogenesis or programmed cell death (Beyersmann and Hartwig 2008). Heavy metals are directly linked to the increased production of reactive oxygen species (ROS) and oxidative stress affecting master regulatory protein and the guardian of the genome p53, thereby increasing susceptibility to cancer and other diseases (Patlolla *et al*., 2009).

2.1 Cadmium

Cadmium is a heavy metal with environmental and occupational concerns. The anthropogenic sources of cadmium into the environment include the production of nickel-cadmium batteries, pigments, nuclear reactors and alloys. The routes of exposure to cadmium are mainly inhalation and ingestion (ASTDR 2008). Cadmium is a severe gastrointestinal and pulmonary irritant, which can be toxic and poisonous when ingested or inhaled. Acute exposures to cadmium are associated with symptoms such as abdominal pain, pulmonary, hepatic or renal injury, muscle cramps, vertigo, convulsions, shock, burning sensation, and nausea (Baselt and Cravey 1995). The chronic exposure to cadmium is associated with hormonal imbalance and pulmonary adenocarcinomas (Waalkes *et al.*, 1996).

2.2 Mercury

Mercury exists in three different forms in the environment (elemental, organic and inorganic) having varying degrees of toxicity. The most frequently encountered organic species in the environment are the methyl mercury formed by the microbial methylation of the inorganic mercury in soil and water (Dopp *et al.*, 2004). Mercury is a widespread environmental pollutant which causes severe alterations in body tissues and induces a range of adverse health effects. The major routes of exposure to mercury are inhalation, skin absorption, ingestion, skin/eye contact. Mercury is a carcinogen that affects the kidney, eye, nervous and reproductive systems (Leaner *et al.*, 2007). The anthropogenic activities leading to the release of mercury into the environment include the industrial production of electrodes, thermometers, barometers, fluorescent lamps and amalgams (Tchounwou *et al.*, 2003).

2.3 Lead

Lead occurs naturally in the environment, but the anthropogenic activities such as mining, burning of fossil fuels and manufacturing of pigments, lead acid accumulators and antiknocks in petrol contribute significantly to the release of lead into the environment. The routes of exposure are mainly inhalation, ingestion, and skin/eye contact (ATSDR 1999). Lead is the most systemic toxicant that affects several organs in the body, including the liver, kidney, central nervous system, haematopoietic system, reproductive system and endocrine system. Children are the most vulnerable to lead poisoning and are prone to nervous system breakdown and mental retardation. The reason is that children absorb lead from contaminated water significantly higher than adults (ATSDR 1999). Decreased sperm count in men and spontaneous abortions are the major effects of lead exposure in the adult population (Hertz-Piccioto 2000).

2.4 Arsenic

Arsenic is found in low concentrations in almost all environmental matrices (ATSDR 2000). It exists in organic form as methylated metabolites and in inorganic form as trivalent arsenite and pentavalent arsenate respectively. Arsenic compounds have a wide application in the medical field for the eradication of many diseases (Centeno *et al.*, 2005). However, the anthropogenic sources of arsenic have serious health consequences. The route of exposure of arsenic compounds to our body system includes inhalation, ingestion, skin and eye contact. Arsenic targets organs such as skin, kidneys, liver, gastrointestinal tract and lungs. It affects our respiratory, nervous and reproductive systems (Tchounwou *et al.*, 2003). The major anthropogenic sources of arsenic into the environment include the industrial production of herbicides, insecticides, fungicides and algicides for agricultural purposes.

2.5 Arsenic

The chemistry and toxicology of chromium are complex and interesting. For example, chromium poses detrimental as well as beneficial properties. It exists in two different states in the environment, Cr (III) and Cr (VI) with contrasting toxicities, bioavailability, and mobility. Cr (III) is not mobile and is required for the normal functioning of our body system. Chromium released into the environment from anthropogenic sources is mainly in hexavalent form. It is a toxic pollutant classified as a human carcinogen by many agencies (USEPA 1992). It moves readily through the soil and aquatic environment and is a strong oxidizing agent capable of being absorbed into our body tissues (Park and Jung 2001). The industries with the largest contributions to chromium release into the environment include tannery facilities, metal processing, chromate production, welding, and pigment production.

3. ADSORBENTS

The increase in the density of a fluid within the vicinity of an interphase is termed as adsorption. Adsorption is mainly controlled by physical forces (such as dipole, dispersive forces, or polarization forces) and chemical forces (such as the interaction of electrons between solid particles and the atoms of the liquid or vapour phase) (F. Rouquerol *et al* 2014). Adsorption on commercial and industrial scale depends solely on the type, quantity, the economic cost and effectiveness of the adsorbent material. The substances used as adsorbents can either be of organic or inorganic origin. Regardless of the origins, each adsorbent material should be: mechanically and thermally stable, highly porous, possessing a high surface area with the balance between the micro and macro-pores, easily regenerative, and high affinity for the adsorbate (Czaja *et al* 2009).

In recent years, adsorptive removal of heavy metals from effluents primarily uses conventional adsorbents such as activated carbon, polymers, activated alumina, silica and zeolite (Mouni *et al.*, 2011; Haladu *et al.*, 2016; Aguado *et al.*, 2009; Motsi *et al.*, 2009). However, of late, the use of modified and non-modified carbon nanotubes (CNT), magnetic nanoparticles (MNP) and metal organic frameworks (MOFs) has received increasing attention for the removal of heavy metals from contaminated water (Gupta *et al.*, 2011; Zagoosh *et al.*, 2013; Khan *et al.*, 2013). The fibrous shape, large external surface area and established mesopores of CNTs are offering higher adsorbing capacity for them. MNPs are very effective and easy to operate; hence they are widely used for the removal of heavy metals. On the other hand, MOFs offer a great potential as their porous surfaces can be easily modified which ultimately leads to the improved selectivity of certain functional groups.

3.1 Activated Carbon

The large surface area and microporous nature of activated carbon, which can be modified by sulfuration, ammonification, oxidation, and coordinated ligand anchorage, has made it a suitable adsorbent for the removal of heavy metals from aqueous solutions (Rivera-Utrilla *et al.*, 2011). Activated carbon produced by carbonizing organic materials is the most widely used adsorbent. Yet, the high cost of the activation process limits its use in wastewater treatment applications. Numerous studies have focused on enhancing the oxygen-containing functional groups on the surface of activated carbon by physical and chemical treatment (Saleh and Danmaliki 2016 [a,b]; Danmaliki and Saleh 2016). Activated carbon is less specific and ineffective in isolating heavy metals from solution when the concentration of heavy metals is too high.

3.2 Nanomaterials

Development of cost-effective and novel nanomaterials for heavy metals treatment is gaining wide interest in recent years. Nanocarbon materials, nanometal particles, and polymer-supported nanoparticles have unique properties that provide high adsorption capacities and selectivity. Nanosized metal oxides, including nanosized cerium oxides, manganese oxides, magnesium oxides, ferric oxides, zinc oxides, aluminium oxides, and titanium oxides provide high surface area and specific affinity for heavy metal removal from aqueous solutions (Hua *et al.*, 2012). Iron oxide nanoparticle based nanomaterials are widely used adsorbents for the remediation of contaminated water due to their favourable physiochemical properties. The removal of chromium was monitored using polypyrrole/Fe_3O_4 magnetic Nano composite. The magnetic nanocomposite was prepared by in situ polymerization of pyrrole monomer and it was tested for its effectiveness in the removal of toxic Cr (VI). The initial concentration of chromium was 200 ppm and there was 100% removal efficiency at pH of 2. Ion exchange and the reduction of the surface of the adsorbent were proposed as the possible mechanism of adsorption (Bhaumika *et al.*, 2011).

The removal of Cd^{2+}, Pb^{2+}, Zn^{2+}, and Cu^{2+} was monitored using a polymer modified magnetic nanoparticles. Fe_3O_4 magnetic nanoparticles modified with 3-aminopropyltriethoxysilane and copolymers of acrylic acid and crotonic acid were evaluated for the removal of heavy metals from aqueous solution. The magnetic nanoparticles removed the metal ions efficiently with high maximum adsorption capacity at pH 5.5 (Ge *et al.*, 2012). Nano zerovalent iron particles have been used as an effective adsorbent for the removal of cadmium in the concentration range of 25-450 ppm. The maximum adsorption capacity and activation energy of the adsorbent were 769.2 mg/g and 54.8 kJ/mol at 297 K respectively (Boparai *et al.*, 2011). The major drawback with nanomaterials is that they are expensive and they require a considerable amount of time for their preparation and functionalization. Additionally, for treating large quantities of waste water an enormous amount of nanomaterials is required.

3.3 Low Cost Adsorbents

Apart from industrial by-products and natural materials, agricultural residues or wastes can also serve as the sustainable source for low-cost adsorbents. Some agricultural residues such as rice husk, wheat bran, sawdust and peanut shell are discarded into the environment as they are available in plenty and having little commercial value (Abdel *et al,*. 2011). Hence, these residues were attempted by researchers as an absorbent for the removal of heavy metals from effluents. Ricordel *et al.* (2001) successfully sequestrated heavy metals from aqueous solutions using activated carbon made from peanut husk and rice husk. The work by Periasamy and Namasivayam (1996) studied employing peanut shell carbon (PHC) for the removal of Cu (II) from waste water. It was interesting to note that PHC exhibited 18 times greater adsorption capacity than commercially available granulated activated carbon (GAC).

Examples of Low-Cost Adsorbent

Activated carbon, Algae, Aragonite shells, Bagasse, Blast furnace slag, Bone chars, Charcoal, Clino-pyrrhotite, Coconut coir, Cow-dung ash, Eggshell, Geothite, Immobilised biomass, Lignite, Natural zeolites, Olive stones, Orange juice residue, Peanut husk, Peat, Rice husk, Saw dust, Sepiolite (Orera), Waste straw dust, Waste tea leaves, Bacteria and Fungi.

3.4 Biomaterials

With the advent of adsorption as a waste water treatment technique, the need to match water treatment efficiency with cost effectiveness and environmental sustainability became paramount. Although a variety of adsorbents are already in use, the lookout for novel adsorbents that meet the above three requirements has resulted in enormous research works. Most commercially available adsorbents are either of coal origin or artificially synthesized. The sudden needs to be environmentally conscious have pushed most of these adsorbents unfit for use. It was in the wake of these needs that the application of biological materials as adsorbents was thought of.

The use of biomaterials as adsorbents is a new, but promising and alternative concept, especially in the removal of heavy metals from water. Biosorption and bioaccumulation have sometimes been used interchangeably to describe the removal of contaminants using materials of biological origin. However, while biosorbents involves the use of passive biomaterials, bioaccumulation describes the removal of the contaminants by metabolically active organisms. Phytoremediation and bioremediation also involve treatment of contaminants using biological organisms. Phytoremediation describes living plants ability to sequester contaminants and degrade them to less toxic elements during their metabolic activity. Conversely, bio-remediation is the process of introducing a microorganism into an effluent as to remove its pollutants.

The scope of this write-up will be limited to the use of biosorbents in the sequestration of heavy metals from water non-metabolically. Biosorption has shown high efficiency in the removal of heavy metals from water especially at low concentrations. The use of readily available, inexpensive biological materials has made it widely accepted. It also results in the generation of low quantity biological sludge. Biosorbents can also be regenerated. The advantages of the use of biomaterials for heavy metals treatment are illustrated in Figure 1.

Similar to any adsorption process, the use of biosorbents in the adsorption of heavy metals requires a solid phase and a liquid phase. The liquid phase is made up

Figure 1. Advantages of biomaterial for heavy water removal

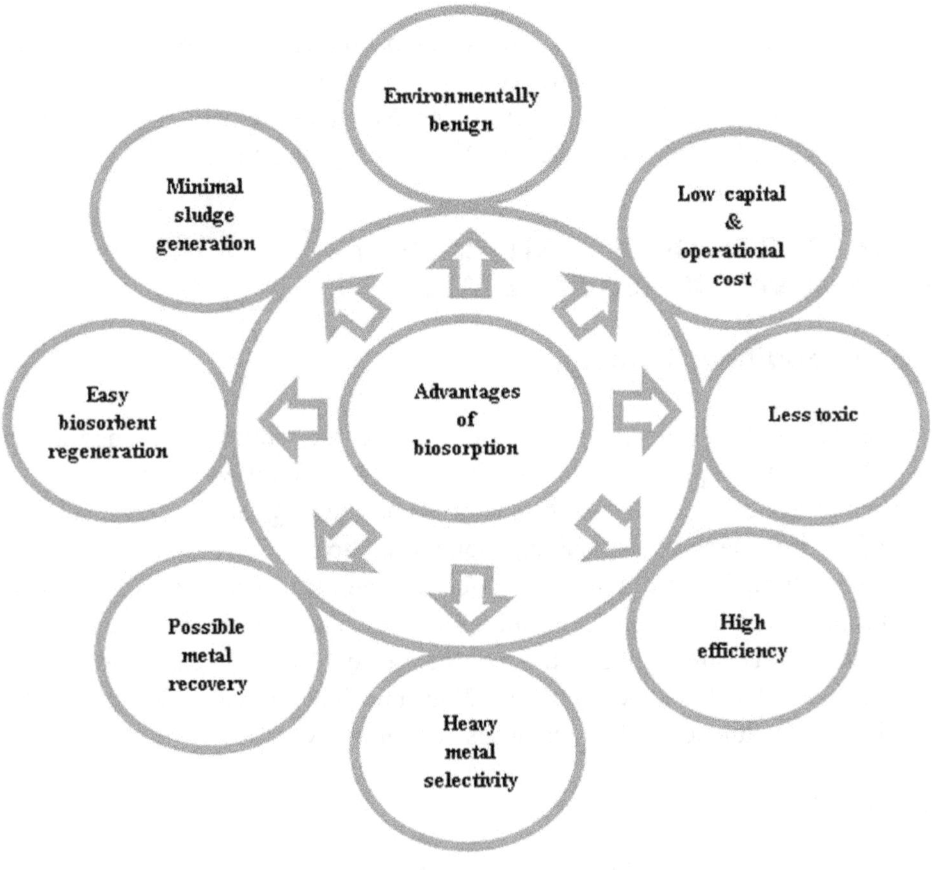

of the contaminants to be sequestered unto the biosorbents. The passive removal of metals from water by biomaterials requires that the substrate shows high metal uptake, suitable mechanical properties, and selectivity (Davis *et al.*, 2003). Biosorbents removal of heavy metals also depends on its composition of proteins, lignin and carbohydrates that make functional groups such as carboxyl, hydroxyl, and amines available for metal sorption (Kuyucak and Volesky, 1988). In recent times, efforts have been directed towards screening efficient biomaterials, investigating different preparation/modification possibilities and the mechanisms involved in the adsorption of heavy metals on biomaterials and the application of biomaterials in water treatment at large scale (Wang and Chen, 2006). Biosorbents can be used directly or after modification. The following section explains the use of biosorbents with and without modification.

Examples of Some Biomaterials:

Aloe, Bagasse, Beat sugar sludge, Chitosan, Coconut shell, Coconut coir, Coffee, Corn cobs, Cottonseed hulls, Distillery waste, Green tea, Japanese coarse tea, Jute stick, Kelp and seaweed, Molasses, Newspaper, Nutshells, Palm tree cobs, Pulp-mill waste, Rice hulls, Sunflower seeds, Tea, Wheat straw and Yuzu.

4. APPLICATION OF BIOMATERIALS IN THE REMOVAL OF HEAVY METALS

4.1 Non-Modified Biomaterials

Biomaterials used with little or no further treatment, alteration or change in their physical or chemical properties is referred to as 'non-modified biomaterials'. It is important to reduce the size of the biomaterials before they can be used as sorbents. As such, within these contexts, grinding biomaterials to powdered form is not referred to as modification because it does not affect its chemical structure. Non-modified biomaterials have been given a particular interest because of their easy availability and application. Despite being untreated, they have adsorption capacities suitable for the removal of different heavy metals. Non-modified biomaterials can be further classified into cellulose or plant materials and animal wastes.

4.2 Plant Waste

Plant cells are made up of cellulose, lignin, hemicellulose, protein, pectic substances, lipids, proteins and other chemical materials (Heredia *et al.,* 1995). These materials provide functional groups that have a high affinity for heavy metals and form chelates or metal complexes, useful for heavy metal adsorption. Cellulosic materials are the most abundant and renewable organic resource on earth (O'Connell, *et al.,* 2008). They are considered inexhaustible because of the huge quantity produced. Photosynthesis aids the production of 10^{11} -10^{12} tons of pure cellulose materials, yearly (O'Connell, *et al.,* 2008). The properties and reactions of cellulose are influenced by factors such as intermolecular interactions, the isolation process employed, chain lengths, cross-linking reactions, crystallinity, distribution of chain-length, functional groups along polymer chains and repeating glucose units (Klemm *et al.,* 2005, Poletto and Zattera, 2013). These factors contribute to the uniqueness of cellulose materials. Cellulose is poly-functional possessing high chain stiffness and is very sensitive to hydrolysis and oxidation of acetal groups (Klemm *et al.,* 2005). This determines its chemistry and method of handling (Klemm *et al.,* 2005). Different cellulosic materials have been used for the removal of heavy metals from aqueous mediums with less modification. In most cases, applying washed, dried and ground wastes of these cellulose materials can be effective enough for the treatment process.

Even micro-organisms such as algae, bacteria, and fungi have been found to adsorb metal ions from solution, forming metal complex effluents. The removal of heavy metals using biomass of dead microorganisms has been given much research attention because of their chemical composition. The sequestration of chromium, lead, and copper from industrial effluents using microorganisms was studied by Illhan *et.al* (Illhan *et al.*, 2004). The influence of pH, temperature and initial concentration on the removal of metal was studied. It was reported that the bacteria *Staphylococcus saprophytics* effectively removed lead and chromium from the effluent. The investigation on the removal of heavy metal from aqueous solution employing *Schizosaccharomyces pombe* (Fission yeast) in free and alginate immobilized cells was performed by Subhashini *et al.* (Subhashini *et al.*, 2011). The parameters such as pH, temperature, and initial concentration were studied. 73% heavy metals removal were achieved using this fungus. The study reported that immobilized cells remove heavy metals better than free cells.

4.3 Animal Waste Biosorption

Animal waste adsorbents are a low-cost option for the removal of heavy metals from solution. The use of animal waste for heavy metal removal from water presents an alternative waste management opportunity. Animal bones (Al-Asheh *et al.,* 2002, Banat *et al.,* 2002, Al-Asheh *et al.,* 1999), chicken feathers (Al-Asheh *et al.,* 2003, Aguayo-Villarreal *et al.,* 2011) and animal horns (Banat *et al.,* 2002) have been successfully applied in removing different heavy metal from aqueous solution. For instance, chicken feathers with minimal treatment were used in the removal of Pb from water (Rosa *et al.,* 2008). Water and ethanol mixed in the ratio of (5:1) were used to rinse the chicken feather and was subsequently dried. When applied in the heavy metal adsorption process, it yielded about 8.3mg/g adsorption capacity at optimal treatment conditions.

The huge generation of animal wastes makes them readily available sorbent for water treatment. Chicken feather for instance constitute about six percent of a bird's total weight (Mittal *et al.,* 2007) while egg shell represents about 11% of an egg's total weight (Ishikawa *et al.,* 2004). They are thus largely generated as waste in poultry industries. The high heavy metal sorption capabilities demonstrated by these materials have been attributed to their keratin content (Banat *et al.,* 2002). They are also hydrophobic as a result of the presence of sterol and biomolecular triphosphate ester layers (Gao *et al.,* 2007). Keratin is a fibrous protein material that forms the main structural components of animal parts such as hooves, teeth, bones, and hair. The structure of Keratin and their corresponding elements that make up the func-

Figure 2. The α-helix and β-pleated sheet structure of Keratin and their corresponding elements (Belamino et al., 2012)

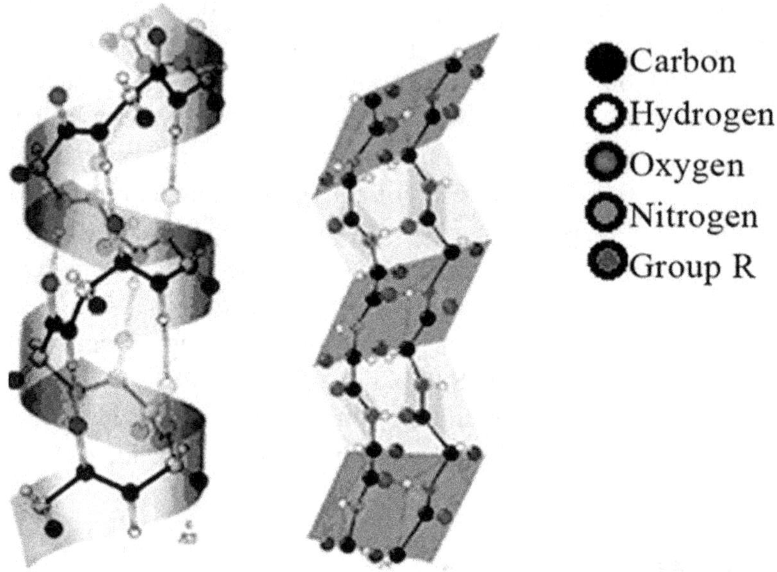

tional groups are shown in Figure 2. Keratin monomers are arranged in bundles forming intermediate, tough filaments. They have high tensile strength, high surface area and are insoluble (Suyama *et al.,* 1996). They are particularly rich in carboxyl, hydroxyl and amines functional group which make them readily made for use as an adsorbent in water treatment. Avian keratin fiber; a keratin protein type contained in bird feathers showed high stability and removal efficiency over a wide pH range in the removal of copper, chromium, lead, uranium and mercury from water (Kar *et al.,* 2004).

4.4 Modified Biomaterials

Numerous factors affect the suitability for use and adsorption capacity of unmodified biomaterials. The presence of the different functional group, heterogeneous/irregular surface areas, pore size distribution and other physical or chemical factors may influence the removal efficiency of a biosorbent (Gautam *et al.,* 2014). In order to increase their sorption properties and enhance their metal selectivity, biomaterials irrespective of their source (plant or animals) are modified (Gautam *et al.,* 2014). Several adsorbents are derived from the modification of biomaterials. This involves the treatment and subsequent alteration in the physical or chemical structure of the materials.

Figure 3. Sorption capacities of alkaline treated, anionic surfactant treated and untreated chicken feather in the removal of Cu from waste water (Al-Asheh et al., 2003)

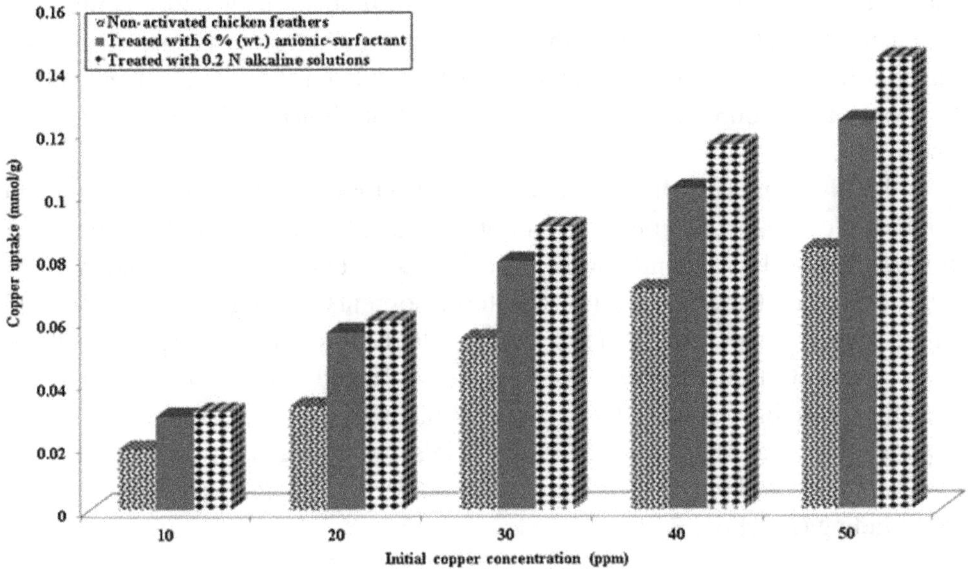

Figure 4. Sorption capacities of alkaline treated, anionic surfactant treated and untreated chicken feather in the removal of Zn from waste water (Al-Asheh et al., 2003)

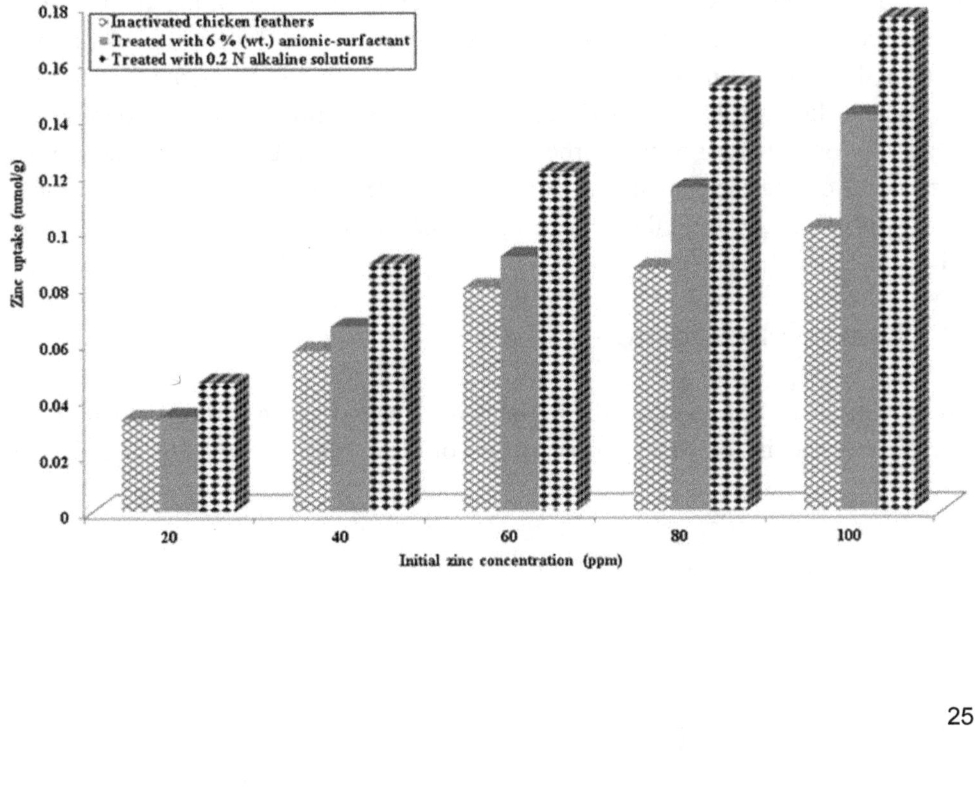

The reason for biomaterial modification has been supported by different researchers that compare the removal efficiency of modified and non-modified biomaterials in heavy metal sorption. For instance, in the removal of copper and zinc from waste water, chicken feathers treated with alkaline solutions showed higher adsorption capacities than those treated with anionic surfactant and untreated chicken feathers (Al-Asheh *et al.,* 2003). The findings of the work are illustrated in Figure 3. and 4 respectively.

Modified and non-modified (native) sugar beet pulp showed similar results in the removal of nickel and copper from water (Reddad *et al.,* 2002). Raw sugar beet pulp was modified by saponification and acid-base extraction. An evaluation of the resulting effect of these modifications on the biosorbents binding properties revealed that that base extracted pulp and saponified pulb displayed higher metal sequestration compared to the raw pulp. The study also showed that the functional group contained in the biomaterials has a tremendous impact on their sorption capacity. The above fact suggests that when biomaterials are modified, different functional groups are added which in turn increases the number of active sites available for metal binding (Lu *et al.,* 2009).

5. MODIFICATION OF BIO-SORBENTS

The objective of the modification is to improve the surface area of biosorbents thereby increasing its sorption capacity. Modification is a kind of pre-treatment on bio sorbents. The modification discards the impurities on the surface of the biosorbents thus converting the surface into active sites readily available for sorption (Huang and Huang, 1996). It also imparts adequate structural stability and improves the adsorption potential of bio-sorbents. Modification can be accomplished either by physical or chemical methods. However, before going for any modification it is necessary to understand the parameters which influence the adsorption capacity of biosorbents.

5.1 Physical Modification

Physical modification of biosorbents involves the activation of the sorbent capability or increase in the adsorption capacity of biosorbents by heating or drying at optimum temperature.

5.2 Chemical Modification

Chemical modification of biosorbents can be broadly classified into two categories. The first one is monomer-grafting, in which specific groups are attached to the substrate. The grafting is again classified into, photografting, high energy radiation grafting and chemical initiation grafting. The other one is a direct modification where the functional groups are attached to the hydroxyl groups of substrate through chemical reactions. The chemical reactions involved in the modification are esterification, etherification, halogenations, oxidation, alkali treatment and silynation (Hokkanen, *et al.*, 2016).

5.3 Surface Functionalization

Surface modification is the process of modifying the surface of biomaterials such that its physical, chemical and biological characteristics are entirely different from its original state. The surface characteristics such as roughness, hydrophilicity, surface charge, surface energy, biocompatibility, and reactivity are altered on the surface of the biomaterials. Surface functionalization is a kind of surface modification. It is the phenomenon of introducing specific functional groups onto the surface of biosorbents to increase the removal efficiency of metal ions (Ihsanullah *et al.*, 2015)

5.4 Nanoparticle Functionalized Biosorbents

Nanoparticles are nano-sized particle possessing very large surface area. The atoms on the nanoparticles are unsaturated and can attach to any atoms (Farnad *et al.*, 2012). Nano-particles due to their surface functionality and high surface area to volume ratio display very high adsorption capacity and efficiency (Fei *et al.*, 2012). It's because of that there is a widespread interest among the researchers to employ nanoparticles in the adsorption of heavy metals from effluents. Of late, the surface of nanoparticles is functionalised to improve their selectivity of metal ions from effluents. Since the selectivity and efficiency are extremely high, there is a growing need for nanoparticle functionalised biosorbents for heavy metals sequestration in aqueous solution.

6. FACTORS INFLUENCING BIOSORPTION

The predominant factors that affect the biosorption processes are pH, temperature, initial metal ion concentration, biomass concentration, contact time and particle size.

6.1 pH

pH is the most important parameter that can significantly influence the biosorption process. It affects the solution chemistry of the metals and disturbs the functional group of the biomass (Das *et al.*, 2008). pH should be neither too high nor too low. It has been found that at a high acidic medium of pH close to 2, the adsorption of metal ions is almost nil. Therefore, the pH should be close to neutral (Ojedokun and Bello, 2016).

6.2 Temperature

The increase in temperature increases the adsorption potential of biomaterials. This could be due to the diffusion of metal ions through the solution as the viscosity of the solution reduces owing to the increase in temperature. It has been reported that at a temperature range of 25-35°C, temperature does not have an influence on biosorption process (Das *et al.*, 2008).

6.3 Adsorbent Concentration

The binding sites for the metal ions are offered by the adsorbent. Hence, the concentration of absorbent has a profound influence on adsorption. Adsorbent concentration in the solution influences the specific uptake. The increase in adsorbent concentration can either increase or decrease the adsorption of metal ions. For some metal ions increase in the biomass favors adsorption whilst for other metal ions decrease in adsorbent concentration supports adsorption (Ihsanullah *et al.*, 2015).

6.4 Ionic Strength of Metal Ions

The Ionic strength of metal ions also can affect the adsorption of heavy metal ions onto the biomass substrate. In general, an increase in ionic strength of metal ions has a negative effect on the sorption process. The detrimental effect could be due to two facts. Firstly, increasing ionic strength reduces the activity coefficients of metal ions, thereby reducing the transfer of metal ions onto the biomaterials. Secondly, at the high ionic strength (metal ions) formation of electrical double layer occurs between cations and biomass substrate hindering adsorption. The adsorption process is also affected by the contention of different metal ions which are trying to accommodate the limited active sites of the substrate (Ihsanullah *et al.*, 2015).

6.5 Contact Time

The contact time also can affect the adsorption of metal ions onto the biomass. For batch adsorption process, it is very important to define the optimum contact time to attain maximum removal of heavy metal ions (Ojedokun and Bello, 2016). At large, longer contact time increases the chances of removal of metal ions from the effluents. However, this holds good until an equilibrium is achieved. The equilibrium is highly determined by the initial concentration of metal ions. By and large, when the initial strength of metal ions is low, the attainment of equilibrium is very rapid. Nonetheless, in few studies, it has also been reported that concentration of metal ions has no effect on adsorption.

6.6 Particle Size

The decrease in particle size of biomaterials increases its adsorption capacity as a reduction in particle size increases the surface area of the substrate (Somasani, 2012).

7. MECHANISM OF HEAVY METAL ADSORPTION BY BIOMATERIALS

Biosorption is the process of removal of metal ions from aqueous solutions by employing certain biomass. The metal biosorption by biomass mainly depends on the functional groups it possesses.

Figure 5. The possible binding combination between functional groups and heavy metal ions

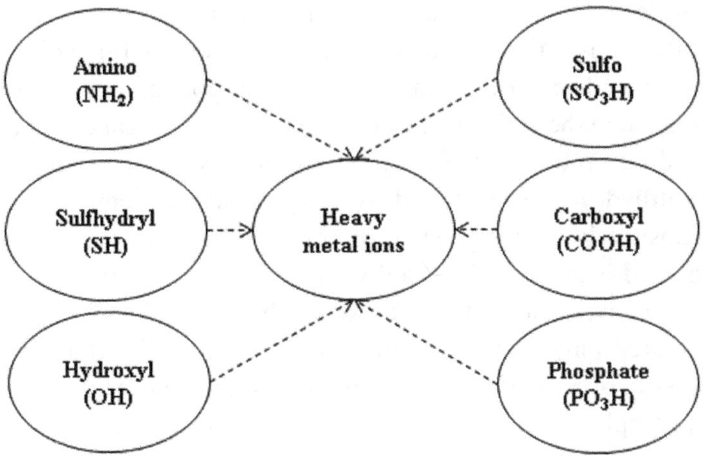

The most important functional groups are carbonyl (ketone), carboxyl, sulfhydryl (thiol), sulfonate, thioether, amine, secondary amine, amide, imine imidazole, phosphonate, and phosphodiester. The possible binding combination between functional groups and heavy metal ions is presented in Figure 5 (Zhang, 2014). Biomass involved in the biosorption can be living or non-living (Prabhakaran, 2006). The use of living organisms is called bio-accumulation. There are some operational difficulties while using live biomass. The industrial effluents, in general, carry toxic metals of high concentration. The pH of the effluents is either too high or low where living organisms cannot sustain. Also, living cells require nutrients and increase the BOC and COD of effluents. Hence, bioaccumulation is not viable for all cases of adsorption of heavy metals. On the other hand, the non-living organisms are not affected by the concentration of heavy metals ions or by the pH of the effluents (Al-Qodah, 2006). Further, dead cells are inexpensive and need less maintenance. Above all, dead cells can be easily regenerated and reused. However, it is a developing technique and is seen as a potential process of removal of heavy metals from industrial effluents.

It is well known that the biosorption is an interaction between a sorbent and sorbate. The sorbent also termed as bio-sorbent or bio-material or substrate is in the solid phase. The sorbate, which is metal ions, in this case, is dissolved in the liquid phase. The liquid phase solvent is water here. The sorbate gets attracted to the sorbent due to the higher affinity between them when they come into contact with each other. The sorption process continues till an equilibrium reaches between sorbent and sorbate. The extent of sorbent affinity for the sorbate decides the sorbate distribution between solid and liquid phases (Das *et al.*, 2008).

The process of metal biosorption in living cells (bio-accumulation) is a two-step phenomenon. In the first step, the metals ions are adsorbed onto the surface of cells by interactions between metal ions and functional groups present on the surface of cells. The metal ion comes across the cell walls. The cell comprises of a wide range of polysaccharides and proteins and hence offers a number of active sites capable of attaching metal ions. Microorganisms *viz.* algae, bacteria, fungi display different adsorption capacity. The variance in the adsorption potential of microorganism is due to the variance in the cell wall composition of them. Algal cells are cellulosic. The metal binding groups in the microbe are carboxylates, amines, imidazoles, phosphates, sulfhydryls, sulfates and hydroxyls. Among them, amines and imidazoles are positively charged, when protonated they show an affinity towards negatively charged metal ions. The cell walls of bacteria are mainly comprised of peptidoglycans. Chiton and chitosan make up the cell walls of fungi. Of the two steps, the first step passive biosorption is independent of metabolism. The step takes place rapidly by a single or a combination of the below binding mechanism: coordination, complexation, ion exchange, physical adsorption (*e.g.* electrostatic) or inorganic micro precipitation. The second step is the active biosorption, where the metal ions penetrate the cell wall and enter into the cells.

Metal uptake by non-living cells is mainly in passive mode. The metal uptake takes place in two steps; passive uptake which takes place rapidly and active uptake which happens slowly. The first stage is more of physical adsorption where adsorption is at the surface of the cell. The adsorption reaches equilibrium within 30-40 min. The passive step is mainly due to the functional groups on the surface of the cell walls. The active mode is dependent on the metabolism and is associated with metal transport and deposition.

8. SUMMARY AND OUTLOOK

Thus, the chapter elaborated on the scope of biomaterials in the context of heavy metals removal from waste water or effluents. The chapter also discussed the modification of biosorbents and parameters influencing the biosorption process. The conventional method of heavy metal removal from effluents is inefficient, non-selective, expensive and time-consuming. On the other hand, the non-conventional techniques are highly efficient and selective. There are numerous non-conventional modes of heavy water removal from effluents. Biosorption is one such alternative method which has been gaining interest recently because of its simple operation, high efficiency, and economics. Also, biosorption offers the flexibility of trying out with different biomass as there is a wide variety of biomaterials availability. Further, a number of modification techniques are available to improve the adsorption capacity of biomaterials. As stated in the article, biosorption is a developing technology and a lot of augmentations are underway to make the technique more dependable. There is no doubt that the technology would be more established in the near future.

ACKNOWLEDGMENT

The author(s) acknowledge the support of the chemistry department, and King Fahd University of Petroleum and Minerals, (KFUPM) Dhahran, Saudi Arabia.

REFERENCES

Agency for Toxic Substances and Disease Registry (ATSDR). (2008). Atlanta, GA: Draft Toxicological Profile for Cadmium.

Agency for Toxic Substances and Disease Registry (ATSDR). (1999). Public Health Service. Atlanta, GA: U.S. Department of Health and Human Services.

Agency for Toxic Substances and Disease Registry (ATSDR) Toxicological Profile for Arsenic TP-92. (2000). *Georgia*. Atlanta, GA: Center for Disease Control.

Aguayo-Villarreal, I. A., & Bonilla-Petriciolet, A. (2011). Batch and column studies of Zn2þ removal from aqueous solution using chicken feathers as sorbents. *Chemical Engineering Journal, 167*, 67–76. doi:10.1016/j.cej.2010.11.107

Al-Asheh, S., Banat, F., & Al-Rousan, D. (2002). Adsorption of copper, zinc and nickel ions from single and binary metal ion mixtures on to chicken feathers. *Adsorption Science and Technology, 20*(9), 849–864. doi:10.1260/02636170260555778

Al-Asheh, S., Banat, F., & Al-Rousan, D. (2003). Beneficial use of chicken feathers in removal of heavy metals from wastewater. *Journal of Cleaner Production, 11*(3), 321–326. doi:10.1016/S0959-6526(02)00045-8

Al-Qodah, A. (2006). Biosorption of heavy metal ions from aqueous solutions by activated sludge. *Desalination, 196*(1-3), 1–3, 164–176. doi:10.1016/j.desal.2005.12.012

Al-sheh, S., Banat, F., & Mohai, F. (1999). Sorption of copper and nickel by spent animal bones. *Chemosphere, 39*(12), 2087–2096. doi:10.1016/S0045-6535(99)00098-3 PMID:10576108

Banat, F., Al-Asheh, S., & Al-Rousan, D. (2002). Comparison between different keratin-composed biosorbents for the removal of heavy metal ions from aqueous solutions. *Adsorption Science and Technology, 20*(4), 393–416. doi:10.1260/02636170260295579

Banat, F., Al-Asheh, S., & Mohai, F. (2002). Multi-metal sorption by spent animal bones. *Separation Science and Technology, 37*(2), 311–327. doi:10.1081/SS-120000790

Baselt, R. C., & Cravey, R. H. (1995). *Disposition of Toxic Drugs and Chemicals in Man* (4th ed.). Chicago, IL: Year Book Medical Publishers.

Beyersmann, D., & Hartwig, A. (2008). Carcinogenic metal compounds: Recent insight into molecular and cellular mechanisms. *Archives of Toxicology, 82*(8), 493–512. doi:10.1007/s00204-008-0313-y PMID:18496671

Bhaumik, M., Maity, A., Srinivasu, V. V., & Onyango, M. S. (2011). Enhanced removal of Cr(VI) from aqueous solution using polypyrrole/Fe_3O_4 magnetic nanocomposite. *Journal of Hazardous Materials, 190*(1-3), 381–390. doi:10.1016/j.jhazmat.2011.03.062 PMID:21497438

Boparai, H. K., Joseph, M., & OCaroll, D. M. (2011). Kinetics and thermodynamics of cadmium ion removal by adsorption onto nano zerovalent iron particles. *Journal of Hazardous Materials*, *186*(1), 458–465. doi:10.1016/j.jhazmat.2010.11.029 PMID:21130566

Centeno, J. A., Tchounwou, P. B., Patlolla, A. K., Mullick, F. G., Murakat, L., Meza, E., & Yedjou, C. G. et al. (2005). Environmental pathology and health effects of arsenic poisoning: a critical review. In R. Naidu, E. Smith, J. Smith, & P. Bhattacharya (Eds.), *Managing Arsenic in the environment: from soil to human health*. Adelaide, Australia: CSIRO Publishing Corporation.

Czaja, A. U., Trkhan, N., & Muller, U. (2009). Industrial applications of metal-organic frameworks. *Chemical Society Reviews*, *38*(5), 1284–1293. doi:10.1039/b804680h PMID:19384438

Danmaliki, G. I., & Saleh, T. A. (2016). Influence of conversion parameters of waste tires to activated carbon on adsorption of dibenzothiophene from model fuels. *Journal of Cleaner Production*, *117*, 50–55. doi:10.1016/j.jclepro.2016.01.026

Das, N., Vimala, R., & Karthika, P. (2008). Bio-sorption of heavy metals- An overview. *Indian Journal of Biotechnology*, *7*, 159–169.

Davis, T. A., Volesky, B., & Mucci, A. (2003). A review of the biochemistry of heavy metal biosorption by brown algae. *Water Research*, *37*(18), 4311–4330. doi:10.1016/S0043-1354(03)00293-8 PMID:14511701

de la Rosa, G., Reynel-Avila, H. E., Bonillia-Petriciolet, A., Cano-Rodriguez, I., Velasco-Santos, C., & Martinez-Hernandez, A. L. (2008). Recycling poultry feathers for Pb removal from wastewater: Kinetic and equilibrium studies. *Proceedings of World Academy of Science, Engineering and Technology*, *30*, 1011–1019.

Demirbas, A. (2008). Heavy metal adsorption onto agro-based waste materials: A review. *Journal of Hazardous Materials*, *157*(2-3), 220–229. doi:10.1016/j.jhazmat.2008.01.024 PMID:18291580

Dimple, L. (2014). Adsorption of heavy metals: A Review. *International Journal of Environmental Research and Development*, *4*, 41–44.

Dopp, E., Hartmann, L. M., Florea, A. M., Rettenmier, A. W., & Hirner, A. V. (2004). Environmental distribution, analysis, and toxicity of organometal (loid) compounds. *Critical Reviews in Toxicology*, *34*(3), 301–333. doi:10.1080/10408440490270160 PMID:15239389

Farnad, N., Farhadi, K., & Voelcker, N. H. (2012). Polydopamine nanoparticles as a new and highly selective biosorbent for the removal of copper (ii) ions from aqueous solution. *Water, Air, and Soil Pollution, 223*(6), 3535–3544. doi:10.1007/s11270-012-1131-7

Feng, N., Guo, X., Liang, S., Zhu, Y., & Liu, J. (2011). Biosorption of heavy metals from aqueous solutions by chemically modified orange peel. *Journal of Hazardous Materials, 185*(1), 49–54. doi:10.1016/j.jhazmat.2010.08.114 PMID:20965652

Förstner, U., & Wittmann, G. T. W. (1981). *Metal Pollution in the Aquatic Environment*. Berlin: Springer-Verlag. doi:10.1007/978-3-642-69385-4

Gao, J., Yu, W., & Pan, N. (2007). Structures and properties of the goose down as a material for thermal insulation. *Textile Research Journal, 77*(8), 617–626. doi:10.1177/0040517507079408

Gautam, R. K., Mudhoo, A., Lofrano, G., & Chattopadhyaya, M. C. (2014). Biomass-derived biosorbents for metal ions sequestration: Adsorbent modification and activation methods and adsorbent regeneration. *Journal of Environmental Chemical Engineering, 2*(1), 239–259. doi:10.1016/j.jece.2013.12.019

Ge, F., Li, M., Ye, H., & Zhao, B. (2012). Effective removal of heavy metal ions Cd^{2+}, Zn^{2+}, Pb^{2+}, Cu^{2+} from aqueous solution by polymer-modified magnetic nanoparticles. *Journal of Hazardous Materials, 211–212*, 366–372. doi:10.1016/j.jhazmat.2011.12.013 PMID:22209322

Gupta, V. K., Agarwal, S., & Saleh, T. A. (2011). Synthesis and characterization of alumina-coated carbon nanotubes and their application for lead removal. *Journal of Hazardous Materials, 185*(1), 17–23. doi:10.1016/j.jhazmat.2010.08.053 PMID:20888691

Gupta, V. K., Ali, I., Saleh, T. A., Nayak, A., & Agarwal, S. (2012). Chemical treatment technologies for waste-water recycling—an overview. *RSC Advances, 2*(16), 6380–6388. doi:10.1039/c2ra20340e

Haladu, S., Muhammad, A. M., Saleh, T. A., & Ali, S. A. (2016). A., A. M. Muhammad, T. A. Saleh and S. A. Ali, "Synthesis of novel cross-linked cyclopolymer bearing polyzwitterion-dianionic moieties and its sorption efficiency for Ni(II) removal from waters. *Chemical Engineering Research & Design, 106*, 337–346. doi:10.1016/j.cherd.2015.12.021

Heredia, A., Jimenez, A., & Guillen, R. (1995). Composition of plant cell walls. *Zeitschrift fur Lebensmittel-Untersuchung und -Forschung, 200*(1), 24–31. doi:10.1007/BF01192903 PMID:7732730

Hertz-Picciotto, I. (2000). The evidence that lead increases the risk for spontaneous abortion. *American Journal of Industrial Medicine*, *38*(3), 300–309. doi:10.1002/1097-0274(200009)38:3<300::AID-AJIM9>3.0.CO;2-C PMID:10940968

Hokkanen, S., Bhatnagar, A., & Sillanpää, M. A. (2016). A review on modification methods to cellulose-based adsorbents to improve adsorption capacity. *Water Research*, *91*, 156–173. doi:10.1016/j.watres.2016.01.008 PMID:26789698

Huang, C., & Huang, C. P. (1996). Application of Aspergillus oryzae and Rhizopus oryzae for Cu(II) removal. *Water Research*, *30*(9), 1985–1990. doi:10.1016/0043-1354(96)00020-6

Igwe, J. C., & Abia, A. A. (2007). Adsorption kinetics and intraparticulate diffusivities for bioremediation of Co(II), Fe(II) and Cu(II) ions from waste water using modified and unmodified maize cob. *International Journal of Physical Sciences*, *2*(5), 119–127.

Ihsanullah, A. (2015). Heavy metal removal from aqueous solution by advanced carbon nanotubes: Critical review of adsorption applications. *Separation and Purification Technology*, *157*(8), 141–161.

Ilhan, S., Nurbas, M., & Nourbaksh, S. (2004). Removal of chromium, lead and copper ions from industrial waste water by *Staphylococcus Saprophyticus*. *Turkish Electronics Journal of Biotechnology*, *2*, 50–57.

Ishikawa, S., Sekine, S., Miura, N., Suyama, K., Arihara, K., & Itoh, M. (2004). Removal of selenium and arsenic by animal biopolymers. *Biological Trace Element Research*, *102*(1), 113–127. doi:10.1385/BTER:102:1-3:113 PMID:15621933

José, A., Arsuaga, J. M., Arencibia, A., Lindo, M., & Gascón, V. (2009). Aqueous heavy metals removal by adsorption on amine-functionalized mesoporous silica. *Journal of Hazardous Materials*, *163*(1), 213–221. doi:10.1016/j.jhazmat.2008.06.080 PMID:18675509

Kar, P., & Misra, M. (2004). Use of keratin fiber for separation of heavy metals from water. *Journal of Chemical Technology and Biotechnology (Oxford, Oxfordshire)*, *79*(11), 1313–1319. doi:10.1002/jctb.1132

Khan, N. A., Hasan, Z., & Jhung, S. H. (2013). Adsorptive removal of hazardous materials using metal-organic frameworks (MOFs): A review. *Journal of Hazardous Materials*, *244-245*, 444–456. doi:10.1016/j.jhazmat.2012.11.011 PMID:23195596

Klemm, D., Heublein, B., Fink, H. F., & Bohn, A. (2005). Cellulose: Fascinating biopolymer and sustainable raw material. *Angewandte Chemie International Edition*, *44*(22), 3358–3393. doi:10.1002/anie.200460587 PMID:15861454

Kumar, G. V., Ali, I., Saleh, T. A., Siddiqui, M. N., & Agarwal, S. (2013). Chromium removal from water by activated carbon developed from waste rubber tires. *Environmental Science and Pollution Research International*, *20*(3), 1261–1268. doi:10.1007/s11356-012-0950-9 PMID:22573097

Kuyucak, N., & Volesky, B. (1988). Biosorbents for recovery of metals from industrial solutions. *Biotechnology Letters*, *10*(2), 137–142. doi:10.1007/BF01024641

Leaner, V. D., Donninger, H., & Birrer, M. J. (2007). Transcription Factors as Targets for Cancer Therapy: AP-1 a Potential Therapeutic Target. *Current Cancer Therapy Reviews*, *3*(1), 1–6. doi:10.2174/157339407780126665

Lotfi, M., Merabet, D., Bouzaza, A., & Belkhiri, L. (2011). Adsorption of Pb(II) from aqueous solutions using activated carbon developed from Apricot stone. *Desalination*, *276*(1-3), 148–153. doi:10.1016/j.desal.2011.03.038

Lu, J. W., Li, Q., Qi, Q. L., Guo, Z. X., & Yu, J. (2009). Surface engineering of poly(D,L-lactic acid) by entrapment of soluble eggshell membrane protein. *Journal of Biomedical Materials Research. Part A*, *91*(3), 701–707. doi:10.1002/jbm.a.32304 PMID:19048638

Ming, H., Zhang, S., Pan, B., Zhang, W., Lu, L., & Zhang, Q. (2012). Heavy metal removal from water/wastewater by nanosized metal oxides: A review. *Journal of Hazardous Materials*, 211–212, 317–331.

Mittal, A., Kurup, L., & Mittal, J. (2007). Freundlich and Langmuir adsorption isotherms and kinetics for the removal of Tartrazine from aqueous solutions using hen feathers. *Journal of Hazardous Materials*, *146*(1-2), 243–248. doi:10.1016/j.jhazmat.2006.12.012 PMID:17222509

Motsi, T., Rowson, N. A., & Simmons, M. J. H. (2009). Adsorption of heavy metals from acid mine drainage by natural zeolite. *International Journal of Mineral Processing*, *92*(1-2), 42–48. doi:10.1016/j.minpro.2009.02.005

OConnell, D. W., Birkinshaw, C., & ODwyer, T. F. (2008). Heavy metal adsorbents prepared from the 1123 modification of cellulose: A review. *Bioresource Technology*, *99*(15), 6709–6724. doi:10.1016/j.biortech.2008.01.036 PMID:18334292

Ojedokun, A. T., & Bello, O. S. (2016). Sequestering heavy metals from wastewater using cow dung. *Water Resources and Industry, 13*, 7–13. doi:10.1016/j.wri.2016.02.002

Park, S., & Jung, W. Y. (2001). Removal of chromium by activated carbon fibers plated with copper metal. *Carbon Science, 2*(1), 15–21.

Patlolla, A., Barnes, C., Yedjou, C., Velma, V., & Tchounwou, P. B. (2009). Oxidative stress, DNA damage and antioxidant enzyme activity induced by hexavalent chromium in Sprague Dawley rats. *Environmental Toxicology, 24*(1), 66–73. doi:10.1002/tox.20395 PMID:18508361

Periasamy, K., & Namasivayam, C. (1996). Removal of copper (II) by adsorption onto peanut hull carbon from water and copper plating industry water. *Chemosphere, 32*(4), 769–789. doi:10.1016/0045-6535(95)00332-0

Poletto, M., & Zattera, A. J. (2013). Materials produced from plant biomass. part III: Degradation kinetics and hydrogen bonding in lignin. *Materials Research, 16*(5), 1065–1070. doi:10.1590/S1516-14392013005000112

Prabhakaran, S. K. (2006). *A comparative study of low-cost biomaterials for the removal of chromium (VI/III) from aqueous solutions* (Master of Engineering Thesis). Department of Chemical and Biomolecular Engineering, National University of Singapore, Singapore.

Reddad, Z., Gerente, C., Andres, Y., Ralet, M. C., Thibault, J. F., & Le Cloirec, P. (2002). Ni(II) and Cu(II) binding properties of native and modified sugar beet pulp. *Carbohydrate Polymers, 49*(1), 23–31. doi:10.1016/S0144-8617(01)00301-0

Ricordel, S., Taha, S., Cisse, I., & Dorange, G. (2001). Heavy metals removal by adsorption onto peanut husks carbon: Characterization, kinetic study and modelling. *Separation and Purification Technology, 24*(3), 389–401. doi:10.1016/S1383-5866(01)00139-3

Rivera-Utrilla, J., Sánchez-Polo, M., Gómez-Serrano, V., Álvarez, P. M., Alvim-Ferraz, M. C. M., & Dias, J. M. (2011). Activated carbon modifications to enhance its water treatment applications. An overview. *Journal of Hazardous Materials, 187*(1-3), 1–23. doi:10.1016/j.jhazmat.2011.01.033 PMID:21306824

Roquerol, F., Roquerol, J., & Sing, K. (1999). *Adsorption by powders and porous solids: Principles, Methodology and Applications* (2nd ed.). Elsevier.

Sai Subhashini, S., Kaliappan, S., & Manickam, V. (2011). Removal of heavy metal from aqueous solution using Schizosaccharomyces Pombo in free and alginate immobilized cells. *2nd International Conference on Environmental Science and Technology*, *6*, 108–111.

Salam, O. E. A., Reiad, N. A., & ElShafei, M. M. (2011). A study of the removal characteristics of heavy metals from wastewater by low-cost adsorbents. *Journal of Advanced Research*, *2*(4), 297–303. doi:10.1016/j.jare.2011.01.008

Saleh, T. A., & Danmaliki, G. I. (2016). Adsorptive desulfurization of dibenzothiophene from fuels by rubber tyres-derived carbons: Kinetics and isotherms evaluation. *Process Safety and Environmental Protection*, *102*, 9–19. doi:10.1016/j.psep.2016.02.005

Saleh, T. A., & Danmaliki, G. I. (2016). Influence of acidic and basic treatments of activated carbon derived from waste rubber tires on adsorptive desulfurization of thiophenes. *JTICE*, *60*, 460–468.

Saleh, T. A., & Gupta, V. K. (2012). Column with CNT/magnesium oxide composite for lead(II) removal from water. *Environmental Science and Pollution Research International*, *19*(4), 1224–1228. doi:10.1007/s11356-011-0670-6 PMID:22124802

Saleh, T. A., Muhammad, A. M., & Ali, S. A. (2016). Synthesis of hydrophobic cross-linked polyzwitterionic acid for simultaneous sorption of Eriochrome black T and chromium ions from binary hazardous waters. *Journal of Colloid and Interface Science*, *468*, 324–333. doi:10.1016/j.jcis.2016.01.057 PMID:26854746

Somasani, S. L. (2012). Removal of heavy metals from drinking water by adsorption onto limestone with a focus on copper and aluminum applications (Masters Theses & Specialist Projects). The Faculty of the Department of Chemistry, Western Kentucky University, Kentucky.

Suyama, K., Fukazawa, Y., & Suzumura, H. (1996). Biosorption of precious metal ions by chicken feather. *Applied Biochemistry and Biotechnology*, *57/67*(1), 67–74. doi:10.1007/BF02941689

Tchounwou, P. B., Ayensu, W. K., Ninashvilli, N., & Sutton, D. (2003). Environmental exposures to mercury and its toxicopathologic implications for public health. *Environmental Toxicology*, *18*(3), 149–175. doi:10.1002/tox.10116 PMID:12740802

Tchounwou, P. B., Patlolla, A. K., & Centeno, J. A. (2003). Carcinogenic and systemic health effects associated with arsenic exposure-a critical review'. *Toxicologic Pathology*, *31*(6), 575–588. PMID:14585726

Tchounwou, P. B., Yedjou, C. G., Patlolla, A. K., & Sutton, D. J. (2012). Heavy metals toxicity and the Environment. *EXS*, *101*, 133–164. doi:10.1007/978-3-7643-8340-4_6 PMID:22945569

UNEP Report. (2008). *Vital water graphics: An overview of the state of the world's fresh and marine waters* (2nd ed.). Available online at: http://www.unep.org/dewa/vitalwater/article141.html

U.S. EPA Environmental Criteria and Assessment Office. (1992). Cincinnati, OH: United States Environmental Protection Agency. Available online at https://cfpub.epa.gov/ncea/iris/iris_documents/documents/subst/0373_summary.pdf

Waalkes, M. P., Misra, R. R., & Chang, L. W. (Eds.). (1996). *Toxicology of Metals*. Boca Raton, FL: CRC Press.

Wan Ngah, W. S., & Hanafiah, M. A. K. M. (2008). Removal of heavy metal ions from wastewater by chemically modified plant wastes as adsorbents: A review. *Bioresource Technology*, *99*(10), 3935–3948. doi:10.1016/j.biortech.2007.06.011 PMID:17681755

Wang, J. L., & Chen, C. (2006). Biosorption of heavy metals by Saccharomyces cerevisiae: A review. *Biotechnology Advances*, *24*(5), 427–451. doi:10.1016/j.biotechadv.2006.03.001 PMID:16737792

Wang, S., & Shi, X. (2001). Molecular mechanisms of metal toxicity and carcinogenesis. *Molecular and Cellular Biochemistry*, *222*(1/2), 1, 3–9. doi:10.1023/A:1017918013293 PMID:11678608

Zargoosh, K., Abedini, H., Abdolmaleki, A., & Molavian, M. R. (2013). Effective removal of heavy metal ions from industrial wastes using Thiosalicylhydrazide-Modified Magnetic Nanoparticles. *Industrial & Engineering Chemistry Research*, *52*(42), 14944–14954. doi:10.1021/ie401971w

Zhang, H. (2014). *Biosorption of heavy metals from aqueous solutions using keratin biomaterials* (Doctoral Thesis). Departament de Quinica, Facultat de Ciencies, Universitat Autonoma de Barcelona, Barcelona, Spain.

Chapter 3
Principles and Advantages of Microwave-Assisted Methods for the Synthesis of Nanomaterials for Water Purification

Tawfik A. Saleh
King Fahd University of Petroleum and Minerals, Saudi Arabia

Shafquat Majeed
University of Kashmir, India

Arunima Nayak
Innovació i Recerca Industrial I Sostenible, Spain & Graphic Era University, India

Brij Bhushan
Innovació i Recerca Industrial I Sostenible, Spain & Graphic Era University, India

ABSTRACT

Nanomaterials are the pillars of nanoscience and nanotechnology and to realize their full potential in various potential applications, synthetic methodologies/routes need to be established that are simple, fast and cost-effective. Wet-chemical approaches for nanomaterial synthesis have proven to be among the most versatile and effective routes to finely tailor nanocrystals with varying compositional and architectural complexity. Microwave-assisted solution route represents an efficient wet-chemical

DOI: 10.4018/978-1-5225-2136-5.ch003

1. INTRODUCTION

approach for the synthesis of nanomaterials that offers additional advantages, such as rapid volumetric heating, high reaction rates, size and shape control by tuning reaction parameters, and energy efficiency. In addition, the homogenous heating of the reactants in microwave synthesis minimizes thermal gradients and provides uniform nucleation and growth conditions that leads to the formation of nanomaterials with uniform size distribution. This chapter deals with the basics of microwave chemistry and its applications towards the synthesis of nanomaterials for catalytic applications.

The prefix '*nano*' which means one billionth of something is derived from the Greek word '*nanos*' meaning dwarf. A nanometre (nm) is an SI (*International System of Units*) unit of length and is equal to one-billionth of a metre, 10^{-9} *m*. Nanoscience and nanotechnology thus deals with the study of phenomena, design and manipulation of matter having dimensions of the order of a billionth of a metre. One of the fundamental issues in nanoscience is to achieve control over size and morphology of nanostructures to tailor their physical and chemical properties. Since the discovery of size-dependent behavior in nanoscale matter, colloidal inorganic nanocrystals have been developed more rapidly than other classes of nanomaterials owing to the high degree of control on geometric and compositional engineering in the synthesis stage and to the versatility with which they can be incorporated into a number of technologically valuable applications (Baghbanzadeh et al. 2011).

Colloidal nanocrystals are synthesized from the reaction of molecular precursors (metal-organic compounds, metal salts and metal–ligand complexes) in a liquid reaction medium heated to a suitable temperature, often in the presence of some stabilizing organic agents, such as ligands and surfactants, that can influence their growth to bring out different morphologies with distinctive properties. Wet-chemical approaches have proven to be among the most versatile and effective routes to finely tailor the size and shape of nanocrystalline materials. The traditional form of heating using heating plates/mantles is rather a slow and inefficient method for transferring energy to a reaction mixture, since it depends on convective currents and on the thermal conductivity of the various compounds or materials that have to be penetrated, and often results in the temperature of the reaction vessel being higher than that of the reaction medium (Galema, 1997). In contrast, microwave (MW) irradiation leads to efficient internal "in core" volumetric heating (that is, the temperature is raised uniformly throughout the whole liquid volume) by direct coupling of MW energy to the molecules present in the reaction medium (Lidström et al. 2001). This volumetric heating has particular implications in nanomaterial synthesis as it leads to a rapid and homogeneous heating of the reaction mixture, that minimizes thermal gradients and provides uniform nucleation and growth conditions.

Moreover, such rapid energy transfer also creates non-equilibrium conditions in the reaction mixture, resulting in high instantaneous temperatures, thus reducing reaction time and improving the crystallinity of the product (Dahal et al. 2012; et al. 2007).

2. MICROWAVE BASICS

"Microwave chemistry" is the science of applying microwave irradiation to chemical reactions and is based on the efficient heating of matter by microwave dielectric heating, i.e., on the ability of a specific material (e.g., solvent and/or reagents) to absorb microwave energy and to convert it into heat. In the liquid phase synthesis of nanomaterials, microwave irradiation triggers heating by two main mechanisms, namely, dipolar polarization and ionic conduction as shown in Figure 1.

In the dipolar mechanism of polarization (or rotation), the dipoles in the reaction mixture (for example, polar solvent molecules or reagents) are involved, whereas charged particles (free ions or ionic species) in a substance contribute to ionic conduction. When irradiated at MW frequencies, the dipoles in the sample try to align themselves in the direction of the applied electric field. As the electric field oscillates, the molecular dipoles try to follow the alternating electric field streamlines and in such a process, energy is lost in the form of heat through molecular friction and dielectric loss (dielectric heating). The amount of heat generated by this process is directly related to the ability of these dipoles to align themselves with the frequency of the applied MW field. If a dipole does not have enough time to realign

Figure 1. Two main heating mechanisms under microwave irradiation: (a) dipolar polarization; (b) ionic conduction mechanism

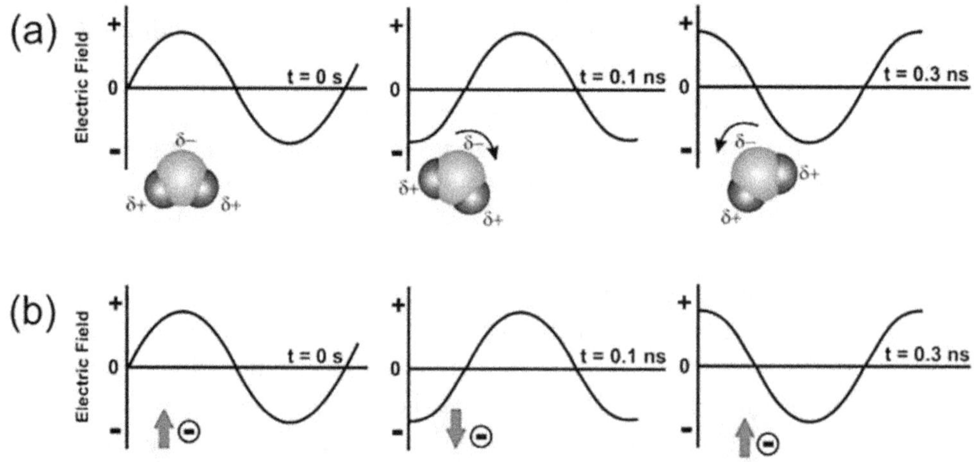

(high-frequency irradiation), or reorients too quickly (low-frequency irradiation) with the applied field, no MW heating will occur. The allocated frequency of 2.45 GHz available in all commercial systems lies between these two extremes and gives the molecular dipoles just enough time to align with the field but does not permit them to follow the alternating field precisely. On the other hand, in the case of ionic conduction the dissolved charged particles in a sample (usually ions) oscillate back and forth under the influence of the MW field, and in the process collide with their neighbouring molecules, generating heat. The ion conductivity mode provides a much stronger heat-generating capacity than the dipolar rotation mechanism and these effects are particularly evident when considering the heating behaviour of ionic liquids in an MW field (Kappe 2004).

The ability of a substance to convert electromagnetic energy (MW energy) into heat at a given frequency and temperature is determined by the so-called loss tangent (*tan* δ) which is defined as the ratio of dielectric loss (ε'') and dielectric constant (ε'). Thus, loss tangent gives a measure of heat generated for a certain degree of polarization. In general, a reaction medium with a high *tan* δ at the standard operating frequency of a MW reactor (2.45 GHz) is required for good absorption and, consequently, for efficient heating. Overall, solvents used in MW chemistry can be classified as high- (*tan* $\delta > 0.5$), medium- (*tan* $\delta \approx 0.1$–0.5), and low-absorbing (*tan* $\delta < 0.1$; Table 1).

On the other hand, polar additives, such as ionic liquids or passive heating elements made out of strongly MW-absorbing materials, may be used on purpose to increase the absorbance level of otherwise low-absorbing reaction media (Kremsner and Kappe 2006).

The dielectric properties (dielectric loss and dielectric constant) of most polar solvents (and other materials in general) vary significantly as a function of tem-

Table 1. Loss tangent (tan δ) values for different solvents

Solvent	tan δ		Solvent	tan δ
Ethylene Glycol	1.350		DMF	0.161
Ethanol	0.941		1,2-dichloroethane	0.127
DMSO	0.825		Water	0.123
2-propanol	0.799		Chlorobenzene	0.101
Formic acid	0.722		Acetonitrile	0.062
Methanol	0.659		Acetone	0.054
1,2-dichlorobenzene	0.280		Tetrahydrofuran	0.047
Dichloromethane	0.042		Toluene	0.040
Acetic acid	0.174		Hexane	0.020

perature (reaction medium) and the frequency of MW radiation and, in turn, loss tangent values are dependent on both temperature and frequency. For example, ethanol is a rather strong MW absorber at room temperature with $tan\ \delta \approx 0.941$. At 100 °C, the $tan\ \delta$ of ethanol has dropped to 0.27, and at 200 °C, it is only 0.08. The reason is that, at 2.45 GHz, most organic solvents, such as ethanol, are heated primarily by the dipolar polarization mechanism and, hence, their ability to absorb MW irradiation decreases with increasing temperature because the molecular friction is reduced due to reduced bulk viscosity. In contrast, the ability of ionic liquid like 1-butyl-3-methylimidazolium-hexafluorophosphate ([bmim][PF6]) to absorb microwaves increases with increasing temperature since it is heated by the ionic conduction mode.

The impact of the associated magnetic-field component should be taken into account when magnetic materials are MW-exposed, in addition to the effects of the electric-field component of MW irradiation on materials

3. MICROWAVE-ASSISTED SYNTHESIS OF INORGANIC NANOMATERIALS

Microwave (MW) chemistry has been widely used in synthetic organic chemistry, with enhanced reaction rates, selectivity, and product yields (Saleh 2016). The technique is also useful for the synthesis of high-quality nanomaterials *via* direct MW heating of the molecular precursors. The microwave-assisted route has been applied for the one-pot synthesis of a large variety of mono- and bimetallic nanostructures and the synthesis generally involves reduction of metal salts either in organic solvents like methanol, ethanol, polyols, etc., or in aqueous medium in presence or absence of surface-directing agents. Ionic liquids, known as excellent microwave-absorbing agents due to their high ionic conductivity and polarizability, have also been used for the preparation of metal nanostructures. By changing the precursor, precursor concentration, solvent and instrumental parameters like microwave power, pressure and temperature metallic nanostructures in various sizes and morphologies have been prepared. Many simple and complex metal oxides have been also synthesized by using the microwave-assisted route. The microwave-hydrothermal method has been used to prepare a large variety of binary and ternary oxides such as ZnO, CuO, PdO, In_2O_3, Tl_2O_3, SnO_2, HfO_2, $BiVO_4$, $ZnAl_2O_4$, $BaTiO_3$, $CaTiO_3$, and $BaZrO_3$ (Huang et a., 2008; Zhao et al., 2004; Wang et al., 2008; Patra and Gedanken 2004; Zhang et al., 2008; Nyutu et al., 2008; Moreira et al., 2009). Organic solvents like ethylene glycol or benzyl alcohol are perfect alternatives to an aqueous system owing to their high boiling points and high dielectric loss factors, and because they combine all the advantages of nonhydrolytic/nonaqueous nanoparticle synthesis to microwave chemistry.

3.1. Polyol Method

The polyol method, for example, was used to synthesise a variety of undoped and doped oxide structures like TiO_2, CeO_2, Mn/$CoFe_2O_4$, $BaZrO_3$ and $PbTiO_3$, Mn/Ni/Co/Cr-doped ZnO and Al- and In-doped ZnO nanoparticles (Wang 2008; Lojkowski et al., 2009). In the polyol route, the solvent, often ethylene glycol or diethylene glycol, can also act as a complexing and capping agent, and thus limits the particle growth and prevent agglomeration (Feldmann 2003). In addition to polyols, benzyl alcohol has been shown to be an excellent solvent for microwave-assisted synthesis and has been used to prepare a variety of oxide structures (Bilecka et al. 2011). In this case, subtle tuning of the crystal size through the initial precursor concentration and the irradiation time is easily possible as well. The strong connection between organic and inorganic reaction pathways and the high sensitivity of organic reactions towards microwave irradiation represents the basis for the size-controlled synthesis of metal oxides in organic solvents. As in the synthesis of metallic nanostructures, ionic liquids have been used as solvents also in the synthesis of metal oxide nanostructures since they have high thermal stability, low vapour pressure and can dissolve many organic, inorganic and organometallic compounds. Microwave irradiation-based routes have also been used for the synthesis of binary and ternary metal chalcogenides, metal phosphates, zeolites, carbon nanotubes and nanofibres, and organic-inorganic hybrid nanomaterials for super-capacitor applications.

3.2. Benefits of Microwave Heating in Nanomaterial Synthesis

Microwave radiation has proved to be a highly effective heating source in inorganic nanomaterial synthesis. It offers the following advantages:

- Efficient source of heating (superheating) leading to increased rates of reaction, enabling faster synthesis;
- Size and shape control by easily tuning instrumental or reaction parameters;
- Selective heating based on the principle that different materials respond differently to microwaves;
- Greater reproducibility of chemical reactions than in conventional heating, owing to uniform heating and better control of process parameters;
- Possible combinations of microwave chemistry with other well-established liquid-phase synthesis routes like solvothermal, sonochemical methods.

3.3. Limitations of Microwave Heating in Nanomaterial Synthesis

On the other hand, some of the limitations of microwave dielectric heating are:

- Its lack of scalability (batch size typically is limited to a few grams),
- Limited applicability (to materials that absorb microwaves), and
- Safety and health hazards related to the use of microwave-heating apparatus.

4. MICROWAVE HEATING IN NANOCATALYSTS

The motivation for the use of MW energy has mainly been to design faster, cleaner, and economically more viable methods of synthesis. Since the early days of MW synthesis, the observed reaction rate acceleration and, sometimes, altered product distribution, compared to those achieved in oil-bath experiments, have led to speculation on the existence of so-called "specific" or "non-thermal" MW effects (Langa et al., 1997; De la Hoz et al., 2005). Historically, such effects were invoked when the outcome of a synthesis performed under MW conditions was remarkably different from that achieved for the conventionally heated counterpart at the same nominal temperature.

5. APPLICATIONS OF NANOCATALYSTS FOR WATER PURIFICATION

The depletion of fresh water resources due to extensive industrialization and unprecedented population growth has caused a spurt in research activities employing new economic and environmentally friendly water and wastewater treatment technologies. Metallic nanoparticles (Ag, Au), metal oxide nano-particles (silicon oxide, aluminium oxide, zinc oxide, iron oxide, titanium oxide etc.) and/or doped onto carbonaceous nano-materials like carbon nano-tubes, carbon nano-sheets, carbon nano-particles serve as adsorbents and efficient photo-catalysts and hence have received great attention in the water purification process (Jimenez et al., 2012; Bootharaju et al., 2010; Yang et al., 2010; Hu et al., 2005; Liang et al., 2004). The unique properties of such nano-particles are associated with their small particle size and high surface area to volume ratio (Wang et al., 2014; Saleh and Gupta 2012; Saleh 2013). The use of microwave heating in the synthesis of such nano-particles has demonstrated faster kinetics and has improved the energy efficiency via increased heating rate and reduced activation energies of the process (Remya 2011). The synthesized nanoparticles have shown enhanced capacity to remove and degrade diverse types of organic and inorganic pollutants like volatile organic compounds, dyes, pesticides, ammonia nitrogen, phenolic compounds, pharmaceuticals, petroleum, pollutants from olive oil processing plants, metal ions etc.

Various carbon based materials like Activated Carbon (AC) and transition metal oxides have been developed into effective nano-catalysts under the influence of MW. The materials that have been used for the synthesis of nano-catalyst and subsequently used in the water purification process are basically good MW absorbers. Examples are AC, carbonaceous materials, transition metal/metal oxide doped AC, mixed metal oxides and metal sulphides (Table-2, 3). The improved performance of such nano-catalysts for pollutant removal by adsorption can be attributed cumulatively to their better adsorption potential as well as good MW absorber properties. The adsorption potential of the nano-catalysts may be attributed to their small particle size and high surface area to volume ratio. Utilising microwave energy for thermal treatment generally leads to more uniform particle size especially in the range of 15–35 nm because of the shorter synthesis time and highly focused local heating when compared to conventional thermal heating. The MW treated nano-catalysts thus exhibited larger specific surface area, micropore volume and total pore volume as compared to virgin materials without MW treatment. The good absorption power of the precursor materials (like AC, carbonaceous adsorbents, transition metal oxides etc.) for MW radiation enhances the rate of increase of temperature leading to ¨hot spots¨. These overheated spots having high temperatures of >1000°C have proved to be responsible for promoting the degradation of the pollutants. For the carbonaceous based nano-catalyst used for the adsorption of pollutants from the aqueous medium, carbon rich precursors like bamboo, maghemite, multiwalled carbon nanotubes, guar gum, rice hull etc. were used. Transition metals, their oxides (Fe_3O_4, TiO_2, ZnO, and WO_3) and their ferrites ($ZnFe_2O4$, $MnFe_2O_4$) etc. doped onto the carbon based materials serve to decrease the activation energy and simultaneously act as active sites for the oxidative degradation of the pollutants. Cu/GAC and Pt/GAC nano-catalyst showed higher p-nitro-phenol degradation and TOC removal than virgin GAC under the influence of MW irradiation. Nano-catalyst synthesized from MW assisted ferrite doped on rice hull based AC (RHM) was used successfully for COD removal from wastewaters. The Chemical Oxygen Demand (COD) removal efficiency increased up to 20% by RHM as compared to rice hull ash. The addition of magnetic Fe_3O_4 improved the degradation capacity of AC-MW system. Nano-Fe_3O_4 displayed higher catalytic activity than the micro-Fe_3O_4.

Transition metal based nano-catalysts can exclusively aid in the catalytic activity for pollution degradation only when the size of the metallic particles is lower than the depth of the MW penetration. This is on account of the fact that MW irradiation is reflected by the metallic particles. In this regard, the higher catalytic activity of the various nano-metal/metal oxide catalysts like nano-NiO_2, nano Ni/ZnO, ZnO, CuO, Fe@Fe_2O_3, Ag–$H_2Ti_3O_7$ nanotubes, Bi-oxy halides, Bi-oxides, $BiVO_4$, WO_3, can be attributed to its MW absorbing property and the oxidation of active oxygen, (OH·) under MW irradiation (Saleh et al., 2011).

Besides the contribution of a higher surface area, generation of localised hot spots and good adsorption potential for photons, the nano-catalysts also exhibit high defect density on account of MW irradiation. This, in turn, tend to promote the transition of electrons excited by the absorption of photons onto the catalyst surface and subsequently help in decreasing the recombination possibility of the electron-hole pair. Greater photo-degradation efficiency is obtained through the non-thermal heating effect of MW on the nano-catalyst which is attributed to changes in the pre-exponential factor A in the Arrhenius equation, $k = Ae^{-E_a/RT}$. Also, the coupling of UV-vis irradiation with MW increases the generation of •OH, which could be demonstrated by electron spin resonance (ESR) studies. A combination of classic UV and visible light along with MW irradiation in the treatment of aqueous pollutants have shown enhancement in the degradation efficiency of various pollutant species present in wastewaters as evident from Table 3. The Table shows the reaction conditions and the outcome of using nano-catalyst derived from MW assistance for the photo-degradation of various organic and inorganic pollutants. Enhancement of the photo-degradation efficiency as seen from the outcome of Table 3 is due to the generation of more •OH and the polarisation effect of highly defect nano-catalyst as a result of MW irradiation. TiO_2, ZnO, $BiVO_4$ doped onto carbon based materials like GO, MWCNT and AC exhibit enhanced photocatalytic activity on account of higher surface area and better light photosensitizing agent of the adsorbent materials. Doping of the nano-catalyst with noble metals (Ag, Pt), semi-conducting materials (oxides of Bi, Zn, Cu, Ti, W etc.) and carbonaceous materials (graphene oxide, MWCNT, AC etc.) help in increasing their band gap and in reducing the recombination of electron-hole charge carriers. Model of graphene and carbon nanotube structures are depicted in Figure 2. The nano-catalysts were successfully used for the destruction of dyes like Methylene blue, methyl orange, rhodamine-B, pharmaceuticals like ciprofloxacin, endocrine disrupters like dimethyl-pthallate, bisphenol, metal ions like Cr(VI) etc. The nano-catalysts were suitable for the complete degradation of pollutants even under low dosage and low light irradiation conditions (Gupta and Saleh 2013).

Several factors can influence the degradation and mineralisation efficiency of MW assisted nano-catalyst ranging from MW power, irradiation time and the temperature. In general, higher MW power and irradiation time has a direct relation with the efficiency of the process. MW energy causes the polarization of molecules leading to electronic vibration, which results in the generation of heat (Remya et al. 2011). Therefore, the reaction temperature increases proportionately with an increase

Figure 2. Carbon nanotube and graphene nanosheet structures

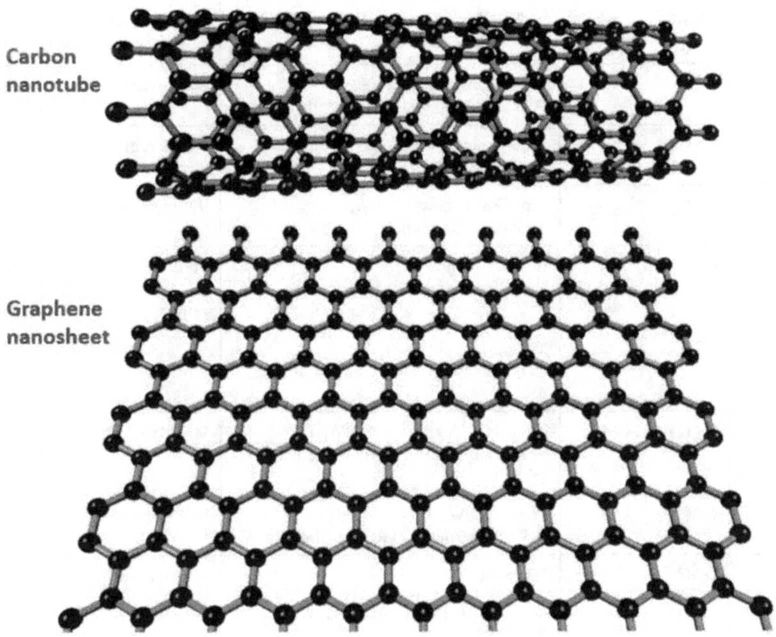

in the MW power input. The treatment time required for the degradation of pollutant could be shortened subsequently by increasing the MW power input. Besides the MW power variable, the properties of the nano-catalyst and operating conditions affect the removal and degradation kinetics of pollutants. The dosage of the nano-catalyst and the pH of the aqueous medium are two important factors that affect the efficiency of the pollutant degradation. pH could affect the surface charge of adsorption, the degree of ionization and speciation of adsorbate during the reaction.

The unique feature of MW irradiation in bringing about distinct advantages on the surface properties of the nano-catalysts has rendered their widespread use in water purification. Satisfactory pollutant degradation and mineralisation can always be achieved due to the special heating effect of irradiation and the possible non-thermal effect. Various disadvantages encountered in traditional and conventional wastewater treatment have been offset by the use of MW assisted nano-catalysts. The unique heating style of MW irradiation viz. fast, selective, volumetric heating and promotion of chemical reactions have the tendency to bring about desired features on the surface of the MW absorbing materials and make them highly efficient for the water purification.

Table 2. Application of nano-catalyst as efficient adsorbents

Pollutant	Nano-catalyst	MW/experimental conditions	Degradation efficiency	Ref
Congo red	AC	Time-1.5min, power-800W; WW conc-50mg/L, pH-8	89.2% removal	[Huang et al., 2008]
Methyl orange	GAC	Time-8mins, power-539W; WW conc-20mg/L,	98.3% removal	[Zhao et al., 2004]
p-nitro-phenol	GAC	Time-8min, power-500W; WW conc-200mg/L	99% removal	[Wang et al., 2008]
carbofuran	GAC	time39min, power-750W; WWconc-100mg/L, pH-8, T-80°C	100% carbofuran removal, 72% COD removal	[Patra and Gedanken 2004]
amoxicillin	GAC	Time-7min, power-480W; COD-1334mg/L after 10times dilution, pH-9	96.4% COD removal	[Jouhannaud et al., 2008]
p-NP	GAC fixed bed	Time-180mins, power-500W; WW conc-1330mg/L, flow rate-6.4ml/min	90% removal	[Zhang et al., 2008]
2,4,5 tri-Cl-biphenyl	GAC	Power-700W; WW conc-2mg/L, flow rate-3ml/min	100% removal	[Zawadzki 2006]
Petroleum refinery	GAC	Power-660W; COD con-5500mg/L	90% removal	[Nyutu et al., 2008]
penta Cl- phenol	GAC	Power-850W, time-10min; PCP conc-198mg/L	100% removal	[Moreira et al., 2009]
Printed circuit board	Rice hull/$MnFe_2O_4$	Time-6mins, power-700W; WW conc-2088mg/L	73.5% COD removal	[Feldmann 2003]
p-NP	Cu/GAC	Time-300mins, power-400W; WW conc-1734mg/L, flow rate-6.4ml/min	91.8% removal	[Pinna 2007]
p-NP penta Cl-phenol	Pt/GAC	Power-400W; Initial conc-1500mg/L, flow rate-6.4ml/min	86%removal-p-NP in 300min 90%removal-penta Cl-P in 240min	[Bilecka et al., 2008]
As(V)	Fe_2O_3/ACF	Power-900W, time-3mins; WW conc-3mg/L, pH-7, Temp-298K	20.33mg/g	[De la Hoz et al., 2005]
Methyl orange	as synthesized ZnO	Power-300W	82% removal	[Bootharaju et al., 2010]
4 Cl-phenol	NiO	Power-100W, time-10mins; pH-7, temp-70°C	100% removal	[Yang et al., 2010]
phenol	NiO	Power-100W; Initial conc-200mg/L, pH-7	100% removal	[Hu et al., 2005]
phenol	Nano-Ni/ZnO	Time-5mins, power-750W; WW conc-100mg/L, pH-7	98%removal	[Liang et al., 2004]
Crystal violet	Nano-NiO_2	Time-5mins, power-750W; WW conc-100mg/L, pH-9	97% removal 81% removal (TOC)	[Javadian 2014]
Rhodamine	Fe@Fe_2O_3/CNT/PTFE	Time-6min, power-na; WW-5mg/L, neutral pH	100% removal	[Wang et al., 2014]

AC-Activated Carbon, GAC-Granular Activated Carbon, ACF-Activated Crabon Fiber, CNT-Carbon Nanotube

Table 3. Application of nano-catalyst for photo-catalytic degradation of pollutants

Pollutant	Nano-catalyst	MW/dosage conditions	Degradation efficiency	Remarks on enhanced photocatalytic activity	Ref
Methylene blue	TiO_2/ MWCNT	1mg, 100mL, 10ppm	100%	Degradation is 40% higher than bare TiO_2. Ascribed due to high surface area and photosensitizing agent of MWCNT	[Tian et al., 2015]
rhodamine	TiO_2/AC	700W, 15min, 0.01g, 50ml, 30mg/L	98% in 30mins	Attributed to high adsorption capacity of mesopore carbon and high TiO_2 content	[Kashinath al., 2015]
rhodamine	TiO_2/AC	0.075g, 50mL, 30mg/L	100% in 10min, 96% mineralization in 20min		[Suresh et al., 2015]
ciprofloxacin	$BiVO_4$/GO	800W, 20mg, 100mL, 10mg/L	100%	Ascribed due to effective seperation of electron-hole pairs	[Zhong et al., 2009]
Textile dyeing WW	Zr-AC, Ni-AC, Zn-AC	750W, 10min, 1g, 1150mg/L	82% COD removal in 120min	Enhanced activity due to synergistic effect of metal oxide and AC	[Zhang et al., 2011]
rhodamine	Bi-oxide	0.05g, 50mL, 10mg/L	100%		[Suchithra et al., 2013]
Cr(VI)	CuO	400W, 30min, 0.02g, 50mL, 100μM	100%		[Yang et al., 2010]
Methylene blue	ZnO	400W, 5min, 1g, 100mL, 15mg/L	99.6% decolorisation; 74.3% TOC removal		[Mohanta et al., 2013]
rhodamine	Ag/ZnO-TiO_2	600W, 0.15g, 90mL, 50mg/L	100%	Due to effective seperation of electron-hole pairs; MW caused efficient particle size distribution	[Liao et al., 2010]
Dimethyl-phthalate	ZrO-ZnO	400W, 1g/L, 100mL, 50mg/L	88% in 30min		[Cao et al., 2015]
ciprofloxacin	Pt/$BiVO_4$	800W 0.15g, 100mL, 10mg/L	91.97% in 60mins	25% enhancement in photocatalytic activity compared to pure $BiVO_4$ photocatalysts	[Adhikari et al., 2014]
Methyl orange	SnS_2	20mg, 50mL, 10mg/L	100%	Enhanced activity due to higher BET surface area of photo-catalyst	[Ahandani et al., 2011]
Methylene blue	Copper doped ZnS	0.1g, 250mL, $2.5x10^{-5}$M	100%	Attributed to high photon absorption and reduced electron-photon pair recombination	[Darwish et al., 2016]
tartrazine	CdS	720W, 5min 25mg, 200mL, 50mg/L	100% in 90mins		[Deng et al., 2013]

ACKNOWLEDGMENT

The author(s) acknowledge the support of the chemistry department, and King Fahd University of Petroleum and Minerals, (KFUPM) Dhahran, Saudi Arabia.

REFERENCES

Adhikari, R., Gyawali, G., kim, T. H., Sekino, T., & Lee, S. W. (2014). EDTA mediated microwave hydrothermal synthesis of WO 3 hierarchical structure and its photoactivity under simulated solar light. *J. Env. Chem. Eng.*, *2*(3), 1365–1370. doi:10.1016/j.jece.2014.02.019

Ahandani, P. Z., & Habibi-Yangjeh, A. (2011). Microwave-assisted preparation of Zn1−xCuxS nanoparticles by a fast, green, and template-free method and photocatalytic. *Desal.*, *271*(1-3), 273–278. doi:10.1016/j.desal.2010.12.042

Baghbanzadeh, M., Carbone, L., Cozzoli, P. D., & Kappe, C. O. (2011). Microwave-assisted synthesis of colloidal inorganic nanocrystals. *Chem. Int. Ed. Engl.*, *50*(48), 11312–11359. doi:10.1002/anie.201101274 PMID:22058070

Bilecka, I., Djerdj, I., & Niederberger, M. (2008). One-minute synthesis of crystalline binary and ternary metal oxide nanoparticles. *Chemical Communications (Cambridge)*, (7): 886–888. doi:10.1039/B717334B PMID:18253537

Bilecka, I., Kubli, M., Amstad, E., & Niederberger, M. (2011). Simultaneous formation of ferrite nanocrystals and deposition of thin films via a microwave-assisted nonaqueous sol–gel process. *Journal of Sol-Gel Science and Technology*, *57*(3), 313–322. doi:10.1007/s10971-010-2165-1

Bootharaju, M. S., & Pradeep, T. (2010). Uptake of Toxic Metal Ions from Water by Naked and Monolayer Protected Silver Nanoparticles: An X-ray Photoelectron Spectroscopic Investigation. *The Journal of Physical Chemistry C*, *114*(18), 8328–8336. doi:10.1021/jp101988h

Cao, W., Chen, L., & Qi, Z. (2015). Microwave-assisted synthesis of Ag/Ag2SO4/ZnO nanostructures for efficient visible-light-driven photocatalysis. *Journal of Molecular Catalysis A Chemical*, *401*, 81–89. doi:10.1016/j.molcata.2015.02.023

Dahal, N., García, S., Zhou, J., & Humphrey, S. M. (2012). Beneficial effects of microwave-assisted heating versus conventional heating in noble metal nanoparticle synthesis. *ACS Nano*, *6*(11), 9433–9446. doi:10.1021/nn3038918 PMID:23033897

Darwish, M., Mohammadi, A., & Assi, N. (2016). Microwave-assisted polyol synthesis and characterization of pvp-capped cds nanoparticles for the photocatalytic degradation of tartrazine. *Research Bulletin (Sun Chiwawitthaya Thang Thale Phuket)*, *74*, 387–396. doi:10.1016/j.materresbull.2015.11.002

De la Hoz, A., Díaz-Ortiz, A., & Moreno, A. (2005). Microwaves in organic synthesis. Thermal and non-thermal microwave effects. *Chemical Society Reviews*, *34*(2), 164–178. doi:10.1039/B411438H PMID:15672180

Deng, C., & Tian, X. (2013). Facile microwave-assisted aqueous synthesis of CdS nanocrystals with their photocatalytic activities under visible lighting. *Materials Research Bulletin*, *48*(10), 4344–4350. doi:10.1016/j.materresbull.2013.07.019

Feldmann, C. (2003). Polyol-Mediated Synthesis of Nanoscale Functional Materials. *Advanced Functional Materials*, *13*(2), 101–107. doi:10.1002/adfm.200390014

Gupta, V. K., & Saleh, T. A. (2013). Sorption of pollutants by porous carbon, carbon nanotubes and fullerene—an overview. *Environmental Science and Pollution Research International*, *20*(5), 2828–2843. doi:10.1007/s11356-013-1524-1 PMID:23430732

Hu, J., Chen, G. H., & Lo, I. M. C. (2005). Removal and recovery of Cr(VI) from wastewater by maghemite nanoparticles. *Water Research*, *39*(18), 4528–4536. doi:10.1016/j.watres.2005.05.051 PMID:16146639

Huang, J., Xia, C., Cao, L., & Zeng, X. (2008). Facile microwave hydrothermal synthesis of zinc oxide one-dimensional nanostructure with three-dimensional morphology. *Materials Science and Engineering B*, *150*(3), 187–193. doi:10.1016/j.mseb.2008.05.014

Javadian, H. (2014). Application of kinetic, isotherm and thermodynamic models for the adsorption of Co(II) ions on polyaniline/polypyrrole copolymer nanofibers from aqueous solution. *Journal of Industrial and Engineering Chemistry*, *20*(6), 4233–4241. doi:10.1016/j.jiec.2014.01.026

Jimenez, I. O., Lopez, X., Arbiol, J., & Puntes, V. (2012). Citrate-coated gold nanoparticles as smart scavengers for mercury(II) removal from polluted waters. *ACS Nano*, *6*(3), 2253–2260. doi:10.1021/nn204313a PMID:22332645

Jouhannaud, J., Rossignol, J., & Stuerga, D. (2008). Rapid synthesis of tin (IV) oxide nanoparticles by microwave induced thermohydrolysis. *Journal of Solid State Chemistry*, *181*(6), 1439–1444. doi:10.1016/j.jssc.2008.02.040

Kappe, C. O. (2004). Controlled microwave heating in modern organic synthesis. *Angewandte Chemie International Edition in English*, *43*(46), 6250–6284. doi:10.1002/anie.200400655 PMID:15558676

Kashinath, L., Namratha, K., & Byrappa, K. (2015). Microwave assisted facile hydrothermal synthesis and characterization of zinc oxide flower grown on graphene oxide sheets for enhanced photodegradation of dyes. *App. Surface Sci.*, *357*, 1849–1856. doi:10.1016/j.apsusc.2015.09.072

Kremsner, J. M., & Kappe, C. O. (2006). Silicon carbide passive heating elements in microwave-assisted organic synthesis. *The Journal of Organic Chemistry*, *71*(12), 4651–4658. doi:10.1021/jo060692v PMID:16749800

Langa, F., de la Cruz, P., de la Hoz, A., Diaz-Ortiz, A., & Diez-Barra, E. (1997). Microwave irradiation: More than just a method for accelerating reactions. *Contemporary Organic Synthesis*, *4*(5), 373–386. doi:10.1039/CO9970400373

Liang, P., Shi, T. Q., & Li, J. (2004). Nanometer-size titanium dioxide separation/preconcentration and FAAS determination of trace Zn and Cd in water sample. *Intern J Environ. Analytical Chemistry*.

Liao, W., Zheng, T., Wang, P., Tu, S., & Pan, W. (2010). Efficient microwave-assisted photocatalytic degradation of endocrine disruptor dimethyl phthalate over composite catalyst ZrO(x)/ZnO. *Journal of Environmental Sciences (China)*, *22*(11), 1800–1806. doi:10.1016/S1001-0742(09)60322-3 PMID:21235170

Lidström, P., Tierney, J., Wathey, B., & Westman, J. (2001). Microwave Reactors: A Brief Review on Its Fundamental Aspects and Applications. *Tetrahedron*, *57*, 9225–9283.

Lojkowski, W., Gedanken, A., Grzanka, E., Opalinska, A., Strachowski, T., Pielaszek, R., & Matysiak, H. et al. (2009). Solvothermal synthesis of nanocrystalline zinc oxide doped with Mn2+, Ni2+, Co2+ and Cr3+ ions. *Journal of Nanoparticle Research*, *11*(8), 1991–2002. doi:10.1007/s11051-008-9559-9

Mohanta, O., Singhbabu, Y. N., Giri, S. K., Dadhich, D., Das, N. N., & Sahu, R. K. (2013). Degradation of Congo red pollutants using microwave derived SrFe12O19: An efficient magnetic photocatalyst under visible light. *Journal of Alloys and Compounds*, *564*, 78–83. doi:10.1016/j.jallcom.2013.02.074

Moreira, M. L., Paris, E. C., do Nascimento, G. S., Longo, V. M., Sambrano, J. R., Mastelaro, V. R., & Longo, E. et al. (2009). Structural and optical properties of CaTiO 3 perovskite-based materials obtained by microwave-assisted hydrothermal synthesis: An experimental and theoretical insight. *Acta Materialia*, *57*(17), 5174–5185. doi:10.1016/j.actamat.2009.07.019

Nyutu, E. K., Chen, C.-H., Dutta, P. K., & Suib, S. L. (2008). Effect of Microwave Frequency on Hydrothermal Synthesis of Nanocrystalline Tetragonal Barium Titanate. *The Journal of Physical Chemistry C*, *112*(26), 9659–9667. doi:10.1021/jp7112818

Patra, C. R., & Gedanken, A. (2004). Rapid synthesis of nanoparticles of hexagonal type In2O3 and spherical type Tl2O3 by microwave irradiation. *New Journal of Chemistry*, *28*(8), 1060–1065. doi:10.1039/b400206g

Pinna, N. (2007). The benzyl alcohol route: An elegant approach towards organic–inorganic hybrid nanomaterials. *Journal of Materials Chemistry*, *17*(27), 2769–2774. doi:10.1039/B702854G

Remya, N., & Lin, J. G. (2011). Current status of microwave application in wastewater treatment—A review. *Chemical Engineering Journal*, *166*(3), 797–813. doi:10.1016/j.cej.2010.11.100

Saleh, T. A. (2016). Nanocomposite of carbon nanotubes/silica nanoparticles and their use for adsorption of Pb (II): From surface properties to sorption mechanism. *Desalination and Water Treatment*, *57*(23), 10730–10744. doi:10.1080/19443994.2015.1036784

Saleh T.A. (2013). *The role of carbon nanotubes in enhancement of photocatalysis*. doi: 10.5772/51050

Saleh, T. A., Agarwal, S., & Gupta, V. K. (2011). Synthesis of MWCNT/MnO2 and their application for simultaneous oxidation of arsenite and sorption of arsenate. *Applied Catalysis B: Environmental*, *106*(1), 46–53.

Saleh, T. A., & Gupta, V. K. (2012). Photo-catalyzed degradation of hazardous dye methyl orange by use of a composite catalyst consisting of multi-walled carbon nanotubes and titanium dioxide. *Journal of Colloid and Interface Science*, *371*(1), 101–106. doi:10.1016/j.jcis.2011.12.038 PMID:22297021

Suchithra, P. S., Shadiya, C. P., Peer Mohamed, A., Velusamy, P., & Ananthakumar, S. (2013). One-pot microwave mediated growth of heterostructured ZnO@AlSi as a potential dual-function eco-catalyst for treating hazardous pollutants in water resources. *Applied Catalysis B: Environmental*, *130*, 44–53. doi:10.1016/j.apcatb.2012.10.010

Suresh, P., Vijaya, J. J., & Kennedy, L. J. (2015). Photocatalytic degradation of textile dyeing wastewater through microwave synthesized Zr–AC, Ni–AC and Zn–AC. *Trans. Nonferrous Met. Soc. China, 25*(12), 4216–4225. doi:10.1016/S1003-6326(15)64072-9

Tian, F., Wu, Z., Chen, Q., Yan, Y., Cravotto, G., & Wu, Z. (2015). Microwave-induced crystallization of AC/TiO2 for improving the performance of rhodamine B dye degradation. *App. Surface Sci., 351*, 104–112. doi:10.1016/j.apsusc.2015.05.133

Wang, K., Huang, T., Liu, H., Zhao, Y., Liu, H., & Sun, C. (2008). Size control synthesis of palladium oxide nanoparticles by microwave irradiation, Colloids Surfaces A Physicochem. *Eng. Asp., 325*(1-2), 21–25. doi:10.1016/j.colsurfa.2008.04.027

Wang, T., Jin, X., Chen, Z., Megharaj, M., & Naidu, R. (2014). Simultaneous removal of Pb (II) and Cr (III) by magnetite nanoparticles using various synthesis conditions. *Journal of Industrial and Engineering Chemistry, 20*(5), 3543–3549. doi:10.1016/j.jiec.2013.12.047

Yang, D. J., Paul, B., Xu, W. J., Yuan, Y., Liu, E. M., Ke, X. B., & Zhu, H. Y. et al. (2010). Alumina nanofibers grafted with functional groups: A new design in efficient sorbents for removal of toxic contaminants from water. *Water Research, 44*(3), 741–750. doi:10.1016/j.watres.2009.10.014 PMID:19889440

Yang, L.-Y., Dong, S.-Y., Sun, J.-H., Feng, J.-L., Wu, Q.-H., & Sun, S.-P. (2010). Microwave-assisted preparation, characterization and photocatalytic properties of a dumbbell-shaped ZnO photocatalyst. *Journal of Hazardous Materials, 179*(1-3), 438–443. doi:10.1016/j.jhazmat.2010.03.023 PMID:20403660

Zawadzki, M. (2008). Preparation and characterization of ceria nanoparticles by microwave-assisted solvothermal process. *Journal of Alloys and Compounds, 454*(1-2), 347–351. doi:10.1016/j.jallcom.2006.12.078

Zhang, H. M., Liu, J. B., Wang, H., Zhang, W. X., & Yan, H. (2008). Rapid microwave-assisted synthesis of phase controlled BiVO4 nanocrystals and research on photocatalytic properties under visible light irradiation. *Journal of Nanoparticle Research, 10*(5), 767–774. doi:10.1007/s11051-007-9310-y

Zhang, L., Cao, X.-F., Chen, X.-T., & Xue, Z.-L. (2011). BiOBr hierarchical microspheres: Microwave-assisted solvothermal synthesis, strong adsorption and excellent photocatalytic properties. *Journal of Colloid and Interface Science, 354*(2), 630–636. doi:10.1016/j.jcis.2010.11.042 PMID:21145062

Zhao, Y., Zhu, J., Hong, J., Bian, N., & Chen, H. (2004). Microwave-Induced Polyol-Process Synthesis of Copper and Copper Oxide Nanocrystals with Controllable Morphology. *Eur. J. Inorg. Chem.*, 4072–4080.

Zhong, H., Shaogui, Y., Yongming, J., & Cheng, S. (2009). Enhancement of visible-light-induced photodegradation over hierarchical porous TiO2 by nonmetal doping and water-mediated dye sensitization. *Journal of Environmental Sciences (China)*, *21*, 268–272. PMID:19402433

Chapter 4
Fundamentals and Sources of Magnetic Nanocomposites and Their Sorption Properties

Victor O. Shikuku
Maseno University, Kenya

Chispin O Kowenje
Maseno University, Kenya

Wilfrida N. Nyairo
Maseno University, Kenya

ABSTRACT

Over the years, adsorption has been the most widely applied technique for pollutants remediation in conventional water and wastewater treatment regimes with commendable results. Consequently, multiple adsorbents have been synthesized, characterized and tested for various pollutants sequestration such as; heavy metals, dyes, pharmaceutically active ingredients, among others, in aqueous media. Unfortunately, most of the sorbents face many inherent limitations such as high production cost, difficult separation of adsorbent from solution, and complex synthesis processes. Therefore, an efficient adsorbent that would be sustainably adopted for industrial application in wastewater treatment requires, among other properties, a simple and efficient recovery step from a continuous flowing system. The regenerated adsorbent must also possess near original properties after several cycles of reuse

DOI: 10.4018/978-1-5225-2136-5.ch004

thereby resulting to low capital investment. To address this challenge, studies conducted in the past few years incorporating magnetism in both natural and synthetic sorbents to improve their removal from water via magnetic separation have yielded stupendous results compared to conventional technologies. This chapter concisely discusses synthesis methods and adsorption capacities and mechanisms of selected magnetic nanocomposite adsorbents under diverse physicochemical conditions for removal of cations, dyes and organic pollutants from wastewater. Magnetic nanocomposites present eco-friendly properties and are potential alternatives for application in water purification processes subject to commercial viability evaluation before practical use.

INTRODUCTION

Nanotechnology aims at designing materials characterized by at least one dimension below 100 nm. When the dimensions of a material are reduced to the nanoscale, the physico-chemical characteristics significantly depart from that of the bulk state. Nanostructured materials exhibit unique chemical, mechanical and magnetic properties that are useful in a wide range of applications. For centuries, adsorption has played an important role in the wastewater treatment processes and it still remains the widely used technique presently. Higher pollutant reduction efficiency and insensitivity to toxic substances compared to the other methods of water purification partly explain its popularity. Several adsorbents widely used for water purification include activated carbon, zeolite, clay minerals etc. However due to their high generation costs, agricultural biomass derived adsorbents have presented an attractive alternative due to their low cost and high abundance (Ngeno *et al.*, 2016). Though these adsorbents have shown stupendous results as effective in the removal of pollutants from wastewater, they suffer one inherent limitation. It is difficult to separate them from wastewater in a continuous flow system. To address this challenge, studies conducted within the past decade have explored the potential of magnetism as an effective means of recovering these suspended adsorbents from water (Dhruv *et al.*, 2015).

The general use of magnetism in water treatment processes such as biological processes and coagulation is an age-old concept dating back to as early as 1873 (Hay, 1873). However, the use of magnetism in adsorption processes is a relatively newer concept; gaining interest among researchers day by day. Magnetic adsorbents are a new class of adsorbents where a base adsorbent is embedded with magnetic particles which are oxides of metals such as Fe, Co, Ni and Cu (Ambashta and Sillanpaa, 2010). In the presence of an external magnetic field, the magnetic adsorbent is quickly and easily segregated from water owing to the presence of the metal oxide component in the adsorbent.

The inclusion of magnetic nanoparticles in composite materials induces responsiveness to magnetic fields, thus synergistically combining the properties of the host and guest particles to generate improved matrices creating new potentials in terms of functionalities and applications in wastewater purification (Dhruv *et al.*, 2015; Alswat et al. 2016). Furthermore, tunability of these magnetic nanocomposites produces materials that can be exploited to overcome the limitations of traditional methods.

This chapter will provide an overview of the fundamentals of magnetism, preparation methods and the applications of magnetic nanocomposites for removal of cations, dyes and organic pollutants from wastewater.

Diverse materials such as; graphene, agricultural biomass, zeolites, clays, and polymers have been shown suitable candidates for consideration as facile adsorbents for constituting the matrix of a nanocomposite (hereby referred to as the host).

The advantages and limitations of the various preparation methods are also outlined. The sorption mechanisms and performance of selected magnetic nanocomposites under different reaction conditions are discussed.

FUNDAMENTALS OF MAGNETIC NANOPARTICLES AND NANOCOMPOSITES

Magnetic effects arise from movements of particles such as electrons, holes, protons, positive or negative ions, which possess both mass and electric charges. A spinning electrically charged particle generates a magnetic dipole, known as *magneton*. A magnetic domain, therefore, designates a volume of ferromagnetic material in which all magnetons are aligned in the same direction by the exchange forces. This concept of magnetic moment, μ, orientation in response to an external magnetic field is the basis for classification of the different forms of magnetism, namely, diamagnetism, paramagnetism, ferromagnetism, antiferromagnetism and ferrimagnetisms (Akbazadeh *et al.*, 2012).

When an external magnetic field is applied, atomic current loops are created resulting from the orbital oscillation of the electrons in opposition to the applied field. This weak repulsion is known as *diamagnetism* and is exhibited in all materials. However, due to the weak nature of diamagnetism, any other type of magnetic characteristics inherent in the material normally quenches these current loops. From the electronic configuration perspective, diamagnetism is observed in materials with completely filled electronic subshells such that the magnetic moments are paired and the effective moment is zero since they cancel each other. Diamagnetic materials have a negative susceptibility ($\chi < 0$). On the other hand, all other magnetic properties are associated with unpaired electrons in the subatomic orbitals more often the $3d$ and $4f$ orbitals (Coey, 2009).

Ferromagnetic materials, such as Fe, Ni and Co, have aligned atomic magnetic moments of equivalent magnitude, and their crystalline structures permit direct coupling interactions between the moments. Additionally, the aligned moments can induce spontaneous magnetization in the absence of an applied external magnetic field. Materials that retain permanent magnetization in the absence of an external field are called hard magnets. Materials possessing atomic magnetic moments of equivalent magnitude arrayed in an antiparallel pattern, resulting to zero net magnetization, are known as *antiferromagnetic* materials.

Above the magnetic ordering temperature (T_N) also called Néel temperature-named after the French physicist Louis Neel, thermal energy becomes large enough to destroy and randomize the alignment of the equal and oppositely aligned atomic moments in the antiferromagnetic material. The material, therefore, displays paramagnetic behavior. If the atomic magnetic moments, below the Neel´s temperature and in the absence of an external field, assume an ordered but unparalleled arrangement, the material is said to possess *ferromagnetic* behavior (e.g. magnetite, Fe_3O_4). Normally, within the magnetic domain, significant net magnetization emanates from the antiparallel alignment of adjacent non-equivalent sub-lattices. The resultant macroscopic characteristics are comparable to ferromagnetism. However, above the Néel temperature, the material remodels to *paramagnetism* (Akbazadeh *et al.*, 2012; Coey, 2009).

Reducing the physical size of materials to the nanoscale often result to properties profoundly divergent from their bulk scale (micro/macro) equivalents, such as inner structure, intrinsic electronic and optical properties including magnetic characteristics (Xiaolei *et al.*, 2013). This is particularly important in the technological application of magnetic nanoparticles and their nanocomposites because these differences include departure from the established laws of magnetic behavior of macroscopic materials. Magnetic nanocomposites are therefore composed of ferromagnetic nanosized particles dispersed either in an amorphous non-magnetic or magnetically soft matrix. The shape, size and distribution of the magnetic particles also significantly affect the properties of the composites. In order to fabricate a nanomagnetic composite of desired physico-chemical properties, an understanding of the structure of the host materials and their tunability is critical (Singh *et al.*, 2011). This is because the integrity of the synthesized composite for water purification will depend on the structural and textural properties of the matrix, which are a function of the interaction mechanisms of the nanoparticle surfaces, and the parent matrix. The interaction mechanisms may be through ionic bonds, covalent bonds, van der Waals forces and hydrogen bonding. The better the interaction and dispersion of the nanoparticles in the matrix, the better the overall properties. In general, the macroscopic properties of the magnetic nanocomposites are determined by several factors including the composition of the matrix, the physicochemical properties of each constituent, the geometry and dispersion of the nanoparticles on the matrix network, guest-guest and guest-host molecular interactions and subsequent modifications of the matrix. It is self-evident that these parameters are closely interlinked. In addition, because of this scale reduction, magnetic nanoparticles possess other specific properties such as coercivity (the field required to bring the magnetization to zero) and superparamagnetism at temperatures above the so-called *blocking temperature* (Lu *et al.*, 2007). As already noted, the orientation of the magnetic moment, μ, of the electrons determines the magnetic properties.

Factors Affecting the Magnetism of Nanoparticles

In this section, we consider in brief a few main factors that influence the magnetic behavior of magnetic nanoparticles. Other intrinsic nanoscale magnetic properties such as spontaneous magnetization, anisotropy (the directional dependence of a material's magnetic properties) and Curie temperature or magnegtostriction have been discussed by Coey (2009) and the details are not reiterated here. A foresight on how the factors could be controlled is key to the design of nanocomposites with desired magnetic properties for varied applications, including water purification. Of particular importance in adsorption, reactions are the particle size.

Effect of Size

As earlier mentioned, the laws of classical physics become less and less applicable and valid with a decrease in the dimensions of a system. Magnetic particles are not an exception to this observation. There are various reasons to account for this characteristic behavior of nanoparticles. Firstly, the finite sizes have a concomitant effect on the quantization of energy states also known as *quantum confinement* relating to the fact that spacing between the electronic energy levels states increases inversely to the square of the nanoparticle size. This gives rise to discrete energy levels in the nanomaterial unlike the continuous sets of levels frequently found in their bulk counterparts (Ortega, 2012). These nanoparticles are sometimes referred to as *quantum dots*. That is, a tiny nanoparticle can be paramagnetic, yet the small nanoparticles are tiny portions of a ferromagnetic bulk material. The domain configuration of a ferromagnetic material, therefore, predetermines the size dependence of its magnetic characteristics. Furthermore, the qualitative change in electronic energy configuration due to quantum confinement also imbues the nanoparticles with unusual catalytic properties remarkably different from those of the macroscopic ones. Secondly, geometric constructions result to an increase in the number of atoms located at the surfaces and interfaces of the nanoparticles, i.e large surface area. For example, a small metal nanocrystal of 1 nm-diameter will have 100% of its atoms on the surface; while on the other hand, a 10 nm diameter has 15% of its atoms on the surface. The material with a higher surface area is expected to be more reactive. This becomes particularly important in adsorption reactions, which mainly occur at the surface, and interfaces. For macroscopic materials, the fraction of surface atoms in bulk materials is negligible compared to the inner atoms while nanostructures possess a large fraction of surface atoms per unit volume, which in turn affects the surface properties of the material[7]. These have a direct influence on the adsorption behavior of the materials. The fraction of atoms at the surface is called dispersion, *F*. The fraction of atoms on the surface of the particle (F_s, percentage) can be estimated using the relation:

$$F_s = 4N^{-1/3} \times 100$$

Where N is the total number of atoms in the particle.

Since the dimensions of the magnetic nanoparticles are significantly smaller than the size of magnetic domains in the bulk material, a new magnetic configuration develops that responds fast to an applied magnetic field with negligible remanence (residual magnetism) and coercivity. Hence, superparamagnetic nanoparticles are of great interest in environmental remediation. So both the magnetic and surface properties can be harnessed to meet for an intended purpose, such as catalytic degradation and adsorption of molecules. Generally, the nanoparticles function best when the dimensions are below a certain critical value which is dependent on the material but is typically between 10-20 nm. It is important to note here that particles in this size range have intrinsic instability over time. The particles tend to agglomerate to reduce the energy associated with the high surface area to volume ratio. Furthermore, metallic nanomagnetic particles are very reactive and easily oxidized when exposed to air, during or after synthesis, losing their magnetism and dispersibility. This challenge is overcome by coating the particles with organic moieties, surfactants, polymers, carbon, silica materials etc. These species not only stabilize the nanoparticles but also are functionalizing agents for a specific application (Figure 1). The matrix also serves to restrain the mobility of the nanoparticles and reduce chain transfer reactions.

Impregnation of materials with nanoparticles, therefore, presents a suitable pathway for controlling the morphological, structural and consequently magnetic properties of water purification adsorbents. Andres et al. (2008) showed that the mean particle sizes could be controlled by varying the reaction conditions, such as reactant type and concentration. The results depict that the average particle size is reduced from 169 to 76 to 33 nm as the concentration of excess OH- is increased from 0.2 to 2 to 20 mM, respectively.

Figure 1. Core-shell nanocomposite structure where the shell serves as stabilizer and functionalizing agent while the magnetic core provides magnetic separation

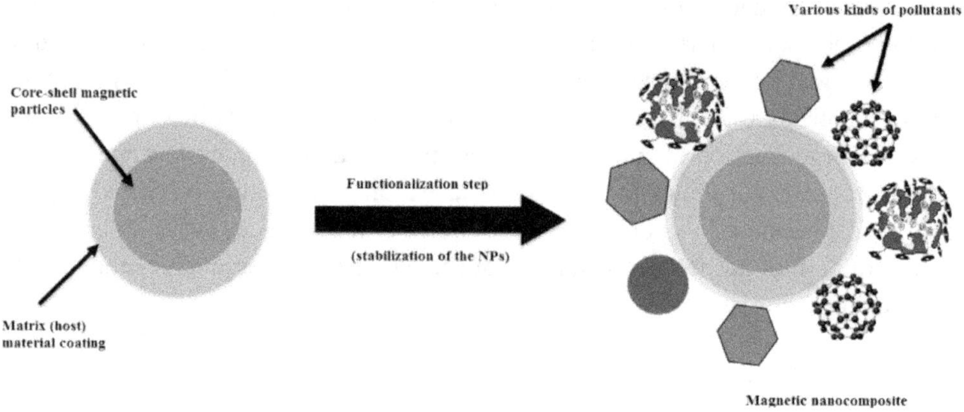

Therefore, properties as high field irreversibility, high saturation field, super-paramagnetism and extra anisotropy contributions, are all traceable to finite-size and corresponding surface area effects.

Natural Sources of Magnetic Nanoparticles

There is a wide range of exceptionally selective naturally occurring magnetic nanomaterials. Magnetite, a ferromagnetic member of the spinel group of minerals, is the most abundant naturally occurring magnetic mineral, often found in large quantities in beach iron sands. It frequently occurs as a constituent of igneous and metamorphic rocks. Magnetites are also formed by serpentinization of peridotite (or dunite) and as the major product of rusting steel surfaces. Massive deposits of magnetite are spread worldwide including places such as Atacama region of Chile, west coast of the North Island of New Zealand and the Valentines region of Uruguay (Tania, 2012). Other related magnetic minerals that occur in nature are maghemite (γ-Fe_2O_3) and hematite (α-Fe_2O_3). Under thermal treatment, magnetite can be converted to maghemite and hematite. Another naturally occurring magnetic nanoparticle is the zero valent iron (ZVI). Zero valent iron based nanocomposites have been used extensively in recent times in water purification regimes. Here, the magnetic nanoparticles are either used directly as adsorbents or as the core material in a core-shell nanoparticle structure where the mixed iron oxide shell provides adsorption sites while the magnetic core realizes magnetic segregation (Shengsen *et al.*, 2016). However, for practical purposes, it is worthwhile to note that nanoscale zero valent iron (nZVI) is more reactive than ZVI, and tends to aggregate due to its high surface energy and strong magnetic attraction, which reduces its effective surface area and contact area for targeted pollutants. A possible mitigation strategy to stabilize nZVI is by immobilizing the nZVI particles onto supporting materials such that the properties of both support materials and nZVI are conserved (Yan *et al.*, 2013).

Advantages of Magnetic Nanocomposites

Magnetic nanocomposites possess the inherent advantages of employing nanoscale particles, that is, high adsorption competence. For example, Yean et al. (2005) observed that decreasing the particle size of nano-magnetite from 300 to 11 nm, its adsorption capacity for arsenic increased 100 fold. This is called the "nanoscale effect" attributed to the change of magnetite surface structure; which both increased the specific surface area and created new adsorption sites. Furthermore, as the particle sizes are decreased further, magnetic particles (e.g nano-maghemite and nano-magnetite) become superparamagnetic, losing permanent magnetic moments while responding to an external magnetic field, thereby allowing untroublesome partitioning and recovery by a low-gradient magnetic field. Therefore, the use of magnetically modified adsorbents makes it possible to perform the separation process directly in crude samples containing suspended solid materials by applying an external magnet without the necessity of filtration, making the segregation step easier, rapid and effective. The fast mass transfer is thus achievable due to a large contact surface area between the adsorbents and pollutants, facilitating fast kinetics and rapid adsorption equilibrium. This is also attributed to the larger specific surface area, a large number of active binding sites (edges, vacancies) and short intraparticle diffusion length typical of a nanomaterial. Magnetite (Fe_3O_4) has been especially widely used as a magnetic material due to its excellent magnetic properties, chemical stability and biocompatibility (Kara *et al.*, 2015). Depending on the properties of the carrier material; the synergistic combination of these materials with magnetic particles offers the composite adsorbent distinguished properties such as good dispersity, large surface area, nontoxicity, strong superparamagnetism and excellent extraction ability. Therefore, manipulation of magnetic adsorbents is much easier in complex multiphase systems with an external magnetic field. Figure 2 demonstrates a magnetic separation. Furthermore, these metal oxides based nanocomposites are easily regenerated by modifying the solution pH (Sharma *et al.*, 2009). Generally, in most cases, the adsorption capacity of nanocomposite remains robust after 3-10 adsorption-desorption cycles (Chang *et al.*, 2016). Moreover, the temperatures required for reclamation of magnetic nanocomposites are very low, the energy consumption for regeneration is greater than 50% of other strategies, low operating pressures are needed with low-pressure drops, and the performance

Figure 2. Magnetic separation process

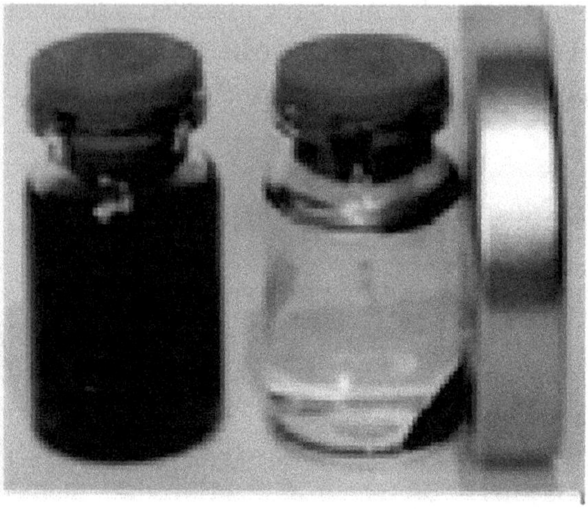

has high flexibility and stamina. The process also exhibits fast adsorption kinetics and does not produce contamination and the adsorbents can be fully recovered with minimum or no damage. However, reduction in adsorption capacity after recycling has also been reported (Mayo *et al.*, 2007). Nevertheless, these advantages prevail over the difficulties normally associated with other powdered adsorbents including non-magnetic nanomaterials. Magnetic nanocomposites are therefore economically and technologically expedient.

Like other adsorbents, the nanocomposites can also be functionalized with target molecules.

Synthesis of Magnetic Nanocomposites

There are various methods of synthesis of magnetic nanocomposites that have been reported in the literature, they include; co-precipitation, microemulsion, sol-gel synthesis, solvothermal synthesis, sonochemical reactions, thermal decomposition, combustion, solid-state, microwave assisted, hydrothermal reactions, electrochemical synthesis, among others (Lu *et al.*, 2007). Figure 3 presents a summary of synthetic routes for magnetic nanocomposites. Factors considered when choosing a method of synthesis of magnetic nanoparticles include; application, particle size distribution and shape, size and crystal structure required. Some of these methods are briefly described below.

Figure 3. Classification of synthetic methods for magnetic nanocomposites

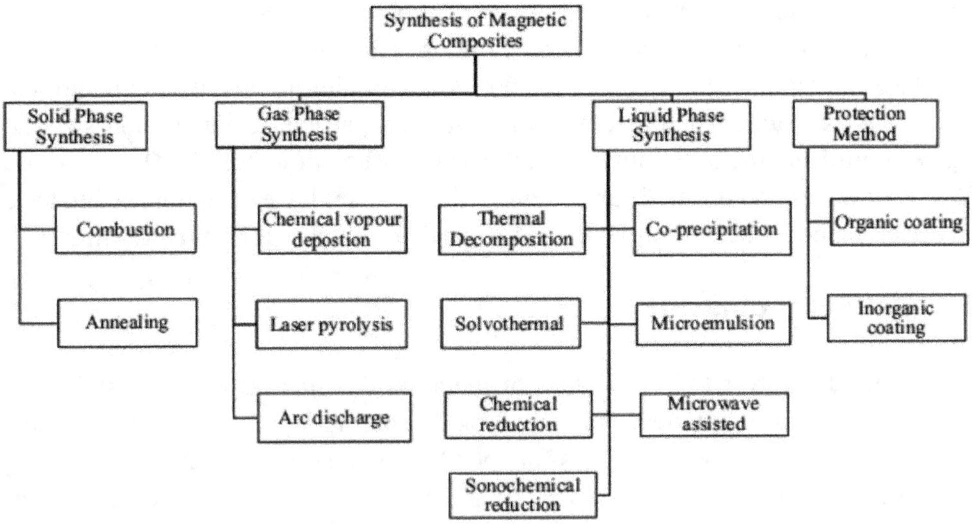

Co-Precipitation

This is the most common method of synthesis of magnetic nanocomposites because it is simple, efficient and it generates high yields (Lu *et al.*, 2007). This method involves a stoichiometric reaction of an aqueous solution of ferrous and ferric ions in an alkaline medium (NaOH or NH_3OH) carried out in an inert atmosphere. The size, shape, and composition of the product can be controlled through careful adjustment of the Fe^{2+}: Fe^{3+} ratio, type of salt used (i.e sulphates, chlorides, nitrates), and the operational conditions such as the pH, temperature and ionic strength of the reactants. The general equation for this reaction is as follows.

$$Fe^{2+} + 2Fe^{3+} + 8OH \rightarrow Fe(OH)_2$$

$$Fe(OH)_2 + 2Fe(OH)_3 \rightarrow Fe_3O_4 + 4H_2O$$

This method is simple and efficient, it does not use harmful materials and it produces large quantities of particles. However, the particles obtained in this method easily agglomerate hence there is broad particle size distribution (Woo *et al.*, 2004).

Microemulsion

It involves dispersal of the nanoparticles in two immiscible liquids (commonly water and oil) in the presence of a surfactant. The surfactant molecules form nanodroplets which encapsulate iron salt solutions separating them from the oil solution. Size control is possible by varying the oil/water ratio, and morphology is achieved by regulating the type and concentration of the surfactant (Li *et al.*, 2011).

The advantages of this method include; use of simple equipment and operational conditions (pressure and temperature), the absence of particle agglomeration, size and shape control and the possibility to synthesize a variety of materials (Hasany *et al.*, 2012). However, the residual surfactants in the particles have been reported to have an effect on the properties of the magnetic particles obtained (Lemine *et al.*, 2012).

Sol-Gel Synthesis

This method involves the fabrication of metal oxides such as titanium and silica onto the iron oxide particles. This metal oxide coating is used to modify the surface properties of the iron particles for specific applications as well as preventing agglomeration. The process takes place in two stages, that is, direct mixing of monomers with the iron oxide nanoparticles followed by a polymerization at mild temperatures. In the first stage, the precursors (metal alkoxides) undergo hydrolysis in which the alkoxides are replaced with hydroxides from water which also generate free alcohol resulting in the formation of a colloidal solution (sol). This then acts as the starting material (monomers) for addition and inorganic polymerization resulting in the formation of a three-dimensional patterned network (gel) (Mou *et al.*, 2012).

Apart from control of size, the sol-gel method is surfactant-free, it is a relatively mild reaction and hence of low cost. However, the three-dimensional patterned network limits its applications (Lemine *et al.*, 2012; Mou *et al.*, 2012).

Hydrothermal Synthesis

This method generates single crystals by use of hot water under high pressure in an autoclave. It only involves the use of one of the ferrous salts as a precursor contrary to the pair salt ions in the co-precipitation method. Size control is achieved by controlling the temperature in the autoclave (Suslick and Price, 1999). Advantages of this method include; cost effectiveness, high yield products, size control and better crystallinity with good morphology.

Solvothermal

This method is similar to hydrothermal procedure except that in the solvothermal method, an organic solvent is used as a dispersion medium instead of water. The organic solvent allows for selection from a wide range of materials and also enables reactions that are not possible with hydrothermal method. This method is useful for purification of wastewater containing organic pollutants. Morphology is dependent on the concentration of reactants while size is controlled by varying operation conditions such as the temperature and reaction time (Li *et al.,* 2011).

The particles obtained in this method are hydrophobic hence there is no adsorption of anions and cations on the nanosurface, size and morphology control is possible.

Sonochemical

Ultrasonication is applied to activate the internal liquid to generate high temperature and high partial pressure which results in oxidation, decomposition, and hydrolysis . The particle size, morphology, and the magnetic behavior depends on the ultrasonication frequency and time as well as the operational temperature (Abu and Gedanken, 2005).

Thermal Decomposition

The organometallic precursors (metal carbonyls or metal acetylacetonates) are injected in high boiling organic solvent containing stabilizing surfactants such as oleic acid (Behrens, 2011; Lu *et al.*, 2007).

MAGNETIC NANOCOMPOSITE ADSORBENTS

Adsorption of Heavy Metal Cations

Heavy metal cations are generally toxic and are potentially carcinogenic under long exposure. Removal of these cations from wastewater still remains a subject of great importance and continues to attract significant attention among researchers. Several

Table 1. Summary of synthesis methods of magnetic nanoparticles

Method of synthesis	Advantages	Disadvantages
Co-precipitation	Simple and produces high yield	Broad size distribution
Micro-emulsion	Size and shape control possible	Low yield
Hydrothermal	Very narrow size distribution	High yield
Solvochemical	Good for adsorption of hydrophobic matter	High yield
Sonochemical	Narrow size distribution	Complicated process-equipment
Thermal decomposition	Very narrow size distribution	High yield

magnetic adsorbents have been successfully designed for various metals and only a case for lead and mercury will be discussed here. A summary of magnetic sorbents, optimal reaction conditions and adsorption capacities for other heavy metals are listed in Table 2.

The upper limit for Lead, 0.05 mg/L, in drinking water by World Health Organization standards demands innovative, sensitive and economical extraction strategies that reach these molecular limits, with possible regeneration of the metal for commercial use.

Magnetic magnetite (Fe_3O_4) and Xanthan Gum (XG) were employed to fabricate a hybrid adsorbent Fe_3O_4@silica–XG, with an excellent affinity for Pb(II) ions with a high extraction efficiency of 100% Pb(II) removal from aqueous media (Peng *et al.*, 2014). Also, the nanocomposite maintained 97% extraction efficiency in the treatment of real wastewater effluent from a battery industry after 21 adsorption-desorption cycles. The Langmuir monolayer maximum adsorption capacity was 21.32 mg/g. However, in the presence of other cations, the removal efficiency for lead decreased in the order Pb(II) > Cu(II) > Zn(II) > Cd(II). The adsorbent was recoverable by the magnetic field and adsorbent regeneration was achieved using hydrochloric acid.

Another novel adsorbent for Pb(II) ions is sodium alginate gel beads embedded with magnetic maghemite nanoparticles (Fe_3O_4@SA-Zr). Here, zirconium (IV) served as the cross-linking ion (Idris *et al.*, 2012). Mechanistically, the –COOH group of sodium alginate and –OH groups connecting to Fe species bound the Pb(II) ions by forming a carboxyl-metal and M–O–metal complexes, respectively. The adsorption is endothermic with 100% removal efficiency for lead concentrations below 200 mg/L at 293 K. This is suitable for ambient temperature operations and the experimental conditions were near real environmental situations.

Magnetic chitosan particles have also shown to be facile carriers for the synthesis of nanocomposites for lead sequestration from the water. Examples include magnetic chitosan impregnated yeast biomass modified with ethylenediamine (EYMC), yeast biomass modified with magnetic chitosan microparticles (YMC), triethylenetetramine grafted magnetic chitosan and ethylenediamine-doped magnetic chitosan microparticles (EMC) (Li *et al.*, 2013). The adsorption reactions were greatly pH dependent with optimum removal observed at pH 4-6. Complexation with NH_2 and ion exchange reactions with –COOH and –OH are conceivable as possible adsorption mechanisms. Water pollution by mercury (Hg) also presents a serious threat to the health of the entire food chain due to its toxicity and bio-accumulative properties which explain the research interests in water purification procedures for complete removal of mercury or reduction to extremely very low concentrations as possible.

Conventional water treatment protocols are generally based on powdered activated carbons (PAC) as adsorbents. However, Faulconer et al. (2012) demonstrated that magnetite coated PACs have the capacity to sorb up to 84% of Hg (II) ions from solution at 100 µg/L metal ion solution. Similar results were obtained using magnetic iron oxide nanoparticles mutated with 2-mercaptobenzothiazole within a wide range of pH (Parham *et al.*, 2012). The hybrid posted stupendous results of complete removal of 200 µg/L of Hg (II) within 4 minutes of contact time. Short residence times are suitable and economical for water purification systems. Noteworthy, the overall performance of the nanocomposite is dependent on the properties of the nanoparticles host material. As Girginova et al. (2010) reported, silica coated magnetite particles (Fe_3O_4/SiO_2) have significantly lower magnetic saturation value compared to pure Fe_3O_4. This is attributed to the fact that the silica shells surrounding the magnetite particles obstruct the interaction between magnetic cores. Consequently, modification of the composite surface by dithiocarbamate groups surface ($Fe_3O_4/SiO_2/NH/CS_2^-$) resulted to increased affinity for Hg (II), from 24% to 74%. Careful choice of the operation pH is critical because of variation of the zeta potential (from negative to positive) of the nanoparticle-mercury complex with pH and Hg (II) concentration.

The adsorption mechanisms are elucidated by varying of solution pH and subsequent characterization of the matrix-metal ion complexes employing tools as FTIR and XRD analysis. For instance, most researchers propose that the adsorption mechanism of Co(II) on hydrous ferric oxide be *via* the reaction with hydroxyl groups, spontaneous precipitation of $Co(OH)_{2,}$ occupation in the vacancy of iron oxide, or reduction by zero valent iron (Xing *et al.*, 2016). Such multi-mechanistic processes of adsorption were proposed by Xing and co-workers whereby the core-shell structure of graphene adsorbent probably experienced dissolution/re-precipitation process as shown in Figure 4.

Table 2. Some magnetic nanocomposites for heavy metals removal from water

Adsorbent	Adsorbate	Adsorption capacity	pH	Temp.	Ref
Magnetic powdered activated carbon	Hg(II)	91%	4.5	30	(Faulconer *et al.* 2012)
EDTA functionalized magnetite	Cu(II)	46.27 mg/g	6	25	(Liu *et al.* 2013)
Magnetic modified sugarcane bagasse	Cd(II)	1mmol/g	5	25	(Yu *et al.* 2013)
Magnetic graphene/iron oxides nanocomposite	U(VI)	69.49	5.5	25	(Zong *et al.* 2013)

Figure 4. Schematic diagram of the adsorption mechanism of Co(II) onto Zero valent iron/graphene nanocomposite. The mechanism proposed by this study was highlighted in the red dotted frame

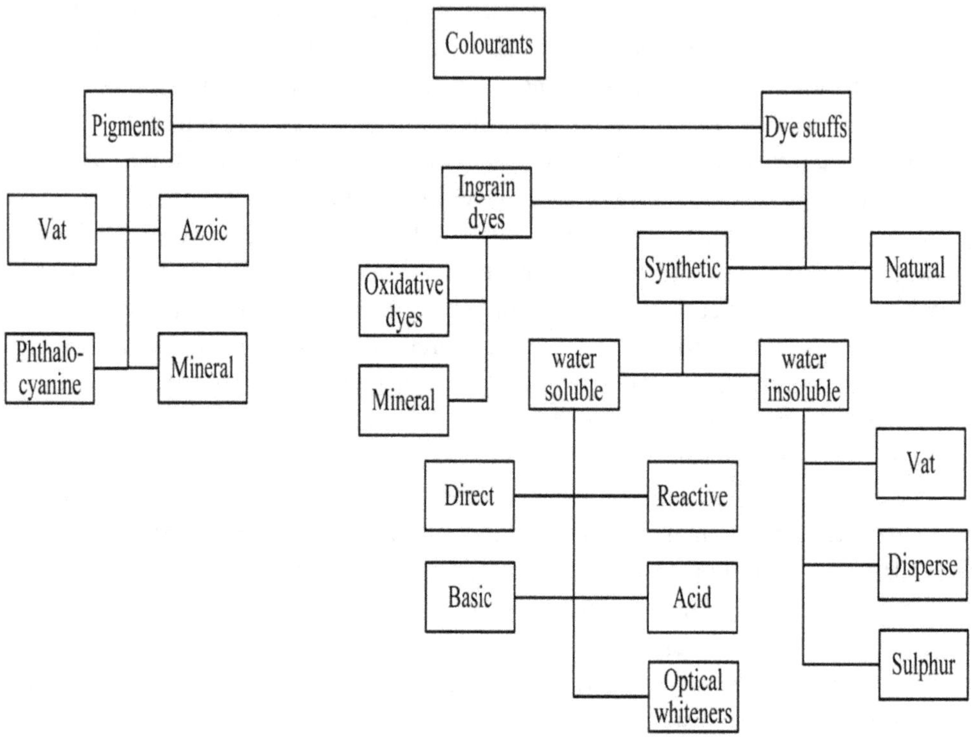

Adsorption of Dyes

Dyes and pigments are coloring agents used for value addition in textile and printing among other finished products. Figure 5 represents the general classification of colorants. The dyes containing effluents from these industries ultimately end into the municipal wastewaters. Besides coloration and making the water odious, dyes are known to possess carcinogenic and mutagenic effects, even at low concentrations. Therefore, environmentally, their elimination from water is of utmost importance.

Dye adsorption studies show that magnetic chitosan nanoparticles coated with ethylene diamine have the capacity to adsorb Acid Orange 7 and Acid Orange 10 dyes (Zhou *et al.*, 2011). The ethylene diamine imbues the composite with amino groups ($-NH_2$) active binding sites that reinforce the adsorptive effectiveness of chitosan nanoparticles. The performance of the adsorbent is however also dependent on the particle size and substrate functional groups. As a result, the composite exhibited a higher affinity for Acid Orange 7 than Acid Orange 10 due to smaller particle size and presence of monovalent sulfonated acid groups on Acid Orange 7 that enhances the availability of more protonated amino groups on the chitosan particle and the ease of packing and reputation of the smaller molecules into the inner pores of the adsorbent. In the same study, desorption of the dyes from dye-sorbent matrix was achieved using NH_4OH/NH_4Cl solution at pH 10. The performance of the recycled nanocomposite was found to be 81.1% and 87.8% for Acid Orange 7 and Acid Orange 10, respectively. The isothermal data confirmed to Langmuir model depicting monolayer adsorption mechanism. Typical of most adsorption reactions involving dyes, the process was strongly pH-dependent indicating that the surface charge of the adsorbent and ionic form of the adsorbates greatly influences the removal efficiency.

In another study, Zhang and co-workers (2007) fabricated copper ferrite/activated carbon ($CuFe_2O_4$@C) nanocomposite of different $CuFe_2O_4$:C ratios by mass

Figure 5. General classification of colorants

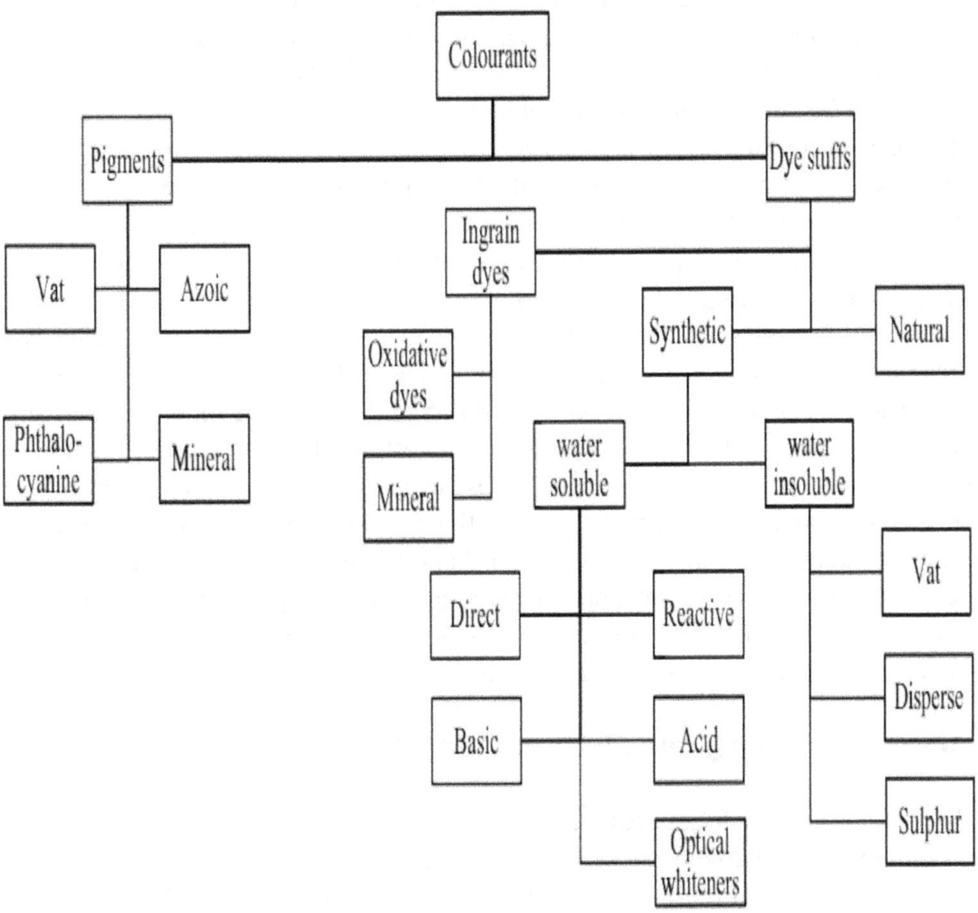

for removal of Acid Orange 7. Saturation magnetization was lower, as expected, for the hybrid will less $CuFe_2O_4$ content. From the isothermal data, it was conceived that the copper ferrite particles homogenized the surface of activated carbon though it was the activated carbon in the matrix that was predominantly responsible for the adsorbed the dye. Therefore, the dual function of the magnetic nanoparticles was surface tuning and easing adsorbent recovery.

Chang et al. (2016) synthesized Fe_3O_4/activated montmorillonite (Fe_3O_4/Mt) nanocomposite for removal of cationic dye methylene blue (MB) from aqueous media. Incorporation of Fe_3O_4 increased the total pore volume and average pore size in the Fe_3O_4/Mt nanocomposite which was beneficial for adsorption efficacy. Impregnation with iron particles is also known to potentially increase the basal spacing than enhances intercalation of adsorbing molecules. Adsorption was mainly attributed to coulombic interactions between the negatively charged adsorbent surface and the cationic dye as evidenced by the high extraction efficiency over a wide range of pH. However, at the lower pH competition between the dye cations and H^+ for the active binding sites becomes significant impeding effective removal efficiency. In such situations, pH values greater than the isoelectric point (pH_{pzc}) of the adsorbent (7.56 in the case of Fe_3O_4/Mt) are recommended. Dynamic equilibrium was achieved in 30minutes which are suitable for system design with 99% dye removal efficiency. Furthermore, the nanocomposite maintained removal efficiency greater than 83% after 5 cycles with concomitant good magnetic separation effect in an external magnetic field, a testament that Fe_3O_4/Mt is plausibly stable and recyclable.

In another study, Zhang and Kong (2011) produced magnetic Fe_3O_4/C core-shell nanoparticles of average size 250 nm for sequestration of two organic dyes, namely methylene blue and cresol red from the water. Magnetic susceptibility investigations showed the composite to be superparamagnetic. The adsorption capacities for methylene blue (MB) and cresol red were 44.38 mg/g and 11.22 mg/g, respectively within the experimental conditions. The higher affinity for MB was attributable to the hydroxyl and carbonyl functional groups on the surface of the composite and the negative potential of the magnetic nanoparticles creating weak coulombic interaction between the cationic dye and the adsorbent. Figure 6 displays the summary of adsorption processes using magnetic nanocomposites for dyes removal from water. A list of available magnetic sorbents, optimal reaction conditions and adsorption capacities for other dyes are listed in Table 3.

Adsorption of Organic Micropollutants

The presence of pharmaceutically active ingredients in water resources has not only raised international environmental concern but has also sparked research on techniques for removal of these compounds from water (Ngumba *et al.*, 2016). This is

Table 3. Some magnetic nanocomposites for dye removal from water

Adsorbent	Adsorbate	Adsorption capacity	pH	Temp.	Ref
Magnetic carbon-iron oxide nanocomposite	Crystal violet	111.8	8.5	50	(Singh et al. 2011)
Magnetic Fe_3O_4 @ graphene	Congo red	33.06	-	-	(Yao et al. 2012)
Superpara-magnetic rectorite nanocomposite	Neutral red	16.0	-	-	(Wu et al. 2011)
Magnetic N-lauryl chitosan nanocomposite	Remazol red 198	267	-	-	(Debrassi et al. 2012b)
Magnetic zeolite/iron oxide nanocomposite	Reactive Orange	16 1.1	-	25 ± 2	(Fungaro et al. 2011)
Magnetic zeolite/iron oxide nanocomposite	Indigo Carmine	0.58	2	25 ± 2	(Fungaro et al. 2011)
Graphene-Fe_3O_4 nanocomposite	Pararosaniline	198.23	3–5	25	(Wu et al. 2013)
Magnetic ferrite nanoparticle alginate composite	Basic Blue 9	106	8	-	(Mahmoodi, 2013)

Figure 6. Summary of sequence of dye removal using magnetic nanocomposites

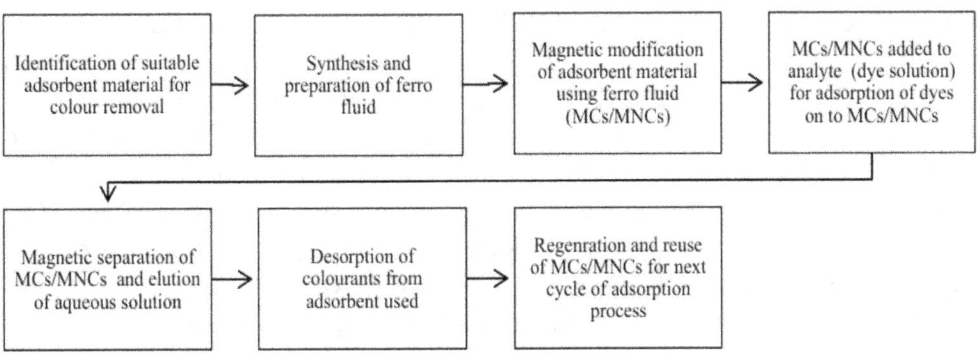

because these compounds are resistant to conventional water treatment approaches and are toxic even at trace concentrations. Pharmaceutically active compounds in aquatic environments have also been shown to induce the development of drug resistant bacterial strains. Magnetically modified adsorbents have since been developed for mitigation of this growing problem. Shikuku (2016) demonstrated that a γ-Fe_2O_3-enriched natural kaolinite clay (Fe-MC) could adsorb sulphonamide antibiotics (sulfachloropyridazine-SCP and sulfadimethoxine-SDM) residues from water at circumneutral pH of 6.4. For both compounds the kinetic data conformed to a pseudo-second order kinetic law which indicates that the adsorption mechanism is dependent on both the adsorbate and adsorbent and the rate-limiting step is a chemical process. Generally, sulfonamides exhibit complexity in adsorption mechanisms because they have two pK_as, hence their molecules may exist as positive, neutral and negative species, which shows different charge property, solubility, and hydrophobicity. The adsorption capacity seemed to be directly proportional to the dipole moment (Debye) of the molecules. Consequently, the dipole moment of sulphonamides is a significant parameter affecting sulfonamide adsorption onto the composite. The higher extraction yield of Fe-MC for SCP (log K_{ow} 0.31) than SDM (log K_{ow} 1.17) depicted that other mechanisms besides hydrophobic interaction were involved in the adsorption process. Furthermore, changes from neutral to anionic form increase the hydrophilicity of the sulfonamides and thus hydrophobic interactions become significantly lessened. Electrostatic interactions are known to influence sorption of ionic compounds. In the study, SCP and SDM molecules (pK_a = 5.7) were mostly anionic under the experimental conditions. The pH of the point of zero charges (pH_{pzc}) of Fe-MC was about 6.1 and therefore the net surface charge of the adsorbent was negative at pH above this value. This testifies that the adsorption of SCP and SDM onto Fe-MC could not be attributed to coulombic attractions.

The presence of multivalent cations in clay materials may promote adsorption of the negatively charged species (SCP^- and SDM^-) by surface bridging with these anions. Therefore, the adsorption of SCP and SDM was partly attributed to cation bridging interactions. Such Fe-MC-adsorbate interactions may be correlated with the molecular dipole moment (Debye). The most polar sulphonamide (SCP) was adsorbed more. This suggested that the adsorption variance of SCP and SDM onto Fe-MC may be closely linked to their polarity and not their kinetic diameters. Also, leaching of iron from the solid phase into the aqueous solution was below the LOQ (0.1ppm) indicating the iron-oxide particles are strongly grafted on the clay matrix, a good characteristic for a longer recycling life of the composite.

Peng et al. (2014) synthesized magnetically impregnated montmorillonite as an adsorbent for the elimination of humic acid from water. A removal efficiency of 96% was reported at pH 6.1. Noteworthy, the adsorption was strongly dependent on pH and ionic strength. The authors realized ligand exchange reaction between -COOH groups of humic acid and the montmorillonite as the adsorption mechanism. The extraction efficiency increased with increase in ionic strength. This phenomenon is a consequence of the compression of the diffuse double layer and the molecular volume of humic acid and charge neutralization.

Carbon nanotubes (CNTs) are also facile matrices for fixation of magnetic nanoparticles. An adsorption study demonstrated that CNTs/Fe_3O_4 nanocomposites have the capacity to remove Bisphenol A (BPA) from wastewater (Ghosh *et al.*, 2013). Fe_3O_4 has a characteristic absorption band in FTIR spectroscopy centered at 584 cm^{-1} assigned to the Fe–O–Fe stretching and bending modes and the band is useful in assessing successful incorporation of Fe_3O_4 in the composite. The BET surface area depends on the CNTs:Fe_3O_4 ratio calculated as 17.9 to 50.6 $m^2 g^{-1}$ for 10% CNTs/Fe_3O_4 and 50% CNTs/Fe_3O_4, respectively. On the contrary, incorporation of the magnetic magnetite nanoparticles decreases the micropore volume of the CNTs. The magnetic saturation value (*Ms*) for Fe_3O_4 was, as expected, higher than that for CNTs/Fe_3O_4 due to the grafting of CNTs onto Fe_3O_4 surface. Furthermore, no significant changes in the coercivity of the central loop of the samples were observed. Such excellent magnetic properties depict a strong magnetic response of CNTs/ Fe_3O_4, which enables the adsorbent to be easily recovered and recycled from aqueous solution in the presence of an external magnetic field. The increase in specific surface area with an increase in CNTs content has a positive influence on the adsorption capability for BPA and other adsorbates as well. This depicts that the additive content of CNTs in the nanocomposite plays a significant driving role in the adsorption process. However, the increase in CNT content resulted in a decrease in the magnetic response of the nanocomposite which is unfavorable for separation from aqueous media. 20%CNTs/ Fe_3O_4 sample, therefore, presents the best ratio with a suitable net combined effect of a relatively high adsorption potential but also a strong magnetic response. The adsorption process was well described by the Freundlich isotherm model indicating a heterogeneous distribution of the adsorption sites.

Research indicates that new magnetic nanocomposites continue to be developed each day and their suitability for target compounds are commendable. Nevertheless, there are still knowledge gaps upon which future research must focus. This includes, but not limited to, studies on the economic feasibility of magnetic nanocomposites on a commercial scale and the performance of these adsorbents in fixed bed systems for real wastewater scenarios containing multi-pollutants such as organic compounds, dyes, and heavy metals.

REFERENCES

Abu, M. R., & Gedanken, S. (2005). Sonochemical synthesis of stable hydrosol of Fe_3O_4 nanoparticles. *Journal of Colloid and Interface Science*, *284*(2), 489–494. doi:10.1016/j.jcis.2004.10.073 PMID:15780286

Akbarzadeh, A., Samiei, M., & Davaran, S. (2012). Magnetic nanoparticles: Preparation, physical properties, and applications in biomedicine. *Nanoscale Research Letters*, *7*(1), 144. doi:10.1186/1556-276X-7-144 PMID:22348683

Alswat, A. A., Ahmad, M. B., & Saleh, T. A. (2016). Zeolite modified with copper oxide and iron oxide for lead and arsenic adsorption from aqueous solutions. *Journal of Water Supply: Research & Technology - Aqua*, *65*(6), 465–479. doi:10.2166/aqua.2016.014

Alswat, A. A., Ahmad, M. B., & Saleh, T. A. (2016). Zeolite modified with copper oxide and iron oxide for lead and arsenic adsorption from aqueous solutions. *Journal of Water Supply: Research & Technology - Aqua*, *65*(6), 465–479. doi:10.2166/aqua.2016.014

Ambashta, R. D., & Sillanpää, M. (2010). Water purification using magnetic assistance: A review. *Journal of Hazardous Materials*, *180*(1-3), 38–49. doi:10.1016/j.jhazmat.2010.04.105 PMID:20488616

Andres, M. V., Costo, R., Roca, A. G., Marco, J. F., Goya, G. F., Serna, C. J., & Morales, M. P. (2008).. . *Journal of Physics. D, Applied Physics*, *41*(13).

Behrens, S. (2011). Preparation of functionalized magnetic nanocomposites and hybrid materials: Recent progress and future directions. *Nanoscale*, *3*(3), 877–892. doi:10.1039/C0NR00634C PMID:21165500

Chang, J., Jianchao, M., Qingliang, M., Duoduo, Z., Nannan, Q., Mengxiao, H., & Hongzhu, M. (2016). Adsorption of methylene blue onto Fe_3O_4/activated montmorillonite nanocomposite. *Applied Clay Science*, *119*, 132–140. doi:10.1016/j.clay.2015.06.038

Coey, J. M. D. (2009). *Magnetism and magnetic materials*. New York, NY: Cambridge University Press.

Debrassi, A., Baccarin, T., Demarchi, C. A., Nedelko, N., Slawska-Waniewska, A., Dhuzewski, P., & Rodrigues, C. A. et al. (2012). Adsorption of Remazol Red198 onto magnetic N-lauryl chitosan particles: Equilibrium, kinetics, reuse and factorial design. *Environ. Sci. Pollut. Res*, *19*(5), 1594–1604. doi:10.1007/s11356-011-0662-6 PMID:22134860

Dhruv, M., Siddharth, M., & Singha, S. K. (2015). Magnetic adsorbents for the treatment of water/wastewater—A review. *Journal of Water Process Engineering*, *7*, 244–265. doi:10.1016/j.jwpe.2015.07.001

Faulconer, E. K., Hoogesteijn von Reitzenstein, N. V., & Mazyck, D. W. (2012). Optimization of magnetic powdered activated carbon for aqueous Hg(II) removal and magnetic recovery. *Journal of Hazardous Materials*, *199-200*, 9–14. doi:10.1016/j.jhazmat.2011.10.023 PMID:22104766

Fungaro, A., Yamaura, M., & Carvalho, T. E. M. (2011). Adsorption of anionic dyes from aqueous solution on zeolite from fly ash-iron oxide magnetic nanocomposite. *J. At. Mol. Sci.*, *2*(4), 305–316.

Ghosh, S., Badruddoza, A.Z., Hidajat, K. & Uddin, M.S. (2013). Adsorptive removal ofemerging contaminants from water using superparamagnetic Fe_3O_4 nanoparticles bearing aminated-cyclodextrin. *J. Environ. Chem. Eng.*, 122–130.

Girginova, P. I., Daniel-da-Silva, A. L., Lopes, C. B., Figueira, P., Otero, M., Amaral, V. S., & Trindade, T. et al. (2010). Silica coated magnetite particles for magnetic removal of Hg^{2+} from water. *Journal of Colloid and Interface Science*, *345*(2), 234–240. doi:10.1016/j.jcis.2010.01.087 PMID:20202646

Hasany, S.F., Ahmed, I., Rajin, I. & Rehman, A. (2012). Systematic Review of the preparation techniques of iron oxide magnetic nanoparticles. *Nanoscience Nanotechnology*, *2*, 148-158

Hay. (1873). *Patent No. 140*. Washington, DC: US Patent Office.

Idris, A., Ismail, N. S., Hassan, N., Misran, E., & Ngomsik, A. F. (2012). Synthesis of magnetic alginate beads based on maghemite nanoparticles for Pb(II) removal in aqueous solution. *Journal of Industrial and Engineering Chemistry*, *18*(5), 1582–1589. doi:10.1016/j.jiec.2012.02.018

Indira, T. K., & Lakshmi, P. K. (2010). Magnetic nanoparticles. A review. *International Journal of Pharmaceutical Sciences and Nanotechnology*, *3*, 1035–1042.

Ishikawa, T., Kataoka, S., & Kandori, K. (1993). The influence of carboxylated ions on the growth of β-FeOOH particles. *Journal of Materials Science*, *28*(10), 2693–2698. doi:10.1007/BF00356205

Kara, A., Demirbel, E., Tekin, N., Osman, B., & Beşirli, N. (2015). Magnetic vinylphenyl boronic acid microparticles for Cr(VI) adsorption: Kinetic, isotherm and thermodynamic studies. *Journal of Hazardous Materials*, *286*, 612–623. doi:10.1016/j.jhazmat.2014.12.011 PMID:25666882

Lemine, O. M., Omri, K., Zhang, B., El Mir, L., Sajieddine, M., Alyamani, A., & Bououdina, M. (2012). Solgel synthesis of 8 nm magnetite (Fe_3O_4) nanoparticles and magnetic properties. *Superlattices and Microstructures*, *52*(4), 793–799. doi:10.1016/j.spmi.2012.07.009

Li, X., Qi, Y., Li, Y., Zhang, Y., He, X., & Wang, Y. (2013). Novel magnetic beads based on sodium alginate gel crosslinked by zirconium(IV) and their effective removal for Pb^{2+} in aqueous solutions by using a batch and continuous systems. *Bioresource Technology, 142*, 611–619. doi:10.1016/j.biortech.2013.05.081 PMID:23771001

Li, X. M., Xu, G., Liu, Y. & He, T. (2011). Magnetic Fe_3O_4 nanoparticles: Synthesis and application in water treatment. *Nanoscience and Nanotechnology-Asia, 1*, 14-24.

Liu, Y., Chen, M., & Hao, Y. (2013). Study on the adsorption of Cu(II) by EDTA-functionalized Fe_3O_4 magnetic nanoparticles. *Chemical Engineering Journal, 218*, 46–54. doi:10.1016/j.cej.2012.12.027

Lu, A. H., Salabas, E. L., & Schuth, F. (2007). Magnetic Nanoparticles: Synthesis, protection, functionalization and application[Review]. *Angewandte Chemie International Edition in English, 46*(8), 1222–1244. doi:10.1002/anie.200602866 PMID:17278160

Mahmoodi, N. M. (2013). Magnetic ferrite nanoparticle–alginate composite: Synthesis, characterization and binary system dye removal. *J. Taiwan Inst. Chem. Eng, 44*(2), 322–330. doi:10.1016/j.jtice.2012.11.014

Majidi, S., Sehrig, F. Z., Farkhani, S. M., Goloujeh, M. S., Akbazadeh, A. (2016). Current methods of synthesis of magnetic nanoparticles. *Artificial Cells, Nanomedicine and Biotechnology, 4*, 722-734

Mayo, J. T., Yavuz, C., Yean, S., Cong, L., Shipley, H., Yu, W., & Colvin, V. L. et al. (2007). The effect of nanocrystalline magnetite size on arsenic removal. *Science and Technology of Advanced Materials, 8*(1-2), 71–75. doi:10.1016/j.stam.2006.10.005

Mou, F., Guan, J., Ma, H., Xu, L., & Shi, W. (2012). Magnetic iron oxide chestnut like hierarchical nanostructures: Preparation and their excellent arsenic removal capabilities. *ACS Application Materials Interfaces, 4*(8), 3987-3993.

Ngeno, E., Orata, F., Lilechi, D. B., Shikuku, V. O., & Kimosop, S. (2016). Adsorption of caffeine and ciprofloxacin onto pyrolytically derived water hyacinth biochar: Isothermal, kinetics and thermodynamics. *Journal of Chemistry and Chemical Engineering, 10*, 185–194.

Parham, H., Zargar, B., & Shiralipour, R. (2012). Fast and efficient removal of mercury from water samples using magnetic iron oxide nanoparticles modified with 2-mercaptobenzothiazole. *Journal of Hazardous Materials, 205-206*, 94–100. doi:10.1016/j.jhazmat.2011.12.026 PMID:22244341

Peng, X., Xu, F., Zhang, W., Wang, J., Zeng, C. & Niu, M. (2014). Magnetic Fe_3O_4@ silica xanthan gum composites for aqueous removal and recovery of Pb^{2+}. *Colloids Surf. A Physicochem. Eng. Aspects,* 27–36.

Puntes, V. F., Krishan, K. M., & Alivisatos, A. P. (2001). Colloidal nanocrystal shape and size control: The case of cobalt. *Science, 291*(5511), 2115–2117. doi:10.1126/science.1057553 PMID:11251109

Sharma, Y. C., Srivastava, V., Singh, V. K., Kaul, S. N., & Weng, C. H. (2009). Nano-adsorbents for the removal of metallic pollutants from water and wastewater. *Environmental Technology, 30*(6), 583–609. doi:10.1080/09593330902838080 PMID:19603705

Shengsen, W., Bin, G., Yuncong, L., Anne, E., & Feng, H. (2016). Adsorptive removal of arsenate from aqueous solutions by biochar supported zero-valent iron nanocomposite: Batch and continuous flow tests. *Journal of Hazardous Materials.* doi:10.1016/j.jhazmat.2016.01.052

Shikuku, V. O. (2016). *Adsorption of sulfonamides by γ-Fe_2O_3 modified natural clay: Equilibrium, kinetics and thermodynamics studies* (Unpublished Doctoral Dissertation). Maseno University, Kenya.

Singh, K. P., Gupta, S., Singh, A. K., & Sinha, S. (2011). Optimizing adsorption of crystal violet dye from water by magnetic nanocomposite using response surface modeling approach. *Journal of Hazardous Materials, 186*(2-3), 1462–1473. doi:10.1016/j.jhazmat.2010.12.032 PMID:21211903

Suslick, K. S., & Price, G. J. (1999). Application of ultrasound to materials chemistry. *Annual Review of Materials Science, 29*(1), 295–326. doi:10.1146/annurev.matsci.29.1.295

Tania, D. (Ed.). (2012). *Nanotechnology for Water Purification.* Brown Walker Press.

Wang, Z., Wu, D., Wu, G., Yang, N., & Wu, A. (2013). Modifying Fe_3O_4 microspheres with rhodamine hydrazide for selective detection and removal of Hg^{2+} ion in water. *Journal of Hazardous Materials, 244-245,* 621–627. doi:10.1016/j.jhazmat.2012.10.050 PMID:23177242

Woo, K., Hung, S., Choi, S., Lee, H. W., Ahn, J. P., Kim, C. S., & Lee, S. W. (2004). Easy synthesis and magnetic properties of iron oxide magnetic nanoparticles. *Chemistry of Materials, 16*(14), 2814–2818. doi:10.1021/cm049552x

Wu, D., Zheng, P., Chang, P. R., & Ma, X. (2011). Preparation and characterization of magnetic rectorite/iron oxide nanocomposites and its application for the removal of the dyes. *Chemical Engineering Journal, 174*(1), 489–494. doi:10.1016/j.cej.2011.09.029

Wu, Q., Feng, C., Wang, C., & Wang, Z. (2013). A facile one-pot solvothermal method to produce superparamagnetic graphene–Fe_3O_4 nanocomposite and its application in the removal of dye from aqueous solution. *Colloids and Surfaces. B, Biointerfaces, 101*, 210–214. doi:10.1016/j.colsurfb.2012.05.036 PMID:23010021

Xiaolei, Q., Pedro, J. J., & Qilin, L. (2013). Applications of nanotechnology in water and wastewater treatment. *Water Research, 47*(12), 3931–3946. doi:10.1016/j.watres.2012.09.058 PMID:23571110

Xing, M., Lejin, X., & Jianlong, W. (2016). Mechanism of Co(II) adsorption by zero valent iron/graphene nanocomposite. *Journal of Hazardous Materials, 301*, 286–296. doi:10.1016/j.jhazmat.2015.09.004 PMID:26368802

Yan, H., Li, H., Yang, H., Li, A., & Cheng, R. (2013). Removal of various cationic dyesfrom aqueous solutions using a kind of fully biodegradable magnetic compositemicrosphere. *Chemical Engineering Journal, 223*, 402–411. doi:10.1016/j.cej.2013.02.113

Yao, X. P., Fu, Z. J., Zhao, Y. G., Wang, L., Fang, L. Y., & Shen, H. Y. (2012). Use of tetraethylenepentamine-functional Fe_3O_4 magnetic polymers for matrix solid phase dispersion extraction and preconcentration of Cr(VI) in water samples at ultra-trace levels. *Talanta, 97*, 124–130. doi:10.1016/j.talanta.2012.04.006 PMID:22841056

Yao, Y., Miao, S., Liu, S., Ma, L. P., Sun, H., & Wang, S. (2012). Synthesis, characterization,and adsorption properties of magnetic Fe_3O_4@graphene nanocomposite. *Chemical Engineering Journal, 184*, 326–332. doi:10.1016/j.cej.2011.12.017

Yu, J. X., Wang, L. Y., Chi, R. A., Zhang, Y. F., Xu, Z. G., & Guo, J. (2013). Competitive adsorption of Pb^{2+} and Cd^{2+} on magnetic modified sugarcane bagasse prepared by two simple steps. *Applied Surface Science, 268*, 163–170. doi:10.1016/j.apsusc.2012.12.047

Zhang, G., Qu, J., Liu, H., Cooper, A. T., & Wu, R. (2007). $CuFe_2O_4$/activated carbon composite: A novel magnetic adsorbent for the removal of acid orange II and catalytic regeneration. *Chemosphere, 68*(6), 1058–1066. doi:10.1016/j.chemosphere.2007.01.081 PMID:17397903

Zhang, X., Zhang, P., Wu, Z., Zhang, L., Zeng, G. & Zhou, C. (2013). Adsorption of methylene blue onto humic acid-coated Fe_3O_4 nanoparticles. *Colloids Surf. A Physicochem. Eng. Aspects*, 85–90.

Zhang, Z., & Kong, J. (2011). Novel magnetic Fe_3O_4@C nanoparticles as adsorbentsfor removal of organic dyes from aqueous solution. *Journal of Hazardous Materials*, *193*, 325–329. doi:10.1016/j.jhazmat.2011.07.033 PMID:21813238

Zhou, L., Jin, J., Liu, Z., Liang, X., & Shang, C. (2011). Adsorption of acid dyes from aqueous solutions by the ethylenediamine-modified magnetic chitosan nanoparticles. *Journal of Hazardous Materials*, *185*(2-3), 1045–1052. doi:10.1016/j.jhazmat.2010.10.012 PMID:21035264

KEY TERMS AND DEFINITIONS

Guest: A nanoparticle embedded on a support material/adsorbent.

Host: A support material on which the magnetic nanoparticles are embedded.

Hydrophobicity: Incapability to effectively combine with water.

Kinetic Diameter: The smallest effective molecular dimension.

Magnetic Nanoparticles: particles whose size are less than 100nm and are responsive to external magnetic field.

Nanocomposite: A chemical matrix consisting of the guest and host materials.

Chapter 5
Advanced Nanomaterials for Water Engineering and Treatment:
Nano-Metal Oxides and Their Nanocomposites

Rabia Nazir
Pakistan Council of Scientific and Industrial Research Laboratories Complex, Pakistan

ABSTRACT

Loading of water with multifarious pollutants has dwindled the availability of quality fresh water and put questions on reliability and efficacy of conventional water treatment technologies. Also the quest for developing robust and cost-effective methods with minimum impact on environment had driven the focus of researchers and technologists on new technological developments. Nanotechnology – better referred as Aqua-nanotechnology in this regard provides scientists a new dimension to deal this big problem with small particles having application in 1) water treatment, 2) remediation, and 3) pollution prevention. This chapter will focus on fabrication and use of advance nanomaterials categorized as nanoadsorbents and nanoatalysts for these three main areas. A range of materials exploited in this regard are single and mixed metal oxides and their composites with polymer, clay, carbon based materials etc. while keeping focus on technological developments taken place over the period in regard with treating water and waste water.

DOI: 10.4018/978-1-5225-2136-5.ch005

INTRODUCTION

The importance of clean and safe water cannot be negated but a famous quote by Jacques Cousteau portrays the true reality about today's world approach that *We forget that the water cycle and the life cycle are one*. As per a report published by World Business Council for Sustainable Development (WBCSD) (Fry, 2005), there is a considerable increase in water consumption with an increase in the industrial revolution. All this development has no doubt made human life easy but on other hand has raised serious concerns over the sustainability of clean water as it caused limited share accounting for only 11% of water available for domestic consumption. Water pollution scenario is creating hype in the world and with its cross-border perspective water pollution is not an issue to be addressed and solved by some countries independently but it requires a consolidated approach to mitigate the pollution jinni. According to United Nations World Water Development report 2014, with depleting world water reservoirs and an anticipated rise of 55% in global water demand by 2050; an appreciable number of people are expected to have no access to clean water and improved sanitation (WWDR, 2014). With the development of new materials, chemicals, and technologies, the pollution scenario is getting worse. Also, the application of stringent laws and requirement of reduced levels of minimum contamination levels have put a question mark on the effectiveness of simple biological processes and conventional water treatment technologies (Zhou et al., 2002). Therefore, with the advent of technological innovations, there is a need to revolutionize methods for water treatment and remediation.

With its growing realm, Nanotechnology provides scientists a new dimension to deal with the ever rising problem of clean water availability (Bora et al., 2014; Qu et al., 2013). Also, the quest for developing robust new methods at lower cost, less energy, minimum use of chemicals and least impact on the environment, had driven the focus of environmentalist, water technologists, and engineers on nanotechnology revolution for rejuvenating the polluted fresh water reservoirs by implication of new materials and technologies. Already, the technology had put in a dramatic impact on water quality research and with its broad and diverse range of applications it offers various opportunities and challenges to further stimulate positively in this area of water engineering and treatment.

BACKGROUND

Access to safe and adequate water is the right of every individual living in this world but still 0.8 billion people according to WHO do not have access to quality drinking water and further 2.5 billion people lack access to improved sanitation. The industrial revolution in addition to bestowing us with amenities of life had also resulted in the production of diverse nature of new chemicals and materials which degradation is becoming a serious concern. Hence, loading of water with these diverse nature of pollutants including disinfectants and their byproducts, inorganic and organic chemicals, radionuclides and even resistant strains of microbes has not only limited the fresh water availability but has greatly impacted the water quality. All this resulted in inadequate drinking water, sanitation, and hygiene, which are estimated to cause more than 0.8 million diarrheal disease deaths per year and contribute substantially to several other diseases. Loading of the above-mentioned diverse range of pollutants and emerging of new contaminants also requires new treatment technologies to address the provision of safe water – the search lead to utilization of nanotechnology for water cleanup and remediation.

AQUA-NANOTECHNOLOGY

"Aqua-nanotechnology" has now been recognized as a branch of nanotechnology that deals with the application of nanomaterials for 1) water treatment, 2) remediation and 3) pollution prevention (Tratnyek et al., 2006). Nanotechnology-enabled treatment of water and wastewater provides the much needed technological revolution desired to facilitate the removal of a large range of contaminants in an easy, timely and affordable manner with much-improved treatment efficiency (Qu et al., 2013). This chapter addresses the development and implication of nanomaterials like metal and mixed metal oxides, polymer composites, clay composites, carbon-based composites etc. that ameliorated the water quality by promoting sustainable water management.

1. Nano Metal Oxides:

A variety of nano metal oxides (NMO) and their composites have been developed to treat water and wastewater for the removal of inorganic, organic and microbial impurities. Among these, titanium oxide, iron oxides, aluminum oxide, and zinc oxide are commonly used ones (Hua et al., 2012; Zhan et al., 2014).

Different synthetic approaches have been used to synthesize these NMO, among which co-precipitation, sol-gel and hydrothermal techniques (Rao, C. N. et al., 2006) have gained attraction because of their ease and low cost. These NMOs have been able to remove multiple contaminants simultaneously ranging from heavy metals and anions to dyes, pesticides, and microbes but in majority of cases their role as adsorbents is manipulated owing to their economic feasibility and ease of modification to enhance their selectivity and adsorption potential as well as considering (Bora et al., 2014).

Titanium dioxide, TiO_2, in anatase phase, had long been recognized as a material of choice for decontamination of water because of its structure and surface properties (Kumar et al., 2014) in addition to being low cost (Ghasemzadeh et al., 2014). Various heavy metals (Pb, Cu, Ni, Cd, Hg etc.) have been removed using TiO_2 showed adsorption capacities in the range of 7 to 166 mg/g mostly obeying pseudo-second order kinetics and Langmuir adsorption model (Ghasemi et al., 2012; Jin et al., 2015; Liang et al., 2000). The adsorption process was observed to be endothermic supported by pH in the range of 5-8 irrespective of the type of metal adsorbed (Ghasemi et al., 2012; Rashidi et al., 2010). At this pH electrostatic attraction between negatively charged TiO_2 NP and positively charged metal ions dominated promoting surface complexation reactions that support metal uptake (Jin et al., 2015). The adsorption process was dependent on crystalline phase both in the case of anions (F^{1-}) and cations (Pb^{2+}) uptake; pure anatase NPs are found more effective (Suriyaraj et al., 2014). Further, the removal efficacy was found to be independent of coexisting ions normally present in water i.e. Na; K, Ca, Mg; Al; Zn, Fe at tolerance levels of 5000; 2000; 1000 and 500 mg/L, respectively (Liang et al., 2000). The photocatalytic potential of these NMOs was exploited for removal/degradation of organic pollutants like dyes etc. (Farooq et al., 2009) and in some cases ensures complete mineralization of organic pollutants. The photocatalytic activity of TiO_2 was initiated by illumination, thereby transferring an electron to conduction band with the generation of a hole in valance band (Ba-Abbad et al., 2012) as shown in Figure 1a depicting the general photocatalysis mechanism. Because of its photocatalytic nature (Gautam et al., 2016; Hua et al., 2012) TiO_2 is found equally effective in degrading microbes and organic pollutant but the process is usually assisted by H_2O_2 or/and UV light (Ba-Abbad et al., 2012; Gautam et al., 2016) to activate TiO_2 to release photo-generated charges that then initiates the degradation (Gautam et al., 2016). This portrays major disadvantage of TiO_2 for its applicability in industrial settings (Abou-Gamra et al., 2015). In this regard few efforts were done to modify the TiO_2 structure to make it visible light responsive (VLR) by undergoing incorporation of other metals (Kaur, T. et al., 2016; Rauf et al., 2011), metal oxides (Wang, X. et al., 2006), metal salts (Hu, C. et al., 2006), non-metals (Long et al., 2011) or a combination of these (Wang, X. et al., 2006). This process led to the introduction of new states in the band gap of TiO_2 resulting in overlapping of these states which consequently improve visible light absorption properties and charge transfer (Long et al., 2011). Enhanced adsorption capacities and degradation rates were observed for these VLR-TiO_2 nanomaterials (Kaur, T. et al., 2016).

Figure 1. Photocatalytic degradation mechanism of pollutants on a) MONP b) ZnO-TiO$_2$ nanocomposite c) ZnO-α-Fe$_2$O$_3$ Nanocomposite for degradation of 2,4-D (Chen, J. et al., 2016; Homaeigohar et al., 2014; Lee et al., 2016; Maya-Treviño et al., 2014; Umar et al., 2013)

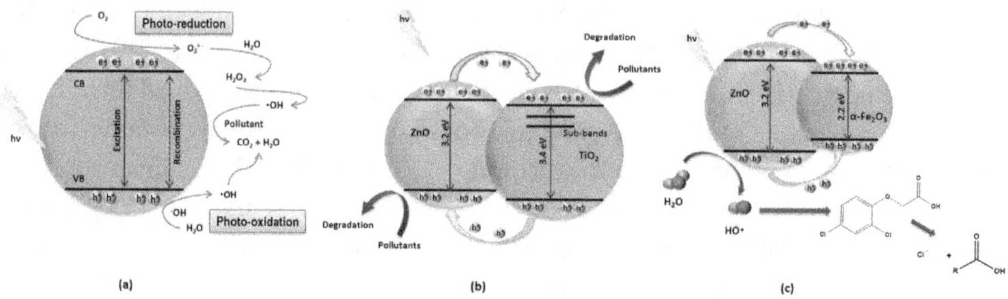

Alumina, Al_2O_3, is another famous material used for water purification purpose as adsorbent because of its amphoteric nature (Szekeres et al., 1998). Enhanced toxicity of alumina and its mixed oxides nanoparticles as compared to their bulk counterparts was reported for algal as well as bacterial strains (Jiang, W. et al., 2009; Sadiq et al., 2011). The activity is affected by the size of NPs as indicated by the studies carried out on gram-negative bacteria i.e. *Pseudomonas* and *Enterobacter* (Riaz et al., 2016). The interaction of alumina NPs with bacterial strain led to changes in cell wall morphology causing cell deactivation and death of the organism (Jiang, W. et al., 2009). In addition to that alumina NPs underwent several other mechanisms that were proposed for the antimicrobial potential of metal oxide nanoparticles in general which are highlighted in Figure 2 (Dizaj et al., 2014). Nano-alumina is also used for scavenging of metals but with low adsorption capacities (Zhan et al., 2014). Efforts were put in to enhance the uptake of metal ions by undergoing modification (Li, J. et al., 2008) but results are not much encouraging except in some cases e.g. Cu^{2+} (167 mg/g) (Mahdavi et al., 2015; Stietiya et al., 2014). On another hand, high adsorption capacities were noticed in the case of dyes on unmodified nano Al_2O_3 (Zhan et al., 2014) while that of pesticides are determined by their nature and availability of positive sites on the adsorbent surface (Clausen et al., 2001).

Zinc oxide, ZnO, antimicrobial properties of its nanoparticles were exploited against various gram-positive and negative bacterial and fungal strains (Dobrucka et al., 2016; He et al., 2011; Yousef et al., 2012) and found to have highest toxicity when compared with their bulk counterparts (Yousef et al., 2012; Zhang, L. et al., 2007) and other metal oxides (Azam et al., 2012; Jiang, W. et al., 2009). The mode of action proposed is typical as for other NPs based on the release of metal ions, electrostatic interactions, generation of reactive oxygen species (ROS) and the col-

Figure 2. Different mechanisms responsible for antimicrobial activity of metal oxide nanoparticles (Dizaj et al., 2014)

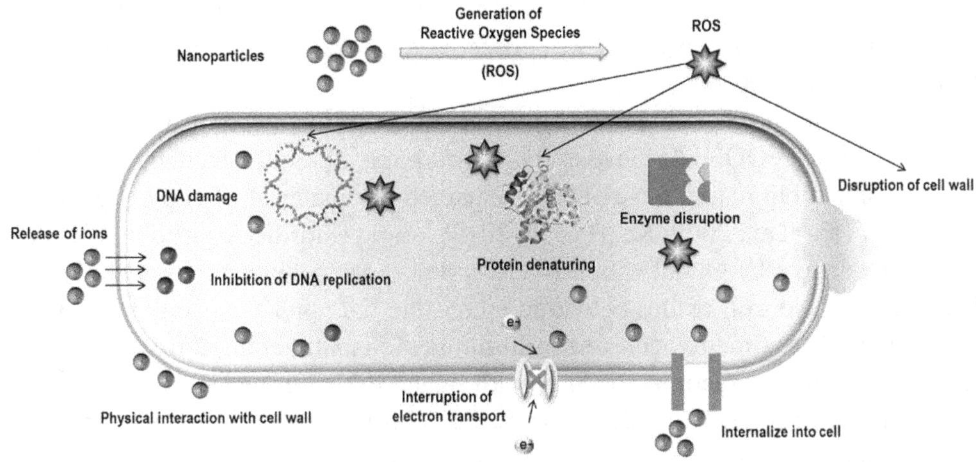

lapse of the bacterial membrane as portrayed in Figure 2 (Dizaj et al., 2014). Antibacterial activity was affected by synthetic approach (Sirelkhatim et al., 2015) as well as water chemistry and showed an inverse dependence on particle size as it directly impacts the release of Zn^{2+} ions (Li et al., 2011; Zhang, L. et al., 2007). Some studies have also marked NP's concentration as a more dominating factor towards antimicrobial activity than the particle size. Moreover, stabilizing agents can augment the activity by directly affecting agglomeration, but no direct impact on the type of stabilizing agents was observed (Zhang, L. et al., 2007).

ZnO system has the advantage of being not UV-assisted (Dobrucka et al., 2016) however use of it can enhance the photocatalytic potential to a considerable extent (Sirelkhatim et al., 2015). High degradation/ removal rates for various dyes (Zhan et al., 2014), pesticides (Khan, S. et al., 2015), organometallics (Olushola et al., 2014), pharmaceutical compounds (Han et al., 2012) and other organic aliphatic and aromatic contaminants (Anju et al., 2012; Shinde et al., 2011) were noticed both in absence and presence of irradiation on treatment with ZnO NP/ thin films/ nanowires/ nanoneedles. The reaction displayed dependence on calcination temperature, irradiation time, pH and ZnO dosage (Hayat et al., 2011), defects formation (Meng et al., 2015), particle size and structure/morphology (Wang, H. et al., 2007; Yang, J. L. et al., 2004). Correspondingly, photocatalytic degradation efficiency of anions like CN^- (Bagabas et al., 2013) and PO_4^{3-} (Elkady et al., 2016) was greatly impacted by the morphology of ZnO. In case of metal adsorption, overall low adsorption capacities were observed for Ni^{2+}, Pb^{2+} and As^{5+} (Mahdavi et al., 2012; Venkatesham et al., 2013; Zhan et al., 2014) while for Cd^{2+}, Cu^{2+}, Zn^{2+} and Hg^{2+} higher removal rates were noticed (Mahdavi et al., 2012; Rafiq et al., 2014; Sheela et al., 2012) which were even greater than the ones shown by iron and copper oxides (Mahdavi et al., 2012) but lower than that of MgO (Rafiq et al., 2014). For Nobel metals, three times improved removal efficacy was observed for nano ZnO as compared to nano TiO_2 (Park et al., 2003).

Oxides of iron, magnetite (Fe_3O_4), maghemite (γ- Fe_2O_3), hematite (α-Fe_2O_3) and goethite (α-FeOOH) because of their low toxicity, unique magnetic properties and biocompatible nature have always remained the material of choice for water treatment (Teja et al., 2009; Xu, P. et al., 2012). Rapid fluoride adsorption had been assisted by nano γ-Fe_2O_3 but with low adsorption capacity (Jayarathna et al., 2015). The use of mixed iron oxides helped in increasing the uptake capacity but system took much longer time to achieve equilibrium (Mohapatra et al., 2011). Chemical and electrostatic interactions together with ligand exchange reactions between an anion and OH^- or water on adsorbent surface played their role in the removal of anions (Mohapatra et al., 2011). The uptake as a general protocol depends on upon the initial pH, adsorbent dose, temperature and initial adsorbate concentration and presence of co-ions like SO_4^{2-} and Cl^{-1} (Mohapatra et al., 2011; Nalbandian et al., 2016). In the case of α-Fe_2O_3 nanofiber, indirect independence of CrO_4^{2-} removal was observed on the diameter of nanofiber which was attributed to change in the surface area solely because of physical nature of adsorption (Nalbandian et al., 2016). In the case of heavy metals, fast removal of on nano γ-Fe_2O_3 was facilitated by the presence of reactive sites on surface hence enabling the system to reach equilibrium in short time (Hu, J. et al., 2006). Contrarily, nano-Fe_3O_4 took a longer time to achieve equilibrium (Giraldo et al., 2013; Mahdavi et al., 2012) probably due to differences in surface area and particle size. Introducing functionalization to Fe_3O_4 had greatly reduced the time to achieve equilibrium and with better adsorption capacity that increased with increase in the concentration of functionalizing moiety (Huang et al., 2009). The uptake efficacy though dependent on cation nature, pH, temperature etc. (Giraldo et al., 2013) but dramatic change in adsorption were observed with change in particle size (Xiao et al., 2015) surface sites, porosity and ionic strength (Giraldo et al., 2013). In contrary to previous studies Cheng *et.al.* showed no effect of a change in ionic strength on adsorption of Arsenic (Cheng et al., 2016). Studies on the effect of competing ions like HCO_3^{1-}, SO_4^{-2}, F^{1-}, Cl^{1-} and NO_3^{1-} indicated no effect on adsorption while HPO_4^{2-} and SiO_3^{2-} impacted to somewhat extent (Cheng et al., 2016). Similar studies performed pointed out that the effect of coexisting ions (i.e. sulphate, phosphate, nitrate, silicate) on removal efficacy vary with change in pH (Tuutijärvi et al., 2012).

The surface adsorption of IONP was more effective and rapid for hydroxyl containing dyes like bromocresol green than the ones that don't have OH group e.g. methyl red (Saha et al., 2011). Complete and fast removal of such dyes was achieved by use of hematite hollow nanofibers (Gao, Q. et al., 2015) and γ-Fe_2O_3 owing to chemisorption (Khan, M. et al., 2016). Simply changing the pH to alkaline media effectively desorbed these IONP ensuring their reuse (Saha et al., 2011). In the case of phenol, the mineralization efficacy depended on the generation of OH radicles and presence of other organic pollutants which suppressed the degradation process (Wang, W. et al., 2013). The process can be speeded up by use of UV irradiation and controlling its irradiation time and laser energy (Bardajee et al., 2013; Gondal et al., 2009).

IONP have proved themselves as effective antimicrobial agents as well against different microbial strains (Ismail et al., 2015). They operate in similar way as other NMO (Figure 2) i.e. on basis on reactive oxygen species generated that later interact with DNA and protein of bacteria leading to damage of cell (Tran et al., 2010) but the activity level achieved by Fe_2O_3 was less than ZnO and CuO NPs (Azam et al., 2012). Gram negative bacteria were more resistant to IONP than gram positive owing to differences in the cell wall, cell physiology and degree of contact (Azam et al., 2012). Positively and negatively charged surface modified Fe_3O_4 NP depicted dissimilar response to bacterial toxicity since different surface charges interacted with the bacterial cell differently; order of toxicity observed was positively charged > bare Fe_3O_4 > negatively charged. Moreover, surface properties of NMO are more decisive in determining antimicrobial activity than the properties of the core itself (Javanbakht et al., 2016).

Nano-Mixed metal oxides (NMMO) of iron, aluminum, titanium and zinc have also been exploited for water treatment. Several methods have been used for their synthesis which includes sol-gel, co-precipitation, wet chemical synthesis, microemulsion etc. Unique properties resulted from the combination of these materials (Kapoor et al., 2004). Among these, photocatalytic activity is one of the most exploited property in water treatment. This in the case of NMMOs helps in reducing electron-hole pair recombination which portrays the major disadvantage of NMO. In addition to that such combinations promote the generation of OH• and ROS thereby enhancing the photocatalysis process. A model system of ZnO-TiO_2 and ZnO-α-Fe_2O_3 is presented in Figure 1 b and c. which are produced by a combination of two NMOs (Chen, J. et al., 2016; Homaeigohar et al., 2014; Lee et al., 2016; Maya-Treviño et al., 2014). The introduction of ZnO into TiO_2 promoted sub-bands production along with generation of defects resulting in a decrease in energy gap and better response to UV illumination (Chen, J. et al., 2016). Hence, owing to these attributes much higher and faster dye degradation was achieved by these NMMO systems when compared to pure NMO. Various combinations used in this aspect are ZnO/TiO_2, γ-Fe_2O_3/SiO_2, Al_2O_3/TiO_2, ZnO/Fe_2O_3, WO_3/TiO_2, ZnO/Fe_3O_4, ZnO/SnO_2, CeO_2/TiO_2, SnO_2/TiO_2, ZrO_2/TiO_2 CuO/ZnO etc. Factors like the type of dopant, doped or molar concentration, irradiation time, precipitation agent during synthesis, sintering temperature, crystallite size, pollutant type, counter ions etc. determined the degradation efficacy in these cases. The change in molar concentration considerably affected the band gap, phase change, purity, morphology, crystallite size (Cun et al., 2002; Hameed et al., 2009; Wang, J. et al., 2009). In the case of mixed metal oxides of IONP presence of competitive ions like heavy metals considerably altered the uptake of dyes depending upon radii and charge they carry. In addition to that low concentrations of humic acid promoted adsorption process because of enhancement in the available surface area but in excess amounts they obstructed the adsorption process by blocking the passage of adsorbate (Chen, D. et al., 2016).

Defluoridation achieved by use of $MgO/\gamma\text{-}Al_2O_3$ was higher than $/\gamma\text{-}Al_2O_3$ and was slightly affected by the presence of counter-ions (Nazari et al., 2015) while in the case of Al_2O_3/Fe_3O_4 NMMO, the presence of other ions had no impact on removal efficacy (Chai et al., 2013). The electrostatic interactions which demarked the main role in adsorption process were affected by the presence of competing ions in order phosphate > carbonate >> sulphate > nitrate > chloride (Azari et al., 2015). TiO_2 doped IONP when used as an adsorbent, the equilibrium was achieved within 5 min for As^{3+} uptake even in dark while for As^{5+} UV illumination was required to remove it from the water. As^{3+} adsorption was predicted to take place by oxidation to As^{5+} which then adsorbed on to the NMMO surface (Yu, L. et al., 2013).

Triple metal combinations were also evaluated by many researchers in the search for having effective and efficient photocatalysts, adsorbents and antimicrobial agents with promising results (Motlagh et al., 2015).

2. Polymer Composites

Metal matrix polymer nanocomposites are usually synthesized by two main techniques

1. *In situ* synthesis, and
2. Direct compounding (Zhang, Q. et al., 2013; Zhao et al., 2011).

Incorporation of these MONP into polymer matrix have not only increased the surface area, reactivity and mechanical strength but also inherent in them the unique combination of properties of both inorganic nanoparticles and organic polymer matrix making them fit for meeting the water treatment process requirements i.e. antifouling, photocatalytic, flux enhancement, hydrophilicity etc. (Baghbanzadeh et al., 2016; Jeon et al., 2010; Zhang, Q. et al., 2013). Further, it facilitates to overcome separation issues of NP (Zhao et al., 2011). Many inorganic materials have been used as inorganic fillers to prepare these polymeric hybrids e.g. metal NP (Palza, 2015), MONP (Baghbanzadeh et al., 2016), metal carbonates (Rong et al., 2006) etc. but here in this chapter the scope will just focus on the ones prepared using MONP.

The addition of metal oxides advocated hydrophilic character of the polymeric material thereby promoting antifouling properties. Moreover, modifications using magnetic or superparamagnetic MONP ensured easy separation and recovery of the adsorbent by application of low-gradient magnetic field (Qu et al., 2013). Highest and fastest recovery achieved in the case of core-shell magnetic nanorods based polymer composites was ~ 25s (Venkateswarlu et al., 2015). Optimization of nano Fe_3O_4 in Polyethersulfone (PES) - nano Fe_3O_4 membrane facilitated in achieving desirable pore size, roughness, porosity, contact angle, water flux and rejection capacity. Using high weight % of IONP put a negative impact on water flux by suppressing microvoids growth and enhancing film skin thickness; the best results were achieved for 10 and 15 wt.% doping (Alam et al., 2013). Likewise, doping of Al_2O_3 into PES, having application as membrane bioreactors, led to decrease in de-ionized water permeation with an increase in polymer ratio while the opposite effect was observed with the rise in NP concentration (Maximous et al., 2009). Correspondingly, varying the amount of SiO_2 NP (30 nm) in PES potentially impacted water flux, hydrophilicity and antifouling abilities of the membrane and hence this can be tailored to achieve desirable properties (Shen et al., 2011).

Magnetic capturing of Cu^{2+} was achieved by controlled coating of polyethylenimine (PEI) onto Fe_3O_4 NP. Increase in PEI coated amount enhanced the uptake which was facilitated by the presence of amino nitrogens in the structure of PEI; this enabled manipulating the adsorbents' colloidal properties to achieve maximum adsorption even at ultra-trace levels of free Cu^{2+} ions selectively (Goon et al., 2010). Contrary to this, uptake capacity increased (341.43 – 489.12 mg/g) with decrease in PVA content (4.55 – 1.43%) in case of PVA/ SiO_2 NC fibers modified by mercapto groups with maximum adsorption reaching to 504.89 mg/g in just 40 min. Higher affinity for Pb^{2+} was detected in hybrid polymers prepared by unmodified or amino functionalized Fe_3O_4 NP and nanorods followed by other metals like Cd^{2+}, Zn^{2+}, Ni^{2+}, Hg^{2+} and As^{3+} (Badruddoza et al., 2013; Ge et al., 2012; Venkateswarlu et al., 2015). Contrarily, better uptake was achieved in the case of Fe_3O_4 NP/carboxymethyl-β-cyclodextrin; attributed to the presence of –COOH and –OH functional groups in adsorbent (Badruddoza et al., 2013). Uptake was slightly affected by the presence of monovalent ions but in the case of divalent strong competition was observed (Ge et al., 2012). Additionally, increase in ionic strength can cause decreases in ions removal (Badruddoza et al., 2013). Ion exchange properties of Fe_3O_4/polypyrrole core-shell NC were exploited for Cr^{6+} that replaced doped Cl^- with chromate ions (Bhaumik et al., 2011). On another hand, Pb^{2+} uptake was expedited by improving porosity and surface roughness of Polyethersulfone /hydrous manganese dioxide (PES/HMO) NC membranes achieved by increasing HMO:PES ratio (Gohari et al., 2013). Type of MONO doped, particle distribution with respect to depth, the amount of doping and dopant to polymer ratio considerably affects the behavior and properties of the composite membranes (Maximous, N. et al., 2010). Desorption of metals like Pb^{2+}, Cd^{2+}, Ni^{2+} and Cu^{2+} from polymer composites was attained through Na_2-EDTA, HNO_3, H_3PO_4 and HCl solutions depending on the polymer system, metal adsorbed and leaching agent nature (Badruddoza et al., 2013; Gohari et al., 2013; Venkateswarlu et al., 2015). Relating these polymer composite systems with activated carbon showed comparable removal efficiencies for Pb^{2+}, little less for Cr^{3+} and much lower for Cd^{2+} (Maximous, N. N. et al., 2010).

High fluoride adsorption capacities were also reported within 60 and 120 min for the magnetically modified polyacrylonitrile and polyaniline hybrids, respectively. Desorption studies performed indicated irreversible nature of adsorption process in former case 85% desorption was achieved for polyaniline NC at pH 12 (Jokar et al., 2014; Nouri et al., 2014). Dye adsorption on to Fe_2O_3/Polyacrylonitrile NC occurred as a result of electrostatic attraction between positively charged IONO and negatively charged dye molecule. The uptake capacity depended on the synthetic approach used with maximum adsorption achieved for the hydrothermally grown nanoparticles i.e. 50.08 mg/g (Patel et al., 2016). Mixed metal oxide polymer (Fe_3O_4/Cu_2O/PANI) NC with different morphologies i.e. flower-like, microsphere with ordered nanosheet and porous dandelion were synthesized by altering aniline/Cu^{2+} ratio and used for dyes and oil adsorption. The high uptake capacity was achieved due to porous structure and in the case of dyes combined effects of electrostatic interactions, coordination of dyes anionic group with metal ions and van der Waals forces. The prepared materials were selective for anionic dyes as cationic dyes face repulsive interactions between positively charged dyes groups and metal ions (Cao, J. et al., 2014). Complete mineralization of phenol was attained under visible light by optimizing the aniline/TiO_2 concentration during *in situ* synthesis of TiO_2-PANI NC. The photocatalyst maintained the activity till 5 recycles and performed much better than neat TiO_2 (Li, X. et al., 2008).

3. Chitosan and Clay Based Nanocomposites

Chitosan is a natural low-cost polymer having excellent chelation properties and easy modification protocols making it suitable for water purification purposes (Ngah et al., 2011; Reddy et al., 2013). The presence of amine groups facilities adsorption of cations, anions or polar molecules by chelation, electrostatic interactions, dipole interactions, van der Waals forces etc. (Rangel-Mendez et al., 2010). Certain disadvantages of these biopolymers i.e. agglomeration, denaturing, separation inefficiency and inadequacy to be used in packed columns limit their applications. Provision of support or modification helps in overcoming these disadvantages but lowers the uptake potential and intraparticle diffusion of these bio-composites (Ngah et al., 2011; Rangel-Mendez et al., 2010; Tanhaei et al., 2015; Zhang, J. et al., 2011).

Using IONP as additives in Chitosan NC enhances chemical stability and provide ion selectivity (Tanhaei et al., 2015). The usual procedure of making these NC is to grow a layer of chitosan on NP making core-shell structure (Figure 3a (Reddy et al., 2013)) that not only prevents agglomeration of chitosan but also increase number of free amino groups (Fan et al., 2011; Zhang, J. et al., 2011). In other studies, hydrogen bonding of MONP with these amino groups and availability of free O-H with partial positive charge was suggested (Shrifian-Esfahni et al., 2015). In the case of Arsenic removal, these magnetically active chitosan materials showed higher selectivity for As^{5+} than As^{5+} but the process was highly pH dependent (Wang, J. et al., 2014). Beads prepared by dispersing of FeOOH NP in chitosan chelates with arsenic either by forming a covalent bond with amine group or iron (Hasan et al., 2014) as contrary to Fe_3O_4/chitosan NC where –OH group on adsorbent surface interacted with arsenic (Wang, J. et al., 2014). Besides, NC of chitosan with single metal dopant e.g. Al_2O_3 (Gandhi et al., 2010), TiO_2 (Miller et al., 2010), CuO (El-wakeel et al., 2015) or with mixed metal oxides like Al_2O_3- TiO_2 (Yamani et al., 2012), Fe-Mn binary oxide (Qi et al., 2015) had been used for heavy metal removal. On comparative account, the bimetallic NC had better adsorption potential than the pure systems (Yamani et al., 2012). The presence of counter ions presents in water impacted differently on the metal uptake. For example As^{5+} uptake was hindered by HPO_4^{2-}, HCO_3^{1-} and SiO_3^{2-} ; not affected by SO_4^{2-} and enhanced by Ca^{2+} and Mg^{2+} ions (Qi et al., 2015) while that of Cr^{6+} was reduced by presence of all the co-ions employed in the experiment i.e. NaCl, Na_2SO_4, $NaNO_3$, $CaSO_4$ and $MgSO_4$ (Gandhi et al., 2010). Effective separation was achieved in the case of these NC but employing ultrasound radiation ensured much faster separation (Liu, X. et al., 2008). These nanocomposites offered the advantage of reuse owing to reversible nature of metal and magnetic chitosan NC interaction hence, the simple weak acid treatment made these adsorbent workable up to 6 cycles (Fan et al., 2011).

Figure 3. a) Magnetic core chitosan shell (Reddy et al., 2013) b) Mechanism of F^{-1} adsorption on Al_2O_3/ Chitosan NC (- - - shows electrostatic adsorption and complexation) (Viswanathan et al., 2010)

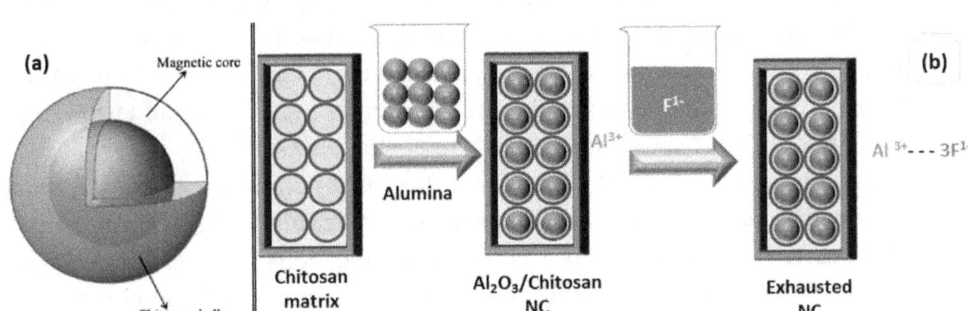

Fluoride removal onto Al_2O_3/chitosan composite was directed by 1) electrostatic attraction between Al^{3+} and F^{1-} and 2) complexation of F^{1-} with positively charged surface as shown in Figure 3b (Viswanathan et al., 2010). However, dyes adsorption on to MONP/chitosan was suggested to be chemisorption in which amino groups on the surface of chitosan NC bind to negatively charged groups of dyes (Salehi et al., 2010; Zhang, J. et al., 2011). Magnetic chitosan NC promoted dye removal as well through spontaneous and endo- or exo-thermic process obeying Langmuir adsorption model and pseudo-second-order kinetic model (Cao, C. et al., 2014; Fan, Zhang, Li, et al., 2012; Fan, Zhang, Luo, et al., 2012; Reddy et al., 2013). The selectivity for a specific dye uptake can be enhanced by using the molecular imprinting technique (Fan, Zhang, Li, et al., 2012). Synergistic effects of mixed metal oxides and chitosan were also exploited to achieve better removal percentage of acidic dyes (Jiang, H. et al., 2013; Zhu, H. et al., 2012). Higher uptake i.e. 216 mg/ g was achieved for Remazol Red 198 in 60min at room temperature by using O-carboxymethylchitosan-N-lauryl doped with γ-Fe_2O_3 NP (Demarchi et al., 2015). Faster (< 20min) and even higher removal were achieved for methyl orange by using Fe_3O_4/Al_2O_3/chitosan and γ-Fe_2O_3/chitosan NC i.e. 416 mg/g and 779 mg/g (Obeid et al., 2013; Tanhaei et al., 2015). Factors like chitosan amount, MONP wt. %, solidifying solutions type and concentration, UV light, adsorbent structure, dye nature etc. affected the degradation efficacy of the chitosan NC (Haldorai et al., 2013; Li, Q. et al., 2008; Zubieta et al., 2008). In the case of dyes as well co-ions present in the water and wastewater like Cl^{1-}, SO_4^{2-} and HCO_3^{1-} affected uptake capacity negatively (Salehi et al., 2010). Concurrently, multifunctional nature of these hybrid materials had been investigated for simultaneous uptake of heavy metal, bactericidal activities and photocatalytic properties and results obtained were encouraging (Haldorai et al., 2013, 2014; Zhang, X. et al., 2011). Also, studies carried out on actual wastewater samples showed effective cleanup by the MONP and chitosan NC (Zhang, X. et al., 2011).

Even better results are obtained by introduction of layer by layer system (i.e. TiO_2/ chitosan supported on glass) that achieved 32 times faster decolourisation rate as compared to single layer system (TiO_2) owing to collective role played by multiple effects of photocatalysis, adsorption in sub-layers and charge-transfer process (Nawi et al., 2012). The rate of degradation strongly depends on the TiO_2 content and number of layers as increase in these parameters led to increasing in film thickness, mobility, and diffusion of dye molecules, penetration of light source and growth in crystallite size of TiO_2 (Nawi et al., 2012; Zainal et al., 2009).

Antimicrobial applications of chitosan against many fungal, bacterial and viral strains are well researched. This broad-spectrum nature, higher bacteriostatic and bactericidal activities and lower toxicity to mammalian cells make it a material of choice. The mode of action is based on interaction with the cell membrane and chelating action (Rabea et al., 2003). With the incorporation of NP, the bactericidal effects of NC are amplified owing to multifarious mechanisms that hamper bacterial growth and promote cell inactivation (Haldorai et al., 2013; Shrifian-Esfahni et al., 2015). The viable bacterial count results showed 99% and 93% *E. coli* inactivation by CuO and MgO impregnated chitosan, respectively as compared to 7% for pristine chitosan (Haldorai et al., 2013, 2014). Likewise, Lower MIC values were observed in the case of susperparamagnetic IONP/chitosan NC *vs* superparamagnetic IONP i.e. 40 μg/mL compared to 70 μg/mL for *P. aeruginosa* and 45 μg/mL compared to 90 μg/mL for *E. coli* respectively (Shrifian-Esfahni et al., 2015).

Clays, another natural source, are hydrous aluminosilicates that act as a natural scavenger of pollutants (Bhattacharyya et al., 2008). Modifications of clays either through intecalation, acid activation, functionalization, incorporation of dopants, the formation of hybrid structures etc. had been practiced for improving their properties (Bhattacharyya et al., 2008). Montmorillonite, bentonite, kaolinite, sepionite, halloysite and zeolite are the common clay minerals that have been used as fillers for preparing NC materials (Motshekga et al., 2013). Dispersing MONP into the clay structure reduce agglomeration and settles the particles on the surface or inside the pores of clay thereby enhancing surface area (Yuan et al., 2009).

Antimicrobial potential of clays is enhanced by harnessing them with the MONP ones making these hybrid materials extremely useful. Comparative account of undoped and doped bentonite against *E. coli* and *E. faecalis* showed that pristine bentonite had no effect on the bacterial growth; ZnO-bentonite showed inhibition zone in the range of 5-9mm and Ag/ZnO-bentonite (6-11mm) had activity comparable to Ag/bentonite (8-11mm) (Motshekga et al., 2013).

Different clays like bentonite, sepiolite, montmorillonite were doped with Fe_3O_4 and utilized for adsorption of heavy metals. 15.3 mg of Cr^{6+} /g of Fe_3O_4/monmarillonite was removed owing to the electrostatic interaction that led to its reduction to Cr^{3+}; the adsorption depended on upon the synthetic approach used to prepare adsorbent (Yuan et al., 2009). Much higher uptake of Cr^{6+} (161.29mg/g) was achieved on $MgFe_2O_4$ /bentonite composite and with easy regeneration using 0.1N NaOH (Kaur, M., Singh, et al., 2015). On another hand, Fe_3O_4 impregnation in bentonite resulted in low uptake capacity (18.76 mg/g) for Co^{2+} (Hashemian et al., 2015). Higher uptakes for As^{3+} i.e. 50.35 mg/g were observed when sepiolite was used instead with good effectiveness on real ground water samples as well. The fast adsorption of As^{3+} into NC was facilitated by the highly porous structure of clay that promoted easy access to iron oxide to result in As(III) – Fe(III) complexes and partial oxidation of As^{3+} to As^{5+} as outlined in Figure 4. The presence of F^{-1} and PO_4^{3-} as counter-ions impact the removal of As (III) to some extent because of competition posed by the similarity in structure and dissolution constants (Tian et al., 2015). Quick uptake of heavy metals (i.e. 2 min) was achieved using Fe_3O_4/talc NC

Figure 4. Proposed mechanism of As^{3+} adsorption on Fe_3O_4/ sepiolite nanofibers (Tian et al., 2015)

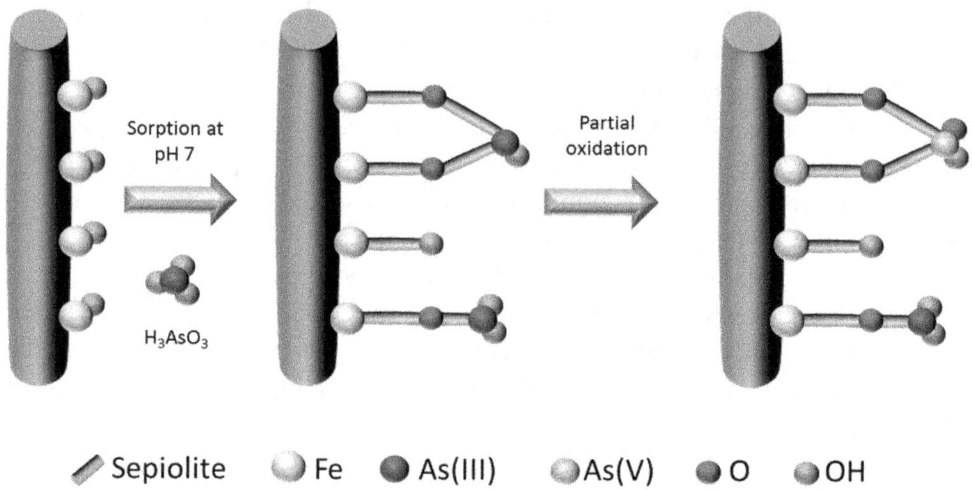

leading to adsorption capacities in order of Pb^{2+} (74.62 mg/g) > Ni^{2+} (33.33 mg/g) > Cu^{2+} (21.05 mg/g) as calculated using Langmuir model (Kalantari et al., 2014).

MONP were supported on different aluminosilicate for evaluating their catalytic degradation pattern towards organic contaminants (Khatamian 2012). In general, the supports provided better degradation potential compared to unsupported ones by augmenting the radicle attack on organic moiety. Moreover, changing its size to nanoscale further enhanced the degradation rate. The type of support also impacted considerably as it not only determines the morphology of NC but also facilitates the binding of adsorbate to its surface.

Among the three type of supports i.e. HZSM-5, clinoptilolite and zeolite HY, the best results were obtained with HZMS-5 because it promoted finely dispersed structure, enhanced hydrophobicity, high adsorption capacity, and delocalization of electrons generated . Reduction in pore size and surface area of support was observed with incorporation of ZnO NP in nanoclinoptilolite or HZSM-5 (Khatamian et al. 2012) while the opposite behavior was observed when Fe_2O_3 NP was doped in montmorillonite (Hristodor et al., 2011) probably due to the differences in synthetic approach used. Also, increase in MONP loading reduced the rate of degradation owing to aggregate formation and surface masking led to a reduction in active sites or due to $OH^{\bullet}$ trapping). Changing the type and concentration of electron scavenging agents e.g. H_2O_2 and $KBrO_3$ influenced degradation rates by controlling $OH^{\bullet}$ availability and electron - hole combination (Hristodor et al., 2011). Incorporating metal ions like Fe^{3+}, Ni^{2+}, La^{2+}, Y^{3+} etc. into TiO_2/Zeolite NC had also played their role in reducing the electron – hole recombination and introducing defects in MONP crystal structure thereby enhancing catalytic degradation rate even higher than the undoped TiO_2/ZSM-5 NC (Khatamian et al., 2012). Further investigation into the mechanism of degradation points out that the pore size of zeolites is usually less than 1nm while the size of organic molecules e.g. dyes is greater than that hence adsorption into the porous structure is not probable just surface interactions of MONP or zeolite play their role in adsorbing these molecules. The size of these molecules sterically hinders adsorption of more molecules on NC surface, therefore, the researchers claim that dye degradation does not take place on the surface of NC directly rather due to $OH^{\cdot}$ generated on the surface of catalyst and holes as well (Wang, C.-C. et al., 2008). Further insight into the adsorption potential of these NC led to important findings i.e. Fe_3O_4 NC with bentonite had a really high affinity (1600 mg/g) for methylene blue which is 2 to 3 times higher than the normal bentonite (Hashem, 2012).

4. Carbon-Based Composite

Graphene (GR), Graphene oxide (GO) and reduced graphene oxide (rGO) are known to be a good adsorbent and used in many water treatment studies (Cohen-Tanugi et al., 2012; Gao, W. et al., 2011; Kemp et al., 2013). Modification of GR or GO were carried out to exploit its properties for water remediation by using metal oxides (Upadhyay et al., 2014). Such incorporation not only averts the possibility of electron-hole recombination and metal leaching but also promotes pollutants' removal by π-π stacking. The arrangement also favors the appearance of new valence bands by promoting hybridization. Therefore, owing to these properties the photocatalytic activity and adsorption of this graphene-based MONP composites have been investigated and exploited for enhanced removal of organic moieties and heavy metals. Figure 5 gives an insight into photocatalytic mechanism operating in these composites and various mechanisms leading to adsorption of pollutants from water (Upadhyay et al., 2014).

TiO_2 doped GO NC membranes were found 100% effective in filtering dye molecules at low concentrations (Xu, C. et al., 2013). Much improved uptake capacity (83.26 mg/g) was achieved by using sheets of reduced GO (rGO) with TiO_2 incorporation which was even higher than its nanotubes (75.36 mg/g), pure sheets of GR (48.73mg/g) and tubular titanate (36.51mg/g) (Nguyen-Phan et al., 2012). Use of $ZnTiO_3$ in modifying rGO resulted in improved methylene blue degradation when assisted with visible light irradiation in comparison to $ZnTiO_3$. The adsorption occurred as a result of π- the π interaction between dye and GO that manipulated with structural changes of the composite, for this reason, less uptake was noticed in the case of the cubic structure as compared to hexagonal one (Gayathri et al.,

Figure 5. Metal oxide-graphene composites a) Photocatalytic activity based on percolation mechanism showing electron transfer from metal oxide's conduction band to graphene b) various interactions operative in adsorption of pollutants i.e. heavy metal ions and dye molecules (Upadhyay et al., 2014)

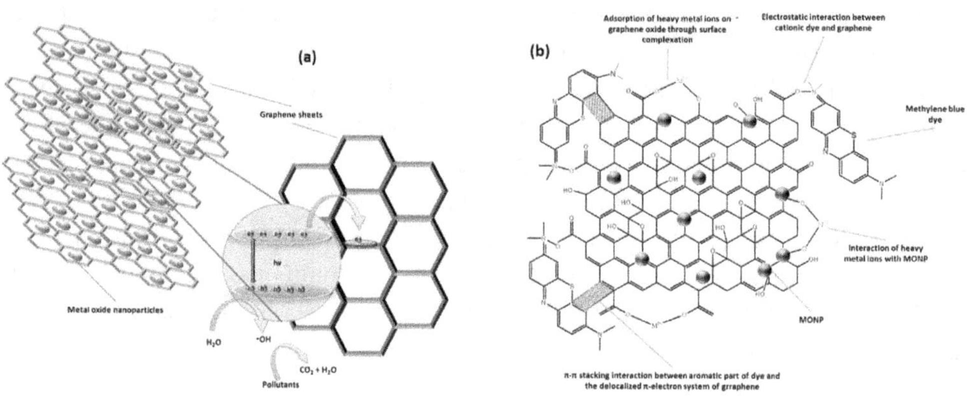

2015). Fe_3O_4/GO magnetically active material had been exploited as a multipurpose water purification material that achieved 97% dye degradation, 77.67% bacterial cell lysis and 69 mg/g Pb (II) adsorption capacity that remains viable up to 5 cycles (Santhosh et al., 2014). The NC was also effective in removal of both anionic (33.66 mg/g) and cationic dyes (13 – 173 mg/g) (Ai et al., 2011; Meral et al., 2014; Sun et al., 2011; Yao et al., 2012), p-chloroaniline (43.86 mg/g) and aniline (375.94 mg/g). Simultaneous removal of Cd^{2+}, methylene blue, and orange G was carried out by using Fe_3O_4-GO leading to reduction in adsorption capacity from 91.92, 64.23, 20.85 to 59.69, 58.08, 20.48 mg/g, respectively as the matrix changed from ultrapure water to tap water owing to presence of interfering ions (Deng et al., 2013).

Fe_3O_4/GO exhibited better adsorption properties for the metals under study at all the pH values following the order Pb(II) > Cu(II) > Cd(II) when compared with MWCNT and powdered activated carbon (AC). Experiments also suggested a decrease in each metal uptake in the multi-component system as compared to the single component system though overall metal adsorption capacity was enhanced in the former case (Hur et al., 2015). Yoon *et.al.* advocated that conversion of GO to rGO was a facile step as a comparison of $As^{3+/5+}$ adsorption on to Fe_3O_4 NP modified GO and rGO displayed a compatible affinity with both the composite materials. Insight into adsorption mechanism suggested that electrostatic interaction played important role in binding As^{5+} to the adsorbent while surface complexation governed As^{3+} attachment (Yoon et al., 2016). Modification of GO with mixed metal oxide NP (Fe_3O_4 and MnO_2) was tried for sorption of $As^{3+/5+}$ but led to the removal of only 14.04 and 12.22 mg/g, respectively (Luo, X. et al., 2012). Metal adsorption on

Fe_3O_4/GO NC depends on the nature of ions and level of affinity with Fe to form an inner or outer shell complex (Tayyebi et al., 2015). In case of double core GR structures containing iron core, Fe_2O_3 inner shell and outer shell of Si-S-O compound the adsorption mechanism was due to synergetic effects of surface complexation and single layer adsorption which is different from the pure GR in which only the later phenomenon plays its role (Zhu, J. et al., 2011). The dispersion of mesoporous Fe_2O_3/TiO_2 nanoparticles on reduced GO helped in enhancing the number of active sites ensuring attainment of equilibrium in 15 min and 99.5 mg As^{5+} and 77.7 mg As^{3+} uptake per gram of adsorbent at neutral pH while maintaining the same sorption up to 4 cycles. The order of interfering anions in this case follows: phosphate > sulphate > nitrate > chloride > fluoride (Babu et al., 2016).

In the case of nanocomposites doping with MONP gained attention and showed considerably high activity as compared to the pristine MONP or Graphene-based compounds (Cao et al., 2013). Diverse nature of bacterial strains was used as a model to determine the effect of these NC using TiO_2, ZnO, and Fe_3O_4 as dopants (Bykkam et al., 2015; Dědková et al., 2015; Upadhyay et al., 2014; Wang, Y.-W. et al., 2014). The usual mechanism responsible for the antimicrobial activity was through the generation of ROS and release of anions (Chang et al., 2015; Wang, Y.-W. et al., 2014). GO role was proposed as the main agent that either proliferates in bacterial cells or caused slicing off a cell resulting in oozing of cell contents (Upadhyay et al., 2014). Factors like weight % of MONP, ROS amount produced, time of irradiation, the viability of microorganisms and production of extracellular protein by microorganisms impacts the antimicrobial potential of these nanocomposites. Among these, the first four are related to each other as Wt.% and time of irradiation directly affects the ROS generation which should be more than the bacterial colonies growth so that bacteria do not have ample time to divide themselves. As far as protein generation is concerned, it is expected to block MONP active sites by adhering onto their surface hence causing a decrease in their activity (Dědková et al., 2015). In addition to that activity was also affected by changing the loading of GR as the excess of it can also hamper the availability of MONP active sites (Cao et al., 2013). Further, the presence of anions e.g. HCO_3^{1-} and HPO_4^{2-} considerably hampered the activity by capturing the holes, reacting with OH•, blocking the active sites of MONP surface by adsorbing on its surface and absorption of exciting light. The same impact but with less intensity was observed for SO_4^{2-} and NO_3^{1-} ions. The presence of Cl^{1-} ions on contrary either had no impact on the activity or assisted the antimicrobial activity by propagating formation of $Cl^{\bullet -}$, a disinfectant. The cations i.e. Na^{1+} and K^{1+} occurrence did not influence on the viable count of bacteria (Chang et al., 2015).

Carbon nanotubes (CNT), single walled (SWCNT) and double walled (MWCNT), properties have been exploited for the water purification with and without modifications (Ong et al., 2010; Rao, G. P. et al., 2007; Upadhyayula et al., 2009). The potential of MWCNT for dye adsorption in absence and presence of light irradiation was accessed by using methyl orange as test contaminant. MWNT synthesized by chemical vapor deposition showed better adsorption capacity (35.4–64.7 mg/g) as compared to the commercial one (25 mg/g) (Yan et al., 2011; Yao et al., 2011). Structural modification of CNT by MONP doping complements the degradation potential of NC (Ahmad et al., 2014; Behnam et al., 2013). Removal efficacy rose in order CNT < MO < MO/CNT and irradiated > non-irradiated systems together with noticeable change in light absorption properties (Ahmad et al., 2014; Saleh et al., 2012) that facilitated enhanced photocatalytic, sonocatalytic and sonophotocatalytic activity as compared to the single component systems (Ahmad et al., 2014). Such systems also accelerated the degradation of contaminant (e.g. diazinone) to less toxic product (i.e. 2-isopropyl-6-methyl-4-pyrimidinol) (Behnam et al., 2013). Varying the amount of either CNT or the dopant directly influenced the uptake potential of NC. Gain in CNT loading from 1-20 wt. % led to increasing in surface area from 49.6 -135.5m^2/g facilitating boost in COD reduction from initial value of 627 to 43 mg/L (Ahmad et al., 2014). The process was governed by slip-induce surface diffusion that assists electron trapping and photon absorption (Nguyen et al., 2016). Also, synergistic effects, enhanced by the homogeneity of the composite, promoted effective charge transfer that impacted the degradation or adsorption potential (Miranda et al., 2014).

Higher uptake potential of modified CNT was also noticed in the case of heavy metals adsorption. Independent studies carried out on NMO/ CNT exhibited improved removal % for metal ions when compared with CNT, charcoal and pure NPs (Chen et al., 2009; Gupta et al., 2011). The studies indicated dependence on agitation time and pH in case of batch studies while on bed thickness subject to column studies (Gupta et al., 2011). The composites were selectively adsorbed Pb^{2+} in mixed systems irrespective of dopant used (Gupta et al., 2011; Peng, Luan, Di, et al., 2005). Ca^{2+} and Mg^{2+} always present in water significantly enhanced the As^{5+} adsorption from 10 to 81.9 and 78.8 mg/g on to CeO_2-CNT owing to the formation of ternary surface complex reaction (Peng, Luan, Ding, et al., 2005). Cd^{2+} removal was aided with MnO_2 doped o-MWCNT (acid oxidized) prepared at different pH and compared with unmodified MWCNT and o-MWCNT; the order followed was MnO_2/o-MWCNT (pH5) > MnO_2/o-MWCNT (pH3) > MnO_2/o-MWCNT (pH7) > o-MWCNT > MnO_2/o-MWCNT (pH1) > raw MWCNT. The reason was attributed to 1) increase in surface roughness with pH, 2) enhancement in uniformity and thinness of MnO_2 layer deposited with an increase in pH and 3) increase in MnO_2 loading with pH that hinders available area for adsorption (Luo, C. et al., 2013). The introduction of 20-30 nm thick diamine functionalized SiO_2 homogeneous layer onto MWCNT was able to remove Pb(II) > Cu(II) > Zn(II) > Ni(II) in larger amounts as compared to SiO_2@MWCNT, o-MWCNT and MWCNT. The intraparticle diffusion model predicts the adsorption process to occur in three stages i.e. adsorption onto the exterior surface of NC, adsorption into the interior surface of pores and diffusion resistance leading to attainment of equilibrium (Yang, W. et al., 2013). Antibacterial activity of TiO_2/CNT thin films (5-30nm) depended on CNT:TiO_2 and annealing temperature that altered the photocatalytic performance owing to indirect impact on the formation of carbonaceous bonds between the CNT and MO (Akhavan et al., 2010)

Carbon encapsulated magnetic nanoparticles (CEMNPs) has emerged as a new class of nano adsorbent consisting of metallic cores completely covered in carbon coatings that are 1-5 nm thick (Pyrzyńska et al., 2010). The magnetic core helps in easy separation and carbon shell ensures effective adsorption (Zhang, D. et al., 2010). Comparative account of these CEMNPs with activated carbon and CNT promoted better uptake of Cu^{2+} and Cd^{2+} i.e. 30-60% *vs.* < 20% owing to differences in surface charge densities of these materials (Pyrzyńska et al., 2010). It was also determined that limiting step in CNMEPs is film diffusion in contrast to intraparticle diffusion (Bystrzejewski et al., 2011). The adsorption capacities reported for these nanoparticles for Cu^{2+}, Cd^{2+}, and Co^{2+} fall in the range of 1.23 – 3.21 mg/g (Bystrzejewski et al., 2009; Pyrzyńska et al., 2010) while for Cr^{6+} it is 0.63 mg/g (Zhang, D. et al., 2010). These NP effectively decontaminated water by adsorbing phenols, phthalates, sulfonamides, tetracyclines and quinolones antibiotics based on hydrogen bonding, π-π and hydrophobic interactions. The adsorption phenomena and uptake of certain compounds were affected by the amount of oxygen-containing species and surface graphitized carbon which can be varied by changing the temperature during synthesis of CNMEP (Niu et al., 2011).

NP impregnated activated charcoal (AC) helps is solving water treatment issues by enhancing the properties of activated charcoal. Incorporation of magnetic nanoparticles assisted in the fast and easy separation of adsorbate making them useful for water purification applications (Fayazi et al., 2015). The addition of these enhanced the basicity as well as adsorption potential of modified charcoal in comparison to pure one. Utilization of these NC in micro-filter membrane system led to 85.3, 80 and 79.6% removal of dissolved organic compounds, turbidity and UV_{254}, respectively, in addition to longer removal operation when compared with other micro-filter systems (Kim et al., 2013). Clustering of $MgFe_2O_4$ into AC structure formed 40-50 nm sized nanocomposite that effectively adsorbed Cr^{6+} through the spontaneous and exothermic process leading to increasing in q_e from 500 (NC) to 333 mg/g (unmodified charcoal) at room temperature. NaOH treatment helped in 80% desorption but with 5% reduction in removal efficiency (Kaur, M., Kaur, et al., 2015). Dyes removal efficacy varies by changing the amount of loading of MONP into AC. The reason can be attributed to resulting differences in particle sizes, surface morphology, and photocatalytic effects. Experiments had also shown a considerable decrease in degradation percentage for the same adsorbent or photocatalyst by changing the systems' exposure mode from UV light to fluorescent lamp (Hu et al., 2011). The efficiency of ZnO/AC NC system for dye adsorption depended on the size of charcoal particles, irradiation time, H_2O_2 amount (Yu, X. et al., 2015) and sonication time (Jamshidi et al., 2016).

In addition to these, there is various nanomaterial i.e. hybrid nanomaterials are used for the water treatment and purification. In this regard NMO-clay-polymer composites, NMO-polymer-CNT, NMO-Chitosan-Graphene oxide, NMO-chitosan-activated carbon etc. have been exploited and some of them proved to have good qualities needed for water purification purposes (Jeon et al., 2010; Larraza et al., 2012; Liu, L. F. et al., 2010; Ray et al., 2015).

CONCLUSION

The water pollution scenario arising from the flow of omnifarious contaminants led to evolution and need of aqua-nanotechnology that through its broad spectrum approach gives rise to various possibilities to solve this problem. In this regard metal oxide nanomaterials had gained attention but owing to their agglomeration and separation issues led researchers and technologist to opt for their nanocomposites. These include composites with carbon-based materials (carbon nanotubes, graphene etc.), natural and synthetic polymers, clays (including zeolite). Efforts were put into synthesizing the materials with high adsorption capacities, photocatalytic potential and recyclable, in essence, to make the water purification procedure fast, cost-effective and environmentally friendly. Among this variety of nanomaterials, the choice of best nanomaterial to be used always depends on the nature and concentration of a contaminant to be removed and matrix composition which needs to be treated in addition to cost involvement. The magnetic clay based composites offer advantages including the combination of materials available owing to its natural source and magnetic nature that makes the material cost effective. In addition, it offers easy separation and multifunctional nature that enables adsorption of heavy metals, decontamination from microbial impurities and photocatalytic degradation of organic pollutants hence offering a single solution to this wide scoped water pollution problem.

REFERENCES

Abou-Gamra, Z. M., & Ahmed, M. A. (2015). TiO_2 nanoparticles for removal of malachite green dye from waste water. *Advances in Chemical Engineering and Science*, *5*(03), 373–388. doi:10.4236/aces.2015.53039

Ahmad, M., Ahmed, E., Hong, Z. L., Ahmed, W., Elhissi, A., & Khalid, N. R. (2014). Photocatalytic, sonocatalytic and sonophotocatalytic degradation of Rhodamine B using ZnO/CNTs composites photocatalysts. *Ultrasonics Sonochemistry*, *21*(2), 761–773. doi:10.1016/j.ultsonch.2013.08.014 PMID:24055646

Ai, L., Zhang, C., & Chen, Z. (2011). Removal of methylene blue from aqueous solution by a solvothermal-synthesized graphene/magnetite composite. *Journal of Hazardous Materials*, *192*(3), 1515–1524. doi:10.1016/j.jhazmat.2011.06.068 PMID:21782325

Akhavan, O., Azimirad, R., Safa, S., & Larijani, M. (2010). Visible light photo-induced antibacterial activity of CNT–doped TiO_2 thin films with various CNT contents. *Journal of Materials Chemistry*, *20*(35), 7386–7392. doi:10.1039/c0jm00543f

Alam, J., Dass, L. A., Ghasemi, M., & Alhoshan, M. (2013). Synthesis and optimization of PES-Fe_3O_4 mixed matrix nanocomposite membrane: Application studies in water purification. *Polymer Composites*, *34*(11), 1870–1877. doi:10.1002/pc.22593

Anju, S., Yesodharan, S., & Yesodharan, E. (2012). Zinc oxide mediated sonophotocatalytic degradation of phenol in water. *Chemical Engineering Journal*, *189*, 84–93. doi:10.1016/j.cej.2012.02.032

Azam, A., Ahmed, A. S., Oves, M., Khan, M. S., Habib, S. S., & Memic, A. (2012). Antimicrobial activity of metal oxide nanoparticles against Gram-positive and Gram-negative bacteria: A comparative study. *International Journal of Nanomedicine*, *7*, 6003–6009. doi:10.2147/IJN.S35347 PMID:23233805

Azari, A., Kalantary, R. R., Ghanizadeh, G., Kakavandi, B., Farzadkia, M., & Ahmadi, E. (2015). Iron–silver oxide nanoadsorbent synthesized by co-precipitation process for fluoride removal from aqueous solution and its adsorption mechanism. *RSC Advances*, *5*(106), 87377–87391. doi:10.1039/C5RA17595J

Ba-Abbad, M. M., Kadhum, A. A. H., Mohamad, A. B., Takriff, M. S., & Sopian, K. (2012). Synthesis and catalytic activity of TiO_2 nanoparticles for photochemical oxidation of concentrated chlorophenols under direct solar radiation. *International Journal of Electrochemical Science*, *7*, 4871–4888.

Babu, C. M., Vinodh, R., Sundaravel, B., Abidov, A., Peng, M. M., Cha, W. S., & Jang, H.-T. (2016). Characterization of reduced graphene oxide supported mesoporous Fe_2O_3/TiO_2 nanoparticles and adsorption of As(III) and As(V) from potable water. *Journal of the Taiwan Institute of Chemical Engineers*, *62*, 199–208. doi:10.1016/j.jtice.2016.02.005

Badruddoza, A. Z. M., Shawon, Z. B. Z., Tay, W. J. D., Hidajat, K., & Uddin, M. S. (2013). Fe_3O_4/cyclodextrin polymer nanocomposites for selective heavy metals removal from industrial wastewater. *Carbohydrate Polymers*, *91*(1), 322–332. doi:10.1016/j.carbpol.2012.08.030 PMID:23044139

Bagabas, A., Alshammari, A., Aboud, M., & Kosslick, H. (2013). Room-temperature synthesis of zinc oxide nanoparticles in different media and their application in cyanide photodegradation. *Nanoscale Research Letters*, *8*(1), 516. doi:10.1186/1556-276X-8-516 PMID:24314056

Baghbanzadeh, M., Rana, D., Lan, C. Q., & Matsuura, T. (2016). Effects of inorganic nano-additives on properties and performance of polymeric membranes in water treatment. *Separation & Purification Reviews*, *45*(2), 141–167. doi:10.1080/15422119.2015.1068806

Bardajee, G., & Hooshyar, Z. (2013). Degradation of 2-chlorophenol from wastewater Using γ-Fe_2O_3 nanoparticles. *International Journal of Nanoscience and Nanotechnology*, *9*(1), 3–6.

Behnam, R., Morshed, M., Tavanai, H., & Ghiaci, M. (2013). Destructive adsorption of diazinon pesticide by activated carbon nanofibers containing Al_2O_3 and MgO nanoparticles. *Bulletin of Environmental Contamination and Toxicology*, *91*(4), 475–480. doi:10.1007/s00128-013-1064-x PMID:23912227

Bhattacharyya, K., & Gupta, S. S. (2008). Adsorption of a few heavy metals on natural and modified kaolinite and montmorillonite: A review. *Advances in Colloid and Interface Science*, *140*(2), 114–131. doi:10.1016/j.cis.2007.12.008 PMID:18319190

Bhaumik, M., Maity, A., Srinivasu, V., & Onyango, M. S. (2011). Enhanced removal of Cr(VI) from aqueous solution using polypyrrole/Fe_3O_4 magnetic nanocomposite. *Journal of Hazardous Materials*, *190*(1), 381–390. doi:10.1016/j.jhazmat.2011.03.062 PMID:21497438

Bora, T., & Dutta, J. (2014). Applications of nanotechnology in wastewater treatment—a review. *Journal of Nanoscience and Nanotechnology*, *14*(1), 613–626. doi:10.1166/jnn.2014.8898 PMID:24730286

Bykkam, S., Narsingam, S., Ahmadipour, M., Dayakar, T., Rao, K. V., Chakra, C. S., & Kalakotla, S. (2015). Few layered graphene Sheet decorated by ZnO Nanoparticles for anti-bacterial application. *Superlattices and Microstructures*, *83*, 776–784. doi:10.1016/j.spmi.2015.03.063

Bystrzejewski, M., & Pyrzyńska, K. (2011). Kinetics of copper ions sorption onto activated carbon, carbon nanotubes and carbon-encapsulated magnetic nanoparticles. *Colloids and Surfaces. A, Physicochemical and Engineering Aspects*, *377*(1), 402–408. doi:10.1016/j.colsurfa.2011.01.041

Bystrzejewski, M., Pyrzyńska, K., Huczko, A., & Lange, H. (2009). Carbon-encapsulated magnetic nanoparticles as separable and mobile sorbents of heavy metal ions from aqueous solutions. *Carbon*, *47*(4), 1201–1204. doi:10.1016/j.carbon.2009.01.007

Cao, B., Cao, S., Dong, P., Gao, J., & Wang, J. (2013). High antibacterial activity of ultrafine TiO_2/graphene sheets nanocomposites under visible light irradiation. *Materials Letters*, *93*, 349–352. doi:10.1016/j.matlet.2012.11.136

Cao, C., Xiao, L., Chen, C., Shi, X., Cao, Q., & Gao, L. (2014). In situ preparation of magnetic Fe_3O_4/chitosan nanoparticles via a novel reduction–precipitation method and their application in adsorption of reactive azo dye. *Powder Technology*, *260*, 90–97. doi:10.1016/j.powtec.2014.03.025

Cao, J., Li, J., Liu, L., Xie, A., Li, S., Qiu, L., & Shen, Y. et al. (2014). One-pot synthesis of novel Fe_3O_4/Cu_2O/PANI nanocomposites as absorbents in water treatment. *Journal of Materials Chemistry A*, *2*(21), 7953–7959. doi:10.1039/c4ta00303a

Chai, L., Wang, Y., Zhao, N., Yang, W., & You, X. (2013). Sulfate-doped Fe_3O_4/Al_2O_3 nanoparticles as a novel adsorbent for fluoride removal from drinking water. *Water Research*, *47*(12), 4040–4049. doi:10.1016/j.watres.2013.02.057 PMID:23602616

Chang, Y.-N., Ou, X.-M., Zeng, G.-M., Gong, J.-L., Deng, C.-H., Jiang, Y., & He, X. et al. (2015). Synthesis of magnetic graphene oxide–TiO_2 and their antibacterial properties under solar irradiation. *Applied Surface Science*, *343*, 1–10. doi:10.1016/j.apsusc.2015.03.082

Chen, C., Hu, J., Shao, D., Li, J., & Wang, X. (2009). Adsorption behavior of multiwall carbon nanotube/iron oxide magnetic composites for Ni(II) and Sr(II). *Journal of Hazardous Materials*, *164*(2), 923–928. doi:10.1016/j.jhazmat.2008.08.089 PMID:18842337

Chen, D., Zeng, Z., Zeng, Y., Zhang, F., & Wang, M. (2016). Removal of methylene blue and mechanism on magnetic γ-Fe_2O_3/SiO_2 nanocomposite from aqueous solution. *Water Resources and Industry*, *15*, 1–13. doi:10.1016/j.wri.2016.05.003

Chen, J., Liao, W., Jiang, Y., Yu, D., Zou, M., Zhu, H., . . . Du, M. (2016). Facile fabrication of ZnO/TiO_2 heterogeneous nanofibres and their photocatalytic behaviour and mechanism towards rhodamine B. *Nanomaterials and Nanotechnology, 6*(9).

Cheng, W., Zhang, W., Hu, L., Ding, W., Wu, F., & Li, J. (2016). Etching synthesis of iron oxide nanoparticles for adsorption of arsenic from water. *RSC Advances*, *6*(19), 15900–15910. doi:10.1039/C5RA26143K

Clausen, L., Fabricius, I., & Madsen, L. (2001). Adsorption of pesticides onto quartz, calcite, kaolinite and α-alumina. *Journal of Environmental Quality*, *30*(3), 846–857. doi:10.2134/jeq2001.303846x PMID:11401273

Cohen-Tanugi, D., & Grossman, J. C. (2012). Water desalination across nanoporous graphene. *Nano Letters*, *12*(7), 3602–3608. doi:10.1021/nl3012853 PMID:22668008

Cun, W., Jincai, Z., Xinming, W., Bixian, M., Guoying, S., Pingan, P., & Jiamo, F. (2002). Preparation, characterization and photocatalytic activity of nano-sized ZnO/SnO_2 coupled photocatalysts. *Applied Catalysis B: Environmental*, *39*(3), 269–279. doi:10.1016/S0926-3373(02)00115-7

Dědková, K., Lang, J., Matějová, K., Peikertová, P., Holešinský, J., Vodárek, V., & Kukutschová, J. (2015). Nanostructured composite material graphite/TiO_2 and its antibacterial activity under visible light irradiation. *Journal of Photochemistry and Photobiology. B, Biology*, *149*, 265–271. doi:10.1016/j.jphotobiol.2015.06.018 PMID:26114221

Demarchi, C. A., Debrassi, A., de Campos Buzzi, F., Nedelko, N., Slawska-Waniewska, A., Dłuzewski, P., & Rodrigues, C. A. (2015). *Adsorption of the dye Remazol Red 198 (RR198) by O-carboxymethylchitosan-N-lauryl/γ-Fe_2O_3 magnetic nanoparticles*. Arabian Journal of Chemistry.

Deng, J.-H., Zhang, X.-R., Zeng, G.-M., Gong, J.-L., Niu, Q.-Y., & Liang, J. (2013). Simultaneous removal of Cd(II) and ionic dyes from aqueous solution using magnetic graphene oxide nanocomposite as an adsorbent. *Chemical Engineering Journal*, *226*, 189–200. doi:10.1016/j.cej.2013.04.045

Dizaj, S. M., Lotfipour, F., Barzegar-Jalali, M., Zarrintan, M. H., & Adibkia, K. (2014). Antimicrobial activity of the metals and metal oxide nanoparticles. *Materials Science and Engineering C*, *44*, 278–284. doi:10.1016/j.msec.2014.08.031 PMID:25280707

Dobrucka, R., & Długaszewska, J. (2016). Biosynthesis and antibacterial activity of ZnO nanoparticles using *Trifolium pratense* flower extract. *Saudi Journal of Biological Sciences*, *23*(4), 517–523. doi:10.1016/j.sjbs.2015.05.016 PMID:27298586

Elkady, M. F., Shokry Hassan, H., & Salama, E. (2016). Sorption profile of phosphorus ions onto ZnO nanorods synthesized via Sonic technique. *Journal of Engineering*, *9*. doi:10.1155/2016/2308560

Elwakeel, K. Z., & Guibal, E. (2015). Arsenic(V) sorption using chitosan/Cu$(OH)_2$ and chitosan/CuO composite sorbents. *Carbohydrate Polymers*, *134*, 190–204. doi:10.1016/j.carbpol.2015.07.012 PMID:26428116

Fan, L., Luo, C., Lv, Z., Lu, F., & Qiu, H. (2011). Preparation of magnetic modified chitosan and adsorption of Zn^{2+} from aqueous solutions. *Colloids and Surfaces. B, Biointerfaces*, *88*(2), 574–581. doi:10.1016/j.colsurfb.2011.07.038 PMID:21868204

Fan, L., Zhang, Y., Li, X., Luo, C., Lu, F., & Qiu, H. (2012). Removal of alizarin red from water environment using magnetic chitosan with Alizarin Red as imprinted molecules. *Colloids and Surfaces. B, Biointerfaces*, *91*, 250–257. doi:10.1016/j.colsurfb.2011.11.014 PMID:22119217

Fan, L., Zhang, Y., Luo, C., Lu, F., Qiu, H., & Sun, M. (2012). Synthesis and characterization of magnetic β-cyclodextrin–chitosan nanoparticles as nano-adsorbents for removal of methyl blue. *International Journal of Biological Macromolecules*, *50*(2), 444–450. doi:10.1016/j.ijbiomac.2011.12.016 PMID:22227306

Farooq, M., Raja, I. A., & Pervez, A. (2009). Photocatalytic degradation of TCE in water using TiO_2 catalyst. *Solar Energy*, *83*(9), 1527–1533. doi:10.1016/j.solener.2009.04.009

Fayazi, M., Ghanei-Motlagh, M., & Taher, M. A. (2015). The adsorption of basic dye (Alizarin red S) from aqueous solution onto activated carbon/γ-Fe_2O_3 nanocomposite: Kinetic and equilibrium studies. *Materials Science in Semiconductor Processing*, *40*, 35–43. doi:10.1016/j.mssp.2015.06.044

Fry, A. (2005). Water: Facts and trends: WBCSD.

Gandhi, M. R., Viswanathan, N., & Meenakshi, S. (2010). Preparation and application of alumina/chitosan biocomposite. *International Journal of Biological Macromolecules*, *47*(2), 146–154. doi:10.1016/j.ijbiomac.2010.05.008 PMID:20471415

Gao, Q., Luo, J., Wang, X., Gao, C., & Ge, M. (2015). Novel hollow α-Fe_2O_3 nanofibers via electrospinning for dye adsorption. *Nanoscale Research Letters*, *10*(1), 1–8. doi:10.1186/s11671-015-0874-7 PMID:25918495

Gao, W., Majumder, M., Alemany, L. B., Narayanan, T. N., Ibarra, M. A., Pradhan, B. K., & Ajayan, P. M. (2011). Engineered graphite oxide materials for application in water purification. *ACS Applied Materials & Interfaces*, *3*(6), 1821–1826. doi:10.1021/am200300u PMID:21568266

Gautam, A., Kshirsagar, A., Biswas, R., Banerjee, S., & Khanna, P. K. (2016). Photodegradation of organic dyes based on anatase and rutile TiO_2 nanoparticles. *RSC Advances*, *6*(4), 2746–2759. doi:10.1039/C5RA20861K

Gayathri, S., Jayabal, P., Kottaisamy, M., & Ramakrishnan, V. (2015). Synthesis of the graphene-$ZnTiO_3$ nanocomposite for solar light assisted photodegradation of methylene blue. *Journal of Physics. D, Applied Physics*, *48*(41), 415305. doi:10.1088/0022-3727/48/41/415305

Ge, F., Li, M.-M., Ye, H., & Zhao, B.-X. (2012). Effective removal of heavy metal ions Cd^{2+}, Zn^{2+}, Pb^{2+}, Cu^{2+} from aqueous solution by polymer-modified magnetic nanoparticles. *Journal of Hazardous Materials*, *211*, 366–372. doi:10.1016/j.jhazmat.2011.12.013 PMID:22209322

Ghasemi, Z., Seif, A., Ahmadi, T. S., Zargar, B., Rashidi, F., & Rouzbahani, G. M. (2012). Thermodynamic and kinetic studies for the adsorption of Hg (II) by nano-TiO 2 from aqueous solution. *Advanced Powder Technology*, *23*(2), 148–156. doi:10.1016/j.apt.2011.01.004

Ghasemzadeh, G., Momenpour, M., Omidi, F., Hosseini, M. R., Ahani, M., & Barzegari, A. (2014). Applications of nanomaterials in water treatment and environmental remediation. *Frontiers of Environmental Science & Engineering*, *8*(4), 471–482. doi:10.1007/s11783-014-0654-0

Giraldo, L., Erto, A., & Moreno-Piraján, J. C. (2013). Magnetite nanoparticles for removal of heavy metals from aqueous solutions: Synthesis and characterization. *Adsorption*, *19*(2-4), 465–474. doi:10.1007/s10450-012-9468-1

Gohari, R. J., Lau, W. J., Matsuura, T., Halakoo, E., & Ismail, A. F. (2013). Adsorptive removal of Pb(II) from aqueous solution by novel PES/HMO ultrafiltration mixed matrix membrane. *Separation and Purification Technology*, *120*, 59–68. doi:10.1016/j.seppur.2013.09.024

Gondal, M. A., Sayeed, M. N., Yamani, Z. H., & Al-Arfaj, A. R. (2009). Efficient removal of phenol from water using Fe_2O_3 semiconductor catalyst under UV laser irradiation. *Journal of Environmental Science and Health. Part A*, *44*(5), 515–521. doi:10.1080/10934520902720033

Goon, I. Y., Zhang, C., Lim, M., Gooding, J. J., & Amal, R. (2010). Controlled fabrication of polyethylenimine-functionalized magnetic nanoparticles for the sequestration and quantification of free Cu2+. *Langmuir*, *26*(14), 12247–12252. doi:10.1021/la101196r PMID:20527930

Gupta, V., Agarwal, S., & Saleh, T. A. (2011). Chromium removal by combining the magnetic properties of iron oxide with adsorption properties of carbon nanotubes. *Water Research*, *45*(6), 2207–2212. doi:10.1016/j.watres.2011.01.012 PMID:21303713

Haldorai, Y., & Shim, J.-J. (2013). Multifunctional chitosan-copper oxide hybrid material: Photocatalytic and antibacterial activities. *International Journal of Photoenergy*.

Haldorai, Y., & Shim, J.-J. (2014). An efficient removal of methyl orange dye from aqueous solution by adsorption onto chitosan/MgO composite: A novel reusable adsorbent. *Applied Surface Science*, *292*, 447–453. doi:10.1016/j.apsusc.2013.11.158

Hameed, A., Montini, T., Gombac, V., & Fornasiero, P. (2009). Photocatalytic decolourization of dyes on NiO–ZnO nano-composites. *Photochemical & Photobiological Sciences*, *8*(5), 677–682. doi:10.1039/b817396f PMID:19424542

Han, J., Liu, Y., Singhal, N., Wang, L., & Gao, W. (2012). Comparative photocatalytic degradation of estrone in water by ZnO and TiO_2 under artificial UVA and solar irradiation. *Chemical Engineering Journal*, *213*, 150–162. doi:10.1016/j.cej.2012.09.066

Hasan, S., Ghosh, A., Race, K., Schreiber, R. Jr, & Prelas, M. (2014). Dispersion of FeOOH on Chitosan Matrix for Simultaneous Removal of As(III) and As(V) from Drinking Water. *Separation Science and Technology*, *49*(18), 2863–2877. doi:10.1080/01496395.2014.949774

Hashem, F. (2012). Adsorption of methylene blue from aqueous solutions using Fe_3O_4/bentonite nanocomposite. *Hydrol Current Res*, *3*(5), 1–6.

Hashemian, S., Saffari, H., & Ragabion, S. (2015). Adsorption of cobalt(II) from aqueous solutions by Fe_3O_4/bentonite nanocomposite. *Water, Air, and Soil Pollution*, *226*(1), 1–10. doi:10.1007/s11270-014-2212-6

Hayat, K., Gondal, M., Khaled, M. M., Ahmed, S., & Shemsi, A. M. (2011). Nano ZnO synthesis by modified sol gel method and its application in heterogeneous photocatalytic removal of phenol from water. *Applied Catalysis A, General*, *393*(1), 122–129. doi:10.1016/j.apcata.2010.11.032

He, L., Liu, Y., Mustapha, A., & Lin, M. (2011). Antifungal activity of zinc oxide nanoparticles against *Botrytis cinerea* and *Penicillium expansum*. *Microbiological Research*, *166*(3), 207–215. doi:10.1016/j.micres.2010.03.003 PMID:20630731

Homaeigohar, S., & Elbahri, M. (2014). Nanocomposite electrospun nanofiber membranes for environmental remediation. *Materials (Basel)*, *7*(2), 1017–1045. doi:10.3390/ma7021017

Hristodor, C. M., Cochecl, L., Pode, R., Copcia, V. E., & Popovici, E. (2011). The Removal of p-cloro-phenol from synthetic waste water on Fe_2O_3-Clays nanocomposites ystems. *REVISTA DE CHIMIE*, *62*(11), 1119–1122.

Hu, C., Lan, Y., Qu, J., Hu, X., & Wang, A. (2006). Ag/AgBr/TiO_2 visible light photocatalyst for destruction of azodyes and bacteria. *The Journal of Physical Chemistry B*, *110*(9), 4066–4072. doi:10.1021/jp0564400 PMID:16509698

Hu, J., Chen, G., & Lo, I. M. (2006). Selective removal of heavy metals from industrial wastewater using maghemite nanoparticle: Performance and mechanisms. *Journal of Environmental Engineering*, *132*(7), 709–715. doi:10.1061/(ASCE)0733-9372(2006)132:7(709)

Hu, X. D., Deng, H., & Du, L. (2011). Preparation and characterization of Fe^{3+}/TiO_2 thin films loaded activated carbon and degradation of methyl orange. *Advanced Materials Research*, *332-334*, 134–137. doi:10.4028/www.scientific.net/AMR.332-334.134

Hua, M., Zhang, S., Pan, B., Zhang, W., Lv, L., & Zhang, Q. (2012). Heavy metal removal from water/wastewater by nanosized metal oxides: A review. *Journal of Hazardous Materials*, *211*, 317–331. doi:10.1016/j.jhazmat.2011.10.016 PMID:22018872

Huang, S.-H., & Chen, D.-H. (2009). Rapid removal of heavy metal cations and anions from aqueous solutions by an amino-functionalized magnetic nano-adsorbent. *Journal of Hazardous Materials*, *163*(1), 174–179. doi:10.1016/j.jhazmat.2008.06.075 PMID:18657903

Hur, J., Shin, J., Yoo, J., & Seo, Y.-S. (2015). Competitive adsorption of metals onto magnetic graphene oxide: Comparison with other carbonaceous adsorbents. *The Scientific World Journal*. PMID:25861683

Ismail, R. A., Sulaiman, G. M., Abdulrahman, S. A., & Marzoog, T. R. (2015). Antibacterial activity of magnetic iron oxide nanoparticles synthesized by laser ablation in liquid. *Materials Science and Engineering C*, *53*, 286–297. doi:10.1016/j.msec.2015.04.047 PMID:26042717

Jamshidi, M., Ghaedi, M., Dashtian, K., Hajati, S., & Bazrafshan, A. (2016). Sonochemical assisted hydrothermal synthesis of ZnO: Cr nanoparticles loaded activated carbon for simultaneous ultrasound-assisted adsorption of ternary toxic organic dye: Derivative spectrophotometric, optimization, kinetic and isotherm study. *Ultrasonics Sonochemistry*, *32*, 119–131. doi:10.1016/j.ultsonch.2016.03.004 PMID:27150752

Javanbakht, T., Laurent, S., Stanicki, D., & Wilkinson, K. J. (2016). Relating the surface properties of superparamagnetic iron oxide nanoparticles (SPIONs) to their bactericidal effect towards a biofilm of *Streptococcus mutans*. *PLoS ONE*, *11*(4), e0154445. doi:10.1371/journal.pone.0154445 PMID:27115356

Jayarathna, L., Bandara, A., Ng, W. J., & Weerasooriya, R. (2015). Fluoride adsorption on $\gamma - Fe_2O_s$ nanoparticles. *Journal of Environmental Health Science and Engineering*, *13*(1), 1–10. doi:10.1186/s40201-015-0210-2 PMID:25763187

Jeon, I.-Y., & Baek, J.-B. (2010). Nanocomposites derived from polymers and inorganic nanoparticles. *Materials (Basel)*, *3*(6), 3654–3674. doi:10.3390/ma3063654

Jiang, H., Chen, P., Luo, S., Luo, X., Tu, X., Cao, Q., & Zhang, W. et al. (2013). Synthesis of novel biocompatible composite Fe_3O_4/ZrO_2/chitosan and its application for dye removal. *Journal of Inorganic and Organometallic Polymers and Materials*, *23*(2), 393–400. doi:10.1007/s10904-012-9792-7

Jiang, W., Mashayekhi, H., & Xing, B. (2009). Bacterial toxicity comparison between nano-and micro-scaled oxide particles. *Environmental Pollution*, *157*(5), 1619–1625. doi:10.1016/j.envpol.2008.12.025 PMID:19185963

Jin, Z., Gao, H., & Hu, L. (2015). Removal of Pb (II) by nano-titanium oxide investigated by batch, XPS and model techniques. *RSC Advances*, *5*(107), 88520–88528. doi:10.1039/C5RA14004H

Jokar, M., Foroutani, R., Safaralizadeh, M. H., & Farhadi, K. (2014). Synthesis and characterization of polyaniline/Fe_3O_4 magnetic nanocomposite as practical approach for fluoride removal process. *Annual Research & Review in Biology*, *4*(21), 3262–3273. doi:10.9734/ARRB/2014/9108

Kalantari, K., Ahmad, M. B., Masoumi, H. R. F., Shameli, K., Basri, M., & Khandanlou, R. (2014). Rapid adsorption of heavy metals by Fe_3O_4/talc nanocomposite and optimization study using response surface methodology. *International Journal of Molecular Sciences*, *15*(7), 12913–12927. doi:10.3390/ijms150712913 PMID:25050784

Kapoor, P. N., Bhagi, A. K., Mulukutla, R. S., & Klabunde, K. J. (2004). *Mixed metal oxide nanoparticle*. Dekker Encyclopedia of Nanoscience and Nanotechnology.

Kaur, M., Kaur, N., Jeet, K., & Kaur, P. (2015). $MgFe_2O_4$ nanoparticles loaded on activated charcoal for effective removal of Cr(VI) – A novel approach. *Ceramics International*, *41*(10), 13739–13750. doi:10.1016/j.ceramint.2015.08.040

Kaur, M., Singh, M., Mukhopadhyay, S. S., Singh, D., & Gupta, M. (2015). Structural, magnetic and adsorptive properties of clay ferrite nanocomposite and its use for effective removal of Cr(VI) from water. *Journal of Alloys and Compounds*, *653*, 202–211. doi:10.1016/j.jallcom.2015.08.265

Kaur, T., Sraw, A., Wanchoo, R. K., & Toor, A. P. (2016). Visible –light induced photocatalytic degradation of fungicide with Fe and Si doped TiO_2 nanoparticles. *Materials Today: Proceedings*, *3*(2), 354–361. doi:10.1016/j.matpr.2016.01.020

Kemp, K. C., Seema, H., Saleh, M., Le, N. H., Mahesh, K., Chandra, V., & Kim, K. S. (2013). Environmental applications using graphene composites: Water remediation and gas adsorption. *Nanoscale*, *5*(8), 3149–3171. doi:10.1039/c3nr33708a PMID:23487161

Khan, M., & Lo, I. M. (2016). Removal of ionizable aromatic pollutants from contaminated water using nano γ-Fe_2O_3 based magnetic cationic hydrogel: Sorptive performance, magnetic separation and reusability. *Journal of Hazardous Materials*. doi:10.1016/j.jhazmat.2016.01.051 PMID:26857338

Khan, S., Suriyaprabha, R., Pathak, B., & Fulekar, M. (2015). Photocatalytic degradation of organophosphate pesticides (Chlorpyrifos) using synthesized zinc oxide nanoparticle by membrane filtration reactor under UV irradiation. *Frontier of Nanoscience and Nanotechnology*, *1*(1), 23–27. doi:10.15761/FNN.1000105

Khatamian, M., & Alaji, Z. (2012). Efficient adsorption-photodegradation of 4-nitrophenol in aqueous solution by using ZnO/HZSM-5 nanocomposites. *Desalination*, *286*, 248–253. doi:10.1016/j.desal.2011.11.031

Kim, S., Kim, J., & Seo, G. (2013). Iron oxide nanoparticle-impregnated powder-activated carbon (IPAC) for NOM removal in MF membrane water treatment system. *Desalination and Water Treatment*, *51*(31-33), 6392–6400. doi:10.1080/19443994.2013.781000

Kumar, R., & Chawla, J. (2014). Removal of cadmium ion from water/wastewater by nano-metal oxides: A review. *Water Quality. Exposure and Health*, *5*(4), 215–226. doi:10.1007/s12403-013-0100-8

Larraza, I., López-Gónzalez, M., Corrales, T., & Marcelo, G. (2012). Hybrid materials: Magnetite–Polyethylenimine–Montmorillonite, as magnetic adsorbents for Cr(VI) water treatment. *Journal of Colloid and Interface Science*, *385*(1), 24–33. doi:10.1016/j.jcis.2012.06.050 PMID:22841705

Lee, K. M., Lai, C. W., Ngai, K. S., & Juan, J. C. (2016). Recent developments of zinc oxide based photocatalyst in water treatment technology: A review. *Water Research*, *88*, 428–448. doi:10.1016/j.watres.2015.09.045 PMID:26519627

Li, J., Shi, Y., Cai, Y., Mou, S., & Jiang, G. (2008). Adsorption of di-ethyl-phthalate from aqueous solutions with surfactant-coated nano/microsized alumina. *Chemical Engineering Journal*, *140*(1), 214–220.

Li, M., Zhu, L., & Lin, D. (2011). Toxicity of ZnO nanoparticles to *Escherichia coli*: Mechanism and the influence of medium components. *Environmental Science & Technology*, *45*(5), 1977–1983. doi:10.1021/es102624t PMID:21280647

Li, Q., Su, H., & Tan, T. (2008). Synthesis of ion-imprinted chitosan-TiO_2 adsorbent and its multi-functional performances. *Biochemical Engineering Journal*, *38*(2), 212–218. doi:10.1016/j.bej.2007.07.007

Li, X., Wang, D., Cheng, G., Luo, Q., An, J., & Wang, Y. (2008). Preparation of polyaniline-modified TiO_2 nanoparticles and their photocatalytic activity under visible light illumination. *Applied Catalysis B: Environmental*, *81*(3), 267–273. doi:10.1016/j.apcatb.2007.12.022

Liang, P., Qin, Y., Hu, B., Li, C., Peng, T., & Jiang, Z. (2000). Study of the adsorption behavior of heavy metal ions on nanometer-size titanium dioxide with ICP-AES. *Fresenius Journal of Analytical Chemistry*, *368*(6), 638–640. doi:10.1007/s002160000546 PMID:11228717

Liu, L. F., Zhang, P. H., & Yang, F. L. (2010). Adsorptive removal of 2,4-DCP from water by fresh or regenerated chitosan/ACF/TiO2 membrane. *Separation and Purification Technology*, *70*(3), 354–361. doi:10.1016/j.seppur.2009.10.022

Liu, X., Hu, Q., Fang, Z., Zhang, X., & Zhang, B. (2008). Magnetic chitosan nanocomposites: A useful recyclable tool for heavy metal ion removal. *Langmuir*, *25*(1), 3–8. doi:10.1021/la802754t PMID:19032060

Long, M., & Cai, W. (2011). Visible light responsive TiO_2 modification with nonmetal elements. *Frontiers of Chemistry in China*, *6*(3), 190–199. doi:10.1007/s11458-011-0243-8

Luo, C., Wei, R., Guo, D., Zhang, S., & Yan, S. (2013). Adsorption behavior of MnO_2 functionalized multi-walled carbon nanotubes for the removal of cadmium from aqueous solutions. *Chemical Engineering Journal, 225*, 406–415. doi:10.1016/j.cej.2013.03.128

Luo, X., Wang, C., Luo, S., Dong, R., Tu, X., & Zeng, G. (2012). Adsorption of As(III) and As(V) from water using magnetite Fe_3O_4-reduced graphite oxide–MnO_2 nanocomposites. *Chemical Engineering Journal, 187*, 45–52. doi:10.1016/j.cej.2012.01.073

Mahdavi, S., Jalali, M., & Afkhami, A. (2012). Removal of heavy metals from aqueous solutions using Fe_3O_4, ZnO and CuO nanoparticles. *Journal of Nanoparticle Research, 14*(8), 1–18. doi:10.1007/s11051-012-0846-0 PMID:22448125

Mahdavi, S., Jalali, M., & Afkhami, A. (2015). Heavy metals removal from aqueous solutions by Al_2O_3 nanoparticles modified with natural and chemical modifiers. *Clean Technologies and Environmental Policy, 17*(1), 85–102. doi:10.1007/s10098-014-0764-1

Maximous, N., Nakhla, G., Wan, W., & Wong, K. (2009). Preparation, characterization and performance of Al_2O_3/PES membrane for wastewater filtration. *Journal of Membrane Science, 341*(1), 67–75. doi:10.1016/j.memsci.2009.05.040

Maximous, N., Nakhla, G., Wan, W., & Wong, K. (2010). Effect of the metal oxide particle distributions on modified PES membranes characteristics and performance. *Journal of Membrane Science, 361*(1), 213–222. doi:10.1016/j.memsci.2010.05.051

Maximous, N. N., Nakhla, G. F., & Wan, W. (2010). Removal of heavy metals from wastewater by adsorption and membrane processes: A comparative study. *International Journal of Environmental, Chemical, Ecological. Geological and Geophysical Engineering, 4*(4), 125–130.

Maya-Treviño, M. L., Guzmán-Mar, J. L., Hinojosa-Reyes, L., Ramos-Delgado, N. A., Maldonado, M. I., & Hernández-Ramírez, A. (2014). Activity of the ZnO–Fe_2O_3 catalyst on the degradation of Dicamba and 2,4-D herbicides using simulated solar light. *Ceramics International, 40*(6), 8701–8708. doi:10.1016/j.ceramint.2014.01.088

Meng, A., Xing, J., Li, Z., & Li, Q. (2015). Cr-doped ZnO nanoparticles: Synthesis, characterization, adsorption property and recyclability. *ACS Applied Materials & Interfaces, 7*(49), 27449–27457. doi:10.1021/acsami.5b09366 PMID:26600320

Meral, K., & Metin, Ö. (2014). Graphene oxide-magnetite nanocomposite as an efficient and magnetically separable adsorbent for methylene blue removal from aqueous solution. *Turkish Journal of Chemistry, 38*(5), 775–782. doi:10.3906/kim-1312-28

Miller, S. M., & Zimmerman, J. B. (2010). Novel, bio-based, photoactive arsenic sorbent: TiO_2-impregnated chitosan bead. *Water Research*, *44*(19), 5722–5729. doi:10.1016/j.watres.2010.05.045 PMID:20594571

Miranda, S. M., Romanos, G. E., Likodimos, V., Marques, R. R., Favvas, E. P., Katsaros, F. K., & Falaras, P. et al. (2014). Pore structure, interface properties and photocatalytic efficiency of hydration/dehydration derived TiO_2/CNT composites. *Applied Catalysis B: Environmental*, *147*, 65–81. doi:10.1016/j.apcatb.2013.08.013

Mohapatra, M., Rout, K., Singh, P., Anand, S., Layek, S., Verma, H., & Mishra, B. (2011). Fluoride adsorption studies on mixed-phase nano iron oxides prepared by surfactant mediation-precipitation technique. *Journal of Hazardous Materials*, *186*(2), 1751–1757. doi:10.1016/j.jhazmat.2010.12.076 PMID:21232851

Motlagh, M. M., Hassanzadeh-Tabrizi, S., & Saffar-Teluri, A. (2015). Sol–gel synthesis of Mn_2O_3/ Al_2O_3/ SiO_2 hybrid nanocomposite and application for removal of organic dye. *Journal of Sol-Gel Science and Technology*, *73*(1), 9–13. doi:10.1007/s10971-014-3487-1

Motshekga, S. C., Ray, S. S., Onyango, M. S., & Momba, M. N. (2013). Microwave-assisted synthesis, characterization and antibacterial activity of Ag/ZnO nanoparticles supported bentonite clay. *Journal of Hazardous Materials*, *262*, 439–446. doi:10.1016/j.jhazmat.2013.08.074 PMID:24076479

Nalbandian, M. J., Zhang, M., Sanchez, J., Choa, Y.-H., Nam, J., Cwiertny, D. M., & Myung, N. V. (2016). Synthesis and optimization of Fe_2O_3 nanofibers for chromate adsorption from contaminated water sources. *Chemosphere*, *144*, 975–981. doi:10.1016/j.chemosphere.2015.08.056 PMID:26433935

Nawi, M., Sabar, S., & Sheilatina, . (2012). Photocatalytic decolourisation of Reactive Red 4 dye by an immobilised TiO_2/chitosan layer by layer system. *Journal of Colloid and Interface Science*, *372*(1), 80–87. doi:10.1016/j.jcis.2012.01.024 PMID:22321991

Nazari, M., & Halladj, R. (2015). Optimization of fluoride adsorption onto a sonochemically synthesized nano-MgO/γ-Al_2O_3 composite adsorbent through applying the L16 Taguchi orthogonal design. *Desalination and Water Treatment*, *56*(9), 2464–2476. doi:10.1080/19443994.2014.961558

Ngah, W. W., Teong, L., & Hanafiah, M. (2011). Adsorption of dyes and heavy metal ions by chitosan composites: A review. *Carbohydrate Polymers*, *83*(4), 1446–1456. doi:10.1016/j.carbpol.2010.11.004

Nguyen, K. C., Ngoc, M. P., & Van Nguyen, M. (2016). Enhanced photocatalytic activity of nanohybrids TiO_2/CNTs materials. *Materials Letters*, *165*, 247–251. doi:10.1016/j.matlet.2015.12.004

Nguyen-Phan, T.-D., Pham, V. H., Kim, E. J., Oh, E.-S., Hur, S. H., Chung, J. S., & Shin, E. W. et al. (2012). Reduced graphene oxide–titanate hybrids: Morphologic evolution by alkali-solvothermal treatment and applications in water purification. *Applied Surface Science*, *258*(10), 4551–4557. doi:10.1016/j.apsusc.2012.01.026

Niu, H., Wang, Y., Zhang, X., Meng, Z., & Cai, Y. (2011). Easy synthesis of surface-tunable carbon-encapsulated magnetic nanoparticles: Adsorbents for selective isolation and preconcentration of organic pollutants. *ACS Applied Materials & Interfaces*, *4*(1), 286–295. doi:10.1021/am201336n PMID:22148634

Nouri, J., Nabizadeh, R., Jahangiri-rad, M., Yunesian, M., & Moattar, F. (2014). Fluoride removal from aqueous solution by functionalized-polyacrylonitrile coated with iron oxide nano particles: Characterization and sorption studies. *Desalination and Water Treatment*, *52*(22-24), 4369–4375. doi:10.1080/19443994.2013.801792

Obeid, L., Bée, A., Talbot, D., Jaafar, S. B., Dupuis, V., Abramson, S., & Welschbillig, M. et al. (2013). Chitosan/maghemite composite: A magsorbent for the adsorption of methyl orange. *Journal of Colloid and Interface Science*, *410*, 52–58. doi:10.1016/j.jcis.2013.07.057 PMID:23998368

Olushola, S., Olalekan, S., Folahan, A., Leslie, F., & Ximba, B. J. (2014). Application of nano zinc oxide (nZnO) for the removal of triphenyltin chloride (TPT) from dockyard wastewater. *Water S.A.*, *40*(4), 659–664. doi:10.4314/wsa.v40i4.11

Ong, Y. T., Ahmad, A. L., Zein, S. H. S., & Tan, S. H. (2010). A review on carbon nanotubes in an environmental protection and green engineering perspective. *Brazilian Journal of Chemical Engineering*, *27*(2), 227–242.

Palza, H. (2015). Antimicrobial polymers with metal nanoparticles. *International Journal of Molecular Sciences*, *16*(1), 2099–2116. doi:10.3390/ijms16012099 PMID:25607734

Park, S., Lee, J., Lee, D., & Lee, J. (2003). Photocatalytic ZnO nanopowders prepared by solution combustion method for noble metal recovery. *Journal of Materials Science*, *38*(22), 4493–4497. doi:10.1023/A:1027329501367

Patel, S., & Hota, G. (2016). Iron oxide nanoparticle-immobilized PAN nanofibers: Synthesis and adsorption studies. *RSC Advances*, *6*(19), 15402–15414. doi:10.1039/C5RA20345G

Peng, X., Luan, Z., Di, Z., Zhang, Z., & Zhu, C. (2005). Carbon nanotubes-iron oxides magnetic composites as adsorbent for removal of Pb(II) and Cu(II) from water. *Carbon*, *43*(4), 880–883. doi:10.1016/j.carbon.2004.11.009

Peng, X., Luan, Z., Ding, J., Di, Z., Li, Y., & Tian, B. (2005). Ceria nanoparticles supported on carbon nanotubes for the removal of arsenate from water. *Materials Letters*, *59*(4), 399–403. doi:10.1016/j.matlet.2004.05.090

Pyrzyńska, K., & Bystrzejewski, M. (2010). Comparative study of heavy metal ions sorption onto activated carbon, carbon nanotubes, and carbon-encapsulated magnetic nanoparticles. *Colloids and Surfaces. A, Physicochemical and Engineering Aspects*, *362*(1), 102–109. doi:10.1016/j.colsurfa.2010.03.047

Qi, J., Zhang, G., & Li, H. (2015). Efficient removal of arsenic from water using a granular adsorbent: Fe–Mn binary oxide impregnated chitosan bead. *Bioresource Technology*, *193*, 243–249. doi:10.1016/j.biortech.2015.06.102 PMID:26141284

Qu, X., Alvarez, P. J., & Li, Q. (2013). Applications of nanotechnology in water and wastewater treatment. *Water Research*, *47*(12), 3931–3946. doi:10.1016/j.watres.2012.09.058 PMID:23571110

Rabea, E. I., Badawy, M. E.-T., Stevens, C. V., Smagghe, G., & Steurbaut, W. (2003). Chitosan as antimicrobial agent: Applications and mode of action. *Biomacromolecules*, *4*(6), 1457–1465. doi:10.1021/bm034130m PMID:14606868

Rafiq, Z., Nazir, R., Shah, M. R., & Ali, S. (2014). Utilization of magnesium and zinc oxide nano-adsorbents as potential materials for treatment of copper electroplating industry wastewater. *Journal of Environmental Chemical Engineering*, *2*(1), 642–651. doi:10.1016/j.jece.2013.11.004

Rangel-Mendez, J. R., Davila-Rodriguez, J. L., & Barrios, V. A. E. (2010). Chitin based biocomposites for removal of contaminants from water: A case study of fluoride adsorption. In M. Elnashar (Ed.), *Biopolymers*. INTECH Open Access Publisher.

Rao, C. N., Müller, A., & Cheetham, A. K. (2006). *The Chemistry of nanomaterials: Synthesis, properties and applications*. Wiley.

Rao, G. P., Lu, C., & Su, F. (2007). Sorption of divalent metal ions from aqueous solution by carbon nanotubes: A review. *Separation and Purification Technology*, *58*(1), 224–231. doi:10.1016/j.seppur.2006.12.006

Rashidi, F., Sarabi, R. S., Ghasemi, Z., & Seif, A. (2010). Kinetic, equilibrium and thermodynamic studies for the removal of lead (II) and copper (II) ions from aqueous solutions by nanocrystalline. *Superlattices and Microstructures*, *48*(6), 577–591. doi:10.1016/j.spmi.2010.09.011

Rauf, M., Meetani, M., & Hisaindee, S. (2011). An overview on the photocatalytic degradation of azo dyes in the presence of TiO_2 doped with selective transition metals. *Desalination*, *276*(1), 13–27. doi:10.1016/j.desal.2011.03.071

Ray, P. Z., & Shipley, H. J. (2015). Inorganic nano-adsorbents for the removal of heavy metals and arsenic: A review. *RSC Advances*, *5*(38), 29885–29907. doi:10.1039/C5RA02714D

Reddy, D. H. K., & Lee, S.-M. (2013). Application of magnetic chitosan composites for the removal of toxic metal and dyes from aqueous solutions. *Advances in Colloid and Interface Science*, *201*, 68–93. doi:10.1016/j.cis.2013.10.002 PMID:24182685

Riaz, S., Faima, S., & Naseem, S. (2016). *Bacteria Removal from Water by using Al_2O_3 and FeO Nanoparticles*. Paper presented at the The 2016 World Congress on Advances in Civil, Environmental, and Materials Research (ACEM16), Jeju Island, Korea.

Rong, M., Zhang, M., & Ruan, W. (2006). Surface modification of nanoscale fillers for improving properties of polymer nanocomposites: A review. *Materials Science and Technology*, *22*(7), 787–796. doi:10.1179/174328406X101247

Sadiq, I. M., Pakrashi, S., Chandrasekaran, N., & Mukherjee, A. (2011). Studies on toxicity of aluminum oxide (Al_2O_3) nanoparticles to microalgae species: *Scenedesmus sp.* and *Chlorella sp. Journal of Nanoparticle Research*, *13*(8), 3287–3299. doi:10.1007/s11051-011-0243-0

Saha, B., Das, S., Saikia, J., & Das, G. (2011). Preferential and enhanced adsorption of different dyes on iron oxide nanoparticles: A comparative study. *The Journal of Physical Chemistry C*, *115*(16), 8024–8033. doi:10.1021/jp109258f

Saleh, T. A., & Gupta, V. K. (2012). Photo-catalyzed degradation of hazardous dye methyl orange by use of a composite catalyst consisting of multi-walled carbon nanotubes and titanium dioxide. *Journal of Colloid and Interface Science*, *371*(1), 101–106. doi:10.1016/j.jcis.2011.12.038 PMID:22297021

Salehi, R., Arami, M., Mahmoodi, N. M., Bahrami, H., & Khorramfar, S. (2010). Novel biocompatible composite (chitosan–zinc oxide nanoparticle): Preparation, characterization and dye adsorption properties. *Colloids and Surfaces. B, Biointerfaces*, *80*(1), 86–93. doi:10.1016/j.colsurfb.2010.05.039 PMID:20566273

Santhosh, C., Kollu, P., Doshi, S., Sharma, M., Bahadur, D., Vanchinathan, M. T., & Grace, A. N. et al. (2014). Adsorption, photodegradation and antibacterial study of graphene-Fe_3O_4 nanocomposite for multipurpose water purification application. *RSC Advances*, *4*(54), 28300–28308. doi:10.1039/c4ra02913e

Sheela, T., Nayaka, Y. A., Viswanatha, R., Basavanna, S., & Venkatesha, T. (2012). Kinetics and thermodynamics studies on the adsorption of Zn (II), Cd (II) and Hg (II) from aqueous solution using zinc oxide nanoparticles. *Powder Technology*, *217*, 163–170. doi:10.1016/j.powtec.2011.10.023

Shen, J., Ruan, H., Wu, L., & Gao, C. (2011). Preparation and characterization of PES–SiO_2 organic–inorganic composite ultrafiltration membrane for raw water pretreatment. *Chemical Engineering Journal*, *168*(3), 1272–1278. doi:10.1016/j.cej.2011.02.039

Shinde, S., Shinde, P., Bhosale, C., & Rajpure, K. (2011). Zinc oxide mediated heterogeneous photocatalytic degradation of organic species under solar radiation. *Journal of Photochemistry and Photobiology. B, Biology*, *104*(3), 425–433. doi:10.1016/j.jphotobiol.2011.04.010 PMID:21592815

Shrifian-Esfahni, A., Salehi, M. T., Esfahani, M., & Ekramian, E. (2015). Chitosan-modified superparamgnetic iron oxide nanoparticles: Design, fabrication, characterization and antibacterial activity. *CHEMIK*, *69*(1), 19–32.

Sirelkhatim, A., Mahmud, S., Seeni, A., Kaus, N. H. M., Ann, L. C., Bakhori, S. K. M., & Mohamad, D. et al. (2015). Review on zinc oxide nanoparticles: Antibacterial activity and toxicity mechanism. *Nano-Micro Letters*, *7*(3), 219–242. doi:10.1007/s40820-015-0040-x

Stietiya, M. H., & Wang, J. J. (2014). Zinc and cadmium adsorption to aluminum oxide nanoparticles affected by naturally occurring ligands. *Journal of Environmental Quality*, *43*(2), 498–506. doi:10.2134/jeq2013.07.0263 PMID:25602651

Sun, H., Cao, L., & Lu, L. (2011). Magnetite/reduced graphene oxide nanocomposites: One step solvothermal synthesis and use as a novel platform for removal of dye pollutants. *Nano Research*, *4*(6), 550–562. doi:10.1007/s12274-011-0111-3

Suriyaraj, S., Vijayaraghavan, T., Biji, P., & Selvakumar, R. (2014). Adsorption of fluoride from aqueous solution using different phases of microbially synthesized TiO_2 nanoparticles. *Journal of Environmental Chemical Engineering*, *2*(1), 444–454. doi:10.1016/j.jece.2014.01.013

Szekeres, M., Tombácz, E., Ferencz, K., & Dékány, I. (1998). Adsorption of salicylate on alumina surfaces. *Colloids and Surfaces. A, Physicochemical and Engineering Aspects*, *141*(3), 319–325. doi:10.1016/S0927-7757(98)00242-8

Tanhaei, B., Ayati, A., Lahtinen, M., & Sillanpää, M. (2015). Preparation and characterization of a novel chitosan/Al_2O_3/magnetite nanoparticles composite adsorbent for kinetic, thermodynamic and isotherm studies of methyl orange adsorption. *Chemical Engineering Journal*, *259*, 1–10. doi:10.1016/j.cej.2014.07.109

Tayyebi, A., Outokesh, M., Moradi, S., & Doram, A. (2015). Synthesis and characterization of ultrasound assisted graphene oxide–magnetite hybrid, and investigation of its adsorption properties for Sr(II) and Co(II) ions. *Applied Surface Science*, *353*, 350–362. doi:10.1016/j.apsusc.2015.06.087

Teja, A. S., & Koh, P.-Y. (2009). Synthesis, properties, and applications of magnetic iron oxide nanoparticles. *Progress in Crystal Growth and Characterization of Materials*, *55*(1), 22–45. doi:10.1016/j.pcrysgrow.2008.08.003

Tian, N., Tian, X., Ma, L., Yang, C., Wang, Y., Wang, Z., & Zhang, L. (2015). Well-dispersed magnetic iron oxide nanocrystals on sepiolite nanofibers for arsenic removal. *RSC Advances*, *5*(32), 25236–25243. doi:10.1039/C5RA01592H

Tran, N., Mir, A., Mallik, D., Sinha, A., Nayar, S., & Webster, T. J. (2010). Bactericidal effect of iron oxide nanoparticles on *Staphylococcus aureus*. *International Journal of Nanomedicine*, *5*, 277–283. PMID:20463943

Tratnyek, P. G., & Johnson, R. L. (2006). Nanotechnologies for environmental cleanup. *Nano Today*, *1*(2), 44–48. doi:10.1016/S1748-0132(06)70048-2

Tuutijärvi, T., Repo, E., Vahala, R., Sillanpää, M., & Chen, G. (2012). Effect of competing anions on arsenate adsorption onto maghemite nanoparticles. *Chinese Journal of Chemical Engineering*, *20*(3), 505–514. doi:10.1016/S1004-9541(11)60212-7

Umar, M., & Aziz, H. A. (2013). Photocatalytic degradation of organic pollutants in water, organic pollutants - Monitoring, risk and treatment. In Organic Pollutants - Monitoring, risk and treatment. InTech Open.

Upadhyay, R. K., Soin, N., & Roy, S. S. (2014). Role of graphene/metal oxide composites as photocatalysts, adsorbents and disinfectants in water treatment: A review. *RSC Advances*, *4*(8), 3823–3851. doi:10.1039/C3RA45013A

Upadhyayula, V. K. K., Deng, S., Mitchell, M. C., & Smith, G. B. (2009). Application of carbon nanotube technology for removal of contaminants in drinking water: A review. *The Science of the Total Environment*, *408*(1), 1–13. doi:10.1016/j.scitotenv.2009.09.027 PMID:19819525

Venkatesham, V., Madhu, G., Satyanarayana, S., & Preetham, H. (2013). Adsorption of lead on gel combustion derived nano ZnO. *Procedia Engineering*, *51*, 308–313. doi:10.1016/j.proeng.2013.01.041

Venkateswarlu, S., & Yoon, M. (2015). Core–Shell ferromagnetic nanorod based on amine polymer composite (Fe_3O_4@ DAPF) for fast removal of Pb(II) from aqueous solutions. *ACS Applied Materials & Interfaces*, *7*(45), 25362–25372. doi:10.1021/acsami.5b07723 PMID:26496966

Viswanathan, N., & Meenakshi, S. (2010). Enriched fluoride sorption using alumina/chitosan composite. *Journal of Hazardous Materials*, *178*(1), 226–232. doi:10.1016/j.jhazmat.2010.01.067 PMID:20144851

Wang, C.-C., Lee, C.-K., Lyu, M.-D., & Juang, L.-C. (2008). Photocatalytic degradation of CI Basic Violet 10 using TiO_2 catalysts supported by Y zeolite: An investigation of the effects of operational parameters. *Dyes and Pigments*, *76*(3), 817–824. doi:10.1016/j.dyepig.2007.02.004

Wang, H., Xie, C., Zhang, W., Cai, S., Yang, Z., & Gui, Y. (2007). Comparison of dye degradation efficiency using ZnO powders with various size scales. *Journal of Hazardous Materials*, *141*(3), 645–652. doi:10.1016/j.jhazmat.2006.07.021 PMID:16930825

Wang, J., Jiang, Z., Zhang, L., Kang, P., Xie, Y., Lv, Y., & Zhang, X. et al. (2009). Sonocatalytic degradation of some dyestuffs and comparison of catalytic activities of nano-sized TiO_2, nano-sized ZnO and composite TiO_2/ZnO powders under ultrasonic irradiation. *Ultrasonics Sonochemistry*, *16*(2), 225–231. doi:10.1016/j.ultsonch.2008.08.005 PMID:18842448

Wang, J., Xu, W., Chen, L., Huang, X., & Liu, J. (2014). Preparation and evaluation of magnetic nanoparticles impregnated chitosan beads for arsenic removal from water. *Chemical Engineering Journal*, *251*, 25–34. doi:10.1016/j.cej.2014.04.061

Wang, W., Mao, Q., He, H., & Zhou, M. (2013). Fe_3O_4 nanoparticles as an efficient heterogeneous Fenton catalyst for phenol removal at relatively wide pH values. *Water Science and Technology*, *68*(11), 2367–2373. doi:10.2166/wst.2013.497 PMID:24334884

Wang, X., Yu, J. C., Chen, Y., Wu, L., & Fu, X. (2006). ZrO_2-modified mesoporous nanocrystalline $TiO_{2-x}N_x$ as efficient visible light photocatalysts. *Environmental Science & Technology*, *40*(7), 2369–2374. doi:10.1021/es052000a PMID:16646476

Wang, Y.-W., Cao, A., Jiang, Y., Zhang, X., Liu, J.-H., Liu, Y., & Wang, H. (2014). Superior antibacterial activity of zinc oxide/graphene oxide composites originating from high zinc concentration localized around bacteria. *ACS Applied Materials & Interfaces*, *6*(4), 2791–2798. doi:10.1021/am4053317 PMID:24495147

WWDR. (2014). *Water and Enegy: Facts and Figures*. Division of Water Sciences, UNESCO.

Xiao, Q., Sun, Y., Zhang, J., & Li, Q. (2015). Size-dependent of chromium (VI) adsorption on nano α-Fe_2O_3 surface. *Applied Surface Science*, *356*, 18–23. doi:10.1016/j.apsusc.2015.08.005

Xu, C., Cui, A., Xu, Y., & Fu, X. (2013). Graphene oxide–TiO_2 composite filtration membranes and their potential application for water purification. *Carbon*, *62*, 465–471. doi:10.1016/j.carbon.2013.06.035

Xu, P., Zeng, G. M., Huang, D. L., Feng, C. L., Hu, S., Zhao, M. H., & Xie, G. X. et al. (2012). Use of iron oxide nanomaterials in wastewater treatment: A review. *The Science of the Total Environment*, *424*, 1–10. doi:10.1016/j.scitotenv.2012.02.023 PMID:22391097

Yamani, J. S., Miller, S. M., Spaulding, M. L., & Zimmerman, J. B. (2012). Enhanced arsenic removal using mixed metal oxide impregnated chitosan beads. *Water Research*, *46*(14), 4427–4434. doi:10.1016/j.watres.2012.06.004 PMID:22743162

Yan, Y., Sun, H., Zhang, L., Zhang, J., Mu, J., & Kang, S.-Z. (2011). Effect of multiwalled carbon nanotubes on the photocatalytic degradation of methyl orange in aqueous solution under visible light irradiation. *Journal of Dispersion Science and Technology*, *32*(9), 1332–1336. doi:10.1080/02604027.2010.505868

Yang, J. L., An, S. J., Park, W. I., Yi, G. C., & Choi, W. (2004). Photocatalysis using ZnO thin films and nanoneedles grown by metal–organic chemical vapor deposition. *Advanced Materials*, *16*(18), 1661–1664. doi:10.1002/adma.200306673

Yang, W., Ding, P., Zhou, L., Yu, J., Chen, X., & Jiao, F. (2013). Preparation of diamine modified mesoporous silica on multi-walled carbon nanotubes for the adsorption of heavy metals in aqueous solution. *Applied Surface Science*, *282*, 38–45. doi:10.1016/j.apsusc.2013.05.028

Yao, Y., Bing, H., Feifei, X., & Xiaofeng, C. (2011). Equilibrium and kinetic studies of methyl orange adsorption on multiwalled carbon nanotubes. *Chemical Engineering Journal*, *170*(1), 82–89. doi:10.1016/j.cej.2011.03.031

Yao, Y., Miao, S., Liu, S., Ma, L. P., Sun, H., & Wang, S. (2012). Synthesis, characterization, and adsorption properties of magnetic Fe_3O_4@ graphene nanocomposite. *Chemical Engineering Journal*, *184*, 326–332. doi:10.1016/j.cej.2011.12.017

Yoon, Y., Park, W. K., Hwang, T.-M., Yoon, D. H., Yang, W. S., & Kang, J.-W. (2016). Comparative evaluation of magnetite–graphene oxide and magnetite-reduced graphene oxide composite for As(III) and As(V) removal. *Journal of Hazardous Materials*, *304*, 196–204. doi:10.1016/j.jhazmat.2015.10.053 PMID:26551223

Yousef, J. M., & Danial, E. N. (2012). In vitro antibacterial activity and minimum inhibitory concentration of zinc oxide and nano-particle zinc oxide against pathogenic strains. *Journal of Health Science*, *2*(4), 38–42.

Yu, L., Peng, X., Ni, F., Li, J., Wang, D., & Luan, Z. (2013). Arsenite removal from aqueous solutions by γ-Fe_2O_3–TiO_2 magnetic nanoparticles through simultaneous photocatalytic oxidation and adsorption. *Journal of Hazardous Materials*, *246*, 10–17. doi:10.1016/j.jhazmat.2012.12.007 PMID:23276789

Yu, X., Qin, A., Liao, L., Du, R., Tian, N., Huang, S., & Wei, C. (2015). Removal of organic dyes by nanostructure ZnO-bamboo charcoal composites with photocatalysis function. *Advances in Materials Science and Engineering*.

Yuan, P., Fan, M., Yang, D., He, H., Liu, D., Yuan, A., & Chen, T. et al. (2009). Montmorillonite-supported magnetite nanoparticles for the removal of hexavalent chromium [Cr (VI)] from aqueous solutions. *Journal of Hazardous Materials*, *166*(2), 821–829. doi:10.1016/j.jhazmat.2008.11.083 PMID:19135796

Zainal, Z., Hui, L. K., Hussein, M. Z., Abdullah, A. H., & Hamadneh, I. M. K. R. (2009). Characterization of TiO_2–Chitosan/Glass photocatalyst for the removal of a monoazo dye via photodegradation–adsorption process. *Journal of Hazardous Materials*, *164*(1), 138–145. doi:10.1016/j.jhazmat.2008.07.154 PMID:18809254

Zhan, H., Jiang, Y., & Ma, Q. (2014). Determination of adsorption characteristics of metal oxide nanomaterials: Application as adsorbents. *Analytical Letters*, *47*(5), 871–884. doi:10.1080/00032719.2013.850090

Zhang, D., Wei, S., Kaila, C., Su, X., Wu, J., Karki, A. B., & Guo, Z. et al. (2010). Carbon-stabilized iron nanoparticles for environmental remediation. *Nanoscale*, *2*(6), 917–919. doi:10.1039/c0nr00065e PMID:20648288

Zhang, J., Zhou, Q., & Ou, L. (2011). Kinetic, isotherm, and thermodynamic studies of the adsorption of methyl orange from aqueous solution by chitosan/alumina composite. *Journal of Chemical & Engineering Data*, *57*(2), 412–419. doi:10.1021/je2009945

Zhang, L., Jiang, Y., Ding, Y., Povey, M., & York, D. (2007). Investigation into the antibacterial behaviour of suspensions of ZnO nanoparticles (ZnO nanofluids). *Journal of Nanoparticle Research*, *9*(3), 479–489. doi:10.1007/s11051-006-9150-1

Zhang, Q., & Jiao, T. (2013). *Polymer-supported organic-inorganic nanomaterials: fabrication, characterization and environmental application. In Materials and processes for energy: communicating current research and technological developments* (pp. 903–912). Formatex Research Center.

Zhang, X., Zhao, X., & Su, H. (2011). Degradation characteristic of TiO_2-chitosan adsorbent on Rhodamine B and purification of industrial wastewater. *Korean Journal of Chemical Engineering*, *28*(5), 1241–1246. doi:10.1007/s11814-010-0501-3

Zhao, X., Lv, L., Pan, B., Zhang, W., Zhang, S., & Zhang, Q. (2011). Polymer-supported nanocomposites for environmental application: A review. *Chemical Engineering Journal*, *170*(2), 381–394. doi:10.1016/j.cej.2011.02.071

Zhou, H., & Smith, D. W. (2002). Advanced technologies in water and wastewater treatment. *Journal of Environmental Engineering and Science*, *1*(4), 247–264. doi:10.1139/s02-020

Zhu, H., Jiang, R., Fu, Y., Guan, Y., Yao, J., Xiao, L., & Zeng, G. (2012). Effective photocatalytic decolorization of methyl orange utilizing TiO_2/ZnO/chitosan nanocomposite films under simulated solar irradiation. *Desalination*, *286*, 41–48. doi:10.1016/j.desal.2011.10.036

Zhu, J., Wei, S., Gu, H., Rapole, S. B., Wang, Q., Luo, Z., & Guo, Z. et al. (2011). One-pot synthesis of magnetic graphene nanocomposites decorated with core@double-shell nanoparticles for fast chromium removal. *Environmental Science & Technology*, *46*(2), 977–985. doi:10.1021/es2014133 PMID:22126606

Zubieta, C. E., Messina, P. V., Luengo, C., Dennehy, M., Pieroni, O., & Schulz, P. C. (2008). Reactive dyes remotion by porous TiO_2-chitosan materials. *Journal of Hazardous Materials*, *152*(2), 765–777. doi:10.1016/j.jhazmat.2007.07.043 PMID:17764835

Chapter 6
Advanced Nanomaterials for the Removal of Chemical Substances and Microbes From Contaminated and Waste Water

Kamlesh Shrivas
Guru Ghasidas Vishwavidyalaya, India

Archana Ghosale
Guru Ghasidas Vishwavidyalaya, India

Pathik Maji
Guru Ghasidas Vishwavidyalaya, India

ABSTRACT

The development of cost-effective, efficient and stable materials helps to provide the affordable solutions to get safe and fresh water to increasing population with health guidelines of emerging contaminants. Nanomaterials (NMs)-based techniques involve the design, synthesis, manipulation, characterization and exploitation of materials for adsorption and separation of target species from the contaminated and waste water. NMs show better adsorption capacity and catalytic for number chemical species and microbes because of their small size and large surface area that favors the purification and treatment of waste or contaminated environmental

DOI: 10.4018/978-1-5225-2136-5.ch006

water. Here, we present the chemical properties, adsorption/removal mechanism and applications of advanced NMs such as magnetic nanoparticles (MNPs), carbon nanotubes (CNTs), graphene and graphene oxide (GO), titanium oxide (TiO_2), silica (SiO_2), silver (Ag), gold (Au) NPs and zeolites in effective and efficient removal of toxic metal ions, organic and inorganic chemical substances and disease-causing microbes from contaminated and wastewater.

1. INTRODUCTION

Water is a most important resource for vitality for life on earth and sustaining human life. The major civilizations of the world originated nearby the water reservoirs or river due to the ample supply of water and fertile land. Water is a universal solvent that dissolves most of the organic and inorganic substances to meet the requirement of the life. Providing clean and safe water is a necessity and it is also challenging attempt to sustain the same level (LVovich 1979; Postel et al., 1996; World Health Organization 2013). About, millions of people die every year because of contaminated drinking water incorporated with water borne microorganism, toxic chemical substances and thousands of children die due to the water-related problems and diseases. At present, more than billions of people in the world lack to access the safe drinking water and a couple of decades the supply of safe water would suppose to be reduced one-third of the present scenario (Hutton et al., 2007; Mara 2003; Ashbolt 2004; Moore et al., 2003). The limited supply of fresh water to the increasing populations throughout the world is due to the drastic exploitation of water resources for domestic, industry and irrigation purposes in many parts of the world. The increased disposal of industrial, domestic and agricultural and mining wastages into surface and ground water bodies cause the contamination of several inorganic, organic and microbes (Ritter et al., 2002; Fawell et al., 2003). The inorganic species such as arsenic (As), mercury (Hg), cadmium (Cd), lead (Pb), chromium (Cr) and selenium (Se), etc.; organic substances such as pesticides, dyes and phenolic compounds etc.; and microbes such as bacteria and viruses could cause contamination of water (Cao et al., 2014; Yu et al., 2014; Walcarius et al., 2010). Today, the providing of safe and clean water to the fast-growing world population is on the high challenge and needs the plan to sustain the life for future generations.

There are several technologies have been developed for the removal of toxic chemical substances from environmental water samples, including chemical oxidation or reduction (Kohler et al., 2006; Theepharaksapan et al., 2011), chemical precipitation, coagulation or flocculation (Kim et al., 2002; RubI et al., 2009), UV or chlorination (Zhang et al., 2015), ozone treatment (Xiao et al., 2015), adsorption (Kojima et al., 2002), biological treatment (Rosal et al., 2010), reverse osmosis (Jian et al., 1999), membrane filtration (Amin et al., 2014) and ion exchange system (Velizarov et al., 2001). The use of conventional chemical oxidation or reduction through the activated sludge or carbon process unable to remove complex materials comprises of pharmaceuticals, surfactants, dyes various industrial additives. Activated sludge and biological trickling filters come under the biological treatment systems are also not efficient in removing the wide range of chemical substances and remain soluble in the effluent (Urase et al., 2005). Coagulation and flocculation process showed an ineffective in removing the variety of pharmaceutical drugs in different investigations (Vieno et al., 2006). The purification of water by chlorination is implemented to prevent the growth of bacteria and pathogens; however, this treatment causes the undesirable tastes and odors with different by-products (Becher, 1999). Ozone and UV treatment are used to stop the growth of microbes but these techniques are found to very expensive (Adams et al., 2002). Purification of water through the membrane filtration such as microfiltration, ultrafiltration, nano-filtration and reverse osmosis are found most effective and efficient techniques. All these filtration techniques are pressure driven process and removal of contaminants are dependent on the pore size of a semi-permeable membrane which is found specific for the removal specific particulates from water. Microfiltration, ultrafiltration, nano-filtration and reverse osmosis are cost-effective because these techniques require a more energy for the removal of contaminants (Walha et al., 2007; Saleh et al., 2016; Saleh et al., 2012).

There is the need for the sustainable development of cost-effective, eco-friendly and stable materials that meets the challenges of providing the fresh water to the increasing world population. Nanotechnology has established an effective way for the removal of harmful toxic chemical substances and microbes from the wastewater or contaminated water (Bottero et al., 2006). In recent years, there have been increasing activities in design and development of nanomaterials (NMs) for versatile applications. It is still a challenging task to synthesize a novel NMs with unique physical, chemical and electrical properties suitable for biochemical probes, sensors, separating and pre-concentrating probes (Obare et al., 2004). NMs can be synthesized using various methods such as chemical, electrochemical, γ-radiation, photochemical, laser ablation, etc. NMs of these substances exhibit an excellent adsorption capacity and photochemical activity because of their high surface to volume ratio and unusual electronic properties (Rickerby et al., 2007). Currently, there are many NMs such as magnetic nanoparticles (MNPs), carbon nanotubes (CNTs), titanium oxide (TiO_2), mesoporous silica (SiO_2), graphene, zeolite, silver (Ag), gold (Au) and other oxide NPs are used for removing the toxic metal ions, organic substances and disease-causing microbes from waste and contaminated water.

2. ADVANCE OF NMS FOR REMOVAL OF CHEMICAL SUBSTANCES AND MICROBES FROM CONTAMINATED AND WASTE WATER

2.1. Magnetic Nanoparticles (MNPs)

Magnetic nanoparticles (MNPs) have many unique properties such as supra magnetic, high coactivity and high magnetic susceptibility that make them extraordinary for distinct function. There are various types of MNPs such as hematite (α-Fe_2O_3), maghemite (γ- Fe_2O_3), magnetite (Fe_3O_4) and wustite (FeO) used for variety applications. Among these Fe_3O_4 MNPs exhibit the most fascinating properties because the presence of non-equivalent iron such as Fe(II) and Fe(III) in the crystal structure (Yantasee et al., 2007; Domingo et al., 1994). The co-precipitation method is used for the synthesis of Fe_3O_4 MNPs in basic condition (Mascolo et al., 2013). MNPs have been used in medicinal applications, catalysis and now specifically used for water purification (Yallapu et al., 2011; Rossi et al., 2014; Ambashta et al., 2010). MNPs act as an efficient adsorbent with ionic exchange capacity for removal of heavy metals and organic substances assisted with magnetic separation property. The magnetic and surface properties of MNPs allow it for the selective adsorption of ions, molecules, macromolecules, cells, colloidal substances from sample solution. The zero-valent of MNPs make it for specific removal of metallic species such as Hg(II), As(III), Cd(II), Se(III), Cr(VI) and Pb(II) from drinking water and wastewater to show its potential for environmental applications (Ambashta et al., 2010; Xu et al., 2012; Singh et al., 2011). The oxidation of zero-valent iron produces ferrous iron to magnetite in presence of water and oxygen. The adsorption of different pollutants on the MNPs from water sample is demonstrated with Figure 1. The adsorption

Figure 1. MNPs used for the separation of pollutants present in water by chemical adsorption and further removal of MNPs by magnetic separation

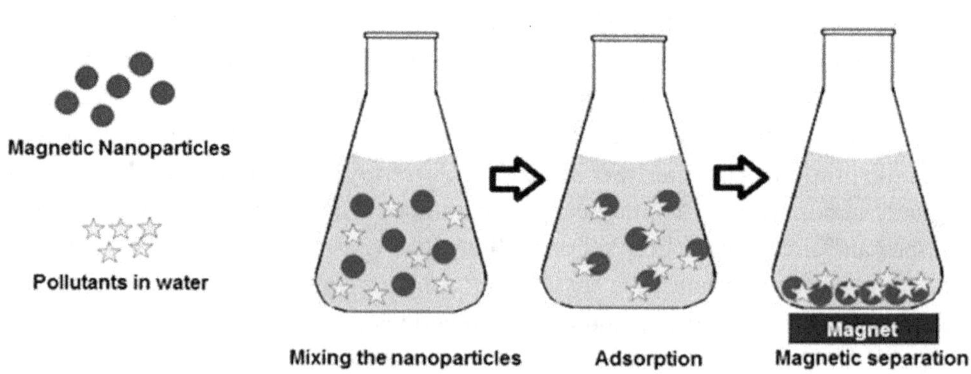

capacity of organic, inorganic and microbes on the surface of MNPs depends on the electrostatic force, van der Waals force and chemical binding. The adsorption and removal of contaminants from water sample is also dependent upon redox conditions and pH (Tang et al., 2013). Many carbonaceous NMs such as graphene, graphene oxide, CNTs, etc., are attached with MNPs for better adsorption and separation for the removal of contaminants (Yang et al., 2012).

2.2. Carbon Nanotubes

Carbon nanotubes (CNTs) are fascinating a new material and have wide range of applications such as reinforcing materials in composites, chemical sensors and catalysts due to their large surface areas, hollow and layered structures and remarkable electrical conductivities (Peng et al., 2014; Saleh et al., 2011; Saleh et al., 2010). CNTs can act as potential solid phase separating probes for removal of metal ions, organic and organometallic compounds from variety of samples. CNTs can be divided into single-walled carbon nanotubes (SWCNT) and multi-walled carbon nanotubes (MWCNT) according to the carbon atom layers in the wall of the nanotubes (Izakmehri et al., 2015; Tawabini et al., 2010). Recently, there is more increase in the applications of CNTs by surface modifications or making the composite materials with NPs. The carboxyl groups or hydroxyl groups can be introduced into CNTs by the treatment with H_2SO_4 and HNO_3. The presence of oxygenated functional groups or other modification helps in binding with a positive species. The removal of dyes using CNTs can be explained with hydrophobic interaction, π-π stacking interaction, covalent and electrostatic interactions. The high π electron density of sp^2 carbons on sidewalls of CNTs provides it high hydrophobic property. The π-π stacking interaction takes place between the bulk π system of CNTs and the C-C or benzene rings (Ai et al., 2011). This is very much similar to the interactions in graphene oxides. Also, the hydrogen bonds formations take place between the functional groups present on the surface of CNTs and the dyes. However, the heavy metals mainly get adsorbed by electrostatic interactions.

2.3. Graphene and Graphene Oxide

Graphene is a two-dimensional sp^2 carbon network with a honeycomb structure in the form of sheets with only one atomic thickness. It is supposed to be next-generation NMs that could be applied for nano-electronics, nano-composites, nano-sensors and nano-devices due to its high electron mobility, high thermal conductivity and mechanical strength. In addition, it has a high specific surface area of 2,630 m^2g^{-1}. Therefore, this superiority allows graphene to be used as a novel adsorbent for the removal of different types of pollutant from water (Liu et al., 2012). The graphite is used as a starting material for preparation of graphene in the presence of $NaNO_3$, H_2SO_4 and $KMnO_4$. The whole solution is stirred for 30 min followed by the addition of 30% H_2O_2 in warm water (Saleh et al., 2010). Graphene oxide (GO) is prepared from graphene by modification with different functional groups such as epoxy (C–O–C), hydroxyl (OH) and carboxyl (COOH) capping groups on the surface (Mkhoyan et al., 2009). Graphene and GO have shown many versatile applications in nano-electronics (Westervelt, 2008), sensors (Ang et al., 2008), and structural composite materials (Liang et al., 2009). The large number of functional groups on GO makes it a smart material for removal of metal ions, biomolecules or cationic substances from sample solution through the electrostatic and coordination interactions. In solution phase, GO can act as a cation exchange resin because of the ionizable carboxyl groups, which allows surface complexation with metal ions or positively charged organic molecules (Dreyer et al., 2010; Zhang et al., 2010). Figure 2 demonstrates the π-π interaction between the sp^2 hybridized GO sheet and the organic molecules responsible for the interaction and separation of organic compounds (Ji et al., 2013). In addition, the removal of positive ions (metal ions) using GO is increased in basic pH due to the base abstracts the protons of the acidic group and hydroxyl groups present on the surface of GO. These take part in attraction with the positive ions and the GO resin by electrostatic interaction. Although the adsorption occurs due to the π-π interactions have no effect on the change of pH of sample solution. Yang and coworkers reported the use of as a GO separating probe for removal of Cu (II) from sample solution considering the electrostatic attraction between the GO and Cu (II) ions as driving force for the separation (Yang et al., 2010). Fan et al. gave a simple route for the fabrication of a magnetic beta-cyclodextrin/GO nanocomposite

Figure 2. Schematic design showing the electrostatic interaction between GO and target positive chemical species from water and also π–π stacking in organic molecules in (a) acidic medium (b) basic medium

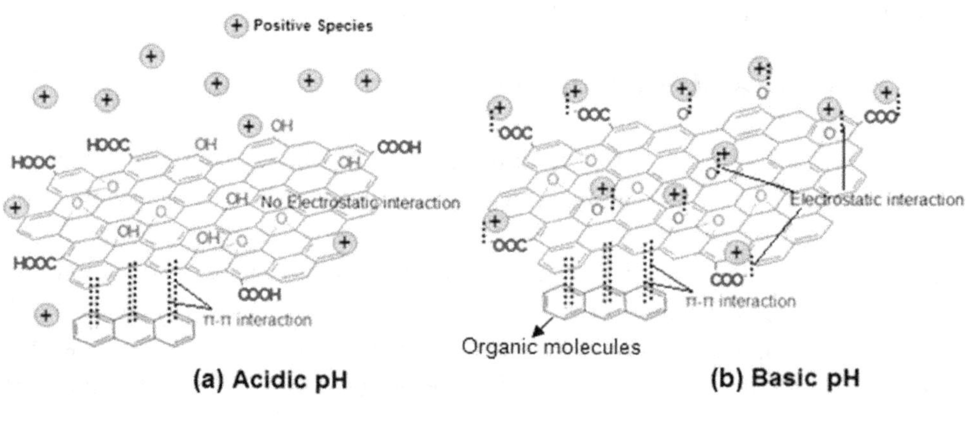

for fast removal of Cr (IV) from water. This material could be a good candidate for efficient separation of Cr (VI) from wastewater and also useful for purification of polluted water (Fan et al., 2012). There are several reports are available for separation of toxic metal ions, organic substances and microbe from water samples. The graphene/GO NPs composite material exhibits an efficient candidate for removal of heavy metals and organic substances in short period of time with good recovery percentage (Cao et al., 2014; Yang et al., 2010; Fan et al., 2012; Teymourian et al., 2013; Sun et al., 2014; Zhu et al., 2012; Sivashankar et al., 2014). Graphite can be also directly used for the removal of organic pollutant and desalination of water with 55% removal of efficiency (Zhou et al., 2015).

2.4. Titanium Oxide Nanoparticles

The surface properties of titanium oxide (TiO_2) NPs are better than that of bulk TiO_2, though it has been used for years in industries and water treatment and purification of water. TiO_2 NPs have better photo-catalytic, adsorption capacity and anti-microbial properties that have been used for removal of toxic substances and microbes from a water sample. The photo-catalytic activity of the TiO_2 NPs causes the removal of pollutants from waste water through the generation of reactive hydroxyl radicals upon the irradiation with UV light (Gupta et al., 2012; Saleh et al., 2011; Daels et al., 2015). Gupta et al. demonstrated the photo-catalytic activity of TiO_2NPs for the degradation of contaminates (Gupta et al., 2012). The interaction of TiO_2 NPs with a photon of light having a comparable band gap of energy results in the separation of conduction and valence bands given in Figure 3. These photo-catalytic reactions involved oxidation and reduction reactions by the photo generated electrons and holes. The hydroxyl radical is a strong oxidizing agent for the organic matters presents on the surface of TiO_2 which further undergo the degradation. This property of TiO_2 makes it an excellent catalytic material for the removal of toxic chemical substances from the environment by gaining solar light as an energy source (Mahlambi et al., 2015). Nabi et al. demonstrated the use of TiO_2 and iron doped-TiO_2 NPs for removal of As(III) and As(V) and iron-doped TiO_2 NPs are found to be better material for the removal of arsenic from contaminated water (Nabi et al., 2009).

2.5. Mesoporous Silica Nanoparticles

Silica (SiO_2) NPs have a high surface area with a number of pore sizes that make them highly reactive and useful for making effective adsorbents in water and wastewater treatment. The mesoporous SiO_2 NPs such as MCM-41 and SBA-15 having a uniform pore, high pore volume, surface area, high adsorption capacity with high hydrothermal and chemical stability (Adjdir et al., 2015; Gandhi et al., 2012). The

Figure 3. Mechanism for the photo-catalysis of organic pollutants on the surface TiO_2 NMs

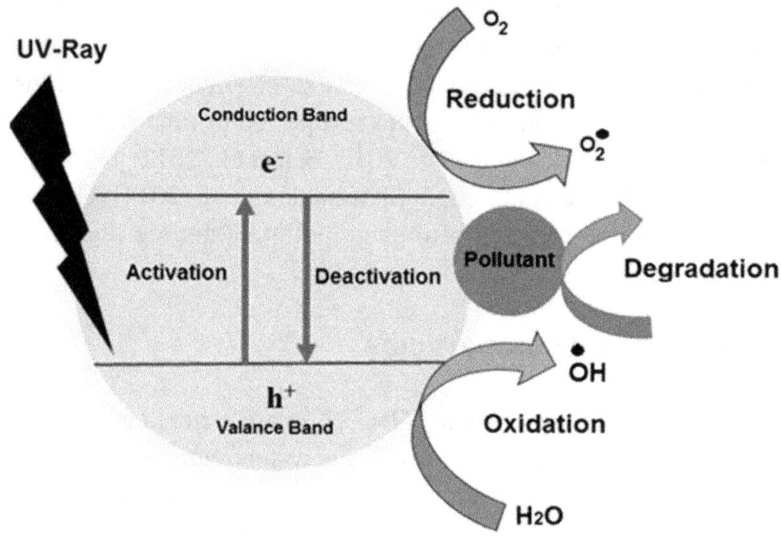

mesoporous SiO_2 NPs is mainly functionalized with NPs or organic compounds with suitable functional groups for better removal of pollutants from water. The amino-functionalized SiO_2 NPs showed an excellent capacity to separate a wide variety of harmful metal ions from water samples through the complexation between the amine group and metal ions. In addition, the regeneration of amino-functionalized adsorbents is found to be easy by proton displacement of the metal ion in acid washing. Chung et al. demonstrated the separation of Pb (II) and Cu(II) using multi-amine grafted mesoporous SiO_2 with MNPs (Chung et al., 2012). MNPs/SiO_2core-shell NPs can be easily dispersed in aqueous solutions due to the hydrophilic nature. In acidic condition, the presence of hydroxyl or silanol groups can be used for attachment and chemical modification to remove the target chemical species from the sample (Emadi et al., 2013). In addition, silica can be grafted with a variety of functional groups, leading to considerable enhancement of their surface properties. Fellenz et al. reported the synthesis and characterization of MNPs embedded with mesoporous MCM-41 (Fellenz et al., 2006). Jing-po et al. accounted the removal and detection of Hg(II) using core-shell structured nanocomposite of SiO_2 NPs (Jing-po et al., 2015).

2.6. Silver and Gold Nanoparticles

Silver nanoparticles (AgNPs) and gold nanoparticles (AuNPs) have received considerable attention for their potential application in catalysis, biological, medicine, pharmaceutical and environmental as adsorbents and disinfectants (Sondi et al., 2004; Agnihotri et al., 2014; Sumesh et al., 2011). The synthesis of these NPs was done by wet chemical methods using a reducing agent and stabilized with certain bulky groups to provide a stability in aqueous solution and also responsible for the interaction with target chemical substances for the separation from contaminated water samples. The colloidal AgNPs have been used as a disinfectant in hospitals and medical centers which are approved by the U.S. Environmental Protection Agency (U.S. EPA, 1993). This is also exploited as a supporting agent for several materials to prevent the microbial formation on the most of the surfaces. AgNPs have good potential antibacterial activities (U.S. EPA, 1993; Gunawan et al., 2011). AgNPs can penetrate into the bacterial membrane of *Escherichia coli* and other gram-negative bacteria. The bactericidal activity of NPs was found to increase with decreasing the size of NPs and also depends on particles shape (Sondi et al., 2004; Agnihotri et al., 2014; Sumesh et al., 2011; Gunawan et al., 2011; Shrivastava et al., 2007). In addition, the inhibition of nitrification in activated sewage sludge is also observed by AgNPs than by the equivalent amount of Ag (mgL^{-1}) as Ag(I) ion (Kaegi et al., 2011). Abbasi et al. reported the synthesis of AgNPs through a greener approach with a garlic extract as stabilizing and reducing agents. The synthesized NPs have demonstrated for the removal of phenanthrene, pyrene and anthracene with 85% removal capacity (Abbasi et al., 2014). Karimi et al. presented the use of activated carbon capped with AgNPs for the removal of methyl orange from water sample (Karimi et al., 2012).

AuNPs is also used as a scavenger for removal of mercury ions from polluted waters in the presence of sodium borohydride. The NPs capped with citrate ions act as a reducing agent to catalyze the Hg(II). The removal of mercury ions from waste water sample is means of amalgamation by mounting onto the colloidal NPs, which are then removed from the AuNPs solution (Ojea-Jimenez et al., 2012). Lee at al. showed the use of nano-composite film of AuNPs/BSA-decorated agarose gel membrane for removal of Hg(II) and Pb(II) from waste water (Lee at al. 2012). Ayati et al. exploited the use of AuNP/TiO_2 NPs for photo-catalytic oxidation of various toxic organic pollutants such as azo dyes and phenols under solar, visible and UV irradiation. The loading amount and particle size of AuNPs are the most important parameters to show the better catalytic activity of AuNP/TiO_2 (Ayati et al., 2014).

2.7. Zeolites

Zeolites are named after two Greek words "zein" and "lithos" which mean "boiling stone" which was first used by a Swedish geologist, in 1756. Zeolites are micro-porous minerals which may be natural or synthetic. They are hydrated aluminosilicates having a general composition of $Al_xSi_yO_{2(x+y)}$ and are insoluble in water (Diale et al., 2011). The structure of zeolites is three-dimensional, micro porous, crystalline solids being used for water treatment because of their ion exchange properties and thermal stability. In addition, the surface modifications and pore size of zeolites make them attractive candidates for a variety of applications (Liu et al., 2013; Rahamani et al., 2006). They act as base exchangers in the removal of metal ions from contaminated water (Hui et al., 2005; Mehdizadeh et al., 2014; Dong et al; 2015).

3. REMOVAL OF INORGANIC CHEMICAL SUBSTANCES FROM WATER USING NMS

Trace metals are essential constituents of our body that are needed for proper physiological and chemical process. However, the excess intake of metals through the drinking water are harmful to humans including Hg, Pb, As, Cr, Se, and Cd. Chronic exposure to these toxic metals can have serious health consequences. Humans are exposed to metals through inhalation of air pollutants, consumption of contaminated drinking water, exposure to contaminated soils or industrial waste, or consumption of contaminated foods. Natural sources such as weathering, erosion of rocks or soils and volcanic emissions also contribute the metal pollution. There are serious effects of exposure of metals on human health including reduced growth, cancer, damage to nervous system and extreme cases death. Therefore, the removal of these toxic metals is important to prevent the entry of these chemical substances into the different compartments of the environment.

3.1. Removal of Arsenic

As is a toxic element with metalloid properties and it exists in soil, water, plants. Arsenic exists in both organic and inorganic compounds in the oxidation states, shown as –3, 0, +3 and +5. The compounds of As have wide application in agriculture as pesticides and semiconductors, transistors, metal adhesives, explosives and pharmaceutical products in various industries (Patel et al., 2005; WHO, 1999; Furst et al., 1983). In term of toxicity, As(III) is more toxic than As(V). The tolerance limit of arsenic is 10 μgL^{-1}. Hyper pigmentation, depigmentation, keratosis, peripheral vascular disorders are the most commonly reported symptoms of chronic exposures of arsenic (Bissen et al., 2000; Chen et al., 1995). Therefore, the separation of arsenic from contaminated water is required to provide the clean water to the world increasing the population.

Recently, there are several NMs have been reported for the removal of As(III) and As(V) from contaminated and waste water. For As removal, NMs exhibited a better response compared to existing conventional technologies. The different NMs are used for separation of As(III) and As(V) from contaminated water, given in Table 1. Martinson et al. synthesized CuO NPs of the surface area of 85 m^2g^{-1} with a diameter of 12–18 nm and employed as an adsorbent for removal of As(III) and As(V) from ground water in the pH range of 6-10. The maximum adsorption capacity of As(III) and As(V) is 26.9 mgg^{-1} for and 22.6 mg/g, respectively (Martinson et al., 2009). Shipley et al. demonstrated the adsorption efficiency of As using MNPs at different NPs concentration, pH, ionic strength, and temperature (Shipley et al. 2009). Jegadeesan et al. synthesized nano-sized amorphous and crystalline TiO_2 and used for specific separation of As (III) and As(V) due to its size and crystalline property of NPs (Jegadeesan et al., 2010). Thanh et al. prepared a nano-composite of γ-Fe_2O_3 and α-MnO_2 through a ball-milled process and illustrated for removal of As(V) from water (Thanh et al., 2011). The adsorption of As(V) on the NMs was dependent on the pH of the sample using pseudo-second-order kinetic model. Selvakumar et al. synthesized a carbonized yeast cells on AgNPs for the removal

Table 1. Different NMs for removal of arsenic at specific particle size, surface area, pH, Adsorption capacity

NMs	Particle Size	Surface area (m^2g^{-1})	pH	Adsorption capacity (mgg^{-1})	Arsenic	Ref.
CuO	12-18 nm	85	–	26.9	As(III)	Martinson et al. 2009
Fe_3O_4	20 nm	60	8	–	As(III)+As(V)	Shipley et al. 2009
Nanocrystalline TiO_2		98	4.8	38.4	As(III)+As(V)	Jegadeesan et al., 2010
Perlite/γ-Fe_2O_3	14.9 µm	27.7	7.1	-	As(V)	Thanh, et al. 2011
Silver/yeast cells	19 nm	–	–	0.975	As(V)	Selvakumar et al., 2011
α-Fe_2O_3–Fe_3O_4–polymer	2 nm	50	–	2.7	As(III)	Savina et al., 2011
P(4-VP)-HCl NPs	20 nm	–	–	31.75	As(V)	Sahiner et al., 2011
Al-doped phenolic bead	0.80 µm	330	–	20	As(V)	Kumar et al., 2011
$Al_2O_{3\,p}$olymer bead	–	–	–	6.56	As(V)	Saha et al., 2012
CMC-Fe-Mn NPs	0.35 µm	~6	–	338	As(III)+As(V)	An et al., 2012
γ-Al_2O_3	3–6 nm	–	–	–	As(III)+As(V)	Patra et al., 2012

of As(V) using yeast strain *Saccharomyces cerevisiae* (Selvakumar et al., 2011). Savina et al. synthesized nanocomposite gels of α-Fe_2O_3 and Fe_3O_4 NPs of pore size up to 100 μm through the polymerization. These nanocomposite materials showed an excellent capability for the removal of 3 mg of As(III) from sample solution (Savina et al., 2011). Similarly, nano p(4-vinylpyridine)-based hydrogels (Sahiner et al., 2011)[106], bi-metal doped nano multifunctional polymeric adsorbents (Kumar et al., 2011), nano-alumina dispersed in chitosan-grafted polyacrylamide (Saha et al., 2012), polysaccharide stabilized Fe-Mn oxide NPs (An et al., 2012) and mesoporous γ-Al_2O_3 NPs demonstrated for efficient removal of different species of As from water (Patra et al., 2012).

3.2. Removal of Chromium

Cr is a transition metal with oxidation states of III and VI, where Cr(VI) is more toxic than Cr (III) due to its high solubility and mobility in biological systems. Cr(III) is an important element needed for proper functions of biological process in human (Agency for Toxic Substances and Disease Registry (ATSDR), 1993; Kotas et al., 2000; Nriagu et al., 1988). The main sources of Cr contamination are through industries like chrome plating, electroplating, pigment fabrication, water-cooling towers and sanitary landfill leaching (Sharma, 2003). The chronic exposure of Cr(VI) causes the chronic ulcer and damage of lining of nose, throat, lungs as well as the gastrointestinal tract (Basu et al., 1983). Thus, the separation of chromium from contaminated water is needed to assure the good public health.

For this, different NMs are showed effective results for the removal of chromium from contaminated water. Xiao et al. used a greener approach for the synthesis of FeNPs for highly effective removal of Cr(VI) from the water. They used fifteen plants extract for the plant-mediated synthesis of FeNPs and then used for removal of Cr(VI) (Xiao et al. 2016). Zhong et al. exploited the use of TiO_2 nanotubes (NTs) coupled with CuNPs for efficient photoelectron-catalytic removal of Cr(VI) (Zhong et al., 2016). Niu et al. reported the application MNPs for the removal Cr(VI) (20 mgL^{-1}) with 100% removal efficiency (Niu et al., 2005). Zhu et al. prepared magnetic graphene nano-composite through a thermal decomposition process to remove Cr(VI) ions from water. The synthesized NMs had double shell-structure where iron oxide as inner shell and amorphous Si-S-O compound as an outer shell. It showed the fast removal of Cr(VI) in few minutes (Zhu et al., 2012).

3.3. Removal of Mercury

Hg is a toxic persistent pollutant that affects the nervous system. There are three main species of mercury such as Hg(II), methyl mercury (MeHg) and dimethylmercury (Me_2Hg). MeHg is found to be more poisonous since it readily undergoes bio-magnifications in the food chain as well as it is found to be water soluble (US Environmental Protection Agency). Mercury has a wide range of industrial applications such as the production of batteries, cathode tubes, medical laboratory chemicals and electronic production of mercury vapor lamps and household cleaners, etc. There are several harmful effects of mercury such as cancer, damages the stomach and large intestine, brain and kidneys, lung and increase blood pressure and heart rate (Stern et al., 2001). Hence, the potential materials are needed for the effective removal of these toxic substances from contaminated water.

For this, Shen et al. synthesized Cys-Fe_3O_4 MNPs functionalized with cysteine using Fe (II) as a precursor by oxidation–precipitation method at room temperature. The MNPs were used for removal of Hg(II) from the water. It showed a fast removal with 95% removal efficiency (Shen et al., 2014). Parham et al. demonstrated the application of MNPs capped with 2-mercaptobenzothiazole as an efficient adsorbent. MNPs showed very high adsorption ability than non-magnetic NMs. It shows fast, simple and effective for treatment of Hg polluted water (Parham et al. 2012). Girginova et al. employed silica coated MNPs for removal of Hg(II) from water with 74% removal capacity (Girginova et al., 2011). Sumesh et al. synthesized mercaptosuccinic acid capped AgNPs supported on activated alumina for the removal of mercuric ions from contaminated water at room temperature (Sumesh et al., 2011).

3.4. Removal of Cadmium

Cd is one of the most toxic metal presents in the form of Cd(II). It enters the environment through mining, smelting, oil and coal combustion. Gastrointestinal absorption of Cd(II) from food or water is the principal source of internally deposited Cd(II) in the human body. The excess intake of Cd(II) through the contaminated water may result in the damage of respiratory system, causing bronchial and pulmonary irritation (Seiler et al., 1988; Forstner et al., 1981). So, the separation of Cd(II) from contaminated water is highly demanded.

For this, Gupta et al. prepared Fe_3O_4 MNPs by aco-precipitation method using agricultural waste for the removal of cadmium ions (Gupta et al., 2012). Vukovic et al. prepared ethylene-diamine modified MWCNTs via chemical modification. The biocompatibility of the material is found to be goad and showed no cytotoxic effect (Vukovic et al., 2010). Skubal et al. synthesized colloids of anatase TiO_2 NPs through the controlled hydrolysis of $TiCl_4$ and surface modification with the bidented thiolactic acid. These nanocolloidsshowed10% removal of cadmium ions from water without undergoing reduction or modification in aerobic condition; however, the removal is increased up to 40% in anoxic condition without undergoing reduction or modification of the metal ion as before (Skubal et al., 2002).

3.5. Removal of Lead

Pb is a highly toxic metal and its compound is found in a variety of commercial products and industrial materials, including paints, storage batteries, insecticides, and ceramics. Presently, the important source of Pb in our environment is anthropogenic activities such as the burning of fossil fuels, mining, and various manufacturing processes. Pb may cause irreversible neurological damage as well as renal diseases, cardiovascular effects, reproductive toxicity, damage the kidney and brain when it is taken in large excess through the contaminated drinking water (Qi et al., 2004; WHO, 1989; Patel et al., 2006). As consequences, smart materials are needed to get rid of the Pb present in contaminated water sample.

Masoumi et al. used rice husk and poly methylmethacrylate-co-maleic anhydride for the removal of Pb(II) ions from water. The system can undergo adsorption and desorption in few cycles of removal process (Masoumi et al., 2016). Mostafa et al. exploited the use of $CoMo(CO_3^{2-})$ NMs for spontaneous and endothermic removal of Pb(II) ions (Mostafa et al., 2016). Awual et al. developed a ligand-based conjugate adsorbent for simultaneous removal and detection of Pb(II) from water samples. The organic ligand 4-dodecyl-6-[{4-(hexyloxy) phenyl}diazenyl] benzene-1,3-diol is immobilized onto mesoporous silica by a direct immobilization approach for removal detection and of lead. The detection of Pb(II) ion is based on charge transfer (π–π transition) transduction mechanism (Awual et al. 2014). Zhang et al. synthesized monodispersed and spherical $Fe_3O_4@SiO_2$–NH_2 core–shell NMs with uniform core-shell structure. The high loading of amino moieties on MNPs helped the material for easy removal of Pb(II) from water (Zhang et al., 2013).

3.6. Removal of Selenium

Se is present in the environment in the form of selenide, selenate, or selenite. A maximum acceptable concentration of Se in drinking water is 0.01 mgL^{-1} (U.S. Environmental Protection Agency, 1980). Se is an important element present in several enzymes needed for the proper function of thyroid gland. The excess intake of inorganic and organic Se through the contaminated drinking water may result in the symptoms such as fatigue, hair loss, and white blotchy nails (Schamberger et al., 1984; Merian, 1984). Hence, the appropriate elimination of this toxic chemical species is required with specific material to get the clean water.

Awual et al. described the use of organic–inorganic based novel composite adsorbent for removal and detection of Se(IV) from contaminated water and showed adsorption capacity of 111.12 mgg^{-1} of Se(IV) is demonstrated. The presence of diverse competing ions did not affect the separation of Se(IV) (Awual et al., 2015). Loyo et al. demonstrated adsorption capacity of SeO_3^{2-} on the surface of Fe_3O_4 and

Fe/Fe_3C NPs. The Fe/Fe_3C NPs showed a better adsorption efficiency than Fe_3O_4 NPs for the removal of Se(IV) from the water samples (Loyo et al. 2008). Gonzalez et al. described $MnFe_2O_4$ NMs for separation of Se from aqueous solutions at pH 6.0 for 5 min of reaction time (Gonzalez et al., 2010).

4. REMOVAL OF ORGANIC CHEMICAL SUBSTANCES FROM CONTAMINATED AND WASTE WATER USING NMS

The disposal of large amounts of domestic, agricultural and industrial waste contains potentially toxic organic chemical substances such as pesticides, phenols, polyaromatic hydrocarbons, and dyes, etc. These toxic substances when entering into drinking water or foods through contamination may cause mutagenic and carcinogenic effects on humans. Recently, NMs are being great attention to using as potential adsorbents to remove toxic and harmful organic substances from contaminated and wastewater.

4.1. Removal of Pesticides

The pesticide is an organic chemical substance used to kill, repel or control certain forms of unwanted weeds or pest in agriculture to prevent the damage of crops. They include insecticides, herbicides, and fungicides. However, both soils and water sources can become contaminated with excess use of these compounds. The acute exposure to these pesticides can disrupt the normal physiological functions of human and cause neurobehavioral disorders, reproductive abnormalities leukemia, malignant lymphoma, nausea, vomiting, hyperventilation and weakness (Hayes et al., 1991; Smith et al., 2002; Corsini et al, 2008). Hence, the removal of the pesticides from drinking water with respect to human health is important concerns.

More recently, Gkementzoglou et al. synthesized a highly specific molecularly imprinted polymer NPs and used for removal of atrazine pesticide from water (Gkementzoglou et al., 2016). Liu et al. prepared graphene-coated silica NPs by mixing GO with acid-treated silica in the presence of hydrazine hydrate. The strong π-bonding network of the benzene rings and interaction of the composite with the electron-donating abilities of the S, P, and N atoms helped in explaining the adsorption and removal mechanism of pesticides from water (Liu et al., 2013). Wu et al. synthesized a graphene-based magnetic nano-composite and used it as an effective adsorbent for the pre-concentration of carbamate pesticides (metolcarb, carbofuran, pirimicarb, isoprocarb and diethofencarb) from environmental water (Wu et al., 2011). Gao et al. utilized a novel nano-structured polyaniline-ionic liquid (i.e. 1-butyl-3-methylimidazolium hexafluorophosphate) composite film coated on steel wire through the electrochemical deposition and used it for the removal of pesticides such as hexachlorocyolohexane, dichlorodiphenyl dichloroethylene, dichlorodiphenyl dichloroethane, dichlorodiphenyl trichloroethane from water samples (Gao et al., 2011).

4.2. Removal of Dyes

Dyes are another class of an organic chemical substance that widely used in many industries such as textile, cosmetics, plastics, leather, paper, plastics for coloring their products. The effluents from various industries are a major source for the disposal of dyes into the environment. Now, thousands of dyes are annually used for coloring the variety of the materials. The discharge of dyes after their use through the industrial waste results in the contamination of groundwater and drinking water supplies that may produce health risks. Dyes give undesirable color to the water and reduce the sunlight penetration and their degradation products being toxic or even mutagenic, carcinogenic (Lin, 1993; Perez-Urquiza, 2000). Therefore, a novel material is highly demanded to eliminate the dyes present in waste water or contaminated water.

Recently, Singh et al. demonstrated the photo-catalytic properties of semiconductor NPs for oxidation and degradation of dyes assisted with UV light (Singh et al., 2013). Patra et al. synthesized polysaccharide (agar)-based adsorbent through the modification of monometallic and bimetallic NPs (Fe, Cu, Pd, Fe/Cu and Fe/Pd) to remove cationic dyes such as methylene blue and rhodamine B. The NPs showed a 90% adsorption capacity for removal of dyes (Patra et al., 2016). Dadfarnia et al. used iron-basedmetal-organic framework loaded on Fe_3O_4 for the removal of methyl red (Dadfarnia et al. 2015). Singla et al. used a novel boron nitride NMs which showed a very high surface area and high adsorption capacity for brilliant green and methyl orange from water samples (Singla et al., 2015). Gautam et al. synthesized a novel bimetallic Fe-Zn NPs via a co-precipitation method and applied it for the removal of a carcinogenic dye such as malachite green and congo red from water samples (Gautam et al., 2015).

4.3. Removal of Phenol and Phenolic Compounds

Phenols and phenolic compounds are highly toxic to the human health at low concentrations. The major sources of these compounds are oil industry, municipal waste and coke plant effluents (Kolthoff et al., 1971; Kuriacose et al., 1987). In chlorinated water, phenol reacts with chlorine to form chlorophenol which can cause toxic effects on human or animals if the contaminated water is taken. The excess intake of phenol and phenolic compounds through the contaminated water may cause protein denaturation and paralysis of the central nervous system and also damages the kidney, liver, and pancreas (Knop et al., 1985). Thus, it is recommended to remove the phenol and phenolic compounds from waste water before entering the water stream.

For this, Mani et al. reported carbon-doped anatase TiO_2 NPs with high surface area employed for the removal of phenol from water under the simulated solar light (Mani et al., 2015). Yang, et al. demonstrated the separation of phenols through the H-bonding interactions between the target molecule and CNTs followed by π-electron polarizability play important roles on the adsorption of phenols from environmental water (Yang, et al., 2008).]. Zuo et al. presented the catalytic oxidation and removal of phenols from waste water using MWCNTs and MNPs in the presence of H_2O_2 (Zuo et al., 2009). Yao et al. reported the combined effect of cobalt oxide (Co_3O_4) and reduced GO NPs for the catalytic performance for the decomposition of phenol. The combined effect of both the NPs showed a better catalytic activity than individual NPs (Yao et al., 2012). Asmaly et al. presented the Fe_3O_4 MNPs impregnated with carbon fibers for the removal of phenol from water using Langmuir and Freundlich model (Asmaly et al., 2015).

5. REMOVAL OF BACTERIA AND VIRUS FROM CONTAMINATED WASTE WATER USING MNS

Bacteria and virus are microscopic organisms present in contaminated and waste water that may cause water borne diseases. Thus, there is a need of novel technology for effective removal and inactivation of waterborne microbial pathogens from water bodies. For this, Zarpelon et al. prepared a thin film of AgNPs crosslinked with glutaraldehyde as a new material to act as a bactericidal agent for the treatment of industrial wastewater. The material showed a good inhibitory effect against *Escherichia coli* and also decreased the growth of coliform in water (Zarpelon et al., 2016). Jin et al. presented the use of Fe_3O_4 NPs modified with cetyltrimethylammoniumbromide [Fe_3O_4-CTAB] NPs as a bactericidal material in a water sample. The prepared Fe_3O_4-CTAB NPs are found to be good in capturing both Gram-negative *Escherichia coli* and Gram-positive *Bacillus subtilis* from water (Jin et al., 2015). Motshekga et al. developed a novel silver–zinc oxide bentonite chitosan nano-composite for complete inactivation of bacteria from water in short period of time (Motshekga et al. 2015). Zhan et al. prepared an amine-functionalized magnetic (Fe_3O_4–SiO_2–NH_2 NPs) and used for the removal of both pathogenic bacteria and viruses from water. The amine-modified NPs showed an excellent response in capturing a wide range of pathogens including bacteriophage, virus *(Poliovirus-1), S. aureus, E. coli, P. aeruginosa, Salmonella and B. Subtilis* (Zhan et al., 2014). Radzig et al. demonstrated the antibacterial action of AgNP on Gram-negative bacteria such as planktonic cells and biofilms. AgNPs stabilized with casein peptides strongly inhibit the biofilms formation and prevent the growth of various bacteria in water (Radzig et al., 2013). Ma et al. fabricated a novel nanofibrous microfiltration membrane system with nanofibers and employed for the effective removal of bacteria and viruses from contaminated water. Behra et al. exploited the use of porous polyethylene glycol microparticles functionalized with sugar ligands for capturing the specific bacteria from the water. Its selective and specific binding capacity is used for the separation of different strains of *Escherichia Coli*.

6. CONCLUSION

The uses of different NMs such as magnetic, carbon nanotubes, graphene/graphene oxide, titanium oxide, silica, silver, gold NPs are elaborately demonstrated for the removal of toxic metal ions, organic substances and disease-causing microbes from contaminated and waste water. The surface area and the functional groups attached to the NMs play an important role the removal of contaminates which depends on the pH, concentration of NMs and contact time of the sample. The electrostatic, π-π interaction and vander wall forces etc. are the main forces responsible for the separation of target chemical substances from water. The use of NMs is found easy, rapid and eco-friendly for selective and efficient removal of contaminants from water. These NMs are very stable and produce less amount of waste generation compared to another traditional method for removal of contaminants. In addition, these described methods are rapid and energy saving for adsorption of target environmental pollutants from contaminated and waste water.

REFERENCES

Abbasi, M., Saeed, F., & Rafique, U. (2014). Preparation of silver nanoparticles from synthetic and natural sources: Remediation model for PAHs. *Materials Science and Engineering*, *60*, 012061.

Adams, C., Wang, Y., Loftin, K., & Meyer, M. (2002). Removal of antibiotics from surface and distilled water in conventional water treatment processes. *Journal of Environmental Engineering*, *128*(3), 253–260. doi:10.1061/(ASCE)0733-9372(2002)128:3(253)

Adjdir, M., Bendeddouche, C. K., Benhaoua, H., Kaid, M., Karmaoui, M., Boudinar, M., & Weidler, P. G. (2015). Increasing the efficiency of silicon and aluminum extraction from Volclay by a water iteration treatment for the synthesis of MCM-41 nanomaterials. *Comptes Rendus. Chimie*, *18*(4), 385–390. doi:10.1016/j.crci.2014.07.003

Agency for Toxic Substances and Disease Registry (ATSDR). (1993). *Toxicological Profile for Chromium*. Atlanta, GA: US Public Health Service, US Public Health Service, US Department of Health and Human Services.

Agnihotri, S., Mukherji, S., & Mukherji, S. (2014). Size-controlled silver nanoparticles synthesized over the range 5–100 nm using the same protocol and their antibacterial efficacy. *RSC Advances*, *4*(8), 3974–3983. doi:10.1039/C3RA44507K

Ai, L., Zhang, C., Liao, F., Wang, Y., Li, M., Meng, L., & Jiang, J. (2011). Removal of methylene blue from aqueous solution with magnetite loaded multi-wall carbon nanotube: Kinetic, isotherm and mechanism analysis. *Journal of Hazardous Materials*, *198*, 282–290. doi:10.1016/j.jhazmat.2011.10.041 PMID:22040800

Ambashta, R. D., & Sillanpaa, M. (2010). Water purification using magnetic assistance: A review. *Journal of Hazardous Materials*, *180*(1-3), 38–49. doi:10.1016/j.jhazmat.2010.04.105 PMID:20488616

Amin, M. T., Alazba, A. A., & Manzoor, U. (2014). A Review of Removal of Pollutants from Water/Wastewater Using Different Types of Nanomaterials. *Advances in Materials Science and Engineering*, 825910–825934.

An, B., & Zhao, D. (2012). Immobilization of As(III) in soil and groundwater using a new class of polysaccharide stabilized Fe–Mn oxide nanoparticles. *Journal of Hazardous Materials*, *211–212*, 332–341. doi:10.1016/j.jhazmat.2011.10.062 PMID:22119304

Ang, P. K., Chen, W., Wee, A. T. S., & Loh, K. P. (2008). Solution-gated epitaxial graphene as pH sensor. *Journal of the American Chemical Society*, *130*(44), 14392–14393. doi:10.1021/ja805090z PMID:18850701

Ashbolt, N. J. (2004). Microbial contamination of drinking water and disease outcomes in developing regions. *Toxicology*, *198*(1-3), 229–238. doi:10.1016/j.tox.2004.01.030 PMID:15138046

Asmaly, H. A., Abussaud, B., Ihsanullah, , Saleh, T. A., Gupta, V. K., & Atieh, M. A. (2015). Ferric oxide nanoparticles decorated carbon nanotubes and carbon nanofibers: From synthesis to enhanced removal of phenol. *J. Saudi Chem. Soc*, *19*(5), 511–520. doi:10.1016/j.jscs.2015.06.002

Awual, M., & Hasan, M. M. (2014). Novel conjugate adsorbent for visual detection and removal of toxic lead(II) ions from water. *Microporous and Mesoporous Materials, 196*, 261–269. doi:10.1016/j.micromeso.2014.05.021

Awual, M., Yaita, T., Suzuki, S., & Shiwaku, H. (2015). Ultimate selenium(IV) monitoring and removal from water using a new class of organic ligand based composite adsorbent. *Journal of Hazardous Materials, 91*, 111–119. doi:10.1016/j.jhazmat.2015.02.066 PMID:25771216

Ayati, A., Ahmadpour, A., Bamoharram, F. F., Tanhaei, B., Manttari, M., & Sillanpa, M. (2014). A review on catalytic applications of Au/TiO_2 nanoparticles in the removal of water pollutant. *Chemosphere, 107*, 163–174. doi:10.1016/j.chemosphere.2014.01.040 PMID:24560285

Basu, D., Blackburn, K., Harris, B., Neal, M., & Stoss, F. (1983). *Health assessment document for chromium*. Springfield, VA: Environmental Protection Agency.

Becher, G. (1999). Drinking water chlorination and health, *Acta Hydrochim. EtHydrobiol., 27*, 100–102.

Bissen, M., & Frimmel, F. H. (2000). Speciation of As(III), As(V), MMA and DMA in contaminated soil extracts by HPLC-ICP/MS. *Journal of Analytical Chemistry, 367*(1), 51–55. doi:10.1007/s002160051597 PMID:11227433

Bottero, J.-Y., Rose, J., & Wiesner, M. R. (2006). Nanotechnologies: Tools for sustainability in a new wave of water treatment processes. *Integrated Environmental Assessment and Management, 2*(4), 391–395. doi:10.1002/ieam.5630020411 PMID:17069181

Cao, Y., & Li, X. (2014). Adsorption of graphene for the removal of inorganic pollutants in water purification: A review. *Adsorption, 20*(5-6), 713–727. doi:10.1007/s10450-014-9615-y

Chen, C. J., Hsueh, Y. M., Lai, M. S., Shyu, P., Chen, S. Y., Wu, M. M., & Tai, T. Y. et al. (1995). Increased prevalence of hypertension and long term arsenic exposure. *Hypertension, 25*(1), 53–60. doi:10.1161/01.HYP.25.1.53 PMID:7843753

Chung, J., Chun, J., Lee, J., Lee, S. H., Lee, Y. J., & Hong, S. W. (2012). Sorption of Pb(II) and Cu(II) onto multi-amine grafted mesoporous silica embedded with nano-magnetite: Effects of steric factors. *Journal of Hazardous Materials, 183*, 239–240. PMID:22985819

Corsini, E., Liesivuori, J., Vergieva, T., Loveren, H. V., & Colosio, C. (2008). Effects of pesticide exposure on the human immune system. *Human and Experimental Toxicology*, *27*(9), 671–680. doi:10.1177/0960327108094509 PMID:19042949

Dadfarnia, S., Shabani, A. M. H., Moradi, S. E., & Emami, S. (2015). Methyl red removal from water by iron based metal-organic frameworks loaded onto iron oxide nanoparticle adsorbent. *Applied Surface Science*, *330*, 85–93. doi:10.1016/j.apsusc.2014.12.196

Daels, N., Radoicic, M., Radetic, M., & Hulle, S. W. H. V. (2015). Electrospunnnanofibre membranes functionalised with TiO_2 nanoparticles: Evaluation of humic acid and bacterial removal from polluted water. *Separation and Purification Technology*, *149*, 488–49. doi:10.1016/j.seppur.2015.06.016

Diale, P. P., & Muzenda, E. (2011). A Study of South African Natural Zeolites Properties and Applications Properties and Applications. *Proceedings of the World Congress on Engineering and Computer Science*, 2.

Domingo, C., Clemente, R. R., & Blesa, M. (1994). Morphological properties of a-FeOOH, g-FeOOH and Fe_3O_4 obtained by oxidation of aqueous Fe (II) solutions. *Journal of Colloid and Interface Science*, *165*(1), 244–252. doi:10.1006/jcis.1994.1225

Dong, L., Yang, H., Liu, S., Wang, X., & Xie, Y. F. (2015). Fabrication and antibiofouling properties of alumina and zeolite nanoparticle embedded ultrafiltration membranes. *Desalination*, *365*, 70–78. doi:10.1016/j.desal.2015.02.023

Dreyer, D. R., Park, S., Bielawski, C. W., & Ruoff, R. S. (2010). The chemistry of graphene oxide. *Chemical Society Reviews*, *39*(1), 228–240. doi:10.1039/B917103G PMID:20023850

Emadi, M., Shams, E., & Amini, M. K. (2013). Removal of zinc from aqueous solutions by magnetite silica core-shell nanoparticles. *J. Chem.*, 1–10.

Fan, L., Luo, C., Sun, M., & Qiu, H. (2012). Synthesis of graphene oxide decorated with magnetic cyclodextrin for fast chromium removal. *Journal of Materials Chemistry*, *22*(47), 24577–24583. doi:10.1039/c2jm35378d

Fawell, J., & Nieuwenhuijsen, M. J. (2003). Contaminants in drinking water. *British Medical Bulletin*, *68*(1), 199–208. doi:10.1093/bmb/ldg027 PMID:14757718

Fellenz, N. A., Marchetti, S. G., Bengoa, J. F., Mercader, R. C., & Stewart, S. J. (2006). Synthesis and magnetic characterization of magnetite particles embedded in mesoporous MCM-41. *J. Magn. Mater.*, *306*(1), 30–34. doi:10.1016/j.jmmm.2006.02.101

Forstner, U., & Wittman, G. T. W. (1981). *Metal pollution in the aquatic environment.* Berlin: Springer-Verlag. doi:10.1007/978-3-642-69385-4

Furst, A., Lederer, E. W. H., Fensterheim, R. J., & Reinhold, V. N. (1983). Arsenic: Industrial, Biomedical, Environmental Prospectives. New York: Academic Press.

Gandhi, M. R., & Meenakshi, S. (2012). Preparation and characterization of silica gel/chitosan composite for the removal of Cu(II) and Pb(II). *International Journal of Biological Macromolecules, 50*(3), 650–657. doi:10.1016/j.ijbiomac.2012.01.012 PMID:22281115

Gao, Z., Li, W., Liu, B., Liang, F., He, H., Yang, S., & Sun, C. (2011). Nano-structured polyaniline-ionic liquid composite film coated steel wire for headspace solid-phase microextraction of organochlorine pesticides in water. *Journal of Chromatography. A, 1218*(37), 6285–6291. doi:10.1016/j.chroma.2011.07.041 PMID:21821255

Gautam, R. K., Rawat, V., Banerjee, S., Sanroman, M. A., Soni, S., Singh, S. K., & Chattopadhyaya, M. C. (2015). Synthesis of bimetallic Fe–Zn nanoparticles and its application towards adsorptive removal of carcinogenic dye malachite green and Congo red in water. *Journal of Molecular Liquids, 212*, 227–236. doi:10.1016/j.molliq.2015.09.006

Girginova, P. I., Daniel-da-Silva, A. L., Lopes, C. B., Figueira, P., Otero, M., Amaral, V. S., & Trindade, T. et al. (2010). Silica coated magnetite particles for magnetic removal of Hg^{2+} from water. *Journal of Colloid and Interface Science, 345*(2), 234–240. doi:10.1016/j.jcis.2010.01.087 PMID:20202646

Gkementzoglou, C., Kotrotsiou, O., Koronaiou, M., & Kiparissides, C. (2016). Development of a sandwich-type filtration unit packed with MIP nanoparticles for removal of atrazine from water source. *Chemical Engineering Journal, 287*, 233–240. doi:10.1016/j.cej.2015.11.018

Gonzalez, C. M., Hernandez, J., Parsons, J. G., & Gardea-Torresdey, J. L. (2010). A study of the removal of selenite and selenate from aqueous solutions using a magnetic iron/manganese oxide nanomaterial and ICP-MS. *Microchemical Journal, 96*(2), 324–329. doi:10.1016/j.microc.2010.05.005

Gunawan, P., Guan, C., Song, X., Zhang, Q., Leong, S. S. J., Tang, C., & Xu, R. et al. (2011). Hollow fiber membrane decorated with Ag/MWNTs: Toward effective water disinfection and biofouling control. *ACS Nano, 5*(12), 10033–10040. doi:10.1021/nn2038725 PMID:22077241

Gupta, V., Jain, R., Mittal, A., Saleh, T. A., & Nayak, A. (2012). Photo-catalytic degradation of toxic dye amaranth on TiO 2/UV in aqueous suspensions. *Materials Sci. Engineer.*, *C 32*, 12-17.

Gupta, V. K., & Nayak, A. (2012). Cadmium removal and recovery from aqueous solutions by novel adsorbents prepared from orange peel and Fe_2O_3 nanoparticles. *Chemical Engineering Journal*, *180*, 81–90. doi:10.1016/j.cej.2011.11.006

Hayes, W. J., Hayes, W., & Lawes, E. R. (1991). *Handbook of Pesticides Toxicology*. San Diego, CA: Academic Press.

Hui, K. S., Chao, C. Y. H., & Kot, S. C. (2005). Removal of mixed heavy metal ions in wastewater by zeolite 4A and residual products from recycled coal fly ash. *Journal of Hazardous Materials*, *127*(1-3), 89–101. doi:10.1016/j.jhazmat.2005.06.027 PMID:16076523

Hutton, G., Haller, L., & Bartram, J. (2007). *Economic and Health Effects of Increasing Coverage of Low Cost Household Drinking Water Supply and Sanitation Intervention*. World Health Organization.

Izakmehri, Z., Ardjmand, M., Ganji, M. D., Babanezhad, E., & Heydarinasab, A. (2015). Removal of dioxane pollutants from water by using Al-doped single walled carbon nanotubes. *RSC Advances*, *5*(59), 48124–48132. doi:10.1039/C5RA07611K

Jawor, A., & Hoek, E. M. V. (2010). Removing cadmium ions from water via nanoparticle-enhanced ultrafiltration. *Environmental Science & Technology*, *44*(7), 2570–2576. doi:10.1021/es902310e PMID:20230021

Jegadeesan, G., Al-Abed, S. A., Sundaram, V., Choi, H., Scheckel, K. G., & Dionysiou, D. D. (2010). Arsenic sorption on TiO_2 nanoparticles: Size and crystallinity effects. *Water Research*, *44*(3), 965–973. doi:10.1016/j.watres.2009.10.047 PMID:20022353

Ji, L., Bai, X., Zhou, L., Shi, H., Chen, W., & Hua, Z. (2013). One-pot preparation of graphene oxide magnetic nanocomposites for the removal of tetrabromobisphenol A. *Front. Environ. Sci. Eng.*, *7*(3), 442–450. doi:10.1007/s11783-013-0515-2

Jian, W., Kitanaka, A., Nishijima, W., Baes, A. U., & Okada, M. (1999). Removal of oil pollutants in seawater as pretreatment of reverse osmosis desalination process. *Water Research*, *33*(8), 1857–1863. doi:10.1016/S0043-1354(98)00404-7

Jin, Y., Deng, J., Liang, J., Shan, C., & Tong, M. (2015). Efficient bacteria capture and inactivation by cetyltrimethylammonium bromide modified magnetic nanoparticles. *Colloids and Surfaces. B, Biointerfaces*, *136*, 659–665. doi:10.1016/j.colsurfb.2015.10.009 PMID:26496472

Jing-po, Y., Fei, L., Jing, C., & Jin-di, Y. (2015). Sensing and magnetic removal of Hg(II) using core–shell structured nanocomposite grafted with fluorescence Off–On probe Microporous. *Mesoporous Mater.*, *202*, 175–182. doi:10.1016/j.micromeso.2014.09.054

Kaegi, R., Voegelin, A., Sinnet, B., Zuleeg, S., Hagendorfer, H., Burkhardt, M., & Siegrist, H. (2011). Behavior of metallic silver nanoparticles in a pilot wastewater treatment Plant. *Environmental Science & Technology*, *45*(9), 3902–3908. doi:10.1021/es1041892 PMID:21466186

Karimi, H., Mousavi, S., & Sadeghian, B. (2012). Silver nanoparticle loaded on activated carbon as efficient adsorbent for removal of methyl orange. *Ind. J. Sci. Technol.*, *5*, 2346–2356.

Kim, Y., Osako, M., & Lee, D. (2002). Removal of hydrophobic organic pollutants by coagulation-precipitation process with dissolved humic matter. *Waste Management & Research*, *20*(4), 341–349. doi:10.1177/0734247X0202000405 PMID:12363094

Knop, A., & Pilato, L. A. (1985). *Phenolic Resins Chemistry, Applications and Performance*. Springer-Verlag. doi:10.1007/978-3-662-02429-4

Kohler, A., Hellweg, S., Escher, B. I., & Hungerbuhler, K. (2006). Organic pollutant removal versus toxicity reduction in industrial wastewater treatment: The example of wastewater from fluorescent whitening agent production. *Environmental Science & Technology*, *40*(10), 3395–3401. doi:10.1021/es060555f PMID:16749712

Kojima, N., Mitomo, A., Itaya, Y., Mori, S., & Yoshida, S. (2002). Adsorption removal of pollutants by active cokes produced from sludge in the energy recycle process of wastes. *Waste Management (New York, N.Y.)*, *22*(4), 399–404. doi:10.1016/S0956-053X(02)00022-3 PMID:12099497

Kolthoff, I. M., Elving, P. J., & Stross, F. H. (1971). Treatise on Analytical Chemistry. Wiley Interscience.

Kotas, J., & Stasicka, Z. (2000). Chromium occurrence in the environment and methods of its speciation. *Environmental Pollution*, *107*(3), 263–283. doi:10.1016/S0269-7491(99)00168-2 PMID:15092973

Kumar, V., Talreja, N., Deva, D., Sankararamakrishnan, N., Sharma, A., & Verma, N. (2011). Development of bi-metal doped micro- and nano multifunctional polymeric adsorbents for the removal of fluoride and arsenic (V) from wastewater. *Desalination*, *282*, 27–38. doi:10.1016/j.desal.2011.05.013

Kuriacose, J. C., & Rajaram, J. (1987). *Chemistry in Engineering and Technology*. New York: McGraw Hill, Publishing Company Ltd.

Lee, Y.-F., Nan, F.-H., Chen, M.-J., Wu, H.-Y., Ho, C.-W., Chen, Y.-Y., & Huan, C.-C. (2012). Detection and removal of mercury and lead ions by using gold nanoparticle-based gel membrane. *Anal. Methods*, *4*(6), 1709–1717. doi:10.1039/c2ay05872c

Liang, J., Huang, Y., Zhang, L., Wang, Y., Ma, Y., Guo, T., & Chen, Y. (2009). Molecular-level dispersion of graphene into poly (vinyl alcohol) and effective reinforcement of their nanocomposites. *Advanced Functional Materials*, *19*(14), 2297–2302. doi:10.1002/adfm.200801776

Lin, S. H. (1993). Adsorption of disperses dye by powdered activated carbon. *Journal of Chemical Technology and Biotechnology (Oxford, Oxfordshire)*, *57*(4), 387–391. doi:10.1002/jctb.280570415

Liu, F., Chung, S., Oh, G., & Seo, T. S. (2012). Three-dimensional graphene oxide nanostructure for fast and efficient water- soluble dye removal. *ACS Applied Materials & Interfaces*, *4*(2), 922–927. doi:10.1021/am201590z PMID:22206476

Liu, M., Xi, B., Hou, L., & Yu, S. (2013). Magnetic multi-functional nano-fly ash-derived zeolite composites for environmental applications. *J. Mater. Chem. A*, *1*(40), 12617–12626. doi:10.1039/c3ta13180g

Liu, X., Zhang, H., Ma, Y., Wu, X., Meng, L., Guo, Y., & Liu, Y. et al. (2013). Graphene-coated silica as a highly efficient sorbent for residual organophosphorus pesticides in water. *J. Mater. Chem. A*, *1*(5), 1875–1884. doi:10.1039/C2TA00173J

Loyo, R. L. A., Nikitenko, S. I., Scheinost, A. C., & Simonoff, M. (2008). Immobilization of selenite on Fe_3O_4 and Fe/Fe_3C ultrasmall particles. *Environmental Science & Technology*, *42*(7), 2451–2456. doi:10.1021/es702579w PMID:18504980

Lvovich, M. I. (1979). World water resources and their future. American Geophysical Union.

Mahlambi, M. M., Ngila, C. J., & Mamba, B. B. (2015). Recent developments in environmental photocatalytic degradation of organic pollutants: The case of titanium dioxide nanoparticles–A review. *Journal of Nanomaterials*, *29*, 790173.

Mani, A. D., Reddy, P. M. K., Srinivaas, M., Ghosal, P., Xanthopoulos, N., & Subrahmanyam, C. (2015). Facile synthesis of efficient visible active C-doped TiO_2 nanomaterials with high surface area for the simultaneous removal of phenol and Cr(VI). *Materials Research Bulletin*, *61*, 391–399. doi:10.1016/j.materresbull.2014.10.051

Mara, D. D. (2003). Water, sanitation and hygiene for the health of developing nations. *Public Health*, *117*(6), 452–456. doi:10.1016/S0033-3506(03)00143-4 PMID:14522162

Martinson, C. A., & Reddy, K. J. (2009). Adsorption of arsenic (III) and arsenic (V) by cupric oxide nanoparticles. *Journal of Colloid and Interface Science*, *336*(2), 406–411. doi:10.1016/j.jcis.2009.04.075 PMID:19477461

Mascolo, M. C., Pei, Y., & Ring, T. A. (2013). Room temperature co-precipitation synthesis of magnetite nanoparticles in a large pH window with different bases. *Materials (Basel)*, *6*(12), 5549–5567. doi:10.3390/ma6125549

Masoumi, A., Hemmati, K., & Ghaemy, M. (2016). Low-cost nanoparticles sorbent from modified rice husk and a copolymer for efficient removal of Pb(II) and crystal violet from water. *Chemosphere*, *146*, 253–262. doi:10.1016/j.chemosphere.2015.12.017 PMID:26735725

Mehdizadeh, S., Sadjadi, S., Ahmadi, S. J., & Outokesh, M. (2014). Removal of heavy metals from aqueous solution using platinum nanopartcles/Zeolite-4A. *J. Environ. Health Sci. Eng.*, *12*(1), 7. doi:10.1186/2052-336X-12-7 PMID:24397886

Merian, E. (1984). Introduction on environmental chemistry and global cycles of chromium, nickel, cobalt, beryllium, arsenic, cadmium and selenium, and their derivatives. *Toxicological and Environmental Chemistry*, *8*(1), 9–38. doi:10.1080/02772248409357038

Mkhoyan, K. A., Contryman, A. W., Silcox, J., Stewart, D. A., Eda, G., Mattevi, C., & Chhowalla, M. et al. (2009). Atomic and electronic structure of graphene-oxide. *Nano Letters*, *9*(3), 1058–1063. doi:10.1021/nl8034256 PMID:19199476

Moore, M., Gould, P., & Keary, B. S. (2003). Global urbanization and impact on health. *International Journal of Hygiene and Environmental Health*, *206*(4-5), 69–278. doi:10.1078/1438-4639-00223 PMID:12971682

Mostafa, M. S., Bakr, A.-S. A., Naggar, A. M. A. E., & Sultan, E.-S. A. (2016). Water decontamination via the removal of Pb (II) using a new generation of highly energetic surface nano-material: Co^{+2}, Mo^{+6} LDH. *Journal of Colloid and Interface Science*, *461*, 261–272. doi:10.1016/j.jcis.2015.08.060 PMID:26402785

Motshekga, S. C., Ray, S. S., Onyango, M. S., & Momba, M. N. B. (2015). Preparation and antibacterial activity of chitosan-based nanocomposites containing bentonite-supported silver and zinc oxide nanoparticles for water disinfection. *Applied Clay Science*, *114*, 330–339. doi:10.1016/j.clay.2015.06.010

Nabi, D., Aslam, I., & Qazi, I. A. (2009). Evaluation of the adsorption potential of titanium dioxide nanoparticles for arsenic removal. *Journal of Environmental Sciences (China)*, *21*(3), 402–408. doi:10.1016/S1001-0742(08)62283-4 PMID:19634455

Niu, S.-F., Liu, Y., Xu, X.-H., & Lou, Z.-H. (2005). Removal of hexavalent chromium from aqueous solution by iron nanoparticles. *Journal of Zhejiang University. Science. B.*, *6*(10), 1022–1027. doi:10.1631/jzus.2005.B1022 PMID:16187417

Nriagu, J. O., & Pacyna, J. M. (1988). Quantitative assessment of worldwide contamination of air, water and soils by trace metals. *Nature*, *333*(6169), 134–139. doi:10.1038/333134a0 PMID:3285219

Obare, S. O., & Meyer, G. J. (2004). Nanostructured materials for environmental remediation of organic contaminants in water, *J. Environ. Sci. Health—Part A*, *39*, 2549–2582.

Ojea-Jimenez, I., Lopez, X., Arbiol, J., & Puntes, V. (2012). Citrate-coated gold nanoparticles As smart scavengers for mercury(II) removal from polluted waters. *ACS Nano*, *6*(3), 2253–2260. doi:10.1021/nn204313a PMID:22332645

Parham, H., Zargar, B., & Shiralipour, R. (2012). Fast and efficient removal of mercury from water samples using magnetic iron oxide nanoparticles modified with 2-mercaptobenzothiazole. *Journal of Hazardous Materials*, *205–206*, 94–100. doi:10.1016/j.jhazmat.2011.12.026 PMID:22244341

Patel, K. S., Shrivas, K., Brandt, R., Jakubowski, N., Corns, W., & Hoffman, P. (2005). Arsenic contamination in water, soil, sediment and rice of central India. *Environmental Geochemistry and Health*, *27*(2), 131–145. doi:10.1007/s10653-005-0120-9 PMID:16003581

Patel, K. S., Shrivas, K., Hoffmann, P., & Jakubowski, N. (2006). A survey of lead pollution in Chhattisgarh State, central India. *Environmental Geochemistry and Health*, *28*(1-2), 11–17. doi:10.1007/s10653-005-9006-0 PMID:16528599

Patra, A. K., Dutta, A., & Bhaumik, A. (2012). Self-assembled mesoporous γ- Al_2O_3 spherical nanoparticles and their efficiency for the removal of arsenic from water. *Journal of Hazardous Materials*, *201–202*, 170–177. doi:10.1016/j.jhazmat.2011.11.056 PMID:22169241

Patra, S., Roy, E., Madhuri, R., & Sharma, P. K. (2016). Agar based bimetallic nanoparticles as high-performance renewable adsorbent for removal and degradation of cationic organic dyes. *Journal of Industrial and Engineering Chemistry*, *33*, 226–238. doi:10.1016/j.jiec.2015.10.008

Peng, L.-M., Zhang, Z., & Wang, S. (2014). Carbon nanotube electronics: Recent advances. *Materials Today*, *17*(9), 433–442. doi:10.1016/j.mattod.2014.07.008

Perez-Urquiza, M., Ferrer, R., & Beltran, J. L. (2000). Determination of sulfonatedazodyes in river water samples by capillary zone electrophoresis. *Journal of Chromatography. A*, *883*(1-2), 277–283. doi:10.1016/S0021-9673(00)00431-3 PMID:10910220

Postel, S. L., Daily, G. C., & Ehrlich, P. R. (1996). Human appropriation of renewable fresh water. *Science*, *271*(5250), 785–788. doi:10.1126/science.271.5250.785

Qi, L., & Xu, Z. (2004). Lead sorption from aqueous solutions on chitosan nanoparticles. *Colloids and Surfaces. A, Physicochemical and Engineering Aspects*, *251*(1-3), 183–190. doi:10.1016/j.colsurfa.2004.10.010

Radzig, M. A., Nadtochenko, V. A., Koksharova, O. A., Kiwi, J., Lipasova, V. A., & Khmel, A. (2013). Antibacterial effects of silver nanoparticles on gram-negative bacteria: Influence on the growth and biofilms formation, mechanisms of action. *Colloids and Surfaces. B, Biointerfaces*, *102*, 300–306. doi:10.1016/j.colsurfb.2012.07.039 PMID:23006569

Rahamani, A. R., & Mahvi, A. H. (2006). Use of ion exchange for removal of ammonium: A biological regeneration of zeolite. *Global NEST Journal*, *8*, 146–150.

Rickerby, D. G., & Morrison, M. (2007). Nanotechnology and the environment: A European perspective. *Science and Technology of Advanced Materials*, *8*(1-2), 19–24. doi:10.1016/j.stam.2006.10.002

Ritter, L., Solomon, K., Sibley, P., Hall, K., Keen, P., Mattu, G., & Linton, B. (2002). Sources, pathways, and relative risks of contaminants in surface water and groundwater: A perspective prepared for the Walkerton inquiry. *Journal of Toxicology and Environmental Health. Part A.*, *65*, 1–142. doi:10.1080/152873902753338572 PMID:11809004

Rosal, R., Rodriguez, A., Perdigon-Melon, J. A., Petre, A., Garcia-Calvo, E., Gomez, M. J., & Fernandez-Alba, A. R. et al. (2010). Occurrence of emerging pollutants in urban wastewater and their removal through biological treatment followed by ozonation. *Water Research*, *44*(2), 578–588. doi:10.1016/j.watres.2009.07.004 PMID:19628245

Rossi, L. M., Costa, N. J. S., Silva, F. P., & Wojcieszak, R. (2014). Magnetic nanomaterials in catalysis: Advanced catalysts for magnetic separation and beyond. *Green Chemistry*, *16*(6), 2906–2933. doi:10.1039/c4gc00164h

Rubi, H., Fall, C., & Ortega, R. E. (2009). Pollutant removal from oily wastewater discharged from car washes through sedimentation-coagulation. *Water Science and Technology*, *59*(12), 2359–2369. doi:10.2166/wst.2009.307 PMID:19542641

Saha, S., & Sarkar, P. (2012). Arsenic remediation from drinking water by synthesized nano-alumina dispersed in chitosan-grafted polyacrylamide. *Journal of Hazardous Materials*, *227–228*, 68–78. doi:10.1016/j.jhazmat.2012.05.001 PMID:22647233

Sahiner, N., Ozay, O., Aktas, N., Blake, D. A., & John, V. T. (2011). Arsenic (V) removal with modifiable bulk and nano p(4-vinylpyridine)-based hydrogels: The effect of hydrogel sizes and quarternization agents. *Desalination*, *79*(1-3), 344–352. doi:10.1016/j.desal.2011.06.028

Saleh, T. A., Gondal, M. A., & Drmosh, Q. A. (2010). Preparation of a MWCNT/ZnO nanocomposite and its photocatalytic activity for the removal of cyanide from water using a laser. *Nanotechnology*, *21*(49), 495705. doi:10.1088/0957-4484/21/49/495705 PMID:21079289

Saleh, T. A., Gondal, M. A., Drmosh, Q. A., Yamani, Z. Y., & Al-Yamani, A. (2011). Enhancement in photocatalytic activity for acetaldehyde removal by embedding ZnO nano particles on multiwall carbon nanotubes. *Chemical Engineering Journal*, *166*(1), 407–412. doi:10.1016/j.cej.2010.10.070

Saleh, T. A., & Gupta, V. (2011). Functionalization of tungsten oxide into MWCNT and its application for sunlight-induced degradation of rhodamine B. *Journal of Colloid and Interface Science*, *362*(2), 337–344. doi:10.1016/j.jcis.2011.06.081 PMID:21788030

Saleh, T. A., & Gupta, V. (2012). Column with CNT/magnesium oxide composite for lead (II) removal from water. *Environ. Sci. Pollution Res.*, *19*(4), 1224–1228. doi:10.1007/s11356-011-0670-6 PMID:22124802

Saleh, T. A., & Gupta, V. (2012). Synthesis and characterization of alumina nano-particles polyamide membrane with enhanced flux rejection performance. *Separation and Purification Technology*, *89*, 245–251. doi:10.1016/j.seppur.2012.01.039

Saleh, T. A., & Gupta, V. (2016). *Nanomaterial and Polymer Membranes: Synthesis, Characterization, and Applications*. Elsevier.

Savina, I. N., English, C. J., Whitby, R. L. D., Zheng, Y., Leistner, A., Mikhalovsky, S. V., & Cundy, A. B. (2011). High efficiency removal of dissolved As(III) using iron nanoparticle-embeddedmacroporous polymer composites. *Journal of Hazardous Materials*, *192*(3), 1002–1008. doi:10.1016/j.jhazmat.2011.06.003 PMID:21715089

Schamberger, R. J., & Frieden, E. (1984). Selenium. In *Biochemistry of the essential ultratrace elements* (pp. 201–237). New York, NY: Plenum Press. doi:10.1007/978-1-4684-4775-0_9

Seiler, H. G., Sigel, H., & Dekker, M. (1988). *Handbook on Toxicity of Inorganic Compounds*. New York.

Selvakumar, R., Jothi, N. A., Jayavignesh, V., Karthikaiselvi, K., Immanual Antony, G., Sharmila, P. R., & Swaminathan, K. et al. (2011). As (V) removal using carbonized yeast cells containing silver nanoparticles. *Water Research*, *45*(2), 583–592. doi:10.1016/j.watres.2010.09.034 PMID:20947119

Sharma, Y. C. (2003). Chromium removal from industrial effluents by adsorption on anindigenous low-cost material. *Colloid Surf. A*, *215*(1-3), 155–162. doi:10.1016/S0927-7757(02)00485-5

Shen, X., Wang, Q., Chen, W. L., & Pang, Y. (2014). One-step synthesis of water-dispersible cysteine functionalized magnetic Fe_3O_4 nanoparticles for mercury (II) removal from aqueous solutions. *Applied Surface Science*, *317*, 1028–1034. doi:10.1016/j.apsusc.2014.09.033

Shipley, H. J., Yean, S., Kan, A. T., & Tomson, M. B. (2009). Adsorption of arsenic to magnetite nanoparticles: Effect of particle concentration, pH, ionic strength, and temperature. *Environmental Toxicology and Chemistry*, *28*(3), 509–515. doi:10.1897/08-155.1 PMID:18939890

Shrivastava, S., Bera, T., Roy, A., Singh, G., Ramachandrarao, P., & Dash, D. (2007). Characterization of enhanced antibacterial effects of novel silver nanoparticles. *Nanotechnology*, *18*(22), 225103. doi:10.1088/0957-4484/18/22/225103

Singh, S., Barick, K. C., & Bahadu, D. (2011). Surface engineered magnetic nanoparticles for removal of toxic metal ions and bacterial pathogens. *Journal of Hazardous Materials*, *192*(3), 1539–1547. doi:10.1016/j.jhazmat.2011.06.074 PMID:21784580

Singh, S., Barick, K. C., & Bahadur, D. (2013). Functional oxide nanomaterials and nanocomposites for the removal of heavy metals and dyes. *Nanomater. Nanotechnology*, *3*, 3–20.

Singla, P., Goel, N., Kumar, V., & Singhal, S. (2015). Boron nitride nanomaterials with different morphologies: Synthesis, characterization and efficient application in dye adsorption. *Ceramics International*, *41*(9), 10565–10577. doi:10.1016/j.ceramint.2015.04.151

Sivashankar, R., Sathya, A. B., Vasantharaj, K., & Sivasubramanian, V. (2014). Magnetic composite an environmental super adsorbent for dye sequestration- A review, *Environ. Nanotechnol. Monitor. Manag., 1–2*, 36–49.

Skubal, L. R., Meshkov, N. K., Rajh, T., & Thurnauer, M. (2002). Cadmium removal from water using thiolactic acid-modified titanium dioxide nanoparticles. *Journal of Photochemistry and Photobiology A Chemistry, 148*(1-3), 393–397. doi:10.1016/S1010-6030(02)00069-2

Smith, A. G., & Gangolli, S. D. (2002). Organochlorine chemicals in seafood: Occurrence and health concerns. *Food and Chemical Toxicology, 40*(6), 767–779. doi:10.1016/S0278-6915(02)00046-7 PMID:11983271

Sondi, I., & Salopek-Sondi, B. (2004). Silver nanoparticles as antimicrobial agent: A case study on *E. coli*as a model for Gram-negative bacteria. *Journal of Colloid and Interface Science, 275*(1), 177–182. doi:10.1016/j.jcis.2004.02.012 PMID:15158396

Stern, A. H., Gochefld, M., Weisel, C., & Burger, J. (2001). Mercury and methyl mercury exposure in the New Jersey Pregnant population. *Archives of Environmental Health, 56*(1), 4–10. doi:10.1080/00039890109604048 PMID:11256856

Sumesh, E., Bootharaju, M. S., Anshup, , & Pradeep, T. (2011). A practical silver nanoparticle-based adsorbent for the removal of Hg^{2+} from water. *Journal of Hazardous Materials, 189*(1-2), 450–457. doi:10.1016/j.jhazmat.2011.02.061 PMID:21398028

Sun, J. Z., Liao, Z. H., Si, R. W., Kingori, G. P., Chang, F. X., Gao, L., & Yong, Y. C. et al. (2014). Adsorption and removal of triphenylmethane dyes from water by magnetic reduced graphene oxide. *Water Science and Technology, 70*(10), 1663–1669. doi:10.2166/wst.2014.427 PMID:25429455

Tang, S. C. N., & Lo, I. M. C. (2013). Magnetic nanoparticles: Essential factors for sustainable environmental applications. *Water Research, 47*(8), 2613–2632. doi:10.1016/j.watres.2013.02.039 PMID:23515106

Tawabini, B., Al-Khaldi, S., Atieh, M., & Khaled, M. (2010). Removal of mercury from water by multi-walled carbon nanotubes. *Water Science and Technology, 61*(3), 591–598. doi:10.2166/wst.2010.897 PMID:20150694

Teymourian, H., Salimi, A., & Khezrian, S. (2013). Fe_3O_4 magnetic nanoparticles/reduced graphene oxide nanosheets as a novel electrochemical and bioeletrochemical sensing platform. *Biosensors & Bioelectronics, 49*, 1–8. doi:10.1016/j.bios.2013.04.034 PMID:23708810

Thanh, D. N., Singh, M., Ulbrich, P., Strnadova, N., & Stepanek, F. (2011). Perlite incorporating γ-Fe_2O_3 and α-MnO_2 nanomaterials: Preparation and evaluation of a new adsorbent for As (V) removal. *Separation and Purification Technology*, *82*, 93–101. doi:10.1016/j.seppur.2011.08.030

Theepharaksapan, S., Chiemchaisri, C., Chiemchaisri, W., & Yamamoto, K. (2011). Removal of pollutants and reduction of bio-toxicity in a full scale chemical coagulation and reverse osmosis leachate treatment system. *Bioresource Technology*, *102*(9), 5381–5388. doi:10.1016/j.biortech.2010.11.091 PMID:21147528

Urase, T., & Kikuta, T. (2005). Separate estimation of adsorption and degradation of pharmaceutical substances and estrogens in the activated sludge process. *Water Research*, *39*(7), 1289–1300. doi:10.1016/j.watres.2005.01.015 PMID:15862328

U.S. Environmental Protection Agency. (1980). *Ambient water quality criteria for selenium.* Report No. EPA-440/5-80-070, Criteria and Standards Division, Office of Water Regulations and Standards.

US Environmental Protection Agency. (n.d.). *Mercury and Compounds*. Retrieved from http://www.epa.gov/opptintr/pbt/mercury.html

U.S. EPA. (1993). Silver: Reregistration eligibility decision fact sheet. prepared by the office of prevention, pesticides and toxic substances. Environmental Protection Agency.

Velizarov, S., Rodrigues, C. M., Reis, M. A., & Crespo, J. G. (2000-2001). Mechanism of charged pollutants removal in an ion exchange membrane bioreactor: Drinking water denitrification. *Biotechnology and Bioengineering*, *71*(4), 245–254. doi:10.1002/1097-0290(2000)71:4<245::AID-BIT1014>3.0.CO;2-F PMID:11291034

Vieno, N., Tuhkanen, T., & Kronberg, L. (2006). Removal of pharmaceuticals in drinking water treatment: Effect of chemical coagulation. *Environmental Technology*, *27*(2), 183–192. doi:10.1080/09593332708618632 PMID:16506514

Vukovic, G. D., Marinkovic, A. D., Colic, M., Ristic, M. D., Aleksic, R., Peric-Grujic, A. A., & Uskokovic, P. S. (2010). Removal of cadmium from aqueous solutions by oxidized and ethylenediamine-functionalized multi-walled carbon nanotubes. *Chemical Engineering Journal*, *157*(1), 238–248. doi:10.1016/j.cej.2009.11.026

Walcarius, A., & Mercier, L. (2010). Mesoporousorganosilica adsorbents: Nanoengineered materials for removal of organic and inorganic pollutants. *Journal of Materials Chemistry*, *20*(22), 4478–4511. doi:10.1039/b924316j

Walha, K., Amar, R. B., Firdaous, L., Quemeneur, F., & Jaouen, P. (2007). Brackish groundwater treatment by nanofiltration, reverse osmosis and electrodialysis in Tunisia: Performance and cost comparison. *Desalination*, *207*(1-3), 95–106. doi:10.1016/j.desal.2006.03.583

Westervelt, R. M. (2008). Graphene nanoelectronics. *Science*, *320*(5874), 324–325. doi:10.1126/science.1156936 PMID:18420920

WHO. (1989). *Environmental health criteria 85: lead*. Geneva: WHO.

WHO. (1999). *Arsenic in drinking water: 210*. Geneva: WHO.

World Health Organization and UNICEF. (2013). *Progress on Sanitation and Drinking-Water, World*. Geneva, Switzerland: Health Organization.

Wu, Q., Zhao, G., Feng, C., Wang, C., & Wang, Z. (2011). Preparation of a graphene-based magnetic nanocomposite for the extraction of carbamate pesticides from environmental water samples. *Journal of Chromatography. A*, *1218*(44), 7936–7942. doi:10.1016/j.chroma.2011.09.027 PMID:21962496

Xiao, J., Xie, Y., & Cao, H. (2015). Organic pollutants removal in wastewater by heterogeneous photocatalyticozonation. *Chemosphere*, *121*, 1–17. doi:10.1016/j.chemosphere.2014.10.072 PMID:25479808

Xiao, Z., Yuan, M., Yang, B., Liu, Z., Huang, J., & Sun, D. (2016). Plant-mediated synthesis of highly active iron nanoparticles for Cr (VI) removal: Investigation of the leading biomolecules. *Chemosphere*, *150*, 357–364. doi:10.1016/j.chemosphere.2016.02.056 PMID:26921588

Xu, P., Zeng, G. M., Huang, D. L., Feng, C. L., Hu, S., Zhao, M. H., & Liu, Z. F. et al. (2012). Use of iron oxide nanomaterials in wastewater treatment: A review. *The Science of the Total Environment*, *424*, 1–10. doi:10.1016/j.scitotenv.2012.02.023 PMID:22391097

Yallapu, M. M., Othman, S. F., Curtis, E. T., Gupta, B. K., Jaggi, M., & Chauhan, S. C. (2011). Multi-functional magnetic nanoparticles for magnetic resonance imaging and cancer therapy. *Biomaterials*, *32*(7), 1890–1905. doi:10.1016/j.biomaterials.2010.11.028 PMID:21167595

Yang, K., Wu, W., Jing, Q. F., & Zhu, L. (2008). Aqueous adsorption of aniline, phenol, and their substitutes by multi-walled carbon nanotubes. *Environmental Science & Technology*, *42*(21), 7931–7936. doi:10.1021/es801463v PMID:19031883

Yang, S.-T., Chang, Y., Wang, H., Liu, G., Chen, S., Wang, Y., & Cao, A. et al. (2010). Folding/aggregation of graphene oxide and its application in Cu^{2+} removal. *Journal of Colloid and Interface Science*, *351*(1), 122–127. doi:10.1016/j.jcis.2010.07.042 PMID:20705296

Yang, X., Chen, C., Li, J., Zhao, G., Ren, X., & Wang, X. (2012). Graphene oxide-iron oxide and reduced graphene oxide-iron oxide hybrid materials for the removal of organic and inorganic pollutants. *RSC Advances*, *2*(23), 8821–8826. doi:10.1039/c2ra20885g

Yantasee, W., Warner, C. L., Sangvanich, T., Addleman, R. S., Carter, T. G., Wiacek, R. J., & Warner, M. G. et al. (2007). Removal of heavymetals from aqueous systems with thiol functionalized superparamagnetic nanoparticles. *Environmental Science & Technology*, *4*(14), 5114–5119. doi:10.1021/es0705238 PMID:17711232

Yao, Y., Yang, Z., Sun, H., & Wang, S. (2012). Hydrothermal synthesis of Co_3O_4–graphene for heterogeneous activation of peroxymonosulfate for decomposition of phenol. *Industrial & Engineering Chemistry Research*, *51*(46), 14958–14965. doi:10.1021/ie301642g

Yu, J.-G., Zhao, X.-H., Yang, H., Chen, X.-H., Yang, Q., Yu, L.-Y., & Chen, X.-Q. et al. (2014). Aqueous adsorption and removal of organic contaminants by carbon nanotubes. *The Science of the Total Environment*, *482–483*, 241–251. doi:10.1016/j.scitotenv.2014.02.129 PMID:24657369

Zarpelon, F., Galiotto, D., Aguzzoli, C., Carli, L. N., Figueroa, C. A., Jacob, I., & Giovanela, M. et al. (2016). Removal of coliform bacteria from industrial wastewaters using polyelectrolytes/silver nanoparticles self-assembled thin films. *J. Environ. Chem. Eng.*, *4*(1), 137–146. doi:10.1016/j.jece.2015.11.013

Zhan, S., Yang, Y., Shen, Z., Shan, J., Li, Y., Yang, S., & Zhu, D. (2014). Efficient removal of pathogenic bacteria and viruses by multifunctional amine-modified magnetic nanoparticles. *Journal of Hazardous Materials*, *274*, 115–123. doi:10.1016/j.jhazmat.2014.03.067 PMID:24769848

Zhang, J., Zhai, S., Li, S., Xiao, Z., Song, Y., An, Q., & Tian, G. (2013). Pb(II) removal of Fe_3O_4@SiO_2–NH_2 core–shell nanomaterials prepared via a controllable sol–gel process. *Chemical Engineering Journal*, *215–216*, 461–471. doi:10.1016/j.cej.2012.11.043

Zhang, K., Dwivedi, V., Chi, C. Y., & Wu, J. S. (2010). Graphene oxide/ferric hydroxide composites for efficient arsenate removal from drinking water. *Journal of Hazardous Materials*, *182*(1-3), 162–168. doi:10.1016/j.jhazmat.2010.06.010 PMID:20580161

Zhang, X., Li, W., Blatchley, E. R. III, Wang, X., & Ren, P. (2015). UV/chlorine process for ammonia removal and disinfection by-product reduction: Comparison with chlorination. *Water Research*, *68*, 804–811. doi:10.1016/j.watres.2014.10.044 PMID:25466638

Zhong, J. S., Wang, Q. Y., Zhou, J., Chen, D. Q., & Ji, Z. G. (2016). Highly efficient photoelectrocatalytic removal of RhB and Cr(VI) by Cu nanoparticles sensitized TiO_2 nanotube arrays. *Applied Surface Science*, *367*, 342–346. doi:10.1016/j.apsusc.2016.01.189

Zhou, G., Li, W., Wang, Z., Wang, X., Li, S., & Zhang, D. (2015). Electrosorption for organic pollutants removal and desalination by graphite and activated carbon fiber composite electrodes. *International Journal of Environmental Science and Technology*, *12*(12), 3735–3744. doi:10.1007/s13762-015-0811-4

Zhu, J., Wei, S., Gu, H., Rapole, S. B., Wang, Q., Luo, Z., & Guo, Z. et al. (2012). One-pot synthesis of magnetic graphenenanocomposites decorated with core@ double-shell nanoparticles for fast chromium removal. *Environmental Science & Technology*, *46*(2), 977–985. doi:10.1021/es2014133 PMID:22126606

Zuo, X., Peng, C., Huang, Q., Song, S., Wang, L., Li, D., & Fan, C. (2009). Design of a carbon nanotube/magnetic nanoparticle-based peroxidase-like nanocomplex and Its application for highly efficient catalytic oxidation of phenols. *Nano Res*, *2*(8), 617–623. doi:10.1007/s12274-009-9062-3

Chapter 7
Biomass-Derived Activated Carbon:
Synthesis, Functionalized, and Photocatalysis Application

Samira Bagheri
University Malaya, Malaysia

Nurhidayatullaili Muhd Julkapli
University Malaya, Malaysia

ABSTRACT

The focus is to highlight the catalytic technologies to convert lignocellulosic biomass into the activated carbon (AC) which can be used in photocatalysis applications. The drawback of carbon production raised by energy assumption and product selectivity has encouraged the development of sustainable carbon synthesis process, where the catalytic approach is considered. This treatment via either homogenous or heterogeneous catalytic approach relative in mild condition provided a bulk, mesoporous and nanostructure AC materials. Those characteristics of AC materials are basic requirements for the efficient photocatalytic system. Due to the excellent oxidizing behavior and stability, semiconductor materials have been widely used in the photocatalytic system. However, they lead to some drawbacks in terms of the separation steps and loss of the photocatalyst. So, attention has been paid to supported semiconductor catalysts which carbon materials were explored. AC reported as a potential support in photocatalytic systems.

DOI: 10.4018/978-1-5225-2136-5.ch007

1. INTRODUCTION

Rising environmental concerns and the need for cost-effective competitive products are becoming two major guidelines in modern material research (Baleizao et al. 2002). Previously developed routes to obtain a periodic porous carbon network were successful, but again did not take into account any criteria of sustainability (Kuzyakos et al. 2009). In the past 20 years, many of laboratories of research institutions have had some in-depth researches on the biomass conversion to carbon-based materials without catalysts (Demirbas 2001). Their studies covered some typical carbon material model compounds, such as glucose, methanol, cellulose, lignin and some real biomass compounds (Hamer et al. 2004). As successful demonstrations accumulated, detailed reaction mechanism, kinetics, and thermodynamics have built a solid foundation for subsequent investigations (Onal et al. 2004). However, to increase the selectivity of carbon materials production, high activation energy is needed for the reaction without a catalyst. The high costs of equipment and operations have undoubtedly become the biggest obstacle to the development of this technology (Manya et al. 2003).

Therefore, the problem of carbon synthesis under sustainable conditions was recently revisited and implemented by several terms, where catalytic treatment of biomass via either homogenous or heterogeneous system relatively in mild condition provided bulk, mesoporous or nanostructured carbon materials (Carlson et al. 2008). The catalytic effect of homogenous catalyst, especially ionic liquid on the biomass conversion has been confirmed by many of open literature. Main characteristics of this catalytic technique are to have a conversion system with minimum energy to confirm the high yield of carbon materials. Compared with the homogeneous catalyst, heterogeneous catalysts have the advantages of high selectivity, recyclable and environmentally friendly. Thus, a heterogeneous catalyst with the wide range of solid acid, ion exchange resin, zeolite and metal oxide has become a research hotspot in this field. Both techniques have been explored extensively, with the need for exploring cheap and sustainable ways to obtain chemicals and carbon from raw materials other than crude oil or natural gas that could lead to a re-exploration of this field. In addition, the implementation of the low-cost pathway to recycle by-products of farmed biomass would additionally represent a way to sequester a significant amount of CO_2 creating a material benefits at the same time (Tessonnier et al. 2009).

This paper will not discuss on the preparation of the catalyst, but instead, focus solely on the use of a catalyst to bring about carbon based materials via a conversion process of biomass, which can be further used as a photocatalysis applications.

Several published papers and review articles have cited the theory and environmental applications of heterogeneous photocatalysis by the use of semiconductors (Chong et al. 2014; Zhang et al. 2014). One of the biggest disadvantages of using these semiconductors is the power requirement due to the use of UV light. However, improving the photocatalyst may increase the reaction rate and thus decrease residence time and reduce energy consumption per unit volume treated. It is generally accepted that this may achieve by deposition of the photocatalyst on a high surface area material (support) that will selectively adsorb the pollutant molecules and concentrate them around the photocatalyst. Unfortunately, strong adsorption of pollutant reduces the rate of diffusion into semiconductor powder, which will thus inhibit the efficiency of the photocatalyst. This limitation has directed many researchers to look for suitable supports for a semiconductor for the treatment of specific pollutants, or to develop new deposition techniques (Kumar et al. 2014). In addition, to the pH of the solution, the type of pollutant and the support play an important role in the success of this process.

By dint of its strong adsorption capacity and because it improves the transfer rate of the interfacial charge and reduces the recombination rate of holes and electrons AC reported as a potential support in photocatalytic systems. Several works have reported the use of activated carbon (AC) as a support for the semiconductor but warned that strong adsorption of pollutants to the adsorbent surface may inhibit diffusion to the catalyst and hence may cause severe problems with the whole process. The use of AC as a support for the titanium oxide (TiO_2) appears to give great advantages over the other supports. These include its ability to rapidly adsorb pollutants, as well as its high adsorption capacity due to its surface area and porosity. As suggested by Quinones et al. and Cherkezova-Zheleva et al., AC will be of the medium surface area and porosity to facilitate the diffusion of pollutants and release of products from the surface (Quinones et al. 2014; Saleh and Al-Saadi 2013). Additionally, the high porosity of AC may prevent the penetration of UV light into small pores, and may cause trapping of pollutants within the pores without being able to diffuse into the external surface for further reaction with the OH radical of pollutants (Saleh and Al-Saadi 2015). Furthermore, some type of pollutant, such as phenol, may undergo polymerization on the carbon surface of the AC, which causes irreversible adsorption (Alansi et al. 2015; Saleh 2014; Saleh et al. 2014).

2. ACTIVATED CARBON

2.1 General View

Carbon materials technology has made remarkable progress in recent years due to its diversity of physicochemical properties, such as lightweight, tunable porosity, exciting electronic properties, chemical and thermal stability, electrical conductivity and the ability to adopt a wide range of morphologies. Therefore, carbon materials

Table 1. Classification of activated carbon

Criteria	Powdered AC	Granule AC	Extruded AC	Beads AC	Impregnated AC
Particle size	< 1.0 mm Diameter: 0.15- 0.25 mm	0.42 to 0.84 mm	0.8 to 130 mm	0.35 to 0.8 mm	0.8 to 200 mm
Properties			Low-pressure drop, high mechanical strength & low dust content	Low-pressure drop, high mechanical strength & low dust content	Porous carbon impregnated with inorganic materials (Iodine, silver, cations) Antimicrobial Antiseptic
Applications	As additives in vessel, wastewater treatment, classifiers, and gravity filter	Vapor and liquid adsorption, water treatment, deodorization, separation of components	Gas phase applications		Pollutant control Purification of domestic water
References	(Shawabkeh et al. 2014)	(Labus et al. 2014)	(Xiang et al. 2013)	(Li et al. 2014)	(Kangwanwatana et al. 2013)

have found a large number of applications in different domains ranging from environmental science, adsorbent, drug delivery, catalyst, electrode materials, stationary phase in a chromatography system, energy storage and others according to its structure, morphological and chemical properties. Nevertheless, for some specific applications, functionalization is required with controllable size and shape. However, the production of such materials usually requires very harsh conditions and has several drawbacks such as the high temperature of the carbonization process in the first steps up to (> 800°C) and followed by physical or chemical activation to transform carbon materials into activated carbon (Saleh et al. 2011; Guptaet al. 2012; Saleh 2011). In addition, it is significantly important here of exploring cheap and sustainable ways to obtain carbon materials from raw materials other than crude oil or natural gas lead to a re-exploration of this field.

2.2 Biomass to Activated Carbon

The utilization of biomass or biomass derived products are becoming vitally crucial for the development of effective and environmentally benign technology and simultaneously solves the problem of agricultural and forestry waste use. Many synthetic methods, such as carbonization, high-voltage-arc electricity, laser ablation or hydrothermal carbonization have been reported for the preparation of amorphous, carbonaceous, porous or crystalline carbon materials with different size, shape and chemical compositions from the biomass feedstocks (Vukcevic et al. 2014).

In addition, the implementation of a low-cost pathway to recycle by-products of formed biomass would additionally represent a way to sequester a significant amount of CO_2 and at the same time, a material benefit would also be created.

Conversion of biomass is very important, however, working with complex biomass feedstocks is difficult and approaches based on the formation of simpler and more stable intermediate derivatives, denoted as platform molecules, have been shown to be effective for efficient biomass conversion to fuels and chemicals, catalytic routes involving deoxygenation and reaction combined with –C-C- coupling processes. The conversion of biomass to carbon materials via catalytic approach generally started with hydrolysis of cellulose chains, dehydration and fragmentation into soluble products of the monomers that come from the hydrolysis of cellulose, polymerization or condensation of the soluble products, aromatization of the polymers thus formed, appearance of short burst of nucleation and finally growth of the nuclei so formed by diffusion and linkage of species from the solution to the surface of the nuclei (Samus et al. 2013; Milbrandt et al. 2014) (Figure 1).

Figure 1. The route of AC derived from biomass

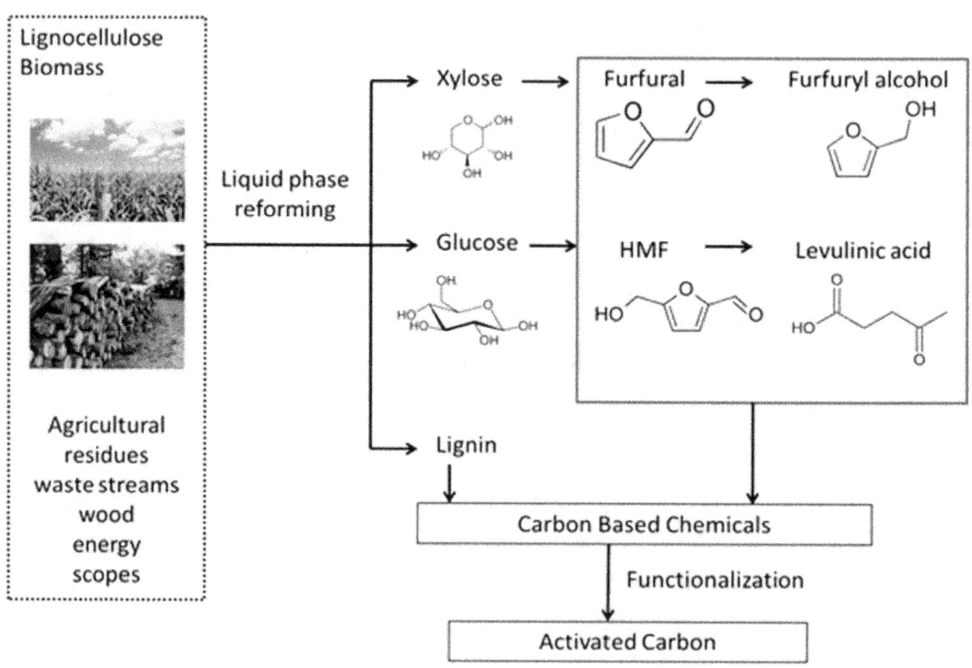

2.3 Conversion of Activated Carbon From Biomass via Catalytic Approach

Nowadays, green chemistry is getting more and more associated with a catalytic process of conversion of biomass to carbon materials.

2.3.1 Homogenous Catalyst

A conventional sulphuric acid solution catalyzed the hydrolysis of cellulose into glucose is a long-time established process. Many large scales units have been built, but the harsh conditions, the recycling and the treatment of the waste sulphuric acid solutions as well as the lack of glucose selectivity, suffer from the complicated separation of products from the solution, toxic and high costs make this process far from the actually sustainable approach.

Therefore, the ionic liquids as solvents of catalyst have received great attention due to its stability, low vapor pressure, and recyclability. Ionic liquids are kinds of the novel green solvents with a relatively low melting point, appears as crystal in normal conditions. Cellulose is stabilized by intra and intermolecular hydrogen bonds so as to form tough bundles, which make it almost impossible to solubilize with water and common organic solvents. It is essential to design solvents for cellulose and to propose a system with minimum energy to confirm the high yield of carbon materials in the biomass conversion. In general, ionic liquid has three important roles; it acts as both a soft template to generate the characterized pore structuring for the formation of a hierarchical (so) porous carbon structure, the solvent of the reaction media and as a catalyst resulting in enhanced isothermal carbon yields. Therefore, It has been demonstrated that the use of ionic liquids, which can dissolve cellulose, as reaction media, can enhance the efficiency of the hydrolysis of cellulose to glucose in the presence of an acid catalyst (Rezan et al. 2009).

Cellulose is dissolved in ionic liquid shows higher reactivity than in what as been demonstrated by Zhang et al. and Iguchi et al. that cellulose dissolved in 1-butyl-3-methylimidazolium chloride [BMIM]Cl (as a model of ionic liquid) was hydrolyzed by sulfuric acid at comparatively low temperature; 100° C (Iguchi et al., 2013). Fan et al. showed that a sulfonated polystyrene resin hydrolyzes cellulose in ionic [BMIM]Cl as ionic liquid and demonstrated that H^+ species was released by sulfonation process by ion-exchange with $[BMIM]^+$ to become a hydrolysis catalyst (Fan et al. 2013). Besides, pioneer studies by Ronald et al. and have investigated on [BMIM]Cl salts as a solvent for cellulose via a solubilizing mechanism to directly convert cellulose into carbons materials such as furans and compounds with longer carbon chains (Ronald et al. 2012). Therefore, the merit of ionic liquid as a catalyst has been mentioned in the dehydration of sugars in the presence of acidic resin and metal chloride. The studies on cellulose catalytic conversions in reaction media other than water are many. In contrast, to other crystalline salts (such as NaCl), the attractive characteristic of the ionic liquid is that it can transform into liquid phase (Gabriele et al. 2011).

In the same context, metal containing ionic liquid have also been employed as good catalyst and solvents for efficient conversion of sugar into carbon materials, with catalytic activity depends on the metal ion employed. The use of another metal containing ionic liquid or secondary additives also enables the development of other useful porous hybrid carbon materials (such as $Fe_3C@C$) that has a great potential use in a variety of applications of polluted water remediation (Hsu et al. 2011). In 2007, Joshua et al. discovered $CrCl_2$ in [EmiM]Cl (1-ethyl-3-methylimidazolium chloride) as an imidazolium type of ionic liquid can efficiently catalyze glucose to 5-Hydroxymethylfurfural (HMF) (Joshua et al. 2011). HMF is a promised platform chemical because it can further transform to a widely used biofuel called 2,5-dimethylfuran (DMF)and other useful materials. Since then, many efforts have worked on the production of HMF from cellulose or glucose in the ionic liquid systems to produced a series of carbon materials presenting high surface areas. Some works such have combined HCl, $CrCl_2$ or $CrCl_3$, DMA/LiCl and 1-ethyl-3-methylimidazolium chloride [EMIM]Cl to convert cellulose to HMF (Joshua et al. 2011). Other works did by Hussein et al. and Furong et al. used $CrCl_2$/$CuCl_2$ as a catalyst in [EMIM]Cl (Hussein et al. 2013) while others discovered IrCl/AuCl and SnCl4 in [EMIM}BF_4 can convert glucose to HMF with the high yield, respectively (Wei et al. 2011). Meanwhile, a novel catalytic system involving $CuCl_2$ (primary metal chloride) paired with a second metal chloride, such as $CrCl_2$, $PdCl_2$, $CrCl_3$ or $FeCl_3$ in [EMIM]Cl ionic liquid solvent has been found to substantially accelerate the rate of cellulose depolymerization under mild conditions (Haixin et al. 2012). Dutta et al. and co-workers reported that $LaCl_3$ could catalyze the degradation of cellulose to HMF in water at 250° C with an approximate yield of 19% (Dutta et al. 2013). Amarasekara & Owereh demonstrated concentrated hydrochloride promoted conversion of cellulose to HMF with up to 84% of yield (Amarasekara et al. 2009). Nawaz et al. showed the transformation of lignocellulosic biomass into furans in the DMA-LiCl system at 140° C with the yield of HMF up to 48% (Nawazet al. 2013). Zhang et al. studied the conversion of lignocellulosic biomass into furans in ionic liquid in the presence of $CrCl_3$ under microwave irradiation, the yield of HMF and furfural could be up to 45 to 52% and 23 to 31%, respectively (Zhang et al. 2009). Yang et al. demonstrated that the common Lewis acid $SnCl_4$ could efficiently convert glucose into HMF in 1-ethyl-3-methylimidazolium tetrafluoroborate [EMIM] BF_4], with the first reported metal chloride in ionic liquid solvents converts sugars to HMF with the yield near to 70% can be obtained (Yang at al., 2011).

However, the limitations of ionic liquid as a catalyst are, it can be easily molten and dissociated into two ionic part at relatively high temperature (> 100° C) which not limits its reaction temperature but also recycle-ability (Lincai et al. 2011). In the principle of green engineering, catalyst recycling is always important, thus study on heterogeneous catalysis is nowadays gain some interest to overcome the disadvantages of homogenous catalytic approaches.

2.3.2 Heterogenous Catalyst

The usage of heterogeneous catalysis with a wide range of design, less expensive, extraordinarily stable at high temperature and chemical inertness against solvents, reactants or by-products is believed to improve catalyst properties and process conditions to obtain high yields of hydrocarbons while minimizing coke formation in the wide range of reaction conditions (Girisuta et al. 2013; Saleh 2015a,b). This approach has the advantage of being very cheap and mild following some important rules of green chemistry since it does not involve organic solvent, with resulting carbon is spherically shaped and the surface is decorated with oxygenated functional groups. This approach also involved simple reaction mechanism for the formation of carbon involves; the dehydration of the carbohydrate into furan-like molecules (mainly 5-(hydroxymethyl)-2-furaldehyde) as the first step and subsequent polymerization and carbonization as a second step (Stephanie et al. 2012). This reaction possesses a great potential for the catalytic upgrading biomass since –C-C- coupling takes place with the simultaneous oxygen removal (the reaction involves the removal of CO_2 and water) from carboxylic acids, the latter of which are common intermediates in biomass conversion processes. In a comparison of homogeneous catalysis, carbon materials derived via heterogeneous catalysis process is composed of an aromatic core containing polyfurane-type units surrounded by oxygen-rich polar functional groups such as –OH, -C-, -O- and COOH makes the materials are more hydrophilic, hygroscopic and have a lower degree of graphitizations. These surface functional groups could act as depositary and premier binder to promote in situ reduce foreign materials (such as metal ions, noble metal nanoparticle etc.) loaded and stabilize on the carbon to form a novel carbon-encapsulated core-shell composites, nanocables and hybrids. Furthermore, heterogeneous catalysis are easy to recover and reused (Saleh et al. 2016; Al-Shalalfeh et al. 2016).

However, the effective degradation of cellulose is the main challenge in heterogeneous catalysis applications, because cellulose itself has a rigid, mainly crystalline, chemically-stable and water insoluble properties, which are induced by the inter-intramolecular hydrogen bonds. The most frequently used heterogeneous catalysis for the biomass conversion are alkali salts, metal (Pt, Ni, Ru, Rh, Pd, Ir, Ag), inorganic oxides (CeO_2, TiO_2, Al_2O_3 and ZrO_3) usually on supports, solid oxide catalysts at moderate temperatures (300 to 425° C) involving deoxygenation reaction combine with –C-C- coupling processes (Cai et al. 2012; Jimenez-Morales et al. 2011; Mario et al. 2013).

2.3.2.1 Solid Acid Catalyst

Conversion of cellulose to water soluble sugars by solid acid catalysts has also received some attention in recent years as solutions to eliminate mineral acids in the production of furfural, easily recovered from the reaction mixture and reused. Furthermore, solid acid catalyst has several advantages over liquid acid catalyst included; it facilitates the separation of product that simultaneously can be recycled without a loss of activity and minimize the costing on post-treatment such as neutralization and effluent processing, it can work at high temperature thus shortening the reaction time. The solid acid catalyst is normally a Lewis acid or Bronsted acid, with a variety of types such as inorganic oxides, ion-exchange resins, Vanadyl phosphate, $ZrO_{2,}$ and zeolites (Andrea et al. 2013).

2.3.2.1.1 Bronsted or Lewis Solid Acid Catalyst

Heterogeneous catalyst with Bronsted or Lewis acidity were synthesized, characterized and evaluated in the conversion of cellulose in hydrothermal conditions into valuable products. These series of catalyst included Zr-P, SiO_2-Al_2O_3, WOX/ ZrO_2, c-Al_2O_3 and HY Zeolite and normally used for aqueous phase dehydration of xylose (Dmitri et al. 2013; Baek et al. 2012; Rick et al. 2012). Some studies have characterized the concentration of both Bronsted and Lewis acid site of these series of the catalyst with Thermal Deposition (TPD) and Fourier Transform Infra Red (FTIR) spectroscopy using gas phase NH_3 and compared the catalytic activity and selectivity with that of homogeneous catalyst for the dehydration of aqueous solution of xylose. The catalyst selectivity is a function of the Bronsted of Lewis acid site ratio of the heterogeneous reaction. Lewis acid sites decrease furfural selectivity by catalyzing a side reaction. In the hot water (190° C), solid Lewis acids promote the cellulose depolymerization and permitted to reduce the formation of soluble oligomers and polymers as regards to the cellulose self-hydrolysis which occurred in the similar conditions (Mingxing et al. 2011). By contrast, in similar conditions, strong Bronsted solid acids did not improve the extent of the depolymerization of cellulose, but has any effect on the products distribution (Ronen et al. 2011). It brought to the meaning that, the usual product of soluble Bronsted acid catalyzed cellulose depolymerization such as glucose and HMF were not produced via the conversion process, but levulinic, formic and lactic acids were favored (Edward et al. 2013). By contrast, Lewis solid acids were capable of improving the extent of cellulose depolymerization but also to favor the direct formation of lactic acid in high yield, close to 30% (Juan et al. 2012). This represents a way of high potential to optimize the cellulose conversion and the selectivity and stability of the Bronsted acid catalyst and to completely understand the reaction pathways.

This also brought an idea for the group of researchers to study the effect of Bronsted acidities of water soluble hetero poly acid catalyst or known as HPA. This heterogeneous micellar HPA catalyst is clean, facile, economical and environmentally begin a process which could be recycled by centrifuge. HPA is known to exhibit activity for the hydrolysis of polysaccharides including disaccharides, starch and cellulose. Among these series of HPA, tungsto- phosphoric acids ($H_3PW_{12}O_{40}$) and tungsto- silicious ($H_4SiW_{12}O_{40}$) are the strongest acid employed as a catalyst in the reaction of hydrocarbons and has been used for example in acylation, alkylation, esterification, isomerization and dehydration of ethanol process (Duan et al. 2013). An acidic cesium salt of $H_3PW_{12}O_{40}$, $Cs_{2.5}H_{0.5}PW_{12}O_{40}$ ($Cs_{2.5}H_{0.5}PW$) with strong acidity is insoluble in water and organic solvents and has micro and mesopores with high surface area (130 m^2gk) (Rick et al. 2012). This $Cs_{2.5}H_{0.5}PW$ has been reported as a solid acid catalyst with desirable and environmentally benign (green) features and is useful in industrial processes such as hydration of ester or olefins and liquid phase dehydration of alcohol. By far, hydrolysis of cellulose into saccharides using a range of HPA catalyst had been investigated for potential applications in comparison to the conventional acid-catalyzed reaction, whereby the yield of glucose is low. It is due to the insolubility of cellulose in any of solvent and difficulty of solid to solid mass transport. In some cases, the acidity of HPA is higher than that of mineral acid. For example, the Hammet acidity function (H0) of $H_5BW_{12}O_{40}$ (0.7 M at 100° C) is -2.1 lower than those of H_2SO_4 and HCl (0.7 M). Hence, HPAs have been expected to be an active and reusable catalyst for the hydrolysis. With that theory, amorphous carbon bearing SO_3H, COOH and OH groups had been reported to exhibit hydrolysis activity (10% of glucose yield) at 100° C for 3 hours. Meanwhile, Yabushita et al. showed that the hydrolysis of cellobiose and cellulose in water promoted by HPAs such as $H_3PW_{12}O_{40}$, $H_4SiW_{12}O_{40}$ and $Sn_{0.75}PW_{12}O_{40}$ record on total yields of reducing sugar was up to 40% at 200° C for only 16 hours of reaction (Yabushita et al. 2014). Hara found that $H_5BW_{12}O_{40}$ showed a good performance for the conversion of crystalline cellulose to give glucose in 77% yield and that various type of HPAs such as $H_3PW_{12}O_{40}$ (glucose yield 8%) and $H_4SiW_{12}O_{40}$ (37%) was less active than $H_5BW_{12}O_{40}$ (Hara 2010). Other research groups also conducted the screening of counter cations of $PW_{12}O_{40}$ for the hydrolysis of cellobiose and found that there was a volcano-type correlation between TOFs for glucose formation and the Lewis acidities (Gurbuz et al. 2013). They also remarked that the acidity and the function of decreasing CrI were the important factors for the catalytic activity of HPAs. In fact, the H0 function was $H_3PW_{12}O_{40} < H_4SiW_{12}O_{40} < H_5BW_{12}O_{40}$ in the same order of catalytic activity. The H0 functions were correlated with the concentration of acids and the number of negative charge anions. Hence, HPAs containing highly negatively charged anions are preferable. The anions also dissociated hydrogen bonding between cellulose molecules to decrease the CrI. Furthermore, protons of

HPAs also weakened the hydrogen bonds of cellulose and a higher concentration of protons was effective for this purpose. Thus, $H_5BW_{12}O_{40}$ worked as a highly active catalyst for the hydrolysis of cellulose. $H_5BW_{12}O_{40}$ was recovered by the extraction and recycled for 10 times (Dora et al. 2012).

2.3.2.1.2 Ion Exchange Resins

Solid acid catalyst with sulfonic acid functionalities of ion exchange resin (such as Amberlyst 15) and sulfonic acid functionalized materials (such as MCM-41) resulted in yield as high as 63 and 76% in pure dimethyl sulfoxide (DMSO) solvent, respectively (Bai et al. 2013). Solid acids such as amorphous carbon materials bearing on SO_3H groups layered transition metal oxide ($HnbMoO_6$) and sulfated resin have been tested for the hydrolysis of cellulose, but glucose yield is still comparatively low. It is hypothesized that side reactions were suppressed using aprotic or organic solvents. Strong acidic resins (Amberlite A 120 and A 200), mixed oxides (silica-alumina and silica-zirconia), niobic acid, silica-niobia and niobium phosphate formed a strong protonic acidity and easy accessibility of saccharides to the active sites of the catalyst surface allow satisfactory reaction rates to be obtained (Carniti et al. 2005).

Sulfonic acidic resins are presented as efficient catalytic systems for the hydrolysis of disaccharides, starch, and cellulose. In general, severe conditions in term of temperature (> 120° C in water and also in supercritical condition) and high saccharide concentration (> 100 g) were used to push the catalytic activities of solid acid toward the achievement of high conversions. Sun & Zhang 2013, found that sulfonated activated-carbon could convert amorphous cellulose into glucose with 41% yield at 200° C for 24 hours. Heinze & Schaller 2000, demonstrated that p-toleunesulfonic acid can catalyze the hydrolysis of cellulose in ionic [EMIM][Cl] system with 28 and 13% yields of THF,10 and 3% yield of mono- and disaccharides, respectively. A sulfonated biomass char can convert natural bamboo, cotton, and stretch up to 20% yield of glucose under microwave assistance. This hydrolysis process (included cellulose and starch) was achieved by layered transition metal oxide, although the yield of glucose was low from glucose. Vyver groups employed sulfonated carbon with the mesoporous structure for hydrolysis of cellulose with 75% yield of glucose, which is the highest recorded yield by a solid acid catalyst (Vyver et al. 2010). In parallel, the use of solid supported acid catalysis for the depolymerization of cellulose in water has also demonstrated some of the interest.

Recent reports describe the hydrolysis of cellulose by solid catalysts such as Amberlyst resin acid modified amorphous carbon, layered transition metal oxides or sulfonated silica/carbon nanocomposites (Geboers et al. 2011). At the same time, the depolymerization of cellulose was also investigated under catalyzed of both Nafion and $FeCl_3$ supported on amorphous silica to be applied to a continuous flow reactor, providing that residual that unreacted cellulose can be easily removed from the system (Potvin et al. 20111). These surfaces acidic species were quite efficient in the conversion of glucose or other small biofuel precursors was limited. Indeed, it is well known that the rate of cellulose hydrolysis depends on the acid strength. As an expansion on previous reports use of Nafion as a supported acid catalyst for the conversion of cellulose to glucose and levulinic acid, some researchers have incorporated the reaction with alkali metal salts to enhance the reaction's yield. Namchot et al. 2014, and Klamrassamee et al. 2013, have recently developed a type of carbon-based solid acid with a high density of Bronsted acid sites (SO_3H and COOH) to carbonize sugar such as glucose, sucrose or cellulose and subsequently sulfonate the as-prepared carbons. It is interesting that these sulfonated carbon materials are very active for the hydrolysis of microcrystalline cellulose to produce water-soluble saccharides even at the low reaction temperature (100°C) with the conventional strong Bronsted acid catalyst such as niobic acid, H-mordenite, and others. The specific surface area of the sulfonated carbon was only around $2m^2g^{-1}$ but the yield of soluble saccharides reached nearly 70% (glucose = 4%; β-1,4-glucan = 64%).Its catalytic performance was attributed to its intrinsic ability to adsorb β-1,4-glucan, which is not adsorbed on the other solid acids. Therefore, it can be concluded that this heterogeneous catalysis are more energy efficient and environmentally-friendly, largely due to a facile product separation as well as catalyst recovery. Takagai et al. reported that acid resins with relatively large pores could effectively depolymerize cellulose in ionic liquid but the major products were cello oligomers, which failed to dissolve in water (Takagai et al. 2011).

2.3.2.1.3 Zeolite

Zeolite catalyst is the workhorse of the petroleum industry efficiently converting petroleum-based feedstocks into targeted fuels and chemicals. Zeolites constitute an important class of crystalline microporous solids due to its widespread application in adsorption, separation and catalysis. It important stems from its unique pore structures, which make them highly selective to adsorbed molecules for separation purposes, or towards product molecules in catalysis. Moreover, it exhibits good thermal and hydrothermal stabilities during heterogeneous catalytic reactions. However, over the last twenty years, there have been dozens of studies that focused on the catalytic conversion of biomass and its derived feedstocks with a range of zeolite catalysts, including ZSM-5, Beta, Y, mordenite, Ferries, FCC, Al-MCM-41, SBA-15, Al-MSU-F FER, MFI, ZSM 23, MCM 22, SSZ 20, ZSM 5, IM 5, TNU 9 and MOR (Guisnet 2013; Stocker 2005). Amongst these series, ZSM 5 and ZSM 11 had the highest aromatic yield and the least amount of coke due to its pore space and steric hindrance. Production of ordered porous carbon material has been previously achieved via nano casting, replication of ordered inorganic of zeolite. Several reactions occur inside the zeolite including dehydration, decarboxylation, isomerization and decarbonylation thereby removing oxygen as water, carbon monoxide, and carbon dioxide and converting the carbon and hydrogen into olefins and aromatics.

In these zeolite catalysts, ZSM-5 has shown the highest aromatic and olefin yields from lignocellulosic biomass. ZSM-5 has a three-dimensional pore system with a pore size of 5.5 to 5.6 A. This small pore size, internal structure, and internal volume make it difficult for larger aromatic coke precursors to form inside the pores. Zapata et al. 2013, have shown that tin-containing zeolite is a highly active catalyst for the isomerization of glucose in water. However, other studies reported that in aqueous environments and at low conversions, mordenite, and faujasite both resulted in high furfural selectivities (80 to 90%) at 200° C with the final yield were lower as the selectivity decreases with increasing conversion (Otomo et al. 2014).

2.3.2.1.4 Metal Ions

HMF and furfural, two of the most important intermediates derived from biomass, were directly produced from the hydrolysis of microcrystalline cellulose with metal ions in ionic liquids such as Cr^{3+}, Mn^{2+}, Fe^{3+}, Fe^{2+}, Co^{2+}as catalyst under mild conditions. Metal ions an acidic supports are a good catalyst for ring opening and hydrogenation of cellulose polymers.

There are reports on carbon produced via hydrolysis of cellulose at moderate temperature (250 to 300°C) metal catalyst in a very short reaction time to prevent the deep degradation of the formed glucose. Gu et al. 2013, and Dreher et al. 2013, used ruthenium (Ru) catalyst to promote the conversion of oligosaccharides and increased yield of glucose (ca, 30%) with the turnover number (TON) of the catalyst was very high (145 based on bulk Ru) by comparison to those of the sulfonated carbon catalysts. Wang et al. 2012, found that the yield of glucose was raised up to 31% by increasing the Ru loading to 10 wt% with the reusable number of catalyst up to five times without loss of the activity or Ru leaching. Other studies reports on the potential of the cellulose hydrolysate generated by Ru catalyst as a carbon sources for bacterial PHA production and found that, the reaction occurs in the aqueous solution that is preferable for feeding to microbes which consequently makes it easily separated from yielded sugar by filtration and facilitating the recycle of the catalyst (Serrano-Ruiz et al. 2013). Those reports confirmed that Ru species was actually in an oxidized state, which acted as the real active site for the hydrolysis of oligosaccharides. In advance, Ru/Co_3O_4 catalyst showed high activity and selectivity as well as good recyclability in the conversion of biomass to carbon (Zu et al. 2014).

Skoulou et al. 2012, and Guan et al. 2012, reported that iron ions and iron oxide nanoparticles could effectively catalyze the hydrothermal carbonization of starch and rice grains under mild conditions (< 200°C) and had a significant influence on the formation of carbon nanomaterials with different shapes. Magnetite catalyst with a sulfonic group such as a composite of mesoporous silica SBA 15 and Fe_3O_4 treated with sulfuric acid and sulfonated $CoFe_2O_4$ embedded silica were also used for the hydrolysis of cellulose. It was easily separable by a magnet after the catalytic reactions. Fe_3O_4-SBA-SO_3H catalyst afforded glucose in 26% yield from microcrystalline cellulose at 200°C for 3 hours, whereas levulinic acid become a major product (42% yield) by prolonging the reaction time up to 12 hours (Aravid and Jong 2012).

2.3.2.1.5 Metal Oxide

Metal oxides play important roles in many catalytic processes. For example, nano metal oxide is a major component of commercially available for methanol synthesis. The extra advantages of this catalyst which make them good candidates for green processes are availability, non-volatility, stability, non-toxicity, reusability and low cost. Studies have examined the decomposition processes associated with methanol, formaldehyde, and formic acid on nano metal oxide surface (Pino et al. 2014). Regarding increased relative surface area of the nanomaterials, it is highly expected that nano metal oxide catalyst would have a better catalytic activity in enhancing the conversion of cellulose to value added products in hydrothermal media with a better separation from the reaction matrix. Layered transition metal oxides containing niobium demonstrated to be particularly active in the disaccharide hydrolysis, suggesting the important to investigate niobium-containing catalyst as an energy inefficient for the biomass conversion.

The use of transition metal oxide such as $HnbMoO_6$was also reported as an efficient solid catalyst to produce glucose from cellulose. In parallel, in the presence of a noble metal-containing a catalyst such as Pt/Al_2O_3, Felice et al. have observed the formation of sugar alcohols in 31 mol% yields from cellulose (sorbitol 25% and mannitol 6%) in hydrothermal conditions. In the other hands, the authors reported that Pt-free catalysts produced only a poor glucose extols amount (Felica et al. 2011).

3. AC IN PHOTOCATALYST SYSTEM

Activated carbon (AC) is a porous, amorphous solid carbon material which derived mainly from the carbonaceous material of plant-based (lignocellulosic) materials. AC adsorption is the most common technique for controlling various environmental pollutants due to its extended surface area, high pore volume and well developed porous structure. Recent interest is the use of AC as a photocatalyst because of its surface area and well-developed porosity. The heterogeneous photocatalytic technology has attracted extensive interest in the past decades due to its excellent properties, such as non-toxicity, chemical stability, and low cost. The advantages of heterogeneous AC photocatalysis methods are examined via its ability to destroy a wide variety of organic compounds, no post-treatment is needed since the organic pollutants can be mineralized into non-toxic by-products such as H_2O, CO_2, and mineral acids, the process can perform at low temperature by promoting of solar radiation, thus resultant in low energy cost. The advantages of AC photocatalyst is also brought from the on-site regeneration of spent adsorbent and destruction of adsorbed organic material is provided which in turn the loss of adsorbents due to attrition and burn off which occurs in thermal regeneration is less. Rather than that, carbons elements in AC compounds are strong light absorbing materials, despite which they have been successfully used as a support of photoactive species. Band gap determination of AC revealed and behave as semiconductor materials and therefore, as photoactive materials in the presence of UV radiation (band gap less than 4 eV). In this regards, there have been recently reported the anomalous response of an AC towards phenol degradation in the aqueous medium by (Velo-Gala et al. 2013). Direct UV irradiation of the solution in the presence of the AC and the absence of any other photoactive species showed that beyond phenol degradation by photolysis, the AC improved the photo-oxidation rate of phenol, compared to bare or immobilized TiO_2. The results suggested the occurrence of interactions between UV light and AC particles. To discriminate whether this behavior is a singular characteristic of that particular AC or other rather an intrinsic property of certain carbon materials, Han & Li 2013, and Jie et al. 2013, have explored the behavior of several porous AC obtained from different precursors and exhibiting varied porosity and composition as exposed to UV irradiation. The aim of these series of works is to unambiguously demonstrate that under UV irradiation, AC is capable of promoting the degradation of organic pollutants in an aqueous medium with the absence of semiconductor. Velo-Gala et al. 2013, have studied on effect on the degradation of sodium diatrizoate using a different type of commercial AC and sixteen gamma radiation-modified AC derived from these. The presence of AC during diatri-zoate degradation markedly accelerates its rate of removal, regardless of the AC used. It is highlighted that the gamma radiation-modified AC shows a higher percentage of carbon atoms with sp2 hybridization, explaining its superior behavior in the photocatalysis process. Meantime, commercial carbon exerts the highest synergistic effect in diatrizoate removal by the UV/AC system, with a synergic contribution is more than 53% after the first minute of treatment. Thus, it can conclude that the synergistic activity of the AC sample is enhanced by gamma radiation treatment with the introduction of a higher percentage of surface oxygen. This makes the activity of the reutilized activated carbons similar to that of the original carbon, the presence of dissolved oxygen enhances the rate of diatrizoate removal by UV/AC system and the UV radiation treatment produces light chemical modifications in the AC. The textural properties of the AC show no clear relationship with their synergistic contribution. There is some other research observed that the gamma radiation treatment reduces the band gap values of the AC which corresponding to higher removal of organic pollutants. Based on this observation, an active mechanism for the photocatalytic removal of diatrizoate in the presence of AC is proposed, in which the AC acts as a photocatalyst, promoting electrons from the valence band to the conduction band and increasing the generating of HO free radical in the pollutant medium.

3.1 Effect Surface Properties of AC Photocatalyst System

The surface properties of AC play an important role in the adsorption of pollutant species. For commercial Acs, their surface properties and textures may differ greatly, depending on the raw material used, carbonization process and activation conditions which may potentially result in distinct photocatalytic performance. The role of surface chemistry and textural features of AC have been reported by many authors. It showed that basic AC with low oxygen content can perform high phenol adsorption. The absence of acidic groups on the carbon surface increases the physisorption and surface polymerization of phenols. Acidic surface groups hinder the ability of AC to adsorb phenolic compounds. The presence of oxygen-containing basic groups on the surface of AC is the key factor in promoting oxidative coupling of phenolic compounds and thus causes irreversible adsorption. Additionally, the graphene layers of an AC increase the π electron density and thus enhance dispersive interactions between π-π electrons in AC and phenols. Thus, regardless of the textural properties, the adsorption capacity of phenol is expected to be higher in carbon of a basic nature. Even there is no clear relationship between carbon basicity and phenol retention, retention of phenol is expected to be less in acidic carbon than in basic carbons.

Yet, the use of AC systems alone has a limitation in that by adsorption, the pollutants are not eliminated, but only transferred to another phase, and the consumed AC becomes a hazardous waste. As such, the regeneration of AC is subsequently required for reusing and lowering disposal costs. Another limitation of AC uses is that the adsorption capacity is a function of inert concentration and is, therefore, relatively low under industrial quality. Meanwhile, the spent AC has to dispose of as hazardous waste or regenerated for reuse. Thermal regeneration is the widely adopted spent AC regeneration technique, but it has several disadvantages because it is, requires off-site regeneration, merely transfers the pollutant from one phase to another phase, results in appreciable carbon lost and may affect the microporous structure of AC. Although chemical regeneration of spent AC could yield high regeneration efficiencies, this technique has associated drawbacks, including non-destructive pollutant removal, significant chemical consumption and generation of secondary waste streams. Other regeneration techniques have been proposed as potential alternatives for regeneration of spent AC such as the ultrasonic regeneration, electrochemical regeneration and microwave regeneration. Through proven effective in bench-scale studies, they generally have a large carbon footprint, consume chemicals or require expensive facilities, and thus, lave limited appeal for industry application. In view of various limitations associated with the current regeneration techniques for the spent AC, it is compelling to develop a sustainable regeneration technique which presents more environmental and economic merit.

Therefore, there is some interest to do the combination the AC and semiconductor to become a single hybrid system. It is expected that the semiconductor elements with a great advanced oxidation process induce some of the destruction hazardous and refractory organic pollutants. This is due to the higher oxidation potential of active oxygen species such as OH and O free radicals, which are generated from irradiated semiconductor catalyst particles. This advanced oxidation process based catalytic process using Fenton reaction and heterogeneous photocatalysis have been studied as regeneration techniques for spent AC. Most of the previous studies on this regeneration technique of the spent AC were about granular AC. For example, Gulyas et al. and Sobana et al. demonstrated that the granular AC adsorption photocatalyst hybrid system resulted in higher removal efficiency of herbicide from water as compared with the photocatalysis system alone. In addition,Yoon et al. 2012, have reported that a rotating adsorbent combines with photocatalysts exhibit higher removal efficiency for formaldehyde as compared with adsorption alone.

3.2 Effect on Hybrid of Transition Metal Oxides: AC in Photocatalysis System

A new generation of a photocatalyst in the near future is expected to possess high internal efficiency related to separation, and set in contact with external pollutant molecules. Therefore, hybrid or composites photocatalyst were designed by immobilization of transition metal oxides on the support materials with large surface areas in order to condense diluted pollutants, to separate easily the catalyst from aqueous solution and to enhance the photocatalytic efficiency. Heterogeneous photocatalysis by use semiconductor metal oxide materials has been applied as an efficient method in the field of environmental purification.

Metal oxides can be impregnated onto the carbon surface by adsorption and impregnation through various methods, nevertheless, they are employed for high-temperature processes or complex experimental procedures. An ideal photocatalyst should be stable, inexpensive, non-toxic and highly proactive. Other primary criteria for the degradation of organic compounds are that the redox potential lies within the bandgap of the semiconductor. Several semiconductors have bandgap energies sufficient for catalyzing a wide range of chemical reactions. These include TiO_2, WO_3, $SrTiO_3$, Fe_2O_3, ZnO and ZnS. Binary metal sulfide semiconductors such as CdS, CdSe or PbS are regarded as insufficiently stable for catalysis, at least in aqueous media as they readily undergo photoanodic corrosion and also toxic. The iron oxide (Fe_3O_4) are not suitable semiconductor as they readily undergo photocathodic corrosion. The band gap for ZnO (3.2 eV) is equal to that of anatase, however, is also unstable in water with $Zn(OH)_2$ being formed on the particle surface. This results in catalyst deactivation.

3.2.1 TiO_2: AC Photocatalyst System

3.2.1.1 TiO_2 Photocatalyst System

Titanium oxide (TiO_2) is the semiconductor most thoroughly investigated in the literature, seems to be the most promising for the photocatalytic destruction of organic pollutants. This semiconductor provides the best compromise between catalytic performance and stability in aqueous media. It is considered usable in environmental clean-up operations and has been of great interest due to its non-toxic nature, photochemical stability, low cost, biocompatibility, and photochemical stability, particularly when the sunlight is used as the source of irradiation. Irradiating UV light with a wavelength around 380 nm on TiO_2 surface creates positive holes and electron pairs that act as powerful oxidizing and reducing agents, respectively. These charge carriers would then lead to the generation of highly reactive oxidants, such as OH radicals, superoxide anions, hydrogen peroxide and other reactive oxygen species and reductive reactions it further involves in the decompositions of organic compounds adsorbed on the TiO_2 surface. The anatase phase of TiO_2 is the material with highest photocatalytic detoxification. In-depth studies show that TiO_2 possesses photodegradation of toxic components (phenols, herbicides, pesticides, dyes, and surfactants) and organic components (sulfosalicylic acid and salicylic acid) that present in wastewater.

However, from an engineering point of view, TiO_2 powders lead to some drawbacks in their use regarding the need for a separation step of the photocatalysis, with the dual purpose of its recovery and to prevent its emission to the atmosphere due to the small particle size, the loss of the photocatalyst if the separation is not good, and the need of powder fluidization in gas phase with the energy cost and the difficulty of scaling involved. Furthermore, since sunlight radiation consists of about 5% of UV light, 47% visible light, and 48% infrared radiation, thus, TiO_2 as the benchmark of UV photocatalysis is inactive under visible light due to its wide band gap. Moreover, the photogenerated electrons and holes in the excited states playing a very important role in pollutant degradation are unstable and can easily recombine, which results in low efficiency of photocatalysis activities. With that, TiO_2 cannot make use of the vast potential of solar photocatalysis.

3.2.1.2 TiO_2 Supported AC photocatalyst system

To achieve rapid and efficient decomposition of organic pollutants and easy manipulation in a total photocatalytic process, it may be effective to load photocatalysis onto suitably fine adsorbents to concentrate the pollutants around the photocatalysis system. Hence, researchers have attempted to support TiO_2 on different types of matrixes such as silica gel, alumina, clay, zeolites and carbon materials such as carbon nanotubes, carbon microspheres, carbon black, carbon nanofibers and AC. Organic pollutants are hydrophobic whose TiO_2 particle when exposed to UV radiation are hydrophilic. Using AC as a support will help to bring the pollutant molecules near to TiO_2 active site for an efficient and effective photodegradation process. AC can generate new adsorption centers to favor approaching organic pollutant which secondary degradation of intermediates to take place in situ. AC is widely used as a support in gas and water remediation because of its good adsorption, high porosity, low cost and supported TiO_2 exhibits a synergism that has marked effects on the kinetics of disappearance of pollutants, each pollutant being more rapidly photodegraded. For example, although the TiO_2 membrane immobilized on the glass surface has the advantages such as high photodegradation efficiency, its major shortcoming is that the adhesion force between the TiO_2 membrane and the glass is very weak, so the TiO_2 is easy to fall off, which causes the declining of the photodegradation efficiency.

A hybrid of TiO_2 with AC support as sensitizers which capable of absorbing visible light has been proven to be a feasible approach to developed visible light responding photocatalyst with high activity. The formation of a heterojunction between TiO_2 and the narrow band gap of AC with more negative conducting band (CB) level can result in the injection of CB electrons from AC to TiO_2, which is helpful for electron-hole separation (Felica et al. 2012). In the meantime, the immobilize of TiO_2 onto the AC support can compensate for the loss of photocatalytic ability of TiO_2 due to the difficulty to effectively disperse in water to full contact with target pollutants. Therefore, many studies have reported that there is an optimum in the TiO_2 loading amount and AC pore structure for achieving a higher photocatalytic activity than that of TiO_2 itself (Lee et al. 2012). Spatial relationships among particles of AC, small particles of TiO_2 photocatalyst and molecules of an organic pollutant are shown schematically in the dark and under UV irradiation. The organic pollutant molecules are assumed to be small enough to be adsorbed into the micropores. In most AC, a large number of micropores exist on the surface of larger pores, mesopores, and macropores, in which a large number of pollutant molecules are accommodated by physical adsorption. On the other hand, only a small number of pollutant molecules are adsorbed on the surface of TiO_2. By depositing TiO_2 particles onto AC, some micropores and mesopores become closed at the entrances and this causes a marked decrease in the apparent surface area. By UV irradiation onto such TiO_2 loaded AC particles, oxidative OH free radicals are formed on the TiO_2 and can degrade pollutant molecules by oxidative. However, it has to be pointed out that these radicals are formed only on the surface of TiO_2 particles that is accessible to UV rays and do not on the surface in the shadow radicals cannot migrate long distances and are confined to a region near active centers in TiO_2 particles. In order for adsorbed pollutant molecules to be photo-catalytically degraded, they have to diffuse along the surface of AC and TiO_2 particles with the interface is not necessarily exposed to UV rays. The driving force for such surface diffusion is mainly a concentration gradient between organic pollutant molecules on the irradiated TiO_2 surface and those on the surface of variable-sized pores in the AC particles. The molecules adsorbed deep inside micropores of AC diffuse with greater difficulty toward TiO_2 particles located on the external surface of the AC particles. Therefore, highly microporous AC are not necessarily advantageous for the TiO_2: AC to have superior photocatalytic performance. The effect of substrate pore structure has been studied using for AC surface area (770 to 1150 m^2g^{-1}) and a dip-hydrothermal method of photocatalyst preparation. Better photocatalytic degradation of methyl orange (MO) has been achieved with TiO_2: AC than with a simple mixture of TiO_2 and AC.

This brought to the design of highly efficient TiO_2:AC hybrid heterojunction photocatalyst, besides the demand of commonsensible matching ability of band potentials between the hybrid components, the smooth and spatially available transfer of charge (electron and hole) at the interface and the high electron and hole mobility of the hybrid system are significant to increase the photocatalytic activity. It has been reported by He & Xia 2012, and Yang et al. 2012, that the surface chemistry and texture of AC exhibited a significant influence on the aggregation of TiO_2 particles and photocatalytic degradation of 4-chlorophenol. In addition, some authors have fabricated TiO_2 with AC microspheres to both maintain dispersion and acceleration separation, because AC microsphere can be suspended by air bubbling and can rapidly settle on the reactor bottom with the aid of gravity after such air bubbling. To overcome this shortcoming, the usage of co-binder promoting the development of an alternative method to obtain a new form of the TiO_2: AC photocatalyst with high activity and better separation performance.

Hsi & Tsai 2012, found that a TiO_2 supported with AC sample heated at 500°C, which mainly consisted of rutile phase, showed the highest photoactivity for degradation and removal of methyl orange from water. Thus, most samples which prepared at high temperature has a very detrimental to photoactivity. Lin et al. 2013, and Kuo et al. 2014, employing different types of AC investigated the associative effects between TiO_2 and AC during the 4-cholorophenol photodegradation and found a clear enhancement of photoactivity due to an increase in electron density of the AC support. It has been stated that the presence of AC in contact with TiO_2 helps because of its strong adsorption capacity and because it improves the transfer rate of the interfacial change and reduce the recombination rate of holes and electrons. This synergistic effect of the combination of AC and TiO_2 has been previously been reported for degradation of some organic compounds in the photocatalytic process. It has been ascribed to a common contact interface between the different solid phases, in which AC acts as an efficient adsorption trap for the organic pollutant. The organic is then more efficiently transferred to the TiO_2 surface, where it is immediately photo-catalytically degraded by a mass transfer to the photoactivated TiO_2.

Therefore, the decomposition rate observed on TiO_2: AC seems to be governed both by the surface diffusion pollutant molecules and by the photocatalytic reaction rate. As the adsorption occurs slowly, the change in relative pollutant concentration with irradiation time depends on both adsorption and photodecomposition, particularly at the onset of UV irradiation. Jung et al. 2013, reported that the adsorption of Ph on the native AC occurs mainly during the first 1 hour and achieves saturation after 3 hours under UV irradiation. For the hybrid of TiO_2: AC, however, adsorption and photodecomposition of Ph occur simultaneously, but the former is reasonably supposed to be the dominant process in the beginning and the latter in the later stage. The observed trend was approximated to consist of two linear processes, the change from one to the other occurring at around 1 hour of irradiation.

3.2.1.2.1 Synthesis of TiO_2: AC Photocatalyst

Various methods are being used for the preparation of TiO_2: AC catalyst, such as chemical vapor deposition (CVD), precipitation, aerosol pyrolysis, hydrothermal, dip-coating, hydrolysis and sol-gel. However, the process of selecting a suitable impregnation method depends on the support used and the pollutant that is to be degraded. It is known that the structure of supported catalyst has a strong influence on the physicochemical properties of TiO_2: AC catalyst and that it depends on the preparation method used (for example calcination treatment). The advantages of physical methods include low-cost, simplicity and the use of commercially available photocatalyst with desired functionalities. For some TiO_2: AC hybrid prepared by common wet chemical methods, the degree of lattice mismatch between these two hybrid components decreases the efficiency of the transfer and separation of photogenerated carriers (electron and hole). Adhesion of TiO_2 particles to AC surface appears to be important for the enrichment of photocatalytic activity and also for the practical application of this hybrid system. To enhance anchorage of TiO_2 on AC, wet method of synthesis is warranted because it will result in physical stable TiO_2:AC hybrid against hydrodynamic shearing method, the surface chemistry and pore structures of AC may have more significant impact on the dispersion of TiO_2 on the support during synthesis, resulting in distinct photocatalytic performance of TiO_2: AC. Nevertheless, TiO_2 photocatalyst usually has low specific surface areas because of the crystalline grain growth that occurs during such heat treatment. To prevent TiO_2 sintering, or to improve its specific surface area, AC could be considered as a good support for photocatalysis materials. Gondal et al. investigated TiO_2: AC hybrid prepared by modified sol-gel method and calcinated at different temperatures. They found that the TiO_2 particle calcinated at 450°C could agglomerate and enter into large pores of AC substrate, including a strong interaction between TiO_2 and carbon matrix. The interaction led to an evident synergy to enhance the photocatalytic ability for the degradation of chromotrope 2R. Wang et al. 2014, prepared nanocrystal anatase TiO_2 particles deposited on AC at low temperature by hydrolysis of titanium butoxide in acidic aqueous solution. They found that phenol pollutants were adsorbed by AC, then migrated continuously onto the surface of TiO_2, which consequently accelerate the photocatalytic oxidation. On the other hands, for the CVD method using tetrabutyl titanate, nano-sized TiO_2 particles were shown to adhere to AC and to do high activity for photodecomposition of methylene orange in the water. The addition of H_2O to titanium tetra isopropoxide vapor was also reported to facilitate the CVD process at a lower temperature and higher deposition rate. The introduction of water vapor during the CVD process and adsorption onto AC in advance were reported to be crucial to obtain anatase-type TiO_2 on the surface of AC particles. Experiments have also shown that the HNO3 treatment results in more efficient TiO_2 loading by CVD, in comparison with other oxidation treatment. TiO_2: AC have also been prepared by dipping the support into sol prepared by alkoxide hydrolysis, and then subjected to heat treatment at 300 to 500°C. In an alternative method, it can be prepared by adding $TiCl_4$ dropwise into an aqueous suspension of AC, followed by heat treatment at 500 °C in an N_2 atmosphere (Tryba et al. 2003). Loading of TiO_2 powder directly onto AC has also been carried out by mixing TiO_2 in an AC aqueous suspension with stirring. Besides, loading onto an AC filter, prepared by gluing granular AC onto the surface of glass cloth, was performed by using a water suspension of 5 mass % TiO_2, and the resultant TiO_2: AC particles in a CCl_4 solution of pitch, followed by heat treatment at 750°C, has also been reported. In the resultant hybrid system, TiO_2 particles on the AC were presumably coated by carbon, formed from the pitch during heat treatment, which might function to fix TiO_2 particles onto the AC surfaces. TiO_2 particle was also able to be loaded onto an AC by a spray-desiccation method, with only a small change in the pore structure of the AC. In another study, loading of TiO_2 onto the surface of AC was carried out by dipping AC particles in a per oxo titanate solution and then heating at 180°C in a Teflon-lined stainless steel vessel, followed by calcination at 300 to 800°C (Gao and Liu H. (2005). By using of AC particles of 0.15 to 0.25 mm, separation of the particles from solution was much easier, and photocatalytic activity for the decomposition of MO remained almost the same even after five cycles (Slimen et al. 2011). Loading of TiO_2 has also been carried out by plugging the pores of AC with paraffin. By removing the paraffin at 250°C in the air after TiO_2 loading, the high surface area of the pristine AC could be recovered and high photocatalytic activity was obtained for the decomposition of methylene blue (MB). TiO_2: AC has been produced also by mixing TiO_2 particles with a liquid or solid state carbon precursor. By hydrolysis of tetraisopropyl orthotitanate, TiO_2 was precipitated on the surface of poly(vinyl butyral) (PVB) and the TiO_2 loaded PVB was carbonized at a high temperature in the flow of CO_2. TiO_2-loaded carbon microspheres with 25 μm diameter have been prepared from the TiO_2-loaded cellulose microspheres, formed by one-step phase separation using cellulose xantate and sodium polyacrylate aqueous solutions with dispersed TiO_2 powder.

3.2.1.2.2 Performance of TiO_2: AC Photocatalyst

Although it is hard to understand how light could penetrate the carbon particles to reach the immobilized photocatalyst, TiO_2: AC composites have shown quite high efficiencies for the photodegradation of a variety of pollutants. For instance, the presence of AC seems to change the photocatalytic activity of TiO_2 towards the degradation of organic pollutants beyond the so-called "synergistic" effect. The synergistic effect of adsorption by AC and photocatalytic decomposition by TiO_2 has been observed in the degradation of several types of organic pollutants. This brought a basic principle of photocatalysis over illuminated TiO_2: AC system. However, the model of a TiO_2: AC photocatalytic process can be quite a complicated matter as it introduces light intensity to the classical aspects of the heterogeneous catalytic system. For example, temporal changes in the concentration of iminoctadine triacetate (IT) which is often used in rice and orchard fields as an insecticide and appears in water paths in the field, are shown for pristine AC and three TiO2: AC (Masahiro et al. 2006). The hybrid system was kept in the dark for 200 hours to saturate IT adsorption and then exposed to UV irradiation. After about 800 hours, the sample was removed from the test solution and dispersed again into a virgin 1.87 X 10^{-4} mol/L solution, and again kept in the dark for about 200 hours, then irradiated by UV. For the pristine AC, the concentration of pollutants was found to be almost constant in the dark and to scatter somewhat under UV irradiation. This brought to the conclusion that, the photocatalyst system of TiO_2: AC could have the similar photocatalytic performance in dark and under UV light. The above mentioned data was supported by Huang et al. who reports on the increment of photocatalytic activity for the decomposition of pollutants via TiO_2: AC catalyst 9 Huang et al. 2011). It is reported that the concentration of 4-chlorophenol solution decreases faster for the TiO_2: AC than TiO_2, revealing that the accelerates photodecomposition can be approximated by a pseudo-first-order equation with a linear relationship between concentration change and time. A similar synergistic effect was observed in the decomposition of Ph and 2,4-dichlorophenoxyacetic acid using the similar AC and TiO_2 mixture. In the other studies, the effect on Ph decomposition was found to depend on the mass ratio of TiO_2 to AC in the range of 5/10 to 75/10. The synergistic effect was thought to be attributable to the rapid transfer of Ph molecules initially adsorbed on the AC to the surface of TiO_2; the driving force for this transfer is presumably the difference in the surface concentration of Ph between AC and TiO_2. A certain amount of Ph was found to remain in the AC, even after the concentration in solution become negligibly small.

3.2.2 ZnO:AC Photocatalyst System

Rather than TiO_2, ZnO supported AC finds wide application. This is due to the some outstanding properties of ZnO such as wide availability, stability, low cost and favorable band gap energy. Problems with the use of ZnO alone as photocatalyst are also well organized; specifically, the difficulty in separating the powder from the solution after reaction is complete, aggregation of particles in suspension especially at high loading and difficulty in application to continuous flow system. For this problem, AC has been determined as a good support for the ZnO photocatalysis system. Owing to the use of dioxygen, light and completely mineralized organic and inorganic substrates, especially bio-recalcitrant, this method is considered to be environmentally friendly for pollution treatment. Spherical AC particles containing ZnS and ZnO were prepared from a cation-exchange resin polystyrene with sulfonate groups and crosslinked by divinylbenzene) and $ZnCl_2$ aqueous solution, followed by carbonization at 500 to 900°C (Cristina et al. 2000). It has reported that ZnO is a suitable alternative to TiO_2 for the photodegradation of Acid Red 14, an azo dye since its photodegradation mechanism has been proven to be similar to that of TiO_2. Recent reports also indicate that simultaneous removal of inorganic pollutants such as Cr(VI) and organic pollutants, such as 4-CP can be realized in ZnO: AC photocatalytic reaction system (Matos et al. 2010). Thus, it is highly desirable that this method can be applied on a large scale for water and waste water treatment.

4.0 OTHERS AC PHOTOCATALYTIC SYSTEM

4.1 Granular and Spherical AC Photocatalytic System

Among all types of AC, the spherical ACs have interesting advantages due to their smooth surface, good fluidity and high mechanical strength over the powdered and the granular AC. For this reason, some approaches have recently been investigated to support TiO2 on spherical AC. However, there are some reports on the granular AC supported TiO_2 photocatalyst that could enhance the removal efficiency of several organic compounds, in regards to various environmental conditions (Yao et al. 2010). Granular AC supports the TiO_2 powder by concentrating the pollutants and intermediates around the TiO_2. These pollutants and intermediate can migrate to the surface of TiO_2 via diffusion. Granular AC also reduces TiO_2 coagulation, which reduces its surface, thereby reducing its catalytic effectiveness. This was brought by the theory that, the absorptivity of AC depends strongly on the molecular morphology and size of the pollutant molecules to the TiO_2 surface occurs directly from solution, and not from the surface of the AC itself. In addition, the TiO_2 can destroy the pollutants, leading to the regeneration of granular AC in situ. However, most of the porous AC still in granules, and the problem of separation and recovery of the photocatalyst from the reaction media still exists.

4.2 ACF Photocatalytic System

AC newly developed the type of photocatalyst support material consisting nano graphite known as activated carbon fiber ACF. Compared to granular AC, ACF has a larger specific surface area, having larger pore volume, more uniform micropores size distribution, a superior rate of adsorption and desorption and a faster attainment of adsorption equilibrium with ACF in the form of flat are preferable in handling than granular support. The surface properties of ACF are known to depend strongly on the preparation process, affect the loading of TiO_2 and consequently the adsorption of pollutant molecules. Nevertheless, the ACF supported TiO_2 photocatalyst has rarely been used for the removal of gas phase environmental pollutants.

4.3 TiO_2:Graphene Photocatalyst System

Meanwhile, the introduction of graphene provides a feasible approach to solve the limitation brought by the TiO_2: AC photocatalyst. Recently, functionalized graphene-based semiconductor photocatalyst have attracted a lot of attention due to their high electron conductivity, large specific surface area and adsorption. Many efforts are put into the synthesis of TiO_2-graphene photocatalysis hybrid system (Liang et al. 2014). Benefiting from the high specific surface area, graphene emerged as a good support to make the loaded nanoparticle metal oxides to achieve a uniform distribution without aggregation. The enhancement of photocatalytic activity for TiO_2-graphene hybrid is ascribed to the giant two-dimensional planar graphene structure favoring dye adsorption and suppressed electron-hole recombination due to high electrical conductivity. Furthermore, incorporation of two or more catalyst particles onto an individual graphene sheet at separate sites can provide greater versatility in carrying out selective catalytic or sensing processes and modify the structure and morphology of photocatalysts to enhance their photocatalytic performances (Zhang et al. 2010).

4.4 TiO_2: CNT Photocatalyst System

Other than graphene, carbon nanotube (CNTs) have been regarded as more attractive catalyst support than activated carbon because of their combination of electronic, adsorption and specific semiconducting characteristics. The studies on TiO_2: CNTs reveal a considerable synergy effect with the metal oxides and carbon phases. Furthermore, researchers have shown that CNTs can increase the adsorption and photocatalytic activity of TiO_2 in the presence of UV and VL irradiation.

Single-walled CNTs exhibited enhanced and selective photocatalytic oxidation of Ph. Heterostructured CNTs consisting of TiO_2-$_xN_x$ and carbon have been prepared by carbonization of electronspun polyacrylonitrile nanofibers containing stabilized titanium oxoacetate (Woan et al. 2009). Therefore, CNTs can be used as a promising material for environmental cleaning and can be adopted to improve the photocatalytic efficiency of TiO_2.

4.5 TiO_2:AC Semiconductor Doped Photocatalyst System

Meanwhile, the means of doping another semiconductor having different corresponding conduction and valence band energy levels with that of TiO_2 on the surface of TiO_2 has been validated to be resultful to improve the photocatalytic efficiency. The mechanism of coupled semiconductors is to increase the photocatalytic efficiency by increasing the charge separation and extending the photo-responding range. Recently, the research about doping another semiconductor on the surface of TiO_2: AC includes, SnO_2, V_2O_5, ZnO, ZrO_2, and CsS. All of them showed higher photodegradation efficiency than TiO_2: AC (Liao et al. 2004)· Furthermore, recent studies have reported the modification of TiO_2: AC by semiconductor doping, which has led to the improvement on the removal of specific contaminants in the gas as a result of changes to the physical and chemical properties of the carbon materials. Iron (Fe_3O_4) dopant TiO_2: AC has been found as a promoter to photocatalysis with a good performance on the degradation of Congo red and methyl orange. Some researchers have combined the Fe_3O_4 compound with TiO_2: AC with aiming in preparing photocatalyst with a magnetic core and photoactive shell. They proved that this hybrid system had magnetic properties and could be separated easily by magnetic materials. Li et al. have pointed out that the magnetic hybrid photocatalyst can also be magnetically agitated by an alternating magnetic field in a suspension system. However Ao et al. found that the photocatalytic activity of the hybrid system declined because the magnetic particles experienced photo dissolution (Ao et al. 2008; Saleh et al. 2016b).

4.6 TiO_2: AC Non-Metal Doped Photocatalyst System

Doping of non-metal; such us N.P.B, C,S,F, Cl, and Br has been also widely used to improve the photocatalytic or to extend the light absorption into visible light of TiO_2: AC system. Among them, the phosphorous (P) doped TiO_2: AC has recently attracted increasing interest due to its enhanced shows a narrow band gap. In addition, it has an adsorption tail in the visible light region. Lin et al. found that F-doped TiO_2: AC prepared by sol-gel method with NaH_2PO_4 as precursor demonstrated a higher photocatalytic activity of acetaldehyde decomposition under visible light irradiation than the pure sample. Huang et al. 2011, synthesize the N-doped TiO_2: AC with high crystallinity and large surface area by hydrothermal method. The methylene blue degradation performance on N-doped sample was significantly increased and superior to that of the commercial P25.

CONCLUSION

AC has been typically employed as an adsorbent for the control of various environmental pollutants due to its high pore volume and large exposed surface area to volume ratio. Thus, many AC studies have dealt primarily with a concentration in lower to higher concentrations than that associated with indoor water quality. The loading of the semiconductor on AC has drawn great attention since the high adsorption capability of AC can help to enrich organic pollutant around the catalyst, promoting the pollutant transfer process and hence increasing the photocatalytic efficiency.

REFERENCES

Al-Shalalfeh, M. M., Saleh, T. A., & Al-Saadi, A. A. (2016). Silver colloid and film substrates in surface-enhanced Raman scattering for 2-thiouracil detection. *RSC Advances*, *6*(79), 75282–75292. doi:10.1039/C6RA14832H

Alansi, A. M., Alkayali, W. Z., Al-Qunaibit, M. H., Qahtan, T. F., & Saleh, T. A. (2015). Synthesis, Surface Morphology and Properties of Polystyrene Modified Synthetic Clay Nanocomposites. *Asian Journal of Chemistry*, *27*(10), 3900–3906. doi:10.14233/ajchem.2015.19070

Amarasekara, A., & Owereh, O. S. (2009). Homogenous phase synthesis of cellulose carbamate silica hybrid materials using 1-n-butyl-3-methylimidazolium chloride ionic liquid. *Carbohydrate Polymers*, *78*(3), 635–636. doi:10.1016/j.carbpol.2009.05.032

Andrea, K., Axel, F., & Titirici, M.-M. (2013). Hydrothermal conversion of biomass to fuels and energetic materials. *Current Opinion in Chemical Biology*, *17*(3), 515–521. doi:10.1016/j.cbpa.2013.05.004 PMID:23707262

Ao, Y., Xu, J., Fu, D., Ba, L., & Yuan, C. (2008). Deposition of anatase titania onto carbon encapsulated magentic nanoparticles. *Nanotechnology*, *19*(40), 405604. doi:10.1088/0957-4484/19/40/405604

Aravid, P. V., & Jong, W. (2012). Evaluation of high temperature gas cleaning options for biomass gasification product gas for solid oxide fuel cells. *Progress in Energy and Combustion Science*, *38*(6), 737–764. doi:10.1016/j.pecs.2012.03.006

Baek, I. G., You, S. J., & Park, E. D. (2012). Direct conversion of cellulose into polyols over Ni/W/SiO2. *Bioresource Technology*, *114*, 684–690. doi:10.1016/j.biortech.2012.03.059 PMID:22497706

Bai, A., Zhao, X., Jin, Y., Yang, G., & Feng, Y. (2013). A novel thermophilic-B-glucosidase from Caldicellulosiruptorbescir: Characterization and its synhergistic catalysis with other cellulose. *Journal of Molecular Catalysis*, *85*, 248–256. doi:10.1016/j.molcatb.2012.09.016

Baleizao, C., Gigante, B., Garcia, H., & Corma, A. (2002). Ionic liquids as green solvents for the asymmetric synthesis of cyanohydrins catalyzed by VOC(salen) complexes. *Green Chemistry*, *4*, 272–274. doi:10.1039/b201497c

Cai, H., Li, C., Wang, A., Xu, G., & Zhang, T. (2012). Zeolite-promoted hydrolusis of cellulose in ionic liquid, insight into the mutual behavior of zeolite, cellulose and ionic liquid. *Applied Catalysis B: Environmental*, *123*, 333–338. doi:10.1016/j.apcatb.2012.04.041

Carlson, T. R., Vispute, T. P., & Huber, G. W. (2008). Green gasoline by catalytic fast pyrolysis of solid biomass derived compounds. *ChemSusChem*, *1*(5), 397–400. doi:10.1002/cssc.200800018 PMID:18702131

Carniti, P., Gervasini, A., Biella, S., & Auroux, A. (2005). Instrinsic and effective acidity study of niobic acid and niobium phosphate by multitechnique approach. *Chemistry of Materials*, *17*(24), 6128–6135. doi:10.1021/cm0512070

Chong, Y., Chen, M., Xu, T., Zhang, Y., & Wang, Q. (2014). Tantalum and aluminum co-doped iron oxide as a robust photocatalyst for water oxidation. *App Catal E*, *147*, 733–740. doi:10.1016/j.apcatb.2013.10.009

Cristina, Y. M., Rodriguez, J., Freer, J., Duran, N., & Mansilla, H. D. (2000). Photocatalytic degradation of cellulose bleaching effluent by supported TiO2 and ZnO. *Chemosphere*, *41*, 1193–1197. doi:10.1016/S0045-6535(99)00551-2 PMID:10901246

Demirbas, A. (2001). Carbonization ranking of selected biomass for charcoal, liquid and gaseous products. *Energy Conv & Mang*, *42*(10), 1229–1238. doi:10.1016/S0196-8904(00)00110-2

Dmitri, A. B., Sergey, B., Pavel, E. P., Yurii, V. S., Valerii, I. B., Sergey, V. K., & Julian, R. H. R. (2013). Vapour phase formic acid decomposition over PdAu/c-Al2O3 catalysts: Effect of composition of metallic particle. *Journal of Catalysis*, *299*, 171–180. doi:10.1016/j.jcat.2012.12.009

Dora, S., Bhasker, T., Singh, R., Naik, D. V., & Adhikari, D. K. (2012). Effective catalytic conversion of cellulose into high yields of methyl glycosides over sulfonated carbon based catalyst. *Bioresource Technology*, *120*, 318–321. doi:10.1016/j.biortech.2012.06.036 PMID:22776237

Dreher, M., Jonson, B., Peterson, A. A., Nachtegaal, M., Wambach, J., & Vogel, F. (2013). Catalysis in supercritical water: Pathway of the methanation reaction and sulfur poisoning over a Ru/C catalyst during the reforming of biomolecules. *Journal of Catalysis*, *301*, 38–45. doi:10.1016/j.jcat.2013.01.018

Dutta, S., De, S., & Saha, B. (2013). Advances in biomass transformation to 5-hydroxymethylfurfural and mechanic aspects. *Biomass & Biores.*, *55*, 355–369. doi:10.1016/j.biombioe.2013.02.008

Edward, W., Yong, T. K., Geoffrey, A. T., Alejandro, F., Kee, S. H., Edward, W. H., & Hubber, G. W. et al. (2013). Conversion of glucose into levulinic acid with solid metal (IV) phosphate catalyts. *Journal of Catalysis*, *304*, 123–134. doi:10.1016/j.jcat.2013.03.023

Fan, G., Liao, C., Fang, T., Wang, M., & Song, G. (2013). Hydrolysis of cellulose catalyzed by sulfonated poly(styrene-co-divinylbenzene) in the ionic liquid 1-n-butyl-3-methylimidazolium bromide. *Fuel Processing Technology*, *116*, 142–148. doi:10.1016/j.fuproc.2013.05.009

Felica, L.-D., Courson, C., Foscoto, P. U., & Kiennemann, A. (2011). Iron and nickel doped alkaline earth catalysis for biomass gasification with simultenous for reformation and CO2 capture. *International Journal of Hydrogen Energy*, *36*(9), 5296–5310. doi:10.1016/j.ijhydene.2011.02.008

Gabriele, C., Paola, L., & Siglinda, P. (2011). Analysis of the alternative routes in the catalytic transformational lignocellulosic materials. *Catalysis Today*, *167*(1), 14–30. doi:10.1016/j.cattod.2010.10.099

Gao, Y., & Liu, H. (2005). Preparation and catalytic property study of a novel kind of suspended photocatalyst of TiO2-activated carbon immobilized on silicon ribber film. *Materials Chemistry and Physics*, *92*(2-3), 604–608. doi:10.1016/j.matchemphys.2005.02.018

Geboers, J. A., Vyver, S. V., Ooms, R., Beeck, B. O., Jacobs, P. A., & Sels, B. F. (2011). Chemocatalytic conversion of cellulose: Opportunities, advances and pitfalls. *Catal Sci Technol.*, *1*(5), 714–720. doi:10.1039/c1cy00093d

Girisuta, B., Dussan, K., Haverty, D., Leahy, J. J., & Hayes, M. H. B. (2013). A kinetic study of acid catalysed hydrolysis of sugar cane bagasse to levulinic acid. *Chemical Engineering Journal*, *217*, 61–70. doi:10.1016/j.cej.2012.11.094

Gu, Q., Long, J., Fan, L., Chen, L., Zhao, L., Lin, H., & Wang, X. (2013). Single site Sn-grafted Ru/TiO2 photocatalysis for biomass reforming synergetic effect of dual co-catalyst and molecular mechanism. *Journal of Catalysis*, *30*, 141–155. doi:10.1016/j.jcat.2013.03.014

Guan, G., Chen, G., Kasai, Y., Lim, E. W. C., Hao, X., Kaewpanha, M., & Tsutmi, A. et al. (2012). Catalytic stream reforming of biomass tar over iron and nickel-based catalyst supported on calcinaed scallop shell. *Applied Catalysis B: Environmental*, *115*, 159–168. doi:10.1016/j.apcatb.2011.12.009

Guisnet, M. (2013). Ideal bifunctional catalysis over Pt acid zeolites. *Catalysis Today*, *218*, 123–134. doi:10.1016/j.cattod.2013.04.028

Gupta, V. K., Ali, I., Saleh, T. A., Nayak, A., & Agarwal, S. (2012). Chemical treatment technologies for waste-water recycling—an overview. *RSC Advances*, *2*(16), 6380–6388. doi:10.1039/c2ra20340e

Gurbuz, E., Bond, J. Q., Domesic, J. A., & Roman-Leshkov, Y. (2013). Role of acid catalysis in the conversion of lignocellulosic biomass to fuels and chemicals. In The role of catalysis for the sustainable production of bio-fuels and bio-chemicals, (pp. 261-288). Academic Press.

Haixin, G., Xinhua, Q., Luyang, L., & Richard, L. S. J. (2012). Hydrolysis of cellulose over functionalized glucose-derived carbon catalyst in ionic liquid. *Bioresource Technology*, *116*, 355–359. doi:10.1016/j.biortech.2012.03.098 PMID:22522013

Hamer, U., Marschner, B., Brodowski, S., & Amelong, W. (2004). Interactive priming of black carbon and glucose mineralisation. *Organic Geochemistry*, *36*(7), 823–830. doi:10.1016/j.orggeochem.2004.03.003

Han, J., & Li, C. J. (2013). Characterization of TiO2/ACF prparation in caprolactom tetrabutyl ammonium bromide ionic liquid. *Adv Mater Res.*, *750*, 1864–1867. doi:10.4028/www.scientific.net/AMR.750-752.1864

Hara, M. (2010). Biomass conversion by a solid acid catalyst. *Energy Envi Sci.*, *3*(5), 601–607. doi:10.1039/b922917e

He, M., & Xia, H. (2012). Photocatalytic performance of activated carbon fibers (ACFs) supported La-S/TiO2 photocatalyst. *Adv Mater Res., 518*(523), 764–767. doi:10.4028/www.scientific.net/AMR.518-523.764

Heinze, T., & Schaller, J. (2000). New water soluble cellulose esters synthesized by on effective acylation procedure. *Macromolecular Chemistry and Physics, 201*(12), 1214–1218. doi:10.1002/1521-3935(20000801)201:12<1214::AID-MACP1214>3.0.CO;2-9

Hsi, H.-C., & Tsai, C.-Y. (2012). Preparation of oxygen-vacant TiO2-x and activated carbon fiber composite using a single step thermal plasma method for low-concentration elemental mercury removal. *Chemical Engineering Journal, 200*, 18–24. doi:10.1016/j.cej.2012.06.021

Hsu, W.-H., Lee, Y.-Y., Peng, W.-H., Kelven, C., & Wu, C.-W. (2011). Cellulosic conversion in ionic liquids (Ils): Effects of H2O/ cellulose molr ratios, temperature, times and different Ils on the production of monosaccharides and 5-hydroxymethylfurfural (HMF). *Catalysis Today, 174*(1), 65–69. doi:10.1016/j.cattod.2011.03.020

Huang, D., Miyamoto, Y., Ding, J., Gu, J., Zhu, S., Liu, Q., & Zhang, D. et al. (2011). A new method prepared high surface-area N-TiO2/activated carbon. *Materials Letters, 65*(2), 326–328. doi:10.1016/j.matlet.2010.10.025

Huang, D., Miyamoto, Y., Matsumoto, T., Tojo, T., Fan, T., Ding, J., & Zhang, D. et al. (2011). Preparation and characterization of high-surface-area TiO2/activated carbon by low-temperature impregnation. *Separation and Purification Technology, 78*(1), 9–15. doi:10.1016/j.seppur.2011.01.010

Hussein, A.-Y., El Barbary, H., & Philip, S. (2013). Rapid conversion of cellulose to 5-hydroxymethylfurfural using single and combined metal chloride catalyst in ionic liquid. *J Fuel Chem & Tech., 41*(2), 214–222. doi:10.1016/S1872-5813(13)60013-4

Iguchi, M., Aida, T. M., Watanabe, M., & Smith, R. L. J. Jr. (2013). Dissolution and recovery of cellulose from 1-butyl-3-methylimidazolium chloride in presence of water. *Carbohydrate Polymers, 92*(1), 651–658. doi:10.1016/j.carbpol.2012.09.021 PMID:23218349

Jimenez-Morales, I., Santamaria-Gonzalez, J., Maireles-Torres, P., & Jimenez-Lopez, A. (2011). Aluminium doped SBA-15 silica as acid catalyst for the methanolysis of sunflower oil. *Applied Catalysis B: Environmental, 105*(1-2), 199–205. doi:10.1016/j.apcatb.2011.04.015

Joshua, P., Erin, S., Jessica, H., Brenton, D. B., & Brett, L. L. (2011). Effect of NaCl on the conversion of cellulose of glucose and levulinic acid via solid supported acid cataylst. *Tetrahend Lett.*, *52*(44), 5891–5893. doi:10.1016/j.tetlet.2011.09.013

Juan, C. S.-R., Antonio, P., Alina, M. B., Rafeal, L., Juan, M. C., Antonio, A. R., & Jose, M. R.-F. (2012). Catalytic transformations of biomass-derived acids into advanced biofuels. *Catalysis Today*, *195*(1), 162–268. doi:10.1016/j.cattod.2012.01.009

Jung, H.-J., Hong, J.-S., & Suh, J.-K. (2013). A comparision of fenton oxidation and photocatalyst reaction efficiency for humic acid degradation. *Journal of Industrial and Engineering Chemistry*, *19*(4), 1325–1330. doi:10.1016/j.jiec.2012.12.036

Kangwanwatana, W., Salwan, C., & Tontiwachwuthikul, P. (2013). Study of CO2 adsorption using adsorbent modified with pipercizine. *Chem Eng Trans*, *35*, 403-408.

Klamrassamee, T., Champreda, V., Reunglek, V., & Laosiripojana, N. (2013). Comparison of homogeneous and heterogeneous acid promoters in single-step aqueous-organosolv fractionation of eucalyptus wood chips. *Bioresource Technology*, *147*, 276–284. doi:10.1016/j.biortech.2013.08.015 PMID:24001557

Kumar, S., Kumar, B., Surender, T., & Shanker, V. (2014). G-C3N4/NaTaO3 organic inorganic hybrid nanocomposite: High performance and recyclable visible light driven photocatalyst. *Materials Research Bulletin*, *49*, 310–318. doi:10.1016/j.materresbull.2013.09.013

Kuo, C. Y., Wu, C. H., & Chen, S. T. (2014). Decolorization of CI Reactive Red 2 by UV/TiO2/PAC and visible light/TiO2/PAC systems. *Desalination and Water Treatment*, *52*(4-6), 834–843. doi:10.1080/19443994.2013.826409

Kuzyakos, Y., Subbotino, I., Chen, H., Bagomolova, I., & Xu, X. (2009). Black carbon decomposition and incorporation into soil microbial biomass estimated by 14C labeling. *Soil Biology & Biochemistry*, *41*(2), 210–219. doi:10.1016/j.soilbio.2008.10.016

Labus, K., Gryglewicz, S., & Machnikowski, J. (2014). Granular KOH-activated carbons from coal based cokes and their CO2 adsorption capacity. *Fuel*, *118*, 9–15. doi:10.1016/j.fuel.2013.10.042

Lee, J.-Y., & Jo, W.-K. (2012). Control of methyl tertiary-butyl ether via carbon-doped photocatalysts under visible light irradiation. *Env Engn Res.*, *17*(4), 179–184. doi:10.4491/eer.2012.17.4.179

Li, T., Ren, Y., & Wei, C. (2014). Study on preparation and properties of PVA-SA-PHB-AC composite carrier for microorganism immobilization. *Journal of Applied Polymer Science*, *131*(3), n/a. doi:10.1002/app.39837

Liang, Y., Wang, H., Casalongue, H. S., Chen, Z., & Dai, H. (2014). TiO2 nanocrystal grown on graphene as advanced photocatalytic hybrid materials. *Nano Res.*, *3*(10), 701–705. doi:10.1007/s12274-010-0033-5

Liao, Z.-H., Chen, J.-J., Yao, K. F., Zhao, F. H., & Li, R.-X. (2004). Preparation and characterization of nanometer-sized magnetic photocatalyst. *J Inorg Mater.*, *04*, 1–10.

Lin, X.-X., Ji, X., Fu, D.-G., & Hao, L.-Y. (2013). Photocatalytic properties of magnetic activated carbon supported F-doped TiO2. *J Inorg Mater.*, *28*(9), 997–1002. doi:10.3724/SP.J.1077.2013.12671

Lincai, P., Lu, L., Junhua, Z., Jianbin, S., & Shijie, L. (2011). Solid acid catalyzed glucose conversion to ethyl levulinate. *App Catal Gen*, *397*(1-2), 259–265. doi:10.1016/j.apcata.2011.03.008

Manya, J. J., Velo, E., & Puigjaner, L. (2003). Kinetics of biomass pyrolysis: A reformulated three-parallel-reaction model. *Industrial & Engineering Chemistry Research*, *42*(3), 433–441. doi:10.1021/ie020218p

Mario, S., Robin, J. W., Tobias, Z., Titirici, M.-M., Markus, A., Regina, R., & Ferdi, S. (2013). Direct methane oxidation over Pt modified nitrogen doped carbons. *Chemical Communications (Cambridge)*, *49*(3), 240–242. doi:10.1039/C2CC36232E PMID:23172370

Masahiro, T., Takashi, Y., Tsumura, T., Yutaka, A., & Michio, I. (2006). Effect of carbon coating on TinO2n-1 for decomposition of imioctadine triacetate in aqeous solution under visible light. *J Adv Oxidation Tech.*, *9*, 1–10.

Matos, J., Garcia-Lopez, E., Palmisano, L., Garcia, A., & Marci, G. (2010). Influence of activated carbon in TiO2 and ZnO mediated photo-assisted degradation of 2-propanol in gas solid regime. *Applied Catalysis B: Environmental*, *99*(1-2), 170–180. doi:10.1016/j.apcatb.2010.06.014

Milbrandt, A. R., Heimiler, D. M., Perry, A. D., & Field, C. B. (2014). Renewable energy potential on marginal lands in the United States. *Renewable & Sustainable Energy Reviews*, *29*, 473–481. doi:10.1016/j.rser.2013.08.079

Mingxing, C., Tian, S., Hongyu, G., Shengtian, W., Xiaohong, W., & Zijiang, J. (2011). Cean production of glucose from polysaccharides using a micellar heteropolyacid as a heterogeous catalyst. *Applied Catalysis B: Environmental*, *107*(1-2), 104–109. doi:10.1016/j.apcatb.2011.07.002

Namchot, W., Panyacharang, N., Jonglertjunya, W., & Sakdarannarong, C. (2014). Hydrolysis of delignified sugarcane bagasse using hydrothermal technique catalyzed by carbonaceous acid catalysts. *Fuel*, *116*, 608–616. doi:10.1016/j.fuel.2013.08.062

Nawaz, H., Pires, P. A. R., & El Seoud, O. A. (2013). Kinetics and mechanism of imidozole-catalyzed acylation of cellulose in LiCl/ N,N-dimethylacetamide. *Carbohydrate Polymers, 92*(2), 997–1005. doi:10.1016/j.carbpol.2012.10.009 PMID:23399120

Onal, Y., Schimpt, S., & Claus, P. (2004). Structure sensitivity and kinetics of e-glucose oxidation to D-gluconic acid over carbon supported gold catalyst. *Journal of Catalysis, 223*(1), 122–133. doi:10.1016/j.jcat.2004.01.010

Otomo, R., Yokoi, T., Kondo, J.N., & Tatsumi, T. (2014). *Dealuminated beta zeolite as effective bifunctional catalyst for direct transformation of glucose to 5-hydroxy-methylfurfural*. Academic Press.

Pino, L., Vita, A., Lagana, M., & Recupero, V. (2014). Hydrogen from biogas: Catalytic tri-reforming process with Ni/L---Ce---O mixed oxides. *Applied Catalysis B: Environmental, 148*, 91–105. doi:10.1016/j.apcatb.2013.10.043

Potvin, J., Sorlien, E., Hegner, J., DeBoet, B., & Luchi, B. L. (2011). Effect of NaCl on the conversion of cellulose to glucose and levulinic acid via solid supported acid catalysis. *Tetrahedron Letters, 52*(44), 5891–5893. doi:10.1016/j.tetlet.2011.09.013

Quinones, D. H., Rey, A., Alvarez, P. M., Beltran, F. J., & Plucinski, P. K. (2014). Enchanced activity and reusability of TiO2 loaded magnetic activated carbon for solar photocatalytic ozonation. *Applied Catalysis B: Environmental, 144*, 96–106. doi:10.1016/j.apcatb.2013.07.005

Rezan, D.-C., Niki, B., Markus, A., & Maria-Magdalena, T. (2009). Carboxylate-rich carbonaceous materials via open-step hydrothermal carbonization of glucose in the presence of arcylic acid. *Chemistry of Materials, 21*(3), 484–490. doi:10.1021/cm802141h

Rick, O., James, R. K., & Joby, M. (2012). Hemicellulose hydrolysis using solid acid catalysts generated from biochar. *Catalysis Today, 190*(1), 89–97. doi:10.1016/j.cattod.2012.02.050

Ronald, G.-B., Xie, Z.-L., Mike, N., Cristina, G., Ruben, L. B., Shiori, K., & Andreas, T. et al. (2012). Synthesis of mesoporous carbon/iron carbide hydrids with unusually high surface areas from ionic liquid precusor[BMIM] [FeCl4]. *CrystEngComm, 14*(15), 4946–4951. doi:10.1039/c2ce25064k

Ronen, W., Geoffrey, A. T., Conner, W. C. J., & George, W. H. (2011). Design of solid acid catalysts for aqueous-phase dehydration of carbohydrates: The role of Lewis and Bronsted acid sites. *Journal of Catalysis, 279*(1), 174–182. doi:10.1016/j.jcat.2011.01.013

Saleh, T. A. (2011). The influence of treatment temperature on the acidity of MW-CNT oxidized by HNO3 or a mixture of HNO3/H2SO4. *Applied Surface Science*, *257*(17), 7746–7751. doi:10.1016/j.apsusc.2011.04.020

Saleh, T. A. (2014) *Activated carbon/aluminum oxide/polyethylenimine composites and methods thereof.* US Patent App. 14/580,311.

Saleh, T. A. (2015a). Mercury sorption by silica/carbon nanotubes and silica/activated carbon: A comparison study. *Journal of Water Supply: Research & Technology - Aqua*, *64*(8), 892–903. doi:10.2166/aqua.2015.050

Saleh, T. A. (2015b). Isotherm, kinetic, and thermodynamic studies on Hg (II) adsorption from aqueous solution by silica-multiwall carbon nanotubes. *Environmental Science and Pollution Research International*, *22*(21), 16721–16731. doi:10.1007/s11356-015-4866-z PMID:26087931

Saleh, T. A., Agarwal, S., & Gupta, V. K. (2011). Synthesis of MWCNT/MnO 2 and their application for simultaneous oxidation of arsenite and sorption of arsenate. *Applied Catalysis B: Environmental*, *106*(1), 46–53.

Saleh, T. A., & Al-Saadi, A. A. (2013). Adsorption of lead ions from aqueous solution using porous carbon derived from rubber tires: Experimental and computational study. *Journal of Colloid and Interface Science*, *396*, 264–269. doi:10.1016/j.jcis.2013.01.037 PMID:23433519

Saleh, T. A., & Al-Saadi, A. A. (2015). Surface characterization and sorption efficacy of tire-obtained carbon: Experimental and semiempirical study of rhodamine B adsorption. *Surface and Interface Analysis*, *47*(7), 785–792. doi:10.1002/sia.5775

Saleh, T. A., Al-Saadi, A. A., & Gupta, V. K. (2014). Carbonaceous adsorbent prepared from waste tires: Experimental and computational evaluations of organic dye methyl orange. *Journal of Molecular Liquids*, *191*, 85–91. doi:10.1016/j.molliq.2013.11.028

Saleh, T. A., Al-Shalalfeh, M. M., & Al-Saadi, A. A. (2016). Graphene Dendrimer-stabilized silver nanoparticles for detection of methimazole using Surface-enhanced Raman scattering with computational assignment. *Scientific Reports*, *6*, 32185. doi:10.1038/srep32185 PMID:27572919

Saleh, T. A., Sarı, A., & Tuzen, M. (2016b). Chitosan-modified vermiculite for As (III) adsorption from aqueous solution: Equilibrium, thermodynamic and kinetic studies. *Journal of Molecular Liquids*, *219*, 937–945. doi:10.1016/j.molliq.2016.03.060

Samus, T., Lang, B., & Rohn, H. (2013). Assesing the natural resource use and the resource efficiency potential of the Desertec Concept. *Solar Eng*, *87*, 176–183. doi:10.1016/j.solener.2012.10.011

Serrano-Ruiz, J. C., Luque, R., & Clark, J. H. (2013). The role of heterogenous catalysis in the biorefinering of the future. In The role of catal for sust poduct biofuel, (pp. 557-576). Academic Press.

Shang, H.-J., Zhou, Y.-L., Zhao, Y., Zhang, G., Zhao, D., Zhang, S.-H., & Zheng, X.-M. (2012). Study on heterogenous photocatalytic degradation of NH4+ N with immobilized AC/TiO2 photocatalyst. *Modern Chem Indust.*, *32*, 59–62.

Shawabkeh, R. A., Al-Harthi, M., & Al-Ghamdi, S. M. (2014). The synthesis and characterization of microporous high surface area activated carbon. *Energy Sources A.*, *36*(1), 93–103. doi:10.1080/15567036.2010.535103

Skoulou, V., & Zabaniotou, A. (2012). Fe catalysis for lignocellulosic biomass conversion to fuels and materials via thermochemical process. *Catal Today, 196*, 56-66.

Slimen, H., Houas, A., & Nogier, J. P. (2011). Elaboration of stable anatase TiO2 through activated carbon addition with light photocatalytic activity under visible light. *J Photochem Photobio A.*, *221*(1), 13–21. doi:10.1016/j.jphotochem.2011.04.013

Stephanie, G. W., David, M. A., Elif, I. G., & James, A. D. (2012). A roadmap for conversion of lignocellulosic biomass to chemicals and fuels. *Current Opinion Chem Eng.*, *1*(3), 218–224. doi:10.1016/j.coche.2012.04.002

Stocker, M. (2005). Gas phase catalysis by zeolites. *Microporous and Mesoporous Materials*, *82*(3), 257–292. doi:10.1016/j.micromeso.2005.01.039

Sun, M., & Zhang, Y. (2013). Study on the preparation of activated carbons with Baiju Vinasse. *App Mech & Mater.*, *448*, 669–693.

Tessonnier, J. P., Villa, A., Majoulet, O., Su, D. S., & Sclogl, R. (2009). Defect-mediated functionalization of carbon nanotubes as a route to design single site basic heterogenous catalysts for biomass conversion. *Ange Chemic Int*, *48*(35), 6543–6546. doi:10.1002/anie.200901658 PMID:19630046

Tryba, B., Morawski, A. W., & Inagaki, M. (2003). Application of TiO2-mounted activated carbon to the removal of phenol from water. *Applied Catalysis B: Environmental*, *41*(4), 427–433. doi:10.1016/S0926-3373(02)00173-X

Velo-Gala, I., Lopez-Penalver, J. J., Sanchez-Polo, M., & Rivera-Utrilla, J. (2013). Activated carbon as photocatalyst of reaction in aqeuous phase. *Applied Catalysis B: Environmental*, *142*, 694–704. doi:10.1016/j.apcatb.2013.06.003

Vukcevic, M., Pejic, B., Kalijadis, A., Pajic-Lijakovic, I., Kostic, M., Lausevic, Z., & Lausevic, M. (2014). Carbon materials from waste short hemp fibers as a sorbent for heavy metal ions matermatical modelling of sorbent structure and ions transport. *Chemical Engineering Journal*, *235*, 284–292. doi:10.1016/j.cej.2013.09.047

Vyver, S. V., Peng, L., Geboers, J., Schepers, H., Clippel, F., Gommes, C. J., & Seis, B. F. et al. (2010). Sulfonated silica/carbon nanocomposites as novel catalyst for hydrolysis of cellulose to glucose. *Green Chemistry*, *12*(9), 1560–1563. doi:10.1039/c0gc00235f

Wang, H., Zhu, L., Peng, S., Peng, F., Yu, H., & Yang, J. (2012). High efficient conversion of cellulose to polyols with Ru/CNTs as catalyst. *Renewable Energy*, *37*(1), 192–196. doi:10.1016/j.renene.2011.06.020

Wang, Y., Ren, P., Feng, C., Zheng, X., Wang, Z., & Li, D. (2014). Photocatalytic behavior and photo-corrosion of visible light active silver carbonate/titanium oxide. *Materials Letters*, *115*, 85–88. doi:10.1016/j.matlet.2013.10.025

Wei, Z., Li, Y., Thurshara, D., Liu, Y., & Ren, Q. (2011). Novel dehydration of carbohydrates to 5-hydroxymethylfurfural catalyzed by Ir and Au chlorides in ionic liquids. *J Taiwan Inst Chem Eng.*, *42*(2), 363–370. doi:10.1016/j.jtice.2010.10.004

Woan, K., Pyrgiotakis, G., & Sigmund, W. (2009). Photocatalytic carbon-nanotube-TiO2 composites. *Advanced Materials*, *21*(21), 2233–2239. doi:10.1002/adma.200802738

Xiang,, Y-Z., Luo,, X., & Han,, W.Y., Yang, J-J., Ma, C.P., Xin, L., & Wang, L.B. (2013). Preparation and catalytic activities of hydrolysis catalyst for carbonyl sulfide in light hydrocarbon. *Modern Chem.*, *33*, 48–51.

Yabushita, M., Kobayashi, H., & Fukuoka, A. (2014). Catalytic transformation of cellulose into platform chemical. *Applied Catalysis*, *145*, 1–9. doi:10.1016/j.apcatb.2013.01.052

Yang, P., Kobayashi, H., & Fukuoka, A. (2011). Recent developments in the catalytic conversion of cellulose into valuable chemicals. *Chinese Journal of Catalysis*, *32*, 1–5. doi:10.1016/j.jcat.2011.08.005

Yang, Q., Liao, Y., & Mao, L. (2012). Kinetics of photocatalytic degradation of gaseous organic compounds on modified TiO2/AC composite photocatalyst. *Chinese Journal of Chemical Engineering*, *20*(3), 572–576. doi:10.1016/S1004-9541(11)60221-8

Yao, S., Jia, Y., Shi, Z., & Zhao, S. (2010). Photoccatalytic oxidation of arsenite by a composite of titanium dioxide and activated carbon fiber. *Photochemistry and Photobiology, 86*(6), 1215–1221. doi:10.1111/j.1751-1097.2010.00813.x PMID:21039575

Yoon, J.-W., Baek, M.-H., Hong, J.-S., Lee, C.-Y., & Suh, J.-K. (2012). Photocatalytic degradation of azo dye using TiO2 supported on spherical activated carbon. *Korean Journal of Chemical Engineering, 29*(12), 1722–1729. doi:10.1007/s11814-012-0076-2

Yu, S., Heather, M. B., Li, G., Zhou, X.-D., Amonette, J. E., John, L. F., & Conrod, Z. et al. (2011). Accelerated cellulose depolymerization catalyzed by paired metal chlorides in ionic liquid solvent. *App Catal Gen., 391*(1-2), 436–442. doi:10.1016/j.apcata.2010.09.021

Zapata, D. A., Huang, Y., Gonzalez-Borja, M. A., & Resosca, D. E. (2013). Silyted hydrophobic zeolites with enchanced tolerance to hot liquid water. *Journal of Catalysis, 308*, 62–97. doi:10.1016/j.jcat.2013.05.024

Zhang, D., Wen, M., Zhang, S., Liu, P., Zhu, W., Li, G., & Li, H. (2014). Au nanoparticles enchanced rutile TiO2 nanorod bundles with high visible light photocatalytic performance for NO oxidation. *Applied Catalysis B: Environmental, 147*, 610–616. doi:10.1016/j.apcatb.2013.09.042

Zhang, X.-Y., Li, H.-P., Cui, X.-L., & Lin, Y. (2010). Graphene/TiO2 nanocomposites synthesis, characterization and application in hydrogen evolution from water photocatalytic splitting. *Journal of Materials Chemistry, 20*(14), 2801–2806. doi:10.1039/b917240h

Zhang, Z., Zongboa, K., & Zhao, K. (2009). Solid acid and microave-assisted hydrolysis of cellulose in ionic liquid. *Carbohydrate Research, 344*(15), 2069–2072. doi:10.1016/j.carres.2009.07.011 PMID:19703684

Zu, Y., Yang, P., Wang, J., Liu, X., Ren, J., Lu, G., & Wang, Y. (2014). Efficient production of the liquid fuel 2,5-dimethylfuran from 5-hydroxymethylfurfural over Ru/CO3O4 catalyst. *Applied Catalysis B: Environmental, 146*, 244–248. doi:10.1016/j.apcatb.2013.04.026

Chapter 8
Polymer Consumption, Environmental Concerns, Possible Disposal Options, and Recycling for Water Treatment

Tawfik A. A. Saleh
King Fahd University of Petroleum and Minerals, Saudi Arabia

Gaddafi I. Danmaliki
King Fahd University of Petroleum and Minerals, Saudi Arabia

ABSTRACT

The increasing awareness of the environment has raised so much concerns in the way we live and the manner to which we dispose our waste. The rapid growth in polymer production has resulted in increasing concerns about the consumption of nonrenewable resources and the environmental impacts associated with its production and disposal. Polymer waste is one of the major components in municipal solid waste and is increasingly becoming a huge burden in industrialized nations. The rise in the use of plastic, coupled with increasing concerns about its disposal, has led to a renewed interest in its recycling and recovery. The technologies involved in the recovery and recycling of polymers are rapidly growing, however, there is no specific pattern of treating polymer waste. The technology depends on the type of material used in the production and consumption pattern. This chapter, therefore, discusses the patterns of polymer consumption, the environmental concerns, and different modes of recycling polymer waste.

DOI: 10.4018/978-1-5225-2136-5.ch008

1. INTRODUCTION

Plastics are lightweight durable and inexpensive materials, which can be converted into a wide array of products that can be used in a wide range of applications. The production of plastics has dramatically increased in the last 60 years due to their wide applications and ease of molding (Jefferson *et al.*, 2009). The amount of polymers used in the industry far exceeds that of metals and is rapidly growing due the industrialization of some developing nations (2003). Polymers are inevitable products in our lives today; they are widely used in place of alternate materials because they are inexpensive and they possess better environmental disposal options due to their light weight, minimal pressure on agricultural resources, and superior insulation abilities. Despite all the advantages, the current level of their usage and disposal pose serious danger to the environment and the health of the populace. In addition, around 4 percent of annual oil and gas production (non-renewable resource) is used as the parent material (precursor for synthesis) and the sole provider of energy for plastic production (British Plastics Federation 2008). Secondly, most of the plastic produced is directed for the production of disposable items that always end up in the landfills and in delicate habitats within short period of time. The impacts of polymer production have been sum up and depicted in Figure 1.

There are so many approaches in use to reduce the impacts of polymer disposal which include: source reduction, composting, degradation, incineration, recycling and landfilling (Richard 1992). Source reduction aim at reducing the volume or toxicity of waste generated and it is the most preferred management option, composting utilized biochemical means with the aid of microorganisms to reduce the volume of plastic waste generated, degradation uses factors such as heat, chemical or light to alter the properties of the polymers and lower its molecular weight for efficient recycling. Incineration which converts combustible waste to inert residue is a viable option because energy is generated but combustion of plastics generates significant amount of pollution. Incineration requires well-designed systems, adequate control and efficient maintenance. Disposal of plastics by landfilling is not a problem so long as space is available, however, most plastics are not biodegradable.

Recycling represents one of the most effective areas in polymer industry; it has significant advantages compared to other means of reducing polymer impacts because it reduces the consumption oil and gas, lowers emissions associated with polymer production, reduces the amount of waste generated and has ecological advantages, but its application is dependent upon the feasibility of elements of waste management i.e. handling, storage, processing, collection, sorting, and transport.

Recycling of plastic materials has received rapid growth in the past few decades however, the quantities been recycled vary with locations according plastic type and application. Developments in the field of recycling are generating wide array of opportunities to people, legislators and government agencies and with proper investments, most plastic wastes been disposed to the landfills will be siphoned and directed for efficient recycling.

The recovery and recycling rate of plastics compared to other materials such as glass, paper, and metals is generally very slow. Plastic recycling even in developed nations with well-developed waste management systems in place is generally slow as compared to other materials because of the huge varieties of plastics, composites and additives used in their production. This complex nature of plastics confer added advantages when compared to other materials but it is generally problematic to recycling.

Figure 1. The impacts of polymer production

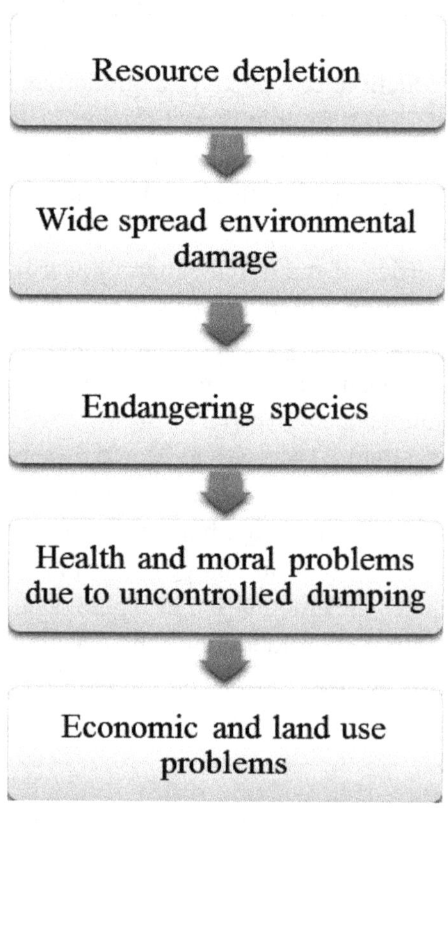

2. POLYMER CONSUMPTION

The major producers of plastics in the world are China (23.9%), Europe (20.4%), North America (20%) and the rest of Asia excluding China (15.8%) (PlasticsEurope 2013) In the Europe, (39.4%) of the whole plastic production is directed for packaging materials, followed by construction (20.3%), automotive accounts for 8.2%, electrical and electronics accounts for 5.5% the plastic demand, agricultural applications account for 4.2% while other applications such as: medical, furniture, household and consumer products and appliances accounts for 22.4% . The world plastic production from 1950-2012 is shown in Figure 2.

There are different types of plastics and they are subdivided into thermoplasts and thermosets. Thermosets are liquids or malleable products that undergo curing with the aid of heat by irradiation or irreversible chemical reaction to produce a hardened thermoset resin that cannot be reformed or tainted. Once a thermosets solidifies they stay solid which makes it difficult to be recycled. Thermosets can only be recycled through a chemical process. Types of thermosets include: epoxy resin, polyester resin, and urea and melamine formaldehydes. Thermoplasts polymer on the other hand do not undergo chemical changes when heated and can be molded again upon cooling. The classes of thermoplastic include: nylon (polyamide), Acrylic (polymethyl methacrylate), Teflon, polyurethane (PU) polyvinyl chloride (PVC), polypropylene (PP), polystyrene (PS), polyethene (PE) which is subdivided

Figure 2. World Plastic Production 1950-2012 (including thermosets, thermoplastics, elastomers, polyurethanes, coatings and sealants, adhesives, and PP-fibers)

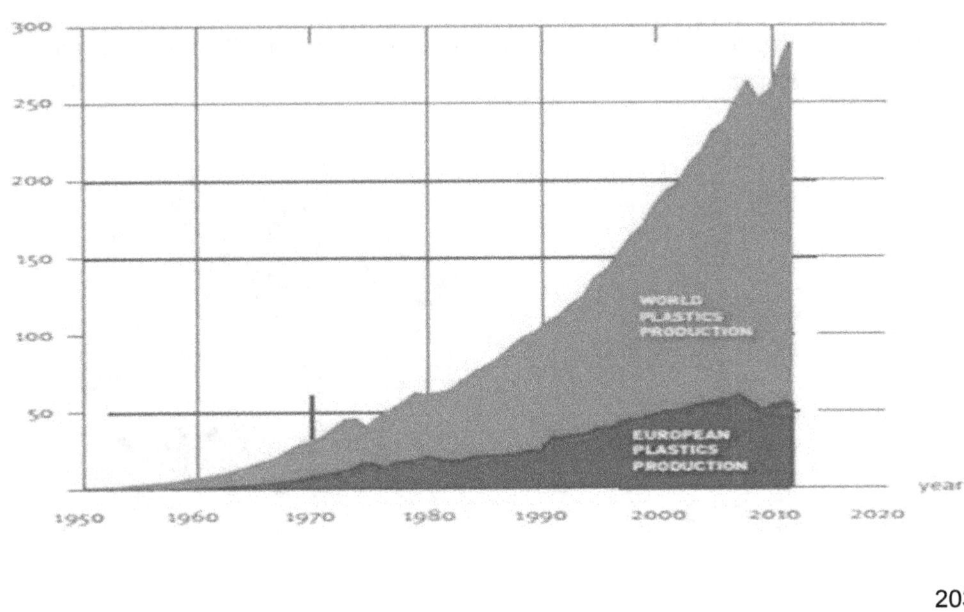

based on the distribution of polymers into: low density polyethene (LDPE), medium-density polyethylene (MDPE), high density polyethene (HDPE), ultra-high molecular weight polyethylene (UHMWPE) and cross-linked polyethylene. The types and classes of these polymers are given in Table 1 and their main applications in EU member states are depicted in Figure 2. The key plastic types used in the EU is depicted in Figure 2. The overall distribution and uses of plastics varies depending on specific conditions. This is evident as many countries have reduced the utilization of PVC in packaging however the same material when combined with PUR and PS serves as the main plastic used in building and construction. Polyolefin dominates the packaging industry.

288 million metric tons of polymer was produced in the year 2012 for all types of polymers including thermoset plastics, thermoplastics, and adhesives and coatings. A 2.8% increase compared to 2011. This shows a growth rate of about 9%/annum. However in Europe, despite the growth in plastic consumption across the globe the production rate decreases by 3% from 2011 to 2012. Thermoplastic resins constitute around 2/3 of this production and their usage is growing at about 5%/annum globally (PlasticsEurope 2013).

The UK plastic industry contributes significantly to the economic strength and development of the geographical location. It accounts for 7% all the manufacturing activity in UK, a percentage much higher than pharmaceutical and automotive industries combined. In 2011, the industry's sales turnover was £19 billion and £6.7bn value of exports a 13% increase from 2010 was recorded. Approximately 8000 firms are engaged in the plastic industry in UK, more than 3000 of this firms are termed primary processors because they are engaged in the processing of plastics

Figure 3: A) European plastic demand by segment 2012 and B) European plastic demand by resin type 2012

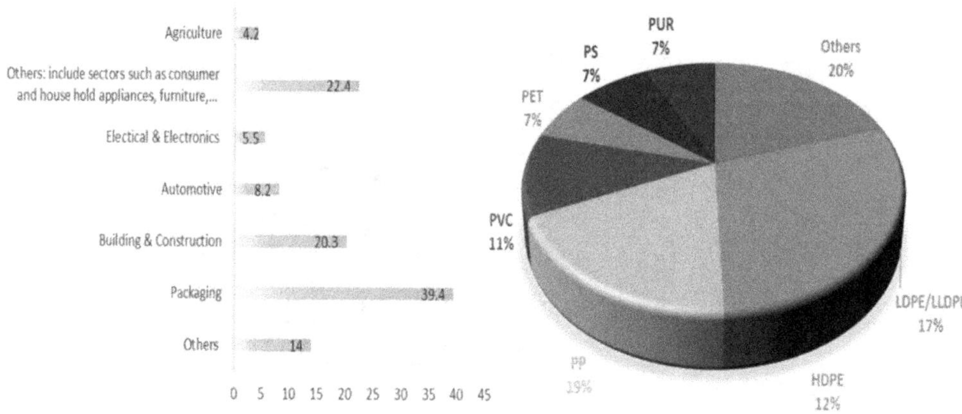

Polymer Consumption

Table 1. The most common types of plastics used, their applications and the symbol which is used in their identification

Polymer Types	Examples of applications	Symbol
Polyethylene Terephthalate	Fizzy drink and water bottles. Salad trays.	1 PET
High Density Polyethylene	Milk bottles, bleach, cleaners and most shampoo bottles.	2 HDPE
Polyvinyl Chloride	Pipes, fittings, window and door frames (rigid PVC). Thermal insulation (PVC foam) and automotive parts.	3 PVC
Low Density Polyethylene	Carrier bags, bin liners and packaging films.	4 LDPE
Polypropylene	Margarine tubs, microwaveable meal trays, also produced as fibers and filaments for carpets, wall coverings and vehicle upholstery.	5 PP
Polystyrene	Yoghurt pots, foam hamburger boxes and egg cartons, plastic cutlery, protective packaging for electronic goods and toys. Insulating material in the building and construction industry.	6 PS
Unallocated References	Any other plastics that do not fall into any of the above categories - for example polycarbonate which is often used in glazing for the aircraft industry	[illegible]

raw materials for goods production. 2.5 million tons of plastics raw materials are approximately produced annually in the UK. The distribution of polymer companies in UK is depicted in Figure and the plastic consumption in the UK by application is given in Figure 4.(BPF 2012)

It is estimated that the UK produces approximately 2·5 million tonnes of plastics raw materials annually. The companies known for manufacturing different forms of polymers and polymer products in UK are: DuPont, lyondellbasell, Sabic and Ineos. UK is among the top five plastic processors in the EU consuming approximately 4.8 million tonnes of materials. A lot of capital and personnel are required in plastic processing to satisfy the individual tasks due inert and assembly work. The number of people currently employed in the UK plastic industry is approximately 180,000.

Figure 4. A) Location of polymer companies in UK and B) Plastic consumption in UK by application

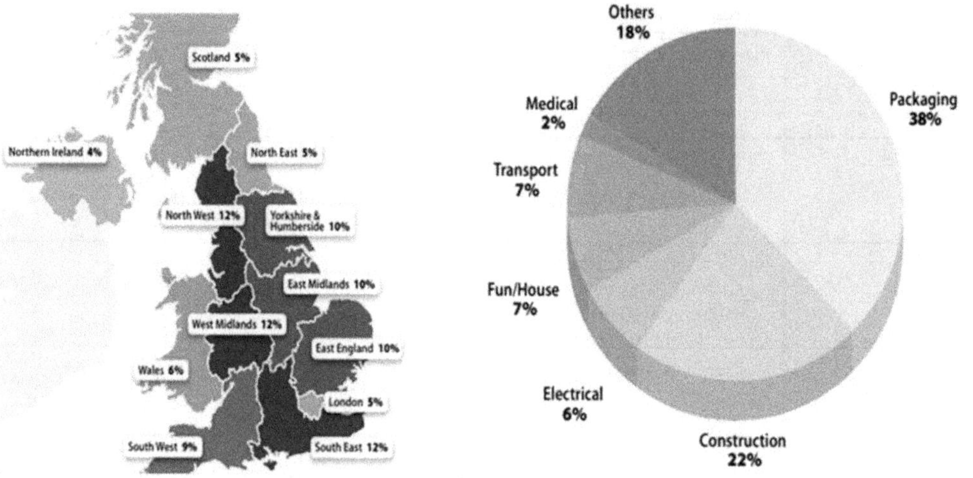

3. LIFECYCLE ASSESSMENT (LCA)

Lifecycle assessment is an analytical tool developed by government for complete evaluation of environmental aspects in a particular process or system and it allows scientist to keep record of a product from the time the raw materials are harnessed to the manufacturing till its final disposal. Public opinion and environmental directives within the last two decades have played a cardinal role in the development of methods to evaluate the environmental impacts of an industrial process. It is believed that LCA is a tool that helps in the development of public policy and design decisions (Subramanian 2000). All aspects of environment are evaluated in a process through all stages of its life cycle. LCA has the ability of creating a tentative inventory elucidating all inputs and outputs including energy sources and environmental emissions of an industrial process. LCA helps in making decisions with regard to tradeoffs in a process and the level of capital to invest in a process. LCA of plastic production measures environmental impacts associated with materials manufacturing, to their packaging and distribution, their final use and disposal.

Life cycle assessment of plastics in Europe from converter demand to disposal and recovery showed that 25.2 Mtonne of plastics in 2012 ended up in the post-consumer waste streams of which 15.6 M tonne were recovered while 9.6 Mtonne were disposed of. Figure 5 gives an elucidation of the LCA of plastic in Europe.

Figure 5. Life cycle assessment of plastic in Europe in 2012

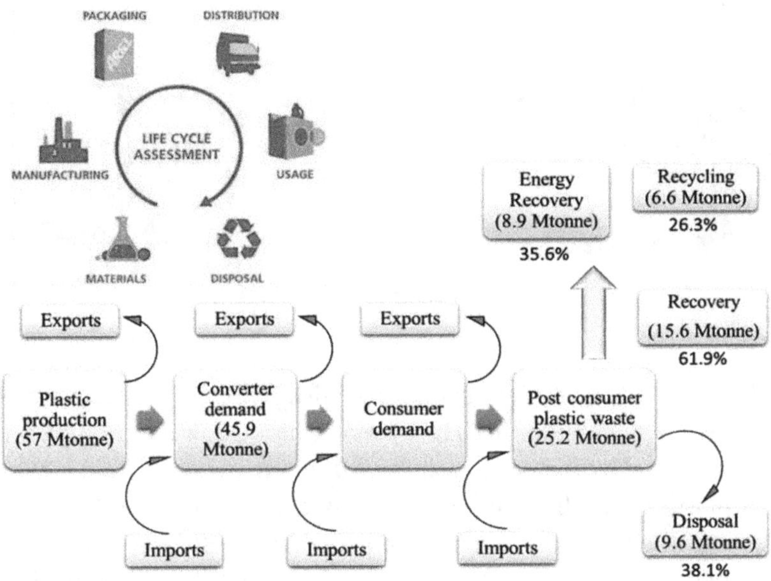

3.1 Polymer Waste

The consumption and generation of plastic solid waste has increased remarkably over decades since industrial revolution. It was estimated that each individual in the world in 1990 produced an average of 250 kg of MSW generating in total 13 Mtons (Al-Salem *et al.*, 2009). Plastic waste currently represents significant amount in MSW in 2012. 32 Mtons and 25.2 Mtons of plastics reached the post-consumer waste stream in America and Europe respectively. Environmental regulations and increasing awareness coupled with inadequate space and cost of disposal to the landfills are forcing for considerations of other environmentally and economically viable alternatives such as recycling and recovery to plastic disposal (Zia et al., 2007; Howard, 2002). Recycling has received much attention in the past few decades.

Plastics make up 32 Mtons almost 13 percent of the total municipal solid waste (MSW) generated in America. Nearly 14 Mtons of plastics were used in packaging and containers production, 11 million tons and 7 million tons as durable goods (appliances) and non-durable goods (cups) respectively. Depending on the parent material, various forms of plastics have different recycling rates. The overall plastics recycling rate in 2012 was 9 percent (28 Mtons) but the recycling rate for some categories of plastic products was higher in the same year e.g. 31 percent of PET

Figure 6. A) Total MSW generation (by material) in 2012 251 million tons before recycling and B) Total MSW recovery (by material)

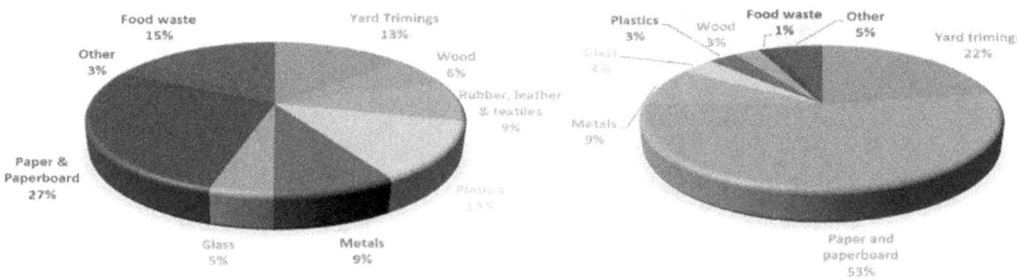

bottles and 28 percent of white translucent bottles (HPDE) were recycled (USEPA 2012). Total MSW generated in 2012 by material is shown in Figure 2 while the total MSW recovered by material is showed in Figure 6.

In 2012, plastics recycling and energy recovery reached 15.6 Mtonne approximately 61.9% of the post-consumer waste generated in Europe. The recycling and recovery of total plastic waste in Europe from 2006 to 2012 has showed increasing trend with an annual growth rate of 4% while the disposal through landfilling continues to diminish at an annual reduction rate of 5.5% which shows a considerable positive development. Europe is one the road towards zero plastic to landfills. Total plastic waste recycling and recovery 2006-2012 is shown in Figure 7.

Figure 7. Total plastic waste recycling and recovery 2006-2012

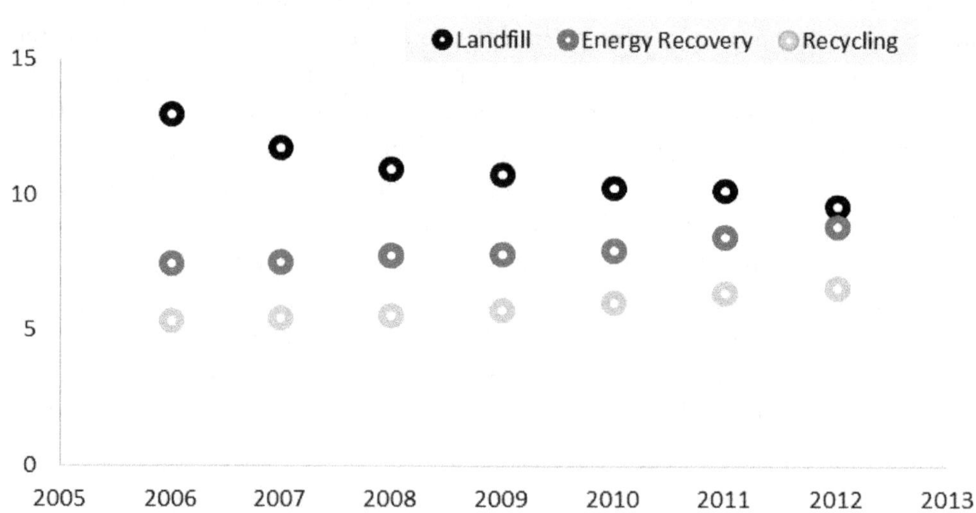

4. ENVIRONMENTAL CONCERNS AND TOXICITY

The consumption and disposal of polymers have enormous effect to the environment and public health. There are evidences that the chemical building blocks of plastics contains substances that affect the environment and human health. This section will discuss the major environmental effects associated with polymer production. These effects include: excessive consumption of nonrenewable resources, emission of toxic substances such as heavy metals from uncontrolled incineration, and pollution of groundwater and surface water by landfill leachates.

There is increasing concern about the consumption of nonrenewable resources generally petroleum for the production of polymers. About 330 Mbarrels of oil (5% of total U.S. oil production) is consumed in plastic manufacturing. Petroleum products such as natural gas, liquid petroleum gas (LPG) and natural gas liquids (NGL) are used as feedstock and fuel for the production of plastics in the United States. About 190 Mbarrels of LPG and NGL equivalent to 2.7% of total U.S. petroleum consumption were used as feedstock and an equivalent of 1 Mbarrels were consumed as fuel in resin and plastic industry in 2010. About 412 billion cubic feet (Bcf) of natural gas in 2010 an equivalent of 1.7% of total U.S. natural gas consumption were utilized in plastic and resin production. In addition to petroleum product consumption, plastic manufacturing consumes a lot of electricity. It was estimated that 65 billion kW of electricity (1.7% of U.S. consumption) were consumed plastic manufacturing (EIA 2014)

Plastic bags presents a significant danger to animal life because they are considered by the sea animals as food. Plastic bags floating in the sea when brightly colored or containing residues of food tend to attract the living species within the water bodies. Marine and land animals choke to death when strangled by plastics, causing so much distress and pain. The animals die of slow death from infection and starvation when they consume plastic bags. Plasticizers such as: Bisphenol A (BPA) and the phthalates are produced in a large quantity and they are regularly detected in aquatic environment due to their continuous release into the environment. Plasticizers used as additives in PVC and other plastics production are known to contain substances that pose adverse health effects to non-mammalian and mammalian species. They impede animal development by disrupting hormones production, interfere with normal reproduction by reducing the normal mitochondrial transport and induce cellular stress by increasing fatty acid oxidation (Justine *et al.*, 2014; Jorg *et al.*, 2008). Some countries still use banned plasticizers as clouding agents in food and beverages which attracts serious international attention (Tzung-Hai *et al.*, 2011).

The growth of the plastic industry has a significant impact on the management of municipal solid waste by incineration and landfilling. Synthetic polymers are regarded as major SW environmental pollutants because they are resistant to physical and chemical degradation. Lack of available space and cost of disposal through the landfills have made incineration a viable alternative in plastic management, however, incineration causes the emission of toxic bottom and fly ash which contain trace levels of heavy metals such as cadmium and lead (Curlee *et al.*, 1991). The presence of these substances brings about surface and groundwater pollution problems. Landfilling is the most common route for disposing of polymer waste. Brominated flame retardants (BFR) used as synthetic additives in plastic production are persistent and pose adverse effects on health. They are released in landfills leachates to the environment and they bio-accumulate (Natsuko *et al.*, 2014; Ying *et al.*, 2014)

5. PLASTIC WASTE MANAGEMENT

These are processes generally taken to minimize the effects of plastics on health, the environment and aesthetics. It is the systematic administration of activities that provide for the collection, source separation, storage, transportation, transfer, processing, treatment and disposal of plastic wastes. Plastic waste management is a distinct practice from resource recovery and it focuses on reducing the rate of consumption of natural resource and disposal of waste by landfilling. Figure 9 gives a graphical illustration of waste management hierarchy.

Philosophies of Waste Disposal and Management

- **Out of site out of mind**: It causes widespread environmental damage. The philosophy persists and continue to pose serious problem to the environment.
- **Dilute and Disperse**: The philosophy believes that the solution to pollution is dilution, it is no longer suitable for waste disposal and it causes widespread marine pollution issues.
- **Concentrate and Contain**: This philosophy is currently the most popular in waste management, however it is very energy intensive and expensive.
- **Resource Recovery**: The philosophy requires large volume of waste and technology to convert wastes into useful materials.
- **Integrated Waste Management Facility (IWMF):** It is a complex set of alternatives that combines source reduction, recycling, composting, incineration and landfilling for efficient waste management.

Figure 8. Plastic waste management hierarchy

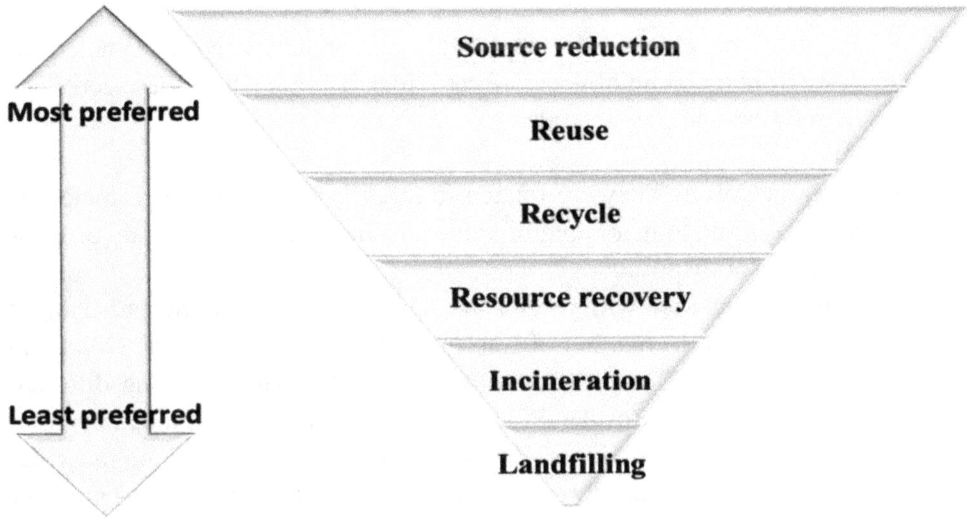

Figure 9. Functional elements of solid waste management

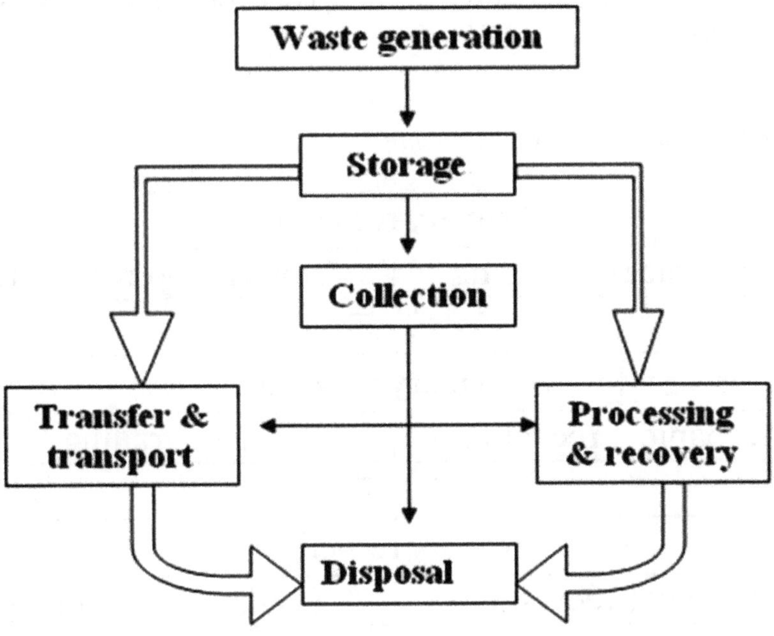

Functional Elements in Solid Waste Management

There are six functional elements in the activities associated with the management of plastic waste from the point of generation to final disposal sites. These activities are depicted in figure and they include:

- **Plastic Generation:** They are related to those activities in which materials are identified as no longer being of value and are either thrown away or gathered together for disposal.
- **Collection:** Concerned with those activities associated with the gathering of plastic wastes and the hauling of wastes to the location where the collection vehicle is emptied. Waste collection methods vary widely among different countries and regions. Residential, commercial and industrial locations have different collection procedures. Plastic wastes are collected through bring-schemes or kerbside procedures. Bring-schemes collection methods are very slow compared to the kerbside methods because they require the willingness and commitment of the public through economic incentives. Kerbside collection methods combine most recyclables together e.g. papers, glass, plastic containers etc. but the recovery rate of plastic using this method is around 30-40% (Jefferson *et al.*, 2009).

Figure 10. Categories of plastic recycling

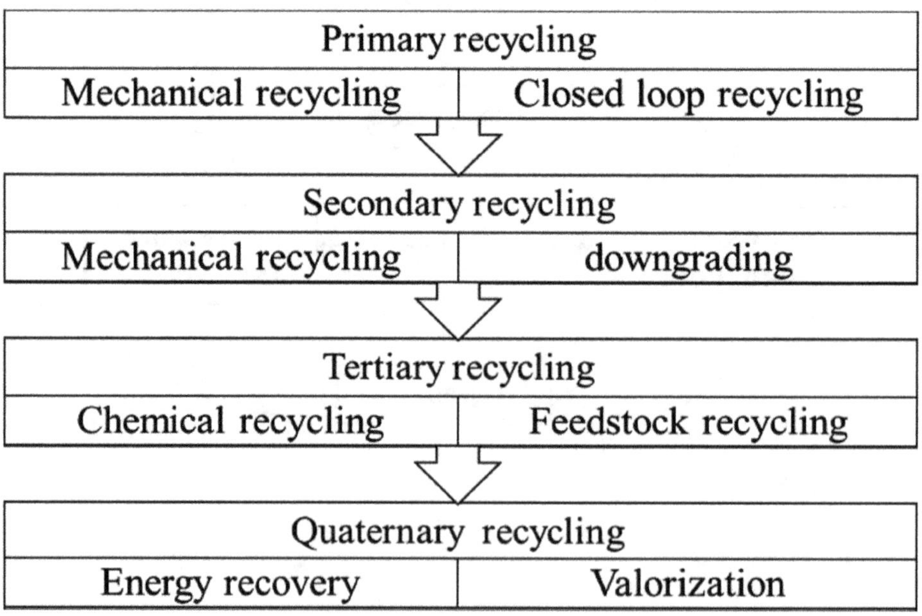

- **On-Site Handling (Sorting, Storage & Processing):** Related to those activities associated with the handling, storage, and processing of plastic wastes at or near the point of generation. Sorting is carried out through manual or automated methods to separate plastics from other recyclables. Manual methods of separation depend on the use of color and physical properties to distinguish between different forms of plastics. Material recovery and recycling facilities use automated methods such as Fourier transform near infrared spectroscopy (FT-NIR) and optical recognition camera techniques to easily separate HDPE and PET bottles out of the stream. X-ray detection methods, ballistic separators, trommels, hydrocyclones, and density based air classification systems are also used for the separation of PVC containers and flexible packaging (Arvanitoyannis 2001; Fisher 2003). Sink/float separation can be used in the separation of polyolefin i.e. PP and all classes of PE from PET, PVC and PS. Air elutriation has been used in separation of labels (low density films) from denser plastics (Chandra 2007). Raman emission spectroscopy, froth flotation, thermal kilns and FT-NIR techniques have been used in reducing the PVC contaminants in PET flake (Marques *et al.*, 2000). Laser-sorting is a new approach that uses emission spectroscopy in the removal of polymers by type and impurities such as nylon and silicones from other forms polymers.
- **Transfer and Transport:** Those activities associated with the transfer of wastes from the smaller collection vehicle to the larger transport equipment and the subsequent transport of the wastes, usually over long distance, to the disposal site.
- **Processing and Recovery:** These are techniques, equipment and facilities used both to improve the efficiency of the other functional elements and to recover useable materials, conversion products, or energy from plastic wastes.
- **Disposal**: They are activities associated with ultimate disposal of plastic wastes.

6. PLASTIC WASTE MANAGEMENT OPTIONS

There are several methods in use to curtail the impacts of plastic waste to our health, species, environment and aesthetics. This section will discuss some of the alternatives for plastic waste management.

Source Reduction

Source reduction means decreasing the amount and toxicity of the waste we dispose. It promotes the use of products that are environmentally benign. Plastic waste is an important target for source reduction because of their rapid growth in MSW. Plastic waste end up as litter that affects aesthetics, health, environment and the economy. Source reduction involves such measures of eliminating unnecessary packaging of products and reducing the amount of item used in packaging will reduce waste volume. The use of alternate materials that will have less ecological effects in place of polymers tend to be a viable economic and environment-friendly approach. Some notable examples are the replacements of paper for plastics in supermarkets and that of synthetic fibers (nylon) for natural degradable fibers (cotton) respectively.

Reuse

The reuse of post-consumer packaging materials is known for a long time. It has been applied for remanufacturing of vehicles and electronic equipment and in industrial scale for the reuse of pallets and containers in haulage (Thompson *et al.*, 2009) Triple layer plastic bags are reused for cowpea storage in thousands of villages along West and Central African countries (Dieudonne *et al.*, 2014). PET bottles are utilized as building units in some parts of Africa replacing the traditional concrete blocks. They are known for their thermal and structural stability (Ashraf *et al.*, 2015). Other countries are known for reusing plastics to store medicine (Muluneh *et al.*, 2014).

Recycling

This is the process of recovering and reprocessing of waste plastic into useful products that maybe entirely different from their original form. Example is evident in the production of tables and chairs by melting down plastic bottles. There are four categories in plastic recycling, these include: primary, secondary, tertiary and quaternary. Figure 11 outlines the categories in plastic recycling

Incineration

Incineration is a waste treatment process that involves the combustion of organic substances contained in waste material. It converts combustible waste into ash, flue gas and heat at a temperature of around 1000°C. Incineration reduces the amount of plastic waste sent to the landfills however, it is associated with the environmental release of halogenated additives such as dioxins, furans and other polychlorinated biphenyls released upon the combustion of PVC (Gilpin *et al.*, 2003). Incineration of plastic waste as a management option is less prevalent in so many countries due to the adverse pollution problems however it is widely used in Asia due to the limited space for landfills. Incineration with energy recovery is one of the several waste to

Figure 11. The waste to energy process

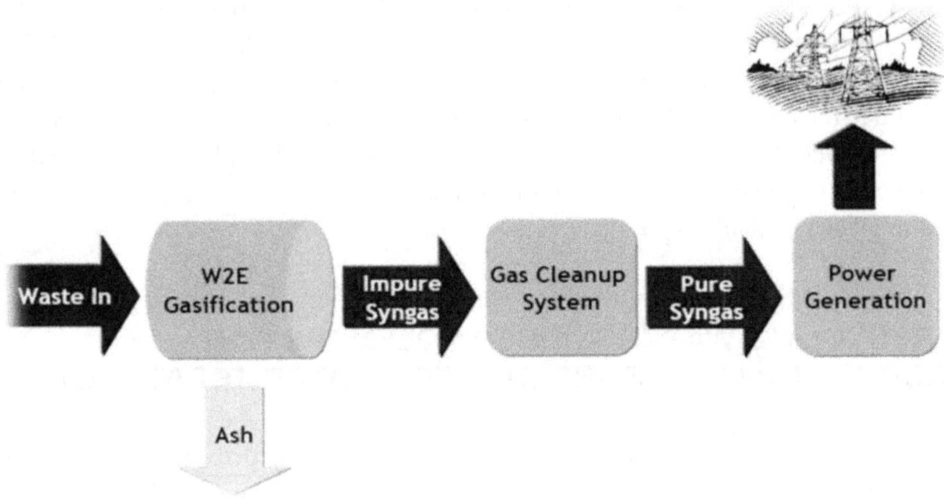

energy technologies such as: gasification, pyrolysis and anaerobic digestion. The waste to energy process is given in Figure 12.

Composting

Composting is a biochemical process of converting organic waste to humus-like material and it helps in the reduction of garbage. Composting can significantly reduce the amount of garbage on a household or municipal scale, and is less expensive. Composting is a natural process of rotting material down to make a soil like product which can be used on the garden. Composting provide fertilizers helps in reduction of waste, save water, and reduces the carbon footprint. Landfill degradation rate is a relatively slow process compared to degradation in compost piles. Biodegradable polymers are decomposed effectively using composting mixtures but it requires controlled conditions in centralized facilities.

Biodegradation

Landfilling

Plastics are difficult to recycle because both the manufacturers and consumers discard different forms of plastics. Private and public agencies are reluctant to investing in plastic recycling because it is mostly non-biodegradable plus the cost of collection, transport, sorting and degreasing. Landfills are the most common methods of waste disposal however disposal of plastic through landfills consumes a lot of space and is associated with long-term risks of contaminating groundwater and soils by persistent organic pollutants released through leachates of additives and degradation by product (Teuten et al. 2009). A major disadvantage of plastic disposal to the landfill is that none of the material used in the manufacturing process can be recovered. Some countries impose taxes on disposal of plastic to the landfill with the ultimate goal of increasing investments in energy recovery, recycling and attaining zero landfill disposal.

7. RECYCLING WASTE RUBBER POLYMERS FOR PRODUCTION OF ADSORBENTS FOR WATER TREATMENT

There are many methods to recycle waste rubber tires which are sub-divided into two categories. The first category involves physical methods of grinding the waste mechanically into smaller pieces to produce fine particles without altering chemical composition while the second category uses mechanochemical means to breakdown the three-dimensional network with the aid of different forms of energy. Thermal, mechanical, biological, chemical, ultrasonic and microwave treatments are used to convert the waste insoluble thermoset into a reprocessable elastomer (De *et al.*, 2005). The best method suitable for tire recycling is pyrolysis (Kaminsky *et al.*, 1995), pyrolysis is the thermal decomposition of rubbers in a closed system for the production of oil, gasses for reuse in petrochemical industries and solid remains such as carbon black for use as filler materials. There is a significant progress in the pyrolysis technology of waste rubber tire, however, despite the progress, pyrolysis is done on a limited scale due to the absence of market for the derived products (Saleh 2013; Saleh et al. 2016; Saleh and Danmaliki, 2016; Alansi et al. 2015; Saleh et al. 2014).

Researchers in recent years have focused more in the recycling and reuse of waste rubber tires among the initiative is the derivation of activated carbon after pyrolysis for various adsorption studies. Activated carbon from waste rubber tire have the ability of adsorbing different compounds however its adsorption capacity is relatively low compared with other adsorbent materials in use. There are various methods of improving the adsorption capacity of the fine powdered particles. Activated carbon from waste rubber tire was synthesized by pyrolysis at 500°C-700°C and later treated with activating agents such as zinc chloride and potassium hydroxide in a study conducted by (Nunes *et al.*, 2011). The adsorbent showed a mesoporous structure containing sulfur and zinc and it showed the possibility of waste water treatment.

The work conducted by (Zabaniotou et al., 2014) derived activated carbon from waste rubber tires by pyrolysis at 800°C in a fixed bed reactor and physical method of activation using CO2 for the removal of pesticide(bromopropylate) by adsorption from aqueous solution. The surface area of the activated carbon was found to reach 432 m2 g−1 and nearly 100% of the pesticide was removed in just 60mins of the experiment. The study of (Edward et al., 2010) derived activated carbon from ELTs by carbonization for the effective removal of dyes from effluents. It was observed that temperatures above 770K resulted in lower yield adsorbent.

Waste rubber tires were pyrolyzed in carbon dioxide and nitrogen atmospheres at various temperatures for char production (Manchon-Vizuete et al., 2005). The char was activated by oxygenation at different temperatures and were used for the removal of lead and copper from aqueous solutions. The study concluded that the optimum temperature for pyrolysis in carbon dioxide and nitrogen atmospheres is 550oC. The study of (Sani et al. 2016; Danmaliki et al. 2017; Alswat et al. 2016) used spectroscopic and computational studies to examine the chemistry of cadmium adsorption by activated from waste rubber tires in aqueous solutions. The adsorption process was found to be fast and equilibrium was achieved after 60mins of interaction. Finally, binding energy of cadmium to carbonyl groups was found to be higher than other functional groups present.

The study conducted by AlSaadi et al., 2013 synthesized activated carbon from ELT by physical activation and used it as an adsorbent for the removal of nickel and lead ions from aqueous solutions. The adsorbent was found to have more affinity to the lead ions (96% removal) and the adsorption process was believed to start from initial surface adsorption followed by intra-particle diffusion. The regeneration ability of the adsorbent was also tested and they believed it may be a substitute to the expensive commercial activated carbon currently in use for pollution clean-up. Danmaliki and Saleh 2017 synthesized different brands of activated carbon from waste rubber tires by microwave assisted chemical treatment and physical heating for the adsorption of phenolic compounds.

Granular particles of tire were carbonized and pyrolyzed by heating at 773K and 1223K respectively in nitrogen atmosphere for the production of active carbon in a study conducted by (Mui *et al.*, 2010). The product produced was activated by carbon di oxide in nitrogen atmosphere and later subjected to acid treatment. The synthesized adsorbent was used in the removal of dyes from waste water and it had a higher surface area and better yield. A study conducted by (Quek et al., 2009) synthesized activated carbon possessing high surface area (414 m2/g) without pyrolysis in a one-step cost effective approach. The study used thermochemical degradation of granular tires in a carbon di oxide environment.

ACKNOWLEDGMENT

The author(s) acknowledge the support of the chemistry department, and King Fahd University of Petroleum and Minerals, (KFUPM) Dhahran, Saudi Arabia.

REFERENCES

Alansi, A. M., Alkayali, W. Z., Al-qunaibit, M. H., Qahtan, T. F., & Saleh, T. A. (2015). Synthesis of exfoliated polystyrene/anionic clay MgAl-layered double hydroxide: Structural and thermal properties. *RSC Advances*, *5*(87), 71441–71448. doi:10.1039/C5RA10545E

Al-Saadi, A. A., Saleh, A. A., & Gupta, V. K. (2013). Spectroscopic and computational evaluation of cadmium adsorption using activated carbon produced from rubber tires. *Journal of Molecular Liquids*, *188*, 136–142. doi:10.1016/j.molliq.2013.09.036

Al-Salem, S. M., Lettieri, P., & Baeyens, J. (2009). Recycling and recovery routes of plastic solid waste (PSW): A review. *Waste Management (New York, N.Y.)*, *29*(10), 2625–2643. doi:10.1016/j.wasman.2009.06.004 PMID:19577459

Alswat, A. A., Ahmad, M. B., & Saleh, T. A. (2016). Zeolite modified with copper oxide and iron oxide for lead and arsenic adsorption from aqueous solutions. *Journal of Water Supply: Research & Technology - Aqua*, *65*(6), 465–479. doi:10.2166/aqua.2016.014

Arvanitoyannis, I., & Bosnea, L. (2001). Recycling of polymeric materials used for food packaging: Current status and perspectives. *Food Reviews International*, *17*(3), 291–346. doi:10.1081/FRI-100104703

Ashraf, M. H. M., & Ali, S. A. (2015). Reusing waste plastic bottles as an alternative sustainable building material. *Energy for Sustainable Development*, *24*, 79–85. doi:10.1016/j.esd.2014.11.001

British Plastic Federation (BPF). (2012). *UK Plastics Industry Capability Guide*. Retrieved from http://www.bpf.co.uk/Industry/Default.aspx

British Plastics Federation. (2008). *Oil consumption*. Retrieved from http://www.bpf.co.uk/Oil_Consumption.aspx

Chanda, M., & Roy, S. (2007). *Plastics technology handbook* (4th ed.). Boca Raton, FL: CRC Press.

Curlee, T. R., & Das, S. (1991). *Plastic Wastes (Management, Control, Recycling and Disposal)*. Noyes Data Corporation.

Danmaliki, G. I., & Saleh, T. A. (2017). Effects of bimetallic Ce/Fe nanoparticles on the desulfurization of thiophenes using activated carbon. *Chemical Engineering Journal*, *307*, 914–927. doi:10.1016/j.cej.2016.08.143

Danmaliki, G. I., Saleh, T. A., & Shamsuddeen, A. A. (2017). Response Surface Methodology Optimization of Adsorptive Desulfurization on Nickel/Activated Carbon. *Chemical Engineering Journal*.

De, Isayev, & Khait. (Eds.). (2005). *Rubber Recycling*. Taylor and Francis.

Baributsa, D., Djibo, K., Lowenberg-DeBoer, J., Moussa, B., & Baoua, I. (2014). J. Lowenberg-DeBoer, Bokar Moussa, Ibrahim Baoua, "The fate of triple-layer plastic bags used for cowpea storage. *Journal of Stored Products Research*, *58*, 97–102. doi:10.1016/j.jspr.2014.02.011

Edward, L. K. (2010). Tyre char preparation from waste tyre rubber for dye removal from effluents. *Journal of Hazardous Materials*, *175*(1-3), 151–158. doi:10.1016/j.jhazmat.2009.09.142 PMID:19854570

Fisher, M. (2003). Plastics recycling. In A. Andrady (Ed.), *Plastics and the environment* (pp. 563–627). Hoboken, NJ: Wiley Interscience.

Gilpin, R., Wagel, D., & Solch, J. (2003). Production, distribution, and fate of polycholorinated dibenzo-p-dioxins, dibenzofurans, and related organohalogens in the environment. In A. Schecter & T. Gasiewicz (Eds.), *Dioxins and health* (2nd ed.). Hoboken, NJ: John Wiley & Sons Inc.

Howard, G. T. (2002). Biodegredation of polyurethane: A review. *International Biodeterioration & Biodegradation*, *49*(1), 245–252. doi:10.1016/S0964-8305(02)00051-3

Independent Statistics and Analysis, US Energy Information Administration. (2014). *Frequently Asked Questions: How much oil is used to make plastic?*. Retrieved from http://www.eia.gov/tools/faqs/faq.cfm?id=34&t=6

Hopewell, J., Dvorak, R., & Kosior, E. (2009). Plastics recycling: Challenges and opportunities. *Phil. Trans. R. Soc. B*, *364*(1526), 2115–2126. doi:10.1098/rstb.2008.0311 PMID:19528059

Oehlmann, J., Oetken, M., & Schulte-Oehlmann, U. (2008). A critical evaluation of the environmental risk assessment for plasticizers in the freshwater environment in Europe, with special emphasis on bisphenol A and endocrine disruption. *Environmental Research*, *108*(2), 140–149. doi:10.1016/j.envres.2008.07.016 PMID:18949832

Mathieu-Denoncourt, Wallace, de Solla, & Langlois. (2014). Plasticizer endocrine disruption: Highlighting developmental and reproductive effects in mammals and non-mammalian aquatic species. *General and Comparative Endocrinology*.

Kaminsky, W., & Sinn, H. (1996). *Book Recycling and Recovery of Plastics* (J. Brandrup, M. Bittner, W. Michaeli, & G. Menges, Eds.). Munich: Hanser Publishers.

Shen, L., & Worrell, E. (2014). Plastic Recycling. In Handbook of Recycling (pp. 179-190). Academic Press.

Manchon-Vizuete, E., Macias-Garcia, A., Gisbert, A. N., & Fernandez-Gonzalez, C. (2000). Use of froth flotation to separate PVC/PET mixtures. *Waste Management (New York, N.Y.)*, *20*(4), 265–269. doi:10.1016/S0956-053X(99)00333-5

Mui, E. L. K., Cheung, W. H., Valix, M., & McKay, G. (2010a). Dye adsorption onto activated carbons from tyre rubber waste using surface coverage analysis. *J. Coll. Interf. Sci.*, *347*(2), 290–300. doi:10.1016/j.jcis.2010.03.061

Ademe, M., & Mekonnen, Z. (2014). Repeated reuse of insulin injection syringes and incidence of bacterial contamination among diabetic patients in Jimma University Specialized Hospital, Jimma, Ethiopia. *Asian Pac J Trop Dis*, *4*(Suppl 2), S712–S716. doi:10.1016/S2222-1808(14)60712-X

Kajiwara, N., Hirata, O., Takigami, H., Noma, Y., Tachifuji, A., & Matsufuji, Y. (2014). Leaching of brominated flame retardants from mixed wastes in lysimeters under conditions simulating landfills in developing countries. *Chemosphere*, *116*, 46–53. doi:10.1016/j.chemosphere.2014.01.025 PMID:24560282

Nunes, M. R., Perez, G. M., Loguercio, L. F., Alves, E. W., Carreno, N. L. V., Martins, J. L., & Garcia, I. T. S. (2011). Active carbon preparation from treads of tire waste for dye removal in waste water. *Journal of the Brazilian Chemical Society*, *22*(11), 2027–2035. doi:10.1590/S0103-50532011001100002

PlasticsEurope. (2013). *Plastics- The facts 2013: an analysis of European latest plastics production, demand and waste*. PlasticsEurope.

Quek, A., & Balasubramanian, R. (2009). Article. *Journal of the Air & Waste Management Association*, *59*, 747. doi:10.3155/1047-3289.59.6.747 PMID:19603742

Richard, S. (2003). *Polymers, Recycling* (3rd ed.). Reference Module in Chemistry, Molecular Sciences and Chemical Engineering, from Encyclopedia of Physical Science and Technology.

Stein, R. S. (1992, February). Polymer recycling: Opportunities and limitations[Colloquium Paper]. *Proceedings of the National Academy of Sciences of the United States of America*, *89*(3), 835–838. doi:10.1073/pnas.89.3.835 PMID:11607263

Saleh, T. A., Tuzen, M., & Sarı, A. (2017). Polyethylenimine modified activated carbon as novel magnetic adsorbent for the removal of uranium from aqueous solution. *Chemical Engineering Research & Design*, *117*, 218–227. doi:10.1016/j.cherd.2016.10.030

Saleh, T. A. (2013). A strategy for integrating basic concepts of nanotechnology to enhance undergraduate nano-education: Statistical evaluation of an application study. *Journal of Nano Education*, *4*(1-2), 1–7.

Saleh, T. A., Muhammad, A. M., & Ali, S. A. (2016). Synthesis of hydrophobic cross-linked polyzwitterionic acid for simultaneous sorption of Eriochrome black T and chromium ions from binary hazardous waters. *Journal of Colloid and Interface Science*, *468*, 324–333. doi:10.1016/j.jcis.2016.01.057 PMID:26854746

Saleh, T. A., & Danmaliki, G. I. (2016). Influence of acidic and basic treatments of activated carbon derived from waste rubber tires on adsorptive desulfurization of thiophenes. *Journal of the Taiwan Institute of Chemical Engineers*, *60*, 460–468. doi:10.1016/j.jtice.2015.11.008

Sani, H., Ahmad, A., & Saleh, T. A. (2016). 2016, Synthesis of Zinc Oxide/Talc Nanocomposite for Enhanced Lead Adsorption from Aqueous Solutions. *RSC Advances*, *6*(110), 108819–108827. doi:10.1039/C6RA24615J

Subramanian, P. M. (2000). Plastics recycling and waste management in the US. *Resources, Conservation and Recycling*, *28*(3-4), 253–263. doi:10.1016/S0921-3449(99)00049-X

Teuten, E. L., Saquing, J. M., Knappe, D. R. U., Barlaz, M. A., Jonsson, S., Bjorn, A., & Takada, H. et al. (2009). Transport and release of chemicals from plastics to the environment and to wildlife. *Phil. Trans. R. Soc. B*, *364*(1526), 2027–2045. doi:10.1098/rstb.2008.0284 PMID:19528054

Thompson, R. C., Moore, C. J., vom Saal, F. S., & Swan, S. H. (2009). Plastics, the environment and human health: Current consensus and future trends. *Phil. Trans. R. Soc. B*, *364*(1526), 2153–2166. doi:10.1098/rstb.2009.0053 PMID:19528062

Yen, T.-H., Lin-Tan, D.-T., & Lin, J.-L. (2011). Food safety involving ingestion of foods and beverages prepared with phthalate-plasticizercontaining clouding agents. *Journal of the Formosan Medical Association, 110*(11), 671–684. doi:10.1016/j.jfma.2011.09.002 PMID:22118310

USEPA. (2008). *Municipal Solid Waste in the United States: 2007 Facts and Figures. Executive summary.* Office of solid waste management and emergency response (5306P), EPA530-R-08-010, November. Available at: www.epa.gov

USEPA. (2012). *Municipal Solid Waste Generation, Recycling, and Disposal in the United States: Facts and Figures for2012*. Available at: www.epa.gov

Li, Y., Li, J., & Deng, C. (2014). Occurrence, characteristics and leakage of polybrominated diphenyl ethers in leachate from municipal solid waste landfills in China. *Environmental Pollution, 184*, 94–100. doi:10.1016/j.envpol.2013.08.027 PMID:24041483

Zabaniotou, A. A., Antoniou, N., & Stavropoulos, G. (2014). Novel sorbent materials for environmental remediation via depolymerization of used tyres. *Desalination and Water Treatment*, 1–10.

Zia, K. M., Bhatti, H. N., & Bhatti, I. A. (2007). Methods for polyurethane and polyurethanevcomposites, recycling and recovery: A review. *Reactive & Functional Polymers, 67*(8), 675–692. doi:10.1016/j.reactfunctpolym.2007.05.004

Chapter 9
Characteristics of Chitosan Nanoparticles for Water and Wastewater Treatment:
Chitosan for Water Treatment

Cayla Cook
Mississippi State University, USA

Veera Gnaneswar Gude
Mississippi State University, USA

ABSTRACT

Chitosan is a naturally occurring biopolymer originating from several microbial species as well as crustacean species, such as shrimp and lobster. Chitosan has excellent physical and chemical properties that allow its use in various environmental applications especially in water treatment. It is a biodegradable polymer, and it is inexpensive providing an environmentally friendly and economic option for water and wastewater treatment. Chitosan offers a myriad of applications through chemical coagulation and flocculation, antimicrobial properties, adsorption capabilities, and nanofiltration and can provide a sustainable route for water and wastewater treatment. This book chapter elaborates the recent developments in chitosan applications in water and wastewater treatment.

DOI: 10.4018/978-1-5225-2136-5.ch009

INTRODUCTION

Water---an innate right---is somehow squandered even in the 21st century. Approximately 10,000 people die as a result of inadequate water supply and related third-world issues every day (UNwater, 2015). Currently, civilizations are growing at twice the rate of the availability of water resources, and it is projected that up to 1.8 billion people will experience absolute water scarcity by 2025 (UNwater, 2015). The importance of economical, clean water cannot be overstated in a society facing environmental issues such as pollution of water resources, droughts, and increasingly dwindling freshwater availability.

Fortunately, modern day science might have the answer in a naturally occurring polymer: chitosan. Chitosan is simply derived by the deacetylation of shellfish exoskeletons. Chitin can also be found in fungi, cell walls, and yeast. Although small, this nanoparticle continues to substantially impact water treatment through its idealistic properties, diverse applications, and economical potential. These diverse applications of chitosan include coagulation and flocculation for water and wastewater treatment, antimicrobial properties, adsorption capabilities of dyes and heavy metals, and desalination through nanofiltration. Many physical and biological properties of chitosan produce a linear polysaccharide ideal for water and wastewater treatment applications.

CHITOSAN PROPERTIES AND PREPARATION

Chitin, as shown in Figure 1, is the acetylated precursor to chitosan. The removal of these acetyl functional groups allows for higher solubility due to the increase in positive charge. Due to its deacetylation, chitosan is considered a pseudo-natural biopolymer which is predominantly characterized by its degree of deacetylation (DD) and molecular weight (MW). Considering that these two properties are directly proportional, the degree of solubility can be seen as inversely proportionate to the MW and directly proportionate to the DD. The degree of deacetylation is a molar percentage which differentiates chitin from chitosan at precisely 50% mol. This 50% DD allows for solubility in acidic aqueous solutions.

Acetylated Chitosan, or Chitin, is known to exist in two allomorphs, α and β. The more abundant α Chitin is found within traditional yeast, fungi, shrimp, crab, krill, lobster, and crawfish. The less abundant β Chitin is found within the protein in squid pens and is not an industrial source. Beyond knowledge of the biopolymer's source, differentiation can be determined through Nuclear Magnetic Resonance (NMR) spectroscopy and infrared in conjunction with X-ray diffraction (Younes & Rinaudo, 2015). Moreover, whereas α Chitin contains two antiparallel molecules

Figure 1. Chitin, chitosan, cellulose (Alvarenga, 2011)

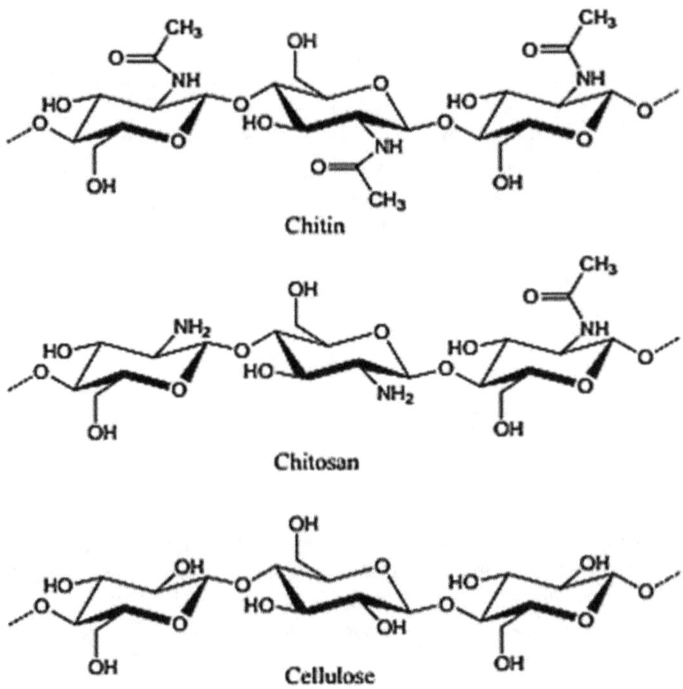

per unit cell, β Chitin contains merely one. β Chitin is more reactive; however, both α and β allomorphs remain insoluble.

Chitosan's solubility can also be affected by the distribution of acetyl compounds as well as ionic concentration. Chitosan's cationic property enables it to strongly adsorb negatively charged toxins. This positive charge can be modified by pH, while chitosan acts with inverse efficacy to its pH for coagulation and flocculation. This capability of changing solubility and cationic charge through pH causes both desirable and undesirable aspects considering that the extent of coagulation and flocculation can also be manipulated to repel particles and prevent agglomeration.

Chitosan Preparation

Chitosan can be prepared through both enzymatic and chemical routes. Chemical route is much more common due to production rates and greater economic viability. Preparation begins with extraction from one of its various sources (Figure 2); however, raw chitin is primarily acquired from the cuticles of shellfish. Both chemical and biological extraction suffice with chemical being the predominant methodology. The extraction process begins with grinding and washing of raw materials followed

Figure 2. Generalized Chitosan Preparation Process

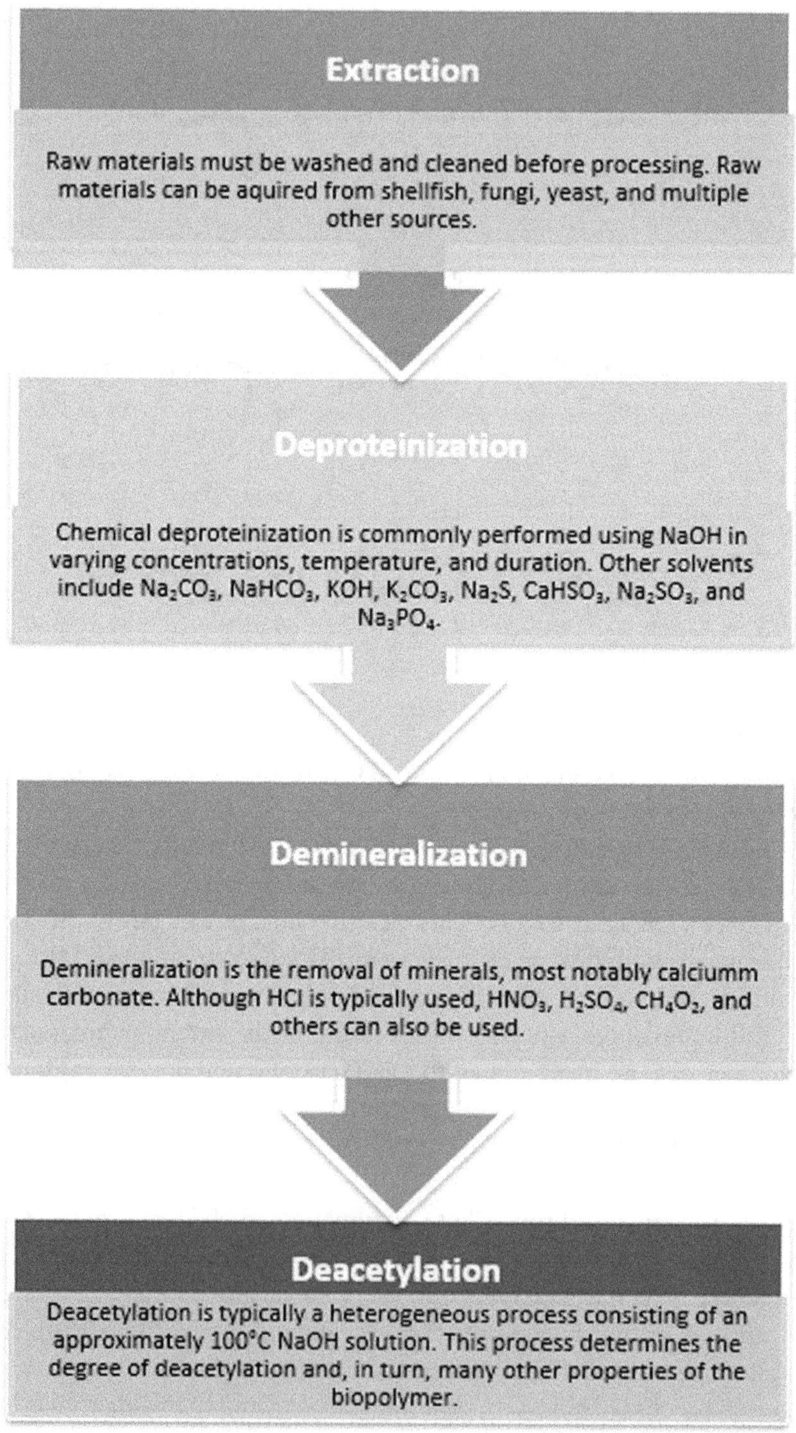

by deproteinization. Deproteinization essentially depolymerizes the biopolymer while removing any protein which is the primary cause of allergic reactions to shellfish. Reagents used for deproteinization include sodium hydroxide (NaOH), sodium bicarbonate ($NaHCO_3$), sodium carbonate (Na_2CO_3), potassium hydroxide (KOH), potassium carbonate (K_2CO_3), and several others. As shown in Table 1, there are multiple concentrations, temperatures, and durations used for NaOH, the most common deproteinization solvent. These variations include 0.125 M to 2.5 M concentration, 25°C to 100°C, and one half hour to overnight soaking in the solution.

Following deproteinization, demineralization removes calcium carbonate as well as other minerals. Although hydrochloric acid (HCl) solution is generally used, nitric acid (HNO_3), sulfuric acid (H_2SO_4), and others will also achieve demineralization. Similar to deproteinization, demineralization's solvent and extent vary with raw materials. Generally two to three hours will suffice; however, much longer incubation times can be used. As the incubation time increases, the potential for degradation of the polymer increases as well. Additionally, the greater the temperature of the process, the more quickly and effectively the solvent will penetrate the chitin matrix. Moreover, the demineralization process may directly affect the physicochemical properties through "high temperatures, longer incubations, high acid concentrations, and granulometry" resulting in potentially undesirable structural changes (Younes & Rinaudo, 2015).

Following deproteinization and demineralization, the chitin must be extracted and dried with acetone. In order to convert the chitin to chitosan, the material must then be deacetylated. This is one of the most important processes in the preparation of chitosan due to its effect on the biopolymer's properties through acetyl group distribution, MW, and viscosity. These properties can be affected by temperature and incubation period. Although this chemical process is performed either heterogeneously or homogeneously, a hot, heterogeneous NaOH solution is most common. In contrast, a homogeneous reaction process would require 12-24 hours of pre-swollen chitin in an alkali solution. The alkali solutions used for either heterogeneous or homogeneous reactions must penetrate the amorphous region and typically leave some crystalline structures intact. These solutions will yield a maximum DD of 75-85%.

These methodologies for preparation directly affect the simple physicochemical properties of the chitosan produced. The differential characteristics enable diverse applications through many physical forms such as chitosan powder, dehydrated chitosan flakes, crystalline chitosan, chitosan nanoparticles, and even chitosan as a modification to pre-existing surfaces. Chitosan surface engineering has proven to be an incredibly complex category, ranging from chitosan biosorbents to chitosan acting as an active layer for nanofiltration surfaces. These various physical forms allow for maximum efficacy of the biopolymer for any desired application.

Table 1.Chitosan preparation procedures

	Deproteinization				Demineralization			
Source	**NaOH Concentration**	**Temperature (°C)**	**Number of Baths**	**Duration (h)**	**HCl Concentration**	**Temperature (°C)**	**Duration (h)**	**References**
12 species of crustaceous and cephalopods	0.3 M	80-85	2 to 7 depending on source	1 h each	0.55 M	Room	15 min to 1 hour repeated 20-5 times depending on source	(Tolaimate, Desbrierews, Rhazi& Alagui, 2003)
Shrimp	0.125 M	100	1	0.5	1.25 M	Room	1	(Madhavan & Nair, 1974)
	0.75 M	100	1	-	-	-	-	
Shrimp	1.25 M	100	1	0.5	1.57 M	20-22	1-3	(Moorjani, Achutha & Khasim, 1975)
Crab	0.5 M	5	1	2	1.57 M	Room	5	(Muzzarelli, Tanfani, Emanuelli & Gentile, 1980)
Crab	1 M	80	1	3	1 M	Room	12	(Mima, Miya, Iwamoto & Yoshikawa, 1983)
Crab	1M	100	1	36	2 M	Room	48	(Shimahara, Ohkouchi & Ikeda, 1992)
Crab	1M	100	3	72	1 M	Room	-	(Hackman, 1954)
Crab	1.25 M	85-90	3	24	1.37 M	Room	24	(Broussignac, 1968)
Crab/Lobster	2.5 M	Room	3	72	11 M	20	4	(BeMiller & Whistler, 1963)
Krill	0.875 M	90-95	1	2	0.6 M	Room	2	(Anderson, de Pablo & Romo, 1978)

continued on following page

Table 1. Continued

	Deproteinization				Demineralization			
Source	NaOH Concentration	Temperature (°C)	Number of Baths	Duration (h)	HCl Concentration	Temperature (°C)	Duration (h)	References
Lobster	1 M	100	5	12	2M	Room	5	(Hakman & Goldberg, 1974)
Squid	2 M	Room	2	One night	1 M	Room	One night	(Kurita et. al, 1993)
	2 M	100		4				
Lobster	10% w/v	100	1	2.5	10% HCl 90% formic	Room	18	(Horowitz, Roseman & Blumental, 1957)
Krill	3.5% w/v	25	1	2	3.5% w/v	20	1.5	(Brzeski, 1982)
Lobster	5% w/v	80-85	2	0.5	5% w/v	70	4	(Blumberg, Southall, van Rensburg & Volckman, 1951)
Crawfish	3.5% w/v	65	1	2	1 M	Room	0.5	(No, Meyers & Lee, 1989)
Crab	1 M	50	1	6	1 M	20	3	(Brine & Austin, 1981)
Shrimp	1% w/v	65	1	1	0.5 M	Room	-	(Wu & Bough, 1978)
Shrimp	3% w/v	100	1	1	1 M	Room	0.5	(Bough, Salter, Wu & Perkins, 1978)
Shrimp	4% w/v	100	1	1	5% w/v	Room	-	(Sluyanarayana, Yashodha, Mahendrakar & Puttarajappa, 1987)

Note. Reprinted from Younes, I., & Rinaudo, M. (2015). Chitin and chitosan preparation from marine sources. Structure, properties and applications. Marine drugs, 13(3), 1133-1174.

CHITOSAN CHARACTERIZATION

Certain methodologies for characterization of chitosan can ensure a thorough deacetylation and transformation from chitin to chitosan. Determination of the DD is a high priority. Methods such as H-NMR Spectroscopy, Titration, and Infrared spectroscopy are primary means of characterization and determination of DD.

Fourier Transform Infrared Spectroscopy

Infrared Spectroscopy (IR) provided below determines the organic functional groups present within the chitosan produced by Czechowska-Biskup et al. (2012). It should be understood that as the acetyl functional group is removed, the number of IR bands decreases to show the loss of the presence of amides. As shown in Figure 3, bands referencing hydroxyl, carbon, carboxyl, and amide bonds are all present. In order to determine the DD, multiple equations can be utilized. However, due to the inexact numerical nature of an FTIR, the calculations themselves are only estimations. For this particular sample, the DD was determined to be 77.84 ±2.36, 66.11 ± 2.36, -12.5 ± 23.30, and 74.60 ±3.60 through the use of FTIR. More importantly, the bands become less pronounced upon increased DD.

Figure 3. Chitosan Fourier transform infrared spectroscopy (Czechowska-Biskup, R., Jarosińska, D., Rokita, B., Ulański, P., & Rosiak, J. M., 2012)

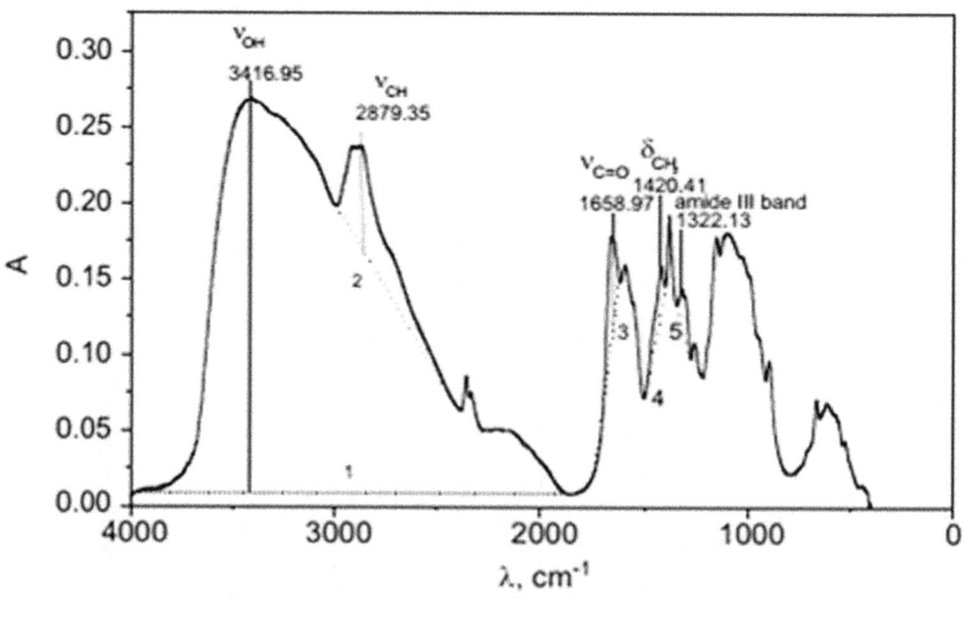

Zeta Potential

Zeta potential is a significant characterization of chitosan nanoparticles not only for the calculation of Degree of Deacetylation but also due to properties dependent upon surface charge interactions, i.e. antimicrobial, adsorption, and coagulation. The greater the positive charge of the chitosan nanoparticles, the greater will be the antimicrobial properties exhibited by modifying the permeability of the cellular membrane. in addition, a higher zeta potential present in the nanoparticle can further attract and coagulate toxins within a solution. Research by Qi et al. (2004) determines a range from 30.8 to 68.9 milliVolts (mV) present in nanoparticles purchased from Chitosan Company of Pan'an. The mean charge is shown to be at 51 mV and the DD was determined to be 85%. The DD directly affects zeta potential by removal of negatively charged acetyl groups while the MW is also important due to the quantity per monomer. In further studies focusing on cationic-anionic ratios, an average zeta potential of 28.0 ±0.6, 28.8 ±0.7, 30.3 ±0.4, 31.4 ±0.5, and 32.4 ± 11.1 mV with increasing particle size was determined, respectively (Abbas, 2010).

GENERAL APPLICATIONS OF CHITOSAN

Being one of the most abundant biopolymers in the world, chitosan's slew of functionalities and properties only add to the vast applications and uses spanning fields from biomedicine to paper finishing. Within the biomedical field, chitosan is being considered for controlled drug release due to its swelling in acidic medium, antacid properties, and prevention of stomach irritation. In ophthalmology, chitosan is being researched for contact lens use due to several properties including "optical clarity, mechanical stability, sufficient optical correction, gas permeability, particularly towards oxygen, wettability, and immunological compatibility" (Kumar, 2000). Chitosan is an ideal polymer for the use in cosmetics and hair care products. More importantly, it also has a spectrum of uses within water treatment and wastewater treatment systems such as coagulation and flocculation, antimicrobial properties and uses as a biosorbent material to remove dyes, metal ions, and salt. Within the coagulation and flocculation functionality alone, many pollutants and suspended particles have successfully been removed and separated including algae, oil-in-water emulsions, brackish water, and lead and chromium wastewater.

Nontoxic and Allergen Considerations

In addition to desirable physicochemical properties for water and wastewater treatment, chitosan is surprisingly nontoxic to those with shellfish allergies and can be used as an antihistamine. Despite its shellfish origins, it remains non-toxic to those with shellfish allergies when applied externally. In 2011, a study was conducted to consider if HemCon, a chitosan bandage mandated for soldiers which stops bleeding, would cause adverse reactions from patients with shellfish allergies. The study concluded that 100% of the patients with predispositions to shellfish allergies were unaffected by the chitosan powder applied externally to wounds. However, this was contradicted when a very small percentage of the public reported delayed onset contact dermatitis to a chitosan cream. Although for some, shellfish increases allergic responses, chitosan could potentially decrease anaphylaxis responses for those with peanut allergies. Currently being tested on mice, this treatment has a 47% efficacy (Bae et. al., 2013). Its antihistamine properties have also resulted in a patent granted in 2011 for topical application.

WATER TREATMENT APPLICATIONS

Due to the interrelated concepts of water treatment and wastewater treatment, coagulation and flocculation are present for both processes as well as for antimicrobial purposes. The following sections will not only discuss studies in which chitosan is used for chemical coagulation and flocculation or antibacterial purposes, but also the mechanisms through which it achieves the goal.

Coagulation and Flocculation in Water Treatment

Among many other preferential processes, chemical coagulation is widely used due to its efficient removal of toxins. This coagulation is due to a chemical's cationic charge which attracts and conglomerates negatively charged toxins. Traditionally, aluminum sulfate, "alum", is used for this process; however, mass mildly toxic chemical addition and sludge generation is a downfall of this option. Considering this, chitosan is a viable option due to biocompatibility which allows degradation of sludge by microorganisms resulting in less landfill usage. As a result, the use of chitosan for many bioeffluents has been explored including food waste, oil-in-water emulsions, effluents containing dyes, lead and chromium wastewater, partially purified sewage, and algal waters. It has been observed that increased MW results in greater coagulation; however, the relationship is nonlinear. Moreover, a greater DD at equivalent MW further optimizes coagulation. This somewhat necessitates a balancing act between MW and DD seeing as an increased DD decreases MW. A greater MW creates greater removal of NTUs and suspended solids (SS) with a reduction in biochemical oxygen demand (BOD) and chemical oxygen demand (COD). Contrastingly, a greater DD which decreases MW results in increased removal of inorganic nutrients and organic compounds. All in all, the DD and MW of chitosan should be tailored to specific applications and purpose. The mechanism of chitosan coagulation-flocculation is shown in Figure 4.

Due to lack of degradation of alum sludge as well as the slight toxicity of alum itself, more biocompatible options such as chitosan must be explored. Recent research has compared traditional alum with high velocity mixing versus proposed chitosan with ultrasound-enhanced flocculation (Fast et al. 2014). Chitosan for coagulation and flocculation creates flocs less vulnerable to shearing although less likely to flocculate again after shearing has occurred while traditional alum is highly vulnerable to shear but can easily agglomerate once more (Fast et. al, 2014). Moreover, the pHs of turbid waters are not a deterrent to alum's treatment efficiency; whereas chitosan only remains soluble in acidic pH solutions. Another significant advantage to the use of chitosan is the much lower dosage necessary to remove turbidity; this, in turn, reduces exposure to consumers compared to alum usage. Ultimately, a 1.5 mL dosage of chitosan using ultrasound was the most effective at removing turbidity (Fast et al. 2014). Further research conducted by Fast & Gude utilized a much wider range of NTU and dosage. Results expressed increased efficiency of chitosan with increased turbidity (2015). Furthermore, the longer the sedimentation time, the lower the final turbidity. In conclusion, optimal conditions were indicated at 40

Figure 4. Coagulation and flocculation of algal turbid water with Chitosan as coagulant

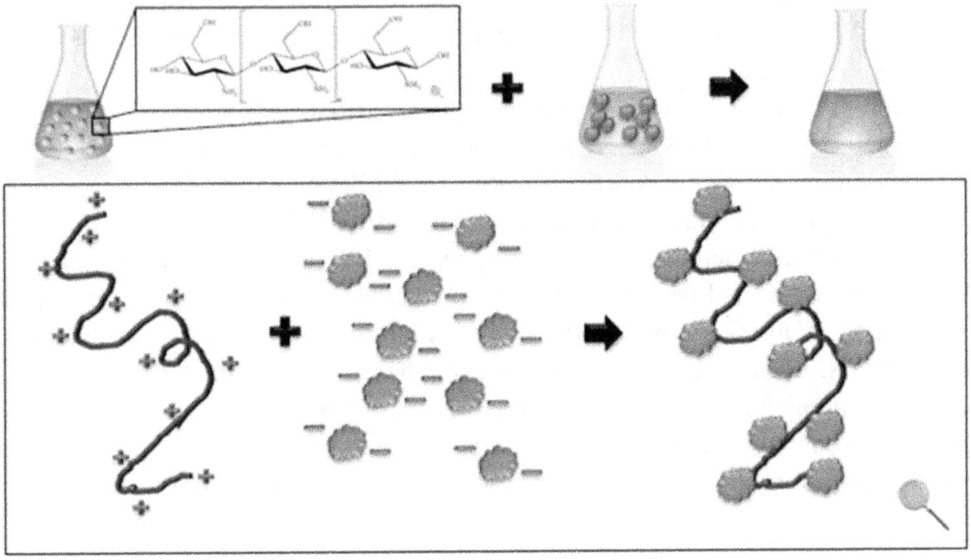

NTUs and 30 minutes flocculation time which produced an optimal 94% removal efficiency which rivals conventional alum at 99%.

Although fresh surface water resources emphasize the removal of algal blooms, salt water constitutes 96.5% of Earth's water with approximately 1% of the global population reliant upon desalination resources (Desalination, n.d.). Future global outlook by the United Nations foresees a massive increase to 14% by 2025 as stated by Christopher Glausson, publisher of Global Water Intelligence. One of many issues desalination industries face is the quality of received saltwater due to high turbidity. Research on chitosan in alkaline medium was conducted in order to determine efficiency as compared to alum at a dosage range from 0-370 mg/L and pH 2-11(Altaher, 2012). Results showed a 97.5% removal of NTUs using significantly less dosage and in turn less toxicity and exposure than alum. The optimum removal efficiency was found at pH 8.1 and 18 mg/L.

Beyond types of water and wastewater which chitosan effectively coagulates, multiple apparatus for which have been studied due to an increase in household water treatment and on-site water treatment interests. In an attempt to address ceramic water filtrations (CWF) lack of virus removal mechanisms, chitosan was tested for *Escherichia coli* K011 and coliphage MS2 contamination with CWF apparatus (Abebe, Chen, & Sobsey, 2016). CWF traditionally uses silver nanoparticles which produce uncertain results. Chitosan hydrochloride, chitosan acetate, and chitosan lactate were tested with 30 minute coagulation, flocculation, and sedimentation. The resulting waters were filtered by gravity flow and sampled. The mean reductions were 4.7 and 7.5 $\log_{10}$ for *Escherichia coli* and 2.8 to 4.5 $\log_{10}$ for MS2. Resulting turbidity present meets both US EPA standards (<1 NTU) and World Health Organization (WHO) guidance. Furthermore, the treated water met viral and bacterial reduction health targets.

Although chitosan is an effective biocoagulant for many water sources, including pre-treated sea water, various types of monomers can be co-grafted onto chitosan to provide further specific functionalization and utilization. Non-ionic co-polymerizations include acrylamide, N-vinyl formamide, and N,N-dimethyl acrylamide. Cationic monomers include (2-Methacryloyloxyethyl) trimethyl ammonium chloride, 3-(Acrylamide) propyl trimethylammonium bromide, N-binyl-2-pyrrolidone, and diallyl dimethyl ammonium chloride. Lastly, anionic grafts include with methacrylic acid and 2-Acrylamidoglycolic acid. However, the most established and commonly grafted nonionic polymer is acrylamide. Overall, the most common chitosan graft would be positively charged copolymerization due to increased efficacy of charge neutralization during coagulation and flocculation (Yang et. al., 2016).

Antimicrobial Properties

One cornerstone property of chitosan is antimicrobial and antifungal abilities. These abilities within a nanoparticle vastly increase possible functionality. It is considered effective in drug delivery schema such as vaccines, insulin adsorption, and even wound treatment as shown in Figure 5 (Qi et al. 2004).

In another study, chitosan nanoparticles are applied to *E. coli*, *S. choleraesuis*, *S. typhimurium*, and *S. aureus* bacteria to determine antibacterial activity. Considering copper nanoparticles also exhibit antibacterial and antimicrobial abilities, copper-loaded chitosan nanoparticles were prepared through ionic gelation. These samples were then left at 121°C with the bacteria at varying concentrations. Minimum Inhibitory Concentration (MIC), or the lowest concentration which causes absorption of the bacteria, was tested as well was Minimum Bactericidal Concentration (MBC). MBC is defined as the lowest concentration of the antibacterial solution which eradicates 99.9% bacteria. The results of this study are shown below.

Although chitosan effectively eradicates 1-$\log_{10}$ bacteria of each strain tested, it is less effective than the copper-loaded chitosan alternative. Due to chitosan's sensitivity to pH, the test was only used in a highly acidic solution of pH 6.5.

Figure 5.

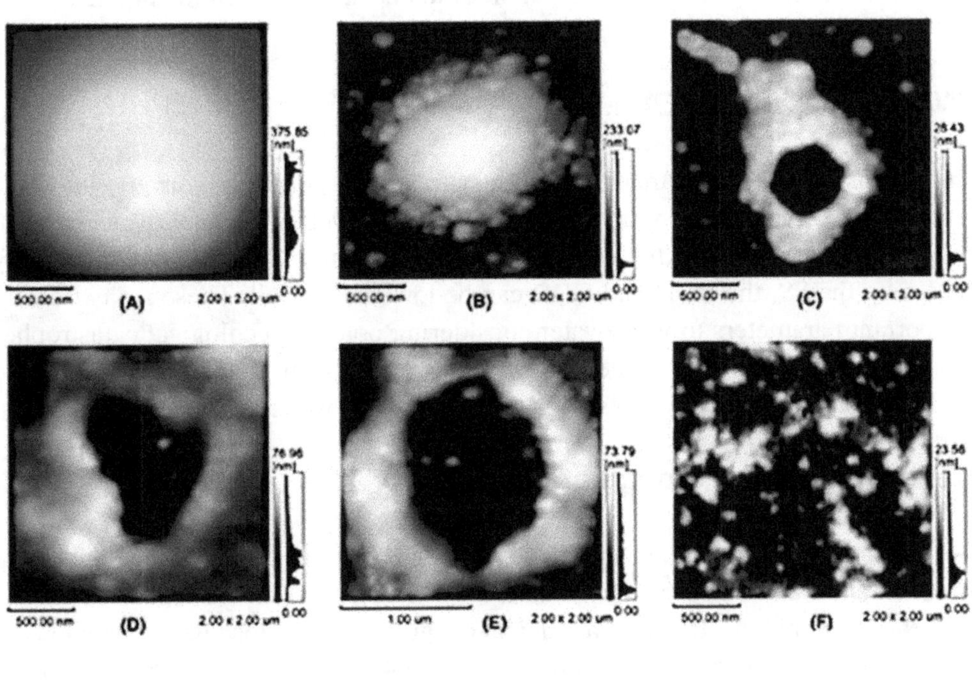

Table 2. Doxycycline, Chitosan Nanoparticles, and Copper-loaded Chitosan Nanoparticle Bactericidal Comparison (Qi, L., Xu, Z., Jiang, X., Hu, C., & Zou, X, 2004)

Bacteria	Doxycycline		Chitosan Nanoparticle Suspension		Copper-loaded Chitosan Nanoparticle Suspension	
	MIC	MBC	MIC	MBC	MIC	MBC
E. coli K88	1	8	1/8	4	1/16	2
E. coli ATCC 25922	2	16	1/16	2	1/16	2
S. chloeraesuis ATCC 50020	4	32	1/8	4	1/16	2
S. typhimurium ATCC 50013	4	64	1/4	8	1/8	4
S. Aureus ATCC 25923	1/4	8	1/4	8	1/8	4

Two mechanisms that facilitate antimicrobial properties are exhibited by the biopolymer. The initial mechanism is believed to create an impermeable layer surrounding the cell which prevents necessary transport from the cell. This effect on microbial cells is only present in slightly acidic solutions and has the potential to be reversed in different solvents. Secondly, the chitosan prevents the synthesis of RNA and protein by invading the cell's nucleus. For the latter effect, DD must be increased to increase electrostatic interaction with the negatively charged microbe. For the former, MW must increase in order to permeate the cell membrane.

WASTEWATER TREATMENT

In contrast to water treatment, wastewater treatment generally converts unusable water into an environmentally safe effluent by reducing the BOD (biochemical oxygen demand), COD (chemical oxygen demand), and SS (suspended solids). By reducing the SS, the BOD and COD can be lowered as well. These are extremely important parameters to wastewater considering potential ecological catastrophes caused by high BOD levels. Due to this concept, coagulation and flocculation are two of the most important processes for water treatment as well as wastewater treatment.

Coagulation and Flocculation in Wastewater Treatment

As previously mentioned, the importance of the coagulation and flocculation step in water and wastewater treatment cannot be overemphasized. What makes chitosan a great potential industrial and municipal coagulation and flocculation agent is not merely its effectiveness, but also its nontoxic nature and the plethora of effluents which are successfully treated with a much lower concentration than is necessary with alum. This process is primarily due to multiple flocculation mechanisms including simple charge neutralization, charge patching, bridging, and sweeping. Simple charge neutralization is inherently the neutralization of positive and negative charges present within the cationic biopolymer and anionic toxin or effluent. Charge patching creates flocs of biopolymer and effluent with the overall particle's charge. Bridging allows larger flocs created by adsorption onto the biopolymer and other primary flocs. Lastly, sweeping allows the entrapment of particles within a net-like structure of biopolymer. These flocculation mechanisms have been tested in food waste, seafood and fish processing wastes, brewery wastewater, surimi wash water, bentonite suspension, organic suspensions, kaolinite, bacterial suspensions, particulate and dissolved contaminants, effluents containing humic substances, effluents containing dyes, pulp and paper mill wastewater, oily wastewater, oil-in-water emulsions, agriculture wastewater, effluents containing metal ions, effluents containing phenol derivatives, partially purified sewage, brackish water, lead and chromium wastewater, poultry waste, coal mine wastewater, algal waters, egg industry effluent, grey wastewater, sugar industry effluent, and even pre-treated sweater as shown in Table 3.

Table 3. Bioeffluents successfully treated with chitosan

Effluent	Purpose	Experimental Details	Results	Reference
Food waste	Coagulation of high fat and protein milk waste	Chitosan-treated sludge is considered for feed supplement of cleaning in place wastewater.	25 mg/L at pH 7.0 yielded optimum efficiency showing that the material could be recycled at no extra cost.	(Chi & Cheng, 2006)
Seafood and fish processing wastes	Crawfish shell chitin tested for pigment extraction of seafood processing streams	150 mg/L Chitosan at pH 6.0 (commercial and synthetic)	97% SS removal and 83% tubidity removal. 150 mg/L at pH 6 resulted in 45% reduction in COD.	(Hong & Meyers, 1989)
Brewery wastewater	Coagulation of wheat dregs in wastewater	120 mg/L in 4.5 pH for optimal removal efficiency; increasing pH decreased turbidity	Maximum removal efficiency was observed at ~85% with a pH of 4.5, pH 7 at ~65%, and pH 8 at ~65%.	(Cheng, Chi, Yu, & Lee, 2005)
Surimi washes water	Protein adsorption and turbidity reduction of surimi wash water	20, 40, and 100 mg/L chitosan-alginate complexes were tested; 100 mg/L for maximum protein removal; 20 mg/L for maximum turbidity removal	Maximum 86% removal of protein, Minimum 73.1%; Maximum 95.7% turbidity removal, minimum 85.5%	(Wibowo, Velazquez, Savant, & Torres, 2007)

continued on following page

Table 3. Continued

Effluent	Purpose	Experimental Details	Results	Reference
Bentonite suspensions	Chitosan coagulation with bentonite suspensions using varying degree of deacetylation (DD), MW, viscosity, and polydispersity	Tap water (TW)–deionized water comparison at varying pH and concentration; maximum efficiency at 0.10 mg/L in TW pH 7.0	The efficiency of coagulation was observed with both TW and DW at pH 5 or 7. With concentration 0.02 mg/L and lower, less than 10 NTU was observed.	Roussy, Cooren, Dempsey, & Guibal, 2005)
Organic suspensions	Varying DD and MW tested using coagulation/ flocculation	Mushroom powder TW mixes were tested for pH 5, 7, and 9	DD was shown to have little effective on coagulation.	(Roussy, Vooren, & Guibal, 2005)
Kaolinite	Chitosan flocculation of kaolinite particles	Distilled water with TW was tested with highly refined kaolinite powder	A maximum turbidity removal was observed as ~85% with a 0.75 mg/L dosage.	(Divakaran & Pillai, 2004)
Bacterial suspensions	To determine antibacterial properties in sterile water	Maleic acid sodium grafted onto hydroxypropyl chitosan was tested using cut plug method and viable cell counting	All concentrations tested yielded maximum 99.90% at removal of *S. aureus* and *E. coli*	(Xie, Xu, Wang, & Liu, 2002)
Particulate and Dissolved Contaminants	Determine removal efficiency of bentonite suspensions, organic suspension, anionic dye solutions, and humic acid solutions	Both sorption and coagulation/flocculation were effective at removal of dyes; charge destabilization effectively coagulated humic acid and kaolin	Multiple coagulation mechanisms were determined for (a) bentonite, (b) organic suspensions, (c) anionic dye solutions, and (d) humic acid solutions which were charge destabilization for a, b, and d. However, sorption was observed for c.	(Guibal, Vooren, Dempsey, & Roussy, 2006)
Effluents containing humic substances	Chitosan hydrochloride (CH) and glutamate (CG) were tested for flocculation in dissolved humic solutions	CH and CG solutions were prepared using air-dry chitosan flakes and tested for color removal	CH yielded maximum effectiveness of >95%	(Bratskaya, Schwarz, & Chervonetsky, 2004)
Effluents containing dyes	Dye removal at differential pH and concentrations	Chitosan samples were ground and dissolved in acetic acid; agitated aqueous solutions of 2% acetic acid, 98% deionized water were incubated overnight. Experiments were done using Jar-Test apparatus at room temperature.	100% efficiency was observed for multiple dye concentrations and chitosan concentrations at pH 3	(Guibal & Roussy, 2007)

continued on following page

Table 3. Continued

Effluent	Purpose	Experimental Details	Results	Reference
Pulp and paper mill wastewater	Coagulation/ flocculation of post-bleaching effluent of the paper industry	Coagulation then photo catalysis determined properties such as biodegradability and through absorbance	100% reduction in absorbance for chromophores at 500 nm	(Rodrigues et. al., 2008)
Oily wastewater	Adsorption of oily wastewater was tested with differential concentrations, sedimentation times, and pH values	Palm oil mill effluent was tested using Jar-Test equipment at ~2000 mg/L of residue	~97-99% removal of residue was obtained at 0.5 g/L of chitosan powder, 15 minute contact time, 5.0 pH, and 30 minute sedimentation time	(Ahmad, Sumathi, & Hameed, 2004)
Oil-in-water emulsions	Flocculation of oil-in-water emulsions was tested with differential pH as well as SDS-stabilized emulsion and surfactant-free emulsion	0.05% diesel oil emulsions with an average droplet diameter of 200 nm were mixed with chitosan using ultrasonic treatment	Removal efficiencies decreased with increasing pH possibly due to chitosan's low solubility; complete phase separation was achieved at pH values greater than 7.5 through SDS-stabilized schema	(Bratskaya, Avramenka, Schwarz, & Phillippova, 2006)
Aquaculture wastewater	Aquaculture wastewaters were tested for turbidity, SS, chemical oxygen demand, ammonia, and phosphates after being treated with varying deacetylation degree, pH, and concentrations of chitosan	Chitosan leaves beneficial bacteria while removing *Edwardsiella itctaluri*	Optimal removal efficacies were 90% turbidity, 61% SS, 69.7% COD, 89.2% NH_3, and 95.6% PO_4^{3-}.	(Chung, 2006)
Effluent containing metal ions	Adsorption of heavy metals Hg(II), Fe(II), Ni(II), Pb(II), Cu(II), and Zn(II) were tested at pH 5, 6, and 7	0.1 g of chitosan mixes were added to 50 mL of wastewater; these samples were left in room temperature for 3 hours followed by a 5 min centrifuging process	Chelation capacities for each heavy metal were 94.1% Fe(II) at pH 5, 22.9% Mn(II) at pH 7, 26.3% Co(II) at pH 7, 62.8% Ni(II) at pH 7, 64.9% Cu(II) at pH 7, 50.3% Zn(II) at pH 7, 15.6% As(II) at pH 5, 19.9% Mo(II) at pH 7, 39.1% Cd(II) at pH 7, 97.5% Hg(II) at pH 7, and 91.0% at pH 7 for Pb(II). Most of these maximums were with a 91.3% DD.	(Gamage & Shahidi, 2007)

continued on following page

Table 3. Continued

Effluent	Purpose	Experimental Details	Results	Reference
Effluent containing phenol derivatives	Chitosan was tested alongside tyrosinase and two synthetic polymers polycondensate and polyethlenemine	Industrial wastewater removal involving a color change from clear to dark brown was tested using tyrosinase and different polymers.	Synthethic polymers were observed to be more effective due to lack of coagulation but precipitation of phenol and aromatic amines. However, all cationic polymer coagulants produced a precipitation. Retention of immobilization and storage stability were 80%.	(Wada, Ichikawa, & Tatsumi, 1995)
Partially purified sewage	Chitosan flocculation composites were compared with traditional flocculation solutions	Raw sewage waters were added to differential dosages and stirred for 10 minutes then transferred to separating funnels and settled for 30 minutes	Chitosan composites were shown to optimize COD by 1.8-23.7%, SS by 50%, and Al^{3+} by 61.2-85.5%.	(Zeng, Wu, & Kennedy, 2007)
Brackish water	Differential pH of 4-9 with a varying SS was tested on flocculation and settling time of river silt.	Turbidity was tested using Jar-Test apparatus with a 0.1 M HCl or $NaCO_3$. Solutions were stirred for 30 minutes and allowed a 30-minute settling time.	Maximum flocculation occurred at a pH of 7.0 while SS vs. NTU relationship was linear. 90% removal efficiency was observed.	(Divakaran & Pillai, 2002)
Lead and chromium wastewater	Comparing chitosan to aluminum sulfate for removal of metal salts and *Daphnia magna* with differential pH in actual conditions and humic acid spiked surface water.	Jar Test apparatus with 1 L high alkalinity water tested from 0.5-10 mg/L chitosan solutions. 1 g/L chlorine solution was added with a 24-hour contact time.	Increased toxicity of *Dapnia magna* was effectively 0% immobilization in some conditions.	(Rizzo, Gennaro, Gallo, & Belgiomo, 2008)
Poultry waste	Chitosan combined with dissolved air flotation were tested at multiple poultry effluent sites for SS and COD.	Poultry processing wastes were coagulated with chitosan and dissolved air flotation at sites such as the chiller and scalder	Composite SS effluents were reduced by 74-94% at different sites; COD of effluents were reduced by up to 82%.	(Bough, Shewfelt, & Salter, 1975)
Coal mine wastewater	Chitosan grafted N-vinyl formamide were tested with varying concentration of N-vinyl formamide, bromate ions, and chitosan.	Experimental setup involved a 1 L beaker of 1% weight coal with stirring apparatus for 15 minutes and 30 minute settling time.	Chitosan had little effect on NTU reduction; however, approximately 10% was the maximum for both coking and non-coking coal.	(Mishra, Tripathy, Srivatstava, Mishra, & Behari, 2008)

continued on following page

Table 3. Continued

Effluent	Purpose	Experimental Details	Results	Reference
Algal waters	Chitosan with ultrasound flocculation was compared with aluminum sulfate rapid mix flocculation of algal turbid waters.	Batch testing with 25 kHz ultrasonic frequencies followed by 50 rpm rapid mix for 20 min at a range of 10 to 40 NTUs.	Turbidity removal efficiencies of chitosan ranged from 84.1%-90.5% for rapid mix and 84%-97% for ultrasound.	(Fast & Gude, 2015)
Egg industry effluent	Differential COD concentration, adsorbent dose, pH, and settling time were tested.	Batch studies performed from 0.7-1.3g/L of wastewater in 100 mL of wastewater with 2-7 pH range. The supernatant was then centrifuged and filtered. A three-level, four-factor Box-Behnken design (BBD) studied each variable.	Optimal pH 4 1.1g/L chitosan with a 40 min settling time was determined. Maximum reductions were 94%, 88%, and 83% for turbidity, COD, and BOD respectively.	(Thirugnansambandham, Sivakumar, & Prakash, 2014)
Grey wastewater	Differential COD concentration, adsorbent dose, pH, and settling time were tested.	Batch studies were carried out using 1-3 min agitation time, 2.5-5.5 pH, 0.3-0.5 g/L chitosan, and 10-20 min settling time. The supernatant was centrifuged and filtered following settling time.	Optimal reduction efficiencies for turbidity, BOD, and COD were 96%, 91%, and 73% respectively.	(Thirugnanasambandham, Sivakumar, Prakash, & Kandasamy, 2014)

One of many effluents successfully treated with chitosan is grey wastewater. Grey wastewater is present within approximately 75% of resident wastewaters and is considered a cleaner effluent of bath tubs, sinks, laundry appliances, and schools. Despite the nonreactive nature of this effluent, it still poses tenable threats to ecosystems and must be treated effectively. Batch testing was used with varying agitation speed, pH, chitosan dose, and settling time were used. As predicted, pH efficiencies increased with increasing pH to 4. The maximum efficiency settling time was determined at 15 minutes due to shear stress and compaction of flocs. Upon optimization, NTU, BOD, and COD were determined to be 96%, 91%, and 73% reduced (Thirugnanasambandham et. al, 2013).

With superior removal and reduction efficiencies of many types of effluent, the natural question is to consider what chitosan cannot effectively treat. In “Influence of Effluent Type on the Performance of Chitosan as a Coagulant”, authors Ritha Pambi and Paul Musonge considered that within the scope of food processing plants. Chitosan with a 75% DD was purchased through Sigma-Aldrich and dissolved in 20 mL of 0.1 M hydrochloric acid. Sugar Processing effluent, Milk Processing Effluent, and Brewery processing effluent were all tested with jar-test apparatus and rapid mixing for 3 minutes at 100 rpm. The COD removal for sugar, milk, and brewery effluents were 11%, 54%, and 50% respectively. The Total Suspended Solids (TSS) were much more significant at 98%, 95%, and 95%. Significant factors which influenced the removal efficiencies include dissolved solids (TDS), initial TSS, and a pH value maximum of 9. The dissolved solids (TDS) were notably higher in the milk effluent. Moreover, the optimal pH value varied with effluent type. It can be concluded from this study that chitosan's coagulation and flocculation efficiency is dependent upon the effluent type and TDS.

Biosorption Properties

Beyond antimicrobial and coagulation-flocculation applications, chitosan and magnetic nanoparticles grafted with chitosan have proven effective at adsorption of Cobalt II, Copper II, Zinc II, and cation or anionic dyes. This adsorption ability is due to amino groups as well as hydroxyl groups. These amino groups are positively charged and attract dyes electrostatically. A wide range of adsorption studies was summarized in Table 4 which includes various types of chitosan and adsorbates.

Dye Adsorption

Due to the inert properties and low concentrations of dyes, they are difficult to be removed effectively and economically. Many other methods are currently being employed; however, they remain ineffective and expensive alternatives. These methods include adsorption, coagulation and flocculation, reverse osmosis, and others. Common chitosan composites that are used to remove dyes are chitosan/montmorillonite composites, chitosan/polyurethane composites, chitosan/activated clay composites, chitosan/bentonite composites, chitosan/oil palm ash composites, and chitosan/kaolin/γ-Fe_2O_3 composites. Each composite type, naturally, has different adsorption properties. For example, a chitosan/montmorillonite composite has an increasing adsorption capacity with respect to the molar ratio of chitosan and the composite, but this property plateaus at a 1:1 ratio. It is also noted that chitosan/montmorillonite composites closely follow a Langmuir isotherm (Wang & Wang, 2007). In contrast, a chitosan/polyurethane composite acts as a matrix due to the high surface area which immobilizes dyes similarly to granular activated carbon. Moreover, chitosan/activated clay was shown to have the highest adsorption capacity when used with the reactive dye RR222. Chitosan/activated clay yielded an absorbance of 1912 mg/g compared to an average of 342 mg/g average adsorption for the other composites.

Table 4. Adsorbent properties of chitosan and chitosan composites (Kamble et. al, 2007) (Bhatnagar & Sillanpää, 2009)

Adsorbent	Adsorbate	Adsorption capacity (mg/g)	Reference
		Dye Adsorbent	
Chitosan/montmorillonite	Congo Red	53.42	(Guzman, Saucedo, Navarro, Revilla, & Guibal, 2002)
Chitosan/polyurethane	Acid Violet 48	30	(Guibal, Milot & Roussy, 1999)
Chitosan/activated clay	Methylene blue	330	(Guibal, Milot & Tobin, 1998)
Chitosan-activated Clay	Reactive dye RRR22	1912	(Moret & Rubio, 2003)
Chitosan/bentonite	Tartrazine	294.1	(Dambies, Roze, & Guibal, 2000)
Chitosan/bentonite	Malachite Green	435	(Chen & Chung, 2006)
Chitosan/palm oil	Reactive blue 19	909.1	
Chitosan flakes	Reactive red 222	293	(Gupta, Chauhan & Sankaramakrishnan, 2009)
Chitosan beads	Reactive red 222	1103	(Adyun & Aksoy, 2009)
Chitosan-ECH	Reactive red 289	1802-1840 (g/kg)	(Schmul, Krieg & Keizer, 2001)
Chitin	Reactive yellow 2	28	(Hasan et. al, 2003)
Chitin	Reactive black 5	65	(Boddu, Abburi, Talbott, & Smith, 2003)
Chitin	synthetic reactive dye wastewater	133-167	(Spinelli, Laranjeira & Favere, 2004)
Chitin modified by sodium hypochlorite	synthetic reactive dye wastewater	59-124	(Rojas, Silva, Flores, Rodriguez, Ly, & Maldonado, 2005)
Quarternary chitosan salt	Reactive orange 16	1060	(Geng, Jin, & Qi, 2009)
Chitosan	Remazol black 13	91.47-130	(Yan, Haijia & Tianwei, 2007)
Powdered cross-linked chitosan grafted with acrylic acid	Remacryl Red TGL	204.2-510.7	(Kim, Benedetti & Boulegue, 2004)
Chitosan beads	Malachite green	74.83-93.55	(Niu & Volesky, 2003)
Chitosan-g-poly(acrylic acid) montmorillonite	Methylene blue	1859	(Ngah & Liang, 1999)
Chitosan-based adsorbent	Basic blue 3	166.5	(Arrascue, Garcia, Horna, & Guibal, 2003)
Disulfonate derivates of chitosan	Basic blue 9	121.9	(Lasko & Hurst, 1999)
Chitosan	Acid green 25	645.1	(Zou, Liu & Liu, 2009)
Chitosan	Acid orange 10	922.9	(Metwally, Elkholy, Salem, & Elsabee, 2009)
Chitosan	Acid orange 12	973.3	(Wang, Liu, Wang, Xie, & Deng, 2009)

continued on following page

Table 4. Continued

Adsorbent	Adsorbate	Adsorption capacity (mg/g)	Reference
Chitosan	Acid red 18	693.2	(Juang, Tseng, Wu & Lin, 1996)
Chitosan	Congo red	93	(Juang, Tseng, Wu & Lee, 1997)
Poly(methylmetacrylate) grafted chitosan	Procion Yellow mx	250	(Annadurai & Krishnan, 1997)
Poly(methylmetacrylate) grafted chitosan	Remazol Brilliant Violet	357	(Annadurai, Chellapandian & Krishnan, 1997)
Poly(methylmetacrylate) grafted chitosan	Reactive Blue H5G	178	(Wu, Tseng & Juang, 2000)
Cross-linked chitosan beads	Metanil yellow	1334	(Wu, Tseng & Juang, 2001)
Cross-linked chitosan beads	Reactive blue 15	722	(Chiou & Li, 2002)
Chitosan grafted with amide groups	Remazol Yellow Gelb 3RS	1211	(Choi & Li, 2003)
Chitosan grafted with carboxyl groups	Basic yellow 37	595	(Chiou, Ho & Li, 2004)
Metal Ion Adsorption			
Chitin	Cd(II)	14	(Akkaya, Uzun, & Guzel, 2007)
Chitosan	Cd(II)	5.93	(Dolphen, Sakkayawong, Thiravetyan & Nakbanpote, 2007)
Chitosan	Hg(II)	815	(Rosa, Larajeira, Riela & Favere, 2008)
Chitosan	Cu(II)	222	(Annadurai, Ling, & Lee, 2008)
Chitosan	Ni(II)	164	
Chitosan	Zn(II)	75	(Chao, Shyu, Lin & Mi, 2004)
Chitosan	Cu, Zn, As, Cr	137, 108, 58, 124	(Lazaridis, Kyzas, Vassilou, & Vikiaris, 2007)
Porous-magnetic chitosan beads	Cd(II)	188-518	(Bekcia, Ozveria, Sekia, & Yurdakoc, 2008)
Chitosan-coated on perlite	Cu(II)	196.07	(Wang, Zhang, & Wang, 2008)
Chitosan coated on perlite	Ni(II)	114.94	(Crini et. al, 2008)
Chitosan from silkworm chyrsalides	Pb(II)	72	(Crini, Martel & Torri, 2008)
Chitosan from silkworm chyrsalides	Cu(II)	87	(Monvisade & Siriphannon, 2009)
Carboxylgrafted chitosan	Cu(II)	318	(Wong, Szeto, Cheung & McKay, 2004)
Amido-grafted chitosan	Cr(VI)	935	(Chatterjee, Chatterjee, Chatterjee & Guha, 2007)
Chitosan coated on polyvinyl chloride beads	Ni(II)	120.5	(Chatterjee, Lee, Lee, & Woo, 2009)

continued on following page

Table 4. Continued

Adsorbent	Adsorbate	Adsorption capacity (mg/g)	Reference
Chitosan coated on polyvinyl chloride beads	Cu(II)	87.9	
Chitosan	Al(III)	45.45	(Azlan & Saime, 2009)
Molybdate-impregnated chitosan gel beads	As(V)	ca. 160 mg/g	
Chitosan beads	As(III)	1.83 mg/g	(Singh, Sharma, Tripathi, & Sanghi, 2009)
Chitosan beads	As(V)	1.94 mg/g	(Wu, Tseng, & Juang, 2001)
Chitosan flakes	Cr(VI)	22.09-102	(Vachoud, Zydowicz& Domard, 2001)
Non-cross-linked chitosan	Cr(VI)	78	(Juang, Wu, & Tseng, 2002)
Cross-linked chitosan	Cr(VI)	50	
Chitosan coated on nonporous ceramic alumina	Cr(VI)	154 mg/g	
Quatenary ammonium salts of chitosan	Cr(VI)	30.2-68.3	(Figueriedo, Loureiro, & Boaventura, 2005)
Cross-linked chitosan	Cr(II)	6	(Chiou & Chang, 2006)
Cross-linked chitosan	Cr(VI)	215	(Kyzas & Lazaridis, 2009)
Aminated chitosan	Ni(II)	30.2	(Zheng, Yang, Jo, & Park, 2004)
Chitosan	Ag	42	(Durshun & Kalayci, 2005)
Chitosan benzoyl thiourea derivative	Co(II)	29.47	(Pang, Ng, Chung, & Wong, 2007)
Chitosan benzoyl thiourea derivative	Eu(III)	34.54	(Nadavala, Swayampukala, Boddu & Abburi, 2009)
Cross-linked chitosan	U(VI)		(Li, Meng, Hu & Du, 2009)
Phenol Adsorption			
Chitin	Phenol	21.5	(Zhang, Xua, Chena, & Zong, 2009)
Magnetite-immobilized chitin	Pentachlorophenol	5.4	(Kumar, Subbaiah, & Reddy, 2009)
Beads of chitosan-sodium alginate	Phenol	108.6	(Chatterjee & Woo, 2009)
Chitosan	Phenol, p-nitrophenol and p-chlorophenol	1.9-2.5	(Chatterjee, Lee, Lee, & Woo, 2009)
Functional chitosan, chemically modified by β-cyclodextrin	Phenol, p-nitrophenol and p-chlorophenol	20.5-179.7	(Sundaram, Viswanthan, & Meenakshi, 2009)
Chitosan-abrus precatorius	Phenol	156	(Sundaram, Viswanthan, Meenakshi, 2009)
Chitosan-abrus precatorius	2-chlorophenol	204	(Jagtap, Thakre, Wanjari, & Kamble, 2009)
Chitosan-abrus precatorius	4-chlorophenol	278	(Viswanathan, Sundaram, & Meenakshi, 2009)

continued on following page

Table 4. Continued

Adsorbent	Adsorbate	Adsorption capacity (mg/g)	Reference
Anion Removal			
Chitosan hydrobeads	Nitrate	92.1	(Viswanathan, Sundaram, & Meenakshi, 2009)
Chitosan beads modified by cross-linked with epichlorohydrin	Nitrate	104	(Sundaram, Viswanathan & Meenakshi, 2009)
Neodymium modified chitosan	Flouride	11.4-22.3	(Xie, Li, Wang, & Liu, 2009)
Magnesia/Chitosan composites	Flouride	11	(Yoshizuka, Lou & Inoue, 2000)
Titanium-incorporated chitosan	Flouride	7.2	(Sismanoglu, 2007)
Protonated chitosan beads	Flouride	4.7-7.3	(Ngah & Musa, 1998)
Nanohydroxypatite/chitin	Flouride	3840	(Zhang & Bai, 2003)
Protonated cross-linked chitosan	Perchlorate	45.45	(Sun, Wang, Liu, Gong, Bao, & Ma, 2008)

Note. Adapted from "Applications of chitin-and chitosan-derivatives for the detoxification of water and wastewater-a short review" by Bhatnaga, & Sillanpää, 2009 Advances in Colloid and Interface Science 152(1), p. 26-38.

Heavy Metal Adsorption

Toxicity of heavy metal ions released daily by industry has not yet accumulated to become a public health concern; however, a long-term risk is present both for the environment and society. Similarly to dye adsorption, chitosan produces the best adsorption as a composite. A few examples of heavy metal adsorption composites are chitosan/ceramic alumina, chitosan/perlite, chitosan/magnetite, chitosan/cotton, chitosan/sand, chitosan/cellulose, and chitosan/montmorillonite. Chitosan composites are much less sensitive to acidic environments than chitosan alone. Magnetite composites, in particular, attract heavy metals magnetically using nanoparticles such as iron oxide and maghemite. These magnetic nanoparticles are not only lower toxicity but can also be produced cheaply. Furthermore, they have been effectively used for removal of Pb(II) through ultrasound enhanced agitation.

Desalination

Foam biosorbents consisting of hemicellulose grafted with diethylene triamine pentaacetic acid (DTPA) and cross-linked with chitosan successfully removed ~0.3 g of salt per gram of biosorbent (Ayoub et al, 2013). The absorption properties of this solution can be attributed to the swelling of the biosorbent and also the effect salinity has upon the polyelectrolyte complexes. As the concentration of NaOH in the solution increases, the gel swells in response. This same biosorbent material was also used for uptake of heavy metals; however, a much different absorption mechanism took place.

A more popular surface modification for desalting and salt rejection purposes employs chitosan for nanofiltration. Nanofiltration is used as a pretreatment for desalination through reverse osmosis in decreasing the pressure necessary. The primary mechanism for rejection of the salt solutions is due to ion valency. Due to the interaction between electrolytes and the nanofiltration surface, different feed solutions produced a range of removal efficiencies varying from 12.7% to 92.8% (Ming, Chen, & Gao, 2005). Many recent studies have been conducted to better understand and assess chitosan's nanofiltration properties through further surface engineering and membrane modification. In studies conducted by Wang et. al, an O-(Carboxymethyl)chitosan nanofiltration membrane was functionalized with graphene oxide to optimize desalination. After characterization and testing, the membranes exhibited better rejection capabilities than "pristine membranes and the commercial [nanofiltration] membranes". These were seen to definitively enhance desalination capabilities of the media (2015). Furthermore, Shenvi, Rashid, Ismail, Kassim, and Isloor synthesized a composite chitosan-poly(1,4-phnylene ether sulfone) (PPEES/CH) membrane which was then crosslinked by glutaraldehyde. Following characterization and water flux studies, the salt rejection was shown as a positive correlation with glutaraldehyde concentration of up to 34% and 53% for NaCL and $MgSO_4$ while simultaneously decreasing flux (2012). In another study, sulfated chitosan was cross-linked with polysulfone ultrafiltration membrane and epichlorohydrin which exhibited salt rejections of 90.8% for Na_2SO_4 and 32.5% for NaCL solutions. Another example of chitosan as an active layer for nanofiltration is exhibited by Zhou, Gao, Li, and Gao (2013) through a composite nanofiltration membrane from O-carboxymethyl chitosan an epoxy chloroprpone with a polyacrylonitrile supporting layer. Optimal concentrations were determined through studying salt rejection and permeation flux. The optimal membrane exhibited 94.36% Na_2SO_4 removal.

ENVIRONMENTAL IMPACT OF CHITOSAN IN WATER TREATMENT

Studies have been done on chitosan to determine potential toxicity after coagulation and flocculation of different surface waters. While some high turbidity levels with an excessive dosage of chitosan can create chitosan flocculation, the appropriate dosage naturally would not increase the NTUs. In addition, higher pH values generally cause a decrease in efficiencies which can sometimes be seen as drastic. Although chitosan's purpose is to decrease toxicity, some studies have shown increased toxicity compared to alum and ferric acid or NOM removal and *Dapnia magna* toxicity removal of humic acid spiked samples (Rizzo et. al., 2008). However, all coagulants decreased the toxicity in unspiked samples. Moreover, chlorination increased toxicity with chitosan being more toxic than traditional coagulants. Thus, surface water type is highly important to determine if chitosan could potentially increase toxicity.

Of course, it must be considered that the lower toxic material in some cases is not always so in others. For example with chitosan use in the well treatment fluid, replacement of phenols and formaldehyde absolutely decrease toxicity to the consumer (Eoff & Reddy, 2004). Moreover, potential cationic polyelectrolytes could possibly contribute to toxicity leading to an increased use in bioflocculants.

COST OF WATER TREATMENT

Many studies show that chitosan can be a less toxic, more economical option compared to traditional coagulants. Chitosan was proven to be more economical than aluminum sulfate and polyaluminum chloride when used as a coagulant for palm oil mill effluent and residue oil (Ahmad, Sumathi, & Hameed, 2006). For the most effective, least toxic alternatives, composites may provide optimal results. Polyaluminum chloride and silicate chitosan composite effectively enhanced COD, SS, and Al^{3+} by 1.8-23.7%, 50%, and 61-85.5% respectively. It also reduced costs by 7-34% (Zeng, Wu, & Kennedy, 2007). Furthermore, chitosan proved an efficient and cost-effective adsorbent for dozens of dyes such as AB25, AB40, AB62, AB113, AB193, AR73, RR24, RY2, RB74, and RY18 in alkali solution (Shen et. al., 2011). It has long been seen as a cost-effective alternative to activated carbon (Gupta & Imran, 2004). Beyond cost-efficiency for water treatment properties, its coagulation capabilities can allow further profitability for companies through recycling materials such as milk and poultry industry effluents (Chi & Cheng, 2006).

SUMMARY

Chitosan and its derivatives provide numerous opportunities for potentially sustainable water and wastewater treatment. Presently, the EPA has approved chitosan for use of point-of-use coagulation of bacteria among other uses with an aim at developing countries as a sustainable and economical alternative. Future research could include cost-benefit analysis as well as the implementation of successful antimicrobial, coagulation and flocculation, and adsorption functionality. Considering the environmental and social need for more drinking water alternatives, cost efficiency for biosorbent capabilities of salts should also be further studied and combined with other leading desalination initiatives such as microbial desalination cells. Environmental and aquatic toxicity of chitosan should be evaluated in detail as well.

ACKNOWLEDGMENT

This work was supported by the research grants from USEPA-P3 program and USGS-MWRRI research program. This work was also supported by the Department of Civil and Environmental Engineering (CEE), the Bagley College of Engineering (BCoE), and the Office of Research and Economic Development (ORED) at Mississippi State University.

REFERENCES

Abbas, A. O. M. (2010). *Chitosan for biomedical applications*. Academic Press.

Abebe, L. S., Chen, X., & Sobsey, M. D. (2016). Chitosan Coagulation to Improve Microbial and Turbidity Removal by Ceramic Water Filtration for Household Drinking Water Treatment. *International Journal of Environmental Research and Public Health*, *13*(3), 269. doi:10.3390/ijerph13030269 PMID:26927152

Ahmad, S., Sumathi, , & Hameed, . (2004). Chitosan: A natural biopolymer for the adsorption of residue oil from oily wastewater. *Adsorption Science and Technology*, *22*(1), 75–88. doi:10.1260/026361704323151015

Akkaya, G., Uzun, İ., & Güzel, F. (2007). Kinetics of the adsorption of reactive dyes by chitin. *Dyes and Pigments*, *73*(2), 168–177. doi:10.1016/j.dyepig.2005.11.005

Ahmad, A. L., Sumathi, S., & Hameed, B. H. (2006). Coagulation of residue oil and suspended solid in palm oil mill effluent by chitosan, alum and PAC. *Chemical Engineering Journal*, *118*(1), 99–105. doi:10.1016/j.cej.2006.02.001

Anderson, C. G., De Pablo, N., & Romo, C. R. (1978). Antarctic krill (Euphausia superba) as a source of chitin and chitosan. *Proceedings of the First International Conference on Chitin/Chitosan. MIT Sea Grant Program,* 54.

Annadurai, G., Chellapandian, M., & Krishnan, M. R. V. (1999). Adsorption of reactive dye on chitin. *Environmental Monitoring and Assessment*, *59*(1), 111–119. doi:10.1023/A:1006072119624

Annadurai, G., & Krishnan, M. R. V. (1997). Adsorption of acid dye from aqueous solution by chitin: Equilibrium studies. *Indian Journal of Chemical Technology, 4*, 217-222.

Aranaz, I., Mengíbar, M., Harris, R., Paños, I., Miralles, B., Acosta, N., & Heras, Á. et al. (2009). Functional characterization of chitin and chitosan. *Current Chemical Biology*, *3*(2), 203–230.

Ayoub, A., Venditti, R. A., Pawlak, J. J., Salam, A., & Hubbe, M. A. (2013). Novel hemicellulose–chitosan biosorbent for water desalination and heavy metal removal. *ACS Sustainable Chemistry & Engineering*, *1*(9), 1102–1109. doi:10.1021/sc300166m

Azlan, K., Wan Saime, W. N., & Lai Ken, L. (2009). Chitosan and chemically modified chitosan beads for acid dyes sorption. *Journal of Environmental Sciences (China)*, *21*(3), 296–302. doi:10.1016/S1001-0742(08)62267-6 PMID:19634439

Bae, M. J., Shin, H. S., Kim, E. K., Kim, J., & Shon, D. H. (2013). Oral administration of chitin and chitosan prevents peanut-induced anaphylaxis in a murine food allergy model. *International Journal of Biological Macromolecules*, *61*, 164–168. doi:10.1016/j.ijbiomac.2013.06.017 PMID:23806320

BeMILLER, J. N., & Whistler, R. L. (1962). Alkaline Degradation of Amino Sugars1. *The Journal of Organic Chemistry*, *27*(4), 1161–1164. doi:10.1021/jo01051a009

Bhatnagar, A., & Sillanpää, M. (2009). Applications of chitin-and chitosan-derivatives for the detoxification of water and wastewater—a short review. *Advances in Colloid and Interface Science*, *152*(1), 26–38. doi:10.1016/j.cis.2009.09.003 PMID:19833317

Blumberg, R., Southall, C. L., Van Rensburg, N. J., & Volckman, O. B. (1951). South african fish products. XXXII.—The rock lobster: A study of chitin production from processing wastes. *Journal of the Science of Food and Agriculture*, *2*(12), 571–576. doi:10.1002/jsfa.2740021210

Bough, W. A., Salter, W. L., Wu, A. C. M., & Perkins, B. E. (1978). Influence of manufacturing variables on the characteristics and effectiveness of chitosan products. I. Chemical composition, viscosity, and molecular-weight distribution of chitosan products. *Biotechnology and Bioengineering*, *20*(12), 1931–1943. doi:10.1002/bit.260201208

Bough, W. A., Shewfelt, A. L., & Salter, W. L. (1975). Use of chitosan for the reduction and recovery of solids in poultry processing waste effluents. *Poultry Science*, *54*(4), 992–1000. doi:10.3382/ps.0540992

Bratskaya, S., Schwarz, S., & Chervonetsky, D. (2004). Comparative study of humic acids flocculation with chitosan hydrochloride and chitosan glutamate. *Water Research*, *38*(12), 2955–2961. doi:10.1016/j.watres.2004.03.033 PMID:15223291

Bratskaya, S., Avramenko, V., Schwarz, S., & Philippova, I. (2006). Enhanced flocculation of oil-in-water emulsions by hydrophobically modified chitosan derivatives. *Colloids and Surfaces. A, Physicochemical and Engineering Aspects*, *275*(1), 168–176. doi:10.1016/j.colsurfa.2005.09.036

Brine, C. J., & Austin, P. R. (1981). Chitin variability with species and method of preparation. *Comparative Biochemistry and Physiology Part B: Comparative Biochemistry*, *69*(2), 283–286. doi:10.1016/0305-0491(81)90242-X

Broussignac, P. (1968). Un haut polymère naturel peu connu dans l'industrie, Le chitosane. *Chim. Ind. Genie Chim*, *99*, 1241–1247.

Brugnerotto, J., Lizardi, J., Goycoolea, F. M., Argüelles-Monal, W., Desbrieres, J., & Rinaudo, M. (2001). An infrared investigation in relation with chitin and chitosan characterization. *Polymer*, *42*(8), 3569–3580. doi:10.1016/S0032-3861(00)00713-8

Brzeski, M. M. (1982). Concept of chitin/chitosan isolation from Antarctic Krill (Euphausia superba) shells on a technique scale.*Proceedings of the 2nd international conference on chitin and chitosan,* 15-29.

Czechowska-Biskup, R., Jarosińska, D., Rokita, B., Ulański, P., & Rosiak, J. M. (2012). Determination of degree of deacetylation of chitosan-comparision of methods. *Prog. Chem. Appl. Chitin Its Deriv*, *17*, 5–20.

Crini, G., Martel, B., & Torri, G. (2008). Adsorption of CI Basic Blue 9 on chitosan-based materials. *International Journal of Environment and Pollution*, *34*(1-4), 451–465. doi:10.1504/IJEP.2008.020809

Chatterjee, S., Chatterjee, S., Chatterjee, B. P., & Guha, A. K. (2007). Adsorptive removal of congo red, a carcinogenic textile dye by chitosan hydrobeads: Binding mechanism, equilibrium and kinetics. *Colloids and Surfaces. A, Physicochemical and Engineering Aspects*, *299*(1), 146–152. doi:10.1016/j.colsurfa.2006.11.036

Chatterjee, S., Lee, D. S., Lee, M. W., & Woo, S. H. (2009). Enhanced adsorption of congo red from aqueous solutions by chitosan hydrogel beads impregnated with cetyl trimethyl ammonium bromide. *Bioresource Technology*, *100*(11), 2803–2809. doi:10.1016/j.biortech.2008.12.035 PMID:19208471

Chatterjee, S., Lee, D. S., Lee, M. W., & Woo, S. H. (2009). Nitrate removal from aqueous solutions by cross-linked chitosan beads conditioned with sodium bisulfate. *Journal of Hazardous Materials*, *166*(1), 508–513. doi:10.1016/j.jhazmat.2008.11.045 PMID:19111975

Chatterjee, S., & Woo, S. H. (2009). The removal of nitrate from aqueous solutions by chitosan hydrogel beads. *Journal of Hazardous Materials*, *164*(2), 1012–1018. doi:10.1016/j.jhazmat.2008.09.001 PMID:18977085

Chi, F. H., & Cheng, W. P. (2006). Use of chitosan as coagulant to treat wastewater from milk processing plant. *Journal of Polymers and the Environment*, *14*(4), 411–417. doi:10.1007/s10924-006-0027-2

Chiou, M. S., Ho, P. Y., & Li, H. Y. (2004). Adsorption of anionic dyes in acid solutions using chemically cross-linked chitosan beads. *Dyes and Pigments*, *60*(1), 69–84. doi:10.1016/S0143-7208(03)00140-2

Chiou, M. S., & Li, H. Y. (2003). Adsorption behavior of reactive dye in aqueous solution on chemical cross-linked chitosan beads. *Chemosphere*, *50*(8), 1095–1105. doi:10.1016/S0045-6535(02)00636-7 PMID:12531717

Chiou, M. S., & Li, H. Y. (2002). Equilibrium and kinetic modeling of adsorption of reactive dye on cross-linked chitosan beads. *Journal of Hazardous Materials*, *93*(2), 233–248. doi:10.1016/S0304-3894(02)00030-4 PMID:12117469

Cheng, W. P., Chi, F. H., Yu, R. F., & Lee, Y. C. (2005). Using chitosan as a coagulant in recovery of organic matters from the mash and lauter wastewater of brewery. *Journal of Polymers and the Environment*, *13*(4), 383–388. doi:10.1007/s10924-005-5533-0

Chung, Y. C., Li, Y. H., & Chen, C. C. (2005). Pollutant removal from aquaculture wastewater using the biopolymer chitosan at different molecular weights. *Journal of Environmental Science and Health*, *40*(9), 1775–1790. doi:10.1081/ESE-200068058 PMID:16134368

Chung, Y. C. (2006). Improvement of aquaculture wastewater using chitosan of different degrees of deacetylation. *Environmental Technology*, *27*(11), 1199–1208. doi:10.1080/09593332708618734 PMID:17203601

Crini, G., Gimbert, F., Robert, C., Martel, B., Adam, O., Morin-Crini, N., & Badot, P. M. et al. (2008). The removal of Basic Blue 3 from aqueous solutions by chitosan-based adsorbent: Batch studies. *Journal of Hazardous Materials*, *153*(1), 96–106. doi:10.1016/j.jhazmat.2007.08.025 PMID:17888569

Crini, G., Martel, B., & Torri, G. (2008). Adsorption of CI Basic Blue 9 on chitosan-based materials. *International Journal of Environment and Pollution*, *34*(1-4), 451–465. doi:10.1504/IJEP.2008.020809

Desalination industry enjoys growth spurt as scarcity starts to bite. (n.d.). Retrieved from globalwaterintel.com

Divakaran, R., & Pillai, V. S. (2004). Mechanism of kaolinite and titanium dioxide flocculation using chitosan assistance by fulvic acids? *Water Research*, *38*(8), 2135–2143. PubMed

Divakaran, R., & Pillai, V. S. (2002). Flocculation of river silt using chitosan. *Water Research*, *36*(9), 2414–2418. doi:10.1016/S0043-1354(01)00436-5 PMID:12108734

Dolphen, R., Sakkayawong, N., Thiravetyan, P., & Nakbanpote, W. (2007). Adsorption of Reactive Red 141 from wastewater onto modified chitin. *Journal of Hazardous Materials*, *145*(1), 250–255. doi:10.1016/j.jhazmat.2006.11.026 PMID:17184913

Domard, A. (2011). A perspective on 30 years research on chitin and chitosan. *Carbohydrate Polymers*, *84*(2), 696–703. doi:10.1016/j.carbpol.2010.04.083

Dursun, A. Y., & Kalayci, C. S. (2005). Equilibrium, kinetic and thermodynamic studies on the adsorption of phenol onto chitin. *Journal of Hazardous Materials*, *123*(1), 151–157. doi:10.1016/j.jhazmat.2005.03.034 PMID:15993297

Elson Santiago de Alvarenga. (2011). *Characterization and Properties of Chitosan, Biotechnology of Biopolymers*. InTech. Retrieved from http://www.intechopen.com/books/biotechnology-of-biopolymers/characterization-and-properties-of-chitosan

Eoff, L. S., & Reddy, B. R. (2004). *U.S. Patent No. 6,764,981*. Washington, DC: U.S. Patent and Trademark Office.

Fast, S. A., & Gude, V. G. (2015). Ultrasound-chitosan enhanced flocculation of low algal turbid waters. *Journal of Industrial and Engineering Chemistry*, *24*, 153–160. doi:10.1016/j.jiec.2014.09.023

Fast, S. A., Kokabian, B., & Gude, V. G. (2014). Chitosan enhanced coagulation of algal turbid waters–Comparison between rapid mix and ultrasound coagulation methods. *Chemical Engineering Journal*, *244*, 403–410. doi:10.1016/j.cej.2014.01.081

Figueiredo, S. A., Loureiro, J. M., & Boaventura, R. A. (2005). Natural waste materials containing chitin as adsorbents for textile dyestuffs: Batch and continuous studies. *Water Research*, *39*(17), 4142–4152. doi:10.1016/j.watres.2005.07.036 PMID:16140355

Gamage, A., & Shahidi, F. (2007). Use of chitosan for the removal of metal ion contaminants and proteins from water. *Food Chemistry, 104*(3), 989–996. doi:10.1016/j.foodchem.2007.01.004

Guibal, E., & Roussy, J. (2007). Coagulation and flocculation of dye-containing solutions using a biopolymer (Chitosan). *Reactive & Functional Polymers, 67*(1), 33–42. doi:10.1016/j.reactfunctpolym.2006.08.008

Guibal, E., Van Vooren, M., Dempsey, B. A., & Roussy, J. (2006). A review of the use of chitosan for the removal of particulate and dissolved contaminants. *Separation Science and Technology, 41*(11), 2487–2514. doi:10.1080/01496390600742807

Gupta, V. K., & Imran, A. M. (2004). Adsorbents for water treatment: development of low-cost alternatives to carbon. Encyclopedia of Surface and Colloid Science, 2004 Update Supplement, 5, 1.

Hackman, R. (1954). Studies on chitin I. Enzymic degradation of chitin and chitin esters. *Australian Journal of Biological Sciences, 7*(2), 168–178. doi:10.1071/BI9540168 PMID:13189822

Hackman, R. H., & Goldberg, M. (1974). Light-scattering and infrared-spectrophotometric studies of chitin and chitin derivatives. *Carbohydrate Research, 38*, 35–45. doi:10.1016/S0008-6215(00)82336-8

No, H. K., & Meyers, S. P. (1989). Crawfish chitosan as a coagulant in recovery of organic compounds from seafood processing streams. *Journal of Agricultural and Food Chemistry, 37*(3), 580–583. doi:10.1021/jf00087a002

Horowitz, S. T., Roseman, S., & Blumenthal, H. J. (1957). The Preparation of Glucosamine Oligosaccharides. I. Separation1, 2. *Journal of the American Chemical Society, 79*(18), 5046–5049. doi:10.1021/ja01575a059

Altaher, H. (2012). The use of chitosan as a coagulant in the pre-treatment of turbid sea water. *Journal of Hazardous Materials, 233*, 97–102. doi:10.1016/j.jhazmat.2012.06.061 PMID:22819482

Jagtap, S., Thakre, D., Wanjari, S., Kamble, S., Labhsetwar, N., & Rayalu, S. (2009). New modified chitosan-based adsorbent for defluoridation of water. *Journal of Colloid and Interface Science, 332*(2), 280–290. doi:10.1016/j.jcis.2008.11.080 PMID:19181328

Juang, R. S., Tseng, R. L., Wu, F. C., & Lee, S. H. (1997). Adsorption behavior of reactive dyes from aqueous solutions on chitosan. *Journal of Chemical Technology and Biotechnology (Oxford, Oxfordshire), 70*(4), 391–399. doi:10.1002/(SICI)1097-4660(199712)70:4<391::AID-JCTB792>3.0.CO;2-V

Juang, R. S., Tseng, R. L., Wu, F. C., & Lin, S. J. (1996). Use of chitin and chitosan in lobster shell wastes for color removal from aqueous solutions. *Journal of Environmental Science & Health Part A*, *31*(2), 325–338. doi:10.1080/10934529609376360

Juang, R. S., Wu, F. C., & Tseng, R. L. (2000). Mechanism of adsorption of dyes and phenols from water using activated carbons prepared from plum kernels. *Journal of Colloid and Interface Science*, *227*(2), 437–444. doi:10.1006/jcis.2000.6912 PMID:10873331

Kamble, S. P., Jagtap, S., Labhsetwar, N. K., Thakare, D., Godfrey, S., Devotta, S., & Rayalu, S. S. (2007). Defluoridation of drinking water using chitin, chitosan and lanthanum-modified chitosan. *Chemical Engineering Journal*, *129*(1), 173–180. doi:10.1016/j.cej.2006.10.032

Kurita, K., Tomita, K., Tada, T., Ishii, S., Nishimura, S. I., & Shimoda, K. (1993). Squid chitin as a potential alternative chitin source: Deacetylation behavior and characteristic properties. *Journal of Polymer Science. Part A, Polymer Chemistry*, *31*(2), 485–491. doi:10.1002/pola.1993.080310220

Kumar, N. S., Suguna, M., Subbaiah, M. V., Reddy, A. S., Kumar, N. P., & Krishnaiah, A. (2010). Adsorption of phenolic compounds from aqueous solutions onto chitosan-coated perlite beads as biosorbent. *Industrial & Engineering Chemistry Research*, *49*(19), 9238–9247. doi:10.1021/ie901171b

Kyzas, G. Z., & Lazaridis, N. K. (2009). Reactive and basic dyes removal by sorption onto chitosan derivatives. *Journal of Colloid and Interface Science*, *331*(1), 32–39. doi:10.1016/j.jcis.2008.11.003 PMID:19026426

Lazaridis, N. K., Kyzas, G. Z., Vassiliou, A. A., & Bikiaris, D. N. (2007). Chitosan derivatives as biosorbents for basic dyes. *Langmuir*, *23*(14), 7634–7643. doi:10.1021/la700423j PMID:17530870

Li, J. M., Meng, X. G., Hu, C. W., & Du, J. (2009). Adsorption of phenol, p-chlorophenol and p-nitrophenol onto functional chitosan. *Bioresource Technology*, *100*(3), 1168–1173. doi:10.1016/j.biortech.2008.09.015 PMID:18930394

Li, Q., Mahendra, S., Lyon, D. Y., Brunet, L., Liga, M. V., Li, D., & Alvarez, P. J. (2008). Antimicrobial nanomaterials for water disinfection and microbial control: Potential applications and implications. *Water Research*, *42*(18), 4591–4602. doi:10.1016/j.watres.2008.08.015 PMID:18804836

Madhavan, P., & Ramachandran Nair, K. G. (1974). Utilization of prawn waste: Isolation of chitin and its conversion to chitosan. *Fishery Technology*, *11*(1), 50–53.

Kumar, M. N. R. (2000). A review of chitin and chitosan applications. *Reactive & Functional Polymers*, *46*(1), 1–27. doi:10.1016/S1381-5148(00)00038-9

Miao, J., Chen, G. H., & Gao, C. J. (2005). A novel kind of amphoteric composite nanofiltration membrane prepared from sulfated chitosan (SCS). *Desalination*, *181*(1), 173–183. doi:10.1016/j.desal.2005.03.001

Mima, S., Miya, M., Iwamoto, R., & Yoshikawa, S. (1983). Highly deacetylated chitosan and its properties. *Journal of Applied Polymer Science*, *28*(6), 1909–1917. doi:10.1002/app.1983.070280607

Mishra, D. K., Tripathy, J., Srivastava, A., Mishra, M. M., & Behari, K. (2008). Graft copolymer (chitosan-g-N- vinyl formamide): Synthesis and study of its properties like swelling, metal ion uptake and flocculation. *Carbohydrate Polymers*, *74*(3), 632–639. doi:10.1016/j.carbpol.2008.04.015

Monvisade, P., & Siriphannon, P. (2009). Chitosan intercalated montmorillonite: Preparation, characterization and cationic dye adsorption. *Applied Clay Science*, *42*(3), 427–431. doi:10.1016/j.clay.2008.04.013

Moorjani, M. N., Achutha, V., & Khasim, D. I. (1975). Parameters affecting the viscosity of chitosan from prawn waste. *Journal of Food Science and Technology*.

Muzzarelli, R. A. A., Tanfani, F., Emanuelli, M., & Gentile, S. (1980). The chelation of cupric ions by chitosan membranes[Callinectes sapidus, blue crab shell]. *Journal of Applied Biochemistry*.

Nadavala, S. K., Swayampakula, K., Boddu, V. M., & Abburi, K. (2009). Biosorption of phenol and o-chlorophenol from aqueous solutions on to chitosan–calcium alginate blended beads. *Journal of Hazardous Materials*, *162*(1), 482–489. doi:10.1016/j.jhazmat.2008.05.070 PMID:18573601

Ngah, W. W., Kamari, A., & Koay, Y. J. (2004). Equilibrium and kinetics studies of adsorption of copper (II) on chitosan and chitosan/PVA beads. *International Journal of Biological Macromolecules*, *34*(3), 155–161. doi:10.1016/j.ijbiomac.2004.03.001 PMID:15225987

Ngah, W. W., Teong, L. C., & Hanafiah, M. A. K. M. (2011). Adsorption of dyes and heavy metal ions by chitosan composites: A review. *Carbohydrate Polymers*, *83*(4), 1446–1456. doi:10.1016/j.carbpol.2010.11.004

No, H. K., Meyers, S. P., & Lee, K. S. (1989). Isolation and characterization of chitin from crawfish shell waste. *Journal of Agricultural and Food Chemistry*, *37*(3), 575–579. doi:10.1021/jf00087a001

Pambi, R. L., & Musonge, P. (2015). The Efficiency of Chitosan as a Coagulant in the Treatment of the Effluents from the Sugar Industry. *Journal of Polymer Materials, 32*(1), 57.

Pambi, R. L. L., & Musonge, P. (2014). *Influence of Effluent Type on the Performance of Chitosan as a Coagulant*. Academic Press.

Pang, K. M., Ng, S., Chung, W. K., & Wong, P. K. (2007). Removal of pentachlorophenol by adsorption on magnetite-immobilized chitin. *Water, Air, and Soil Pollution, 183*(1-4), 355–365. doi:10.1007/s11270-007-9384-2

Qi, L., Xu, Z., Jiang, X., Hu, C., & Zou, X. (2004). Preparation and antibacterial activity of chitosan nanoparticles. *Carbohydrate Research, 339*(16), 2693–2700. doi:10.1016/j.carres.2004.09.007 PMID:15519328

Renault, F., Sancey, B., Badot, P. M., & Crini, G. (2009). Chitosan for coagulation/flocculation processes–an eco- friendly approach. *European Polymer Journal, 45*(5).

Rizzo, L., Di Gennaro, A., Gallo, M., & Belgiorno, V. (2008). Coagulation/chlorination of surface water: A comparison between chitosan and metal salts. *Separation and Purification Technology, 62*(1), 79–85. doi:10.1016/j.seppur.2007.12.020

Rodrigues, A. C., Boroski, M., Shimada, N. S., Garcia, J. C., Nozaki, J., & Hioka, N. (2008). Treatment of paper pulp and paper mill wastewater by coagulation–flocculation followed by heterogeneous photocatalysis. *Journal of Photochemistry and Photobiology A Chemistry, 194*(1), 1–10. doi:10.1016/j.jphotochem.2007.07.007

Rosa, S., Laranjeira, M. C., Riela, H. G., & Fávere, V. T. (2008). Cross-linked quaternary chitosan as an adsorbent for the removal of the reactive dye from aqueous solutions. *Journal of Hazardous Materials, 155*(1), 253–260. doi:10.1016/j.jhazmat.2007.11.059 PMID:18180101

Roussy, J., Van Vooren, M., & Guibal, E. (2005). Influence of chitosan characteristics on coagulation and flocculation of organic suspensions. *Journal of Applied Polymer Science, 98*(5), 2070–2079. doi:10.1002/app.22411

Roussy, J., Van Vooren, M., Dempsey, B. A., & Guibal, E. (2005). Influence of chitosan characteristics on the coagulation and the flocculation of bentonite suspensions. *Water Research, 39*(14), 3247–3258. doi:10.1016/j.watres.2005.05.039 PMID:16019052

Sánchez-Duarte, R. G., Sánchez-Machado, D. I., López-Cervantes, J., & Correa-Murrieta, A. (2012). Adsorption of allura red dye by cross-linked chitosan from shrimp waste. *Water Science and Technology, 65*(4), 618–623. doi:10.2166/wst.2012.900 PMID:22277220

Shenvi, S. S., Rashid, S. A., Ismail, A. F., Kassim, M. A., & Isloor, A. M. (2013). Preparation and characterization of PPEES/chitosan composite nanofiltration membrane. *Desalination, 315*, 135–141. doi:10.1016/j.desal.2012.09.009

Roberts, G. A. (1992). Preparation of chitin and chitosan. In *Chitin chemistry* (pp. 54–84). Macmillan Education UK. doi:10.1007/978-1-349-11545-7_2

Sievers, D. M., Jenner, M. W., & Hanna, M. (1994). Treatment of dilute manure wastewaters by chemical coagulation. *Transactions of the ASAE. American Society of Agricultural Engineers*, *37*(2), 597–601. doi:10.13031/2013.28118

Singh, V., Sharma, A. K., Tripathi, D. N., & Sanghi, R. (2009). Poly (methylmethacrylate) grafted chitosan: An efficient adsorbent for anionic azo dyes. *Journal of Hazardous Materials*, *161*(2), 955–966. doi:10.1016/j.jhazmat.2008.04.096 PMID:18547715

Sismanoglu, T., & Pura, S. (2001). Adsorption of aqueous nitrophenols on clinoptilolite. *Colloids and Surfaces. A, Physicochemical and Engineering Aspects*, *180*(1), 1–6. doi:10.1016/S0927-7757(00)00751-2

Rao, S. S., Yashodha, K. P., Mahendrakar, N. S., & Puttarajappa, P. (1987). deacetylation of chitin at low-temperature by a novel alkali impregnation technique. *Indian Journal of Technology*, *25*(4), 194–196.

Sun, X. F., Wang, S. G., Liu, X. W., Gong, W. X., Bao, N., Gao, B. Y., & Zhang, H. Y. (2008). Biosorption of Malachite Green from aqueous solutions onto aerobic granules: Kinetic and equilibrium studies. *Bioresource Technology*, *99*(9), 3475–3483. doi:10.1016/j.biortech.2007.07.055 PMID:17855080

Sundaram, C. S., Viswanathan, N., & Meenakshi, S. (2009). Defluoridation of water using magnesia/chitosan composite. *Journal of Hazardous Materials*, *163*(2), 618–624. doi:10.1016/j.jhazmat.2008.07.009 PMID:18723280

Shen, C., Shen, Y., Wen, Y., Wang, H., & Liu, W. (2011). Fast and highly efficient removal of dyes under alkaline conditions using magnetic chitosan-Fe (III) hydrogel. *Water Research*, *45*(16), 5200–5210. doi:10.1016/j.watres.2011.07.018 PMID:21839488

Thirugnanasambandham, K., Sivakumar, V., & Prakash, M. J. (2014). Treatment of egg processing industry effluent using chitosan as an adsorbent. *Journal of the Serbian Chemical Society*, *79*(6), 743–757. doi:10.2298/JSC130201053T

Thirugnanasambandham, K., Sivakumar, V., Maran, J. P., & Kandasamy, S. (2014). Chitosan based grey wastewater treatment—A statistical design approach. *Carbohydrate Polymers*, *99*, 593–600. doi:10.1016/j.carbpol.2013.08.058 PMID:24274548

Tolaimate, A., Desbrieres, J., Rhazi, M., & Alagui, A. (2003). Contribution to the preparation of chitins and chitosans with controlled physico-chemical properties. *Polymer*, *44*(26), 7939–7952. doi:10.1016/j.polymer.2003.10.025

UNwater. (n.d.). Retrieved from UNwater Statistics: http://www.unwater.org/statistics/en/

Vachoud, L., Zydowicz, N., & Domard, A. (2001). Sorption and desorption studies on chitin gels. *International Journal of Biological Macromolecules*, *28*(2), 93–101. doi:10.1016/S0141-8130(00)00151-3 PMID:11164225

Vitali, L., Justi, K. C., Laranjeira, M., & Fávere, V. T. (2006). Impregnation of chelating agent 3, 3-bis-N, N bis- (carboxymethyl) aminomethyl-o-cresolsulfonephthalein in biopolymer chitosan: Adsorption equilibrium of Cu (II) in aqueous medium. *Polímeros*, *16*(2), 116–122. doi:10.1590/S0104-14282006000200011

Viswanathan, N., Sundaram, C. S., & Meenakshi, S. (2009). Removal of fluoride from aqueous solution using protonated chitosan beads. *Journal of Hazardous Materials*, *161*(1), 423–430. doi:10.1016/j.jhazmat.2008.03.115 PMID:18479817

Wada, S., Ichikawa, H., & Tastsumi, K. (1995). Removal of phenols and aromatic amines from wastewater by a combination treatment with tyrosinase and a coagulant. *Biotechnology and Bioengineering*, *45*(4), 304–309. doi:10.1002/bit.260450404 PMID:18623183

Wang, L., Zhang, J., & Wang, A. (2008). Removal of methylene blue from aqueous solution using chitosan-g-poly (acrylic acid)/montmorillonite superadsorbent nanocomposite. *Colloids and Surfaces. A, Physicochemical and Engineering Aspects*, *322*(1), 47–53. doi:10.1016/j.colsurfa.2008.02.019

Wibowo, S., Velazquez, G., Savant, V., & Torres, J. A. (2007). Effect of chitosan type on protein and water recovery efficiency from surimi wash water treated with chitosan–alginate complexes. *Bioresource Technology*, *98*(3), 539–545. doi:10.1016/j.biortech.2006.02.014 PMID:16580193

Wong, Y. C., Szeto, Y. S., Cheung, W., & McKay, G. (2004). Adsorption of acid dyes on chitosan—equilibrium isotherm analyses. *Process Biochemistry*, *39*(6), 695–704. doi:10.1016/S0032-9592(03)00152-3

Wu, A. C. M., & Bough, W. A. (1978). A study of variables in the chitosan manufacturing process in relation to molecular-weight distribution, chemical characteristics and waste-treatment effectiveness. *Proceedings of the First International Conference on Chitin/Chitosan*, 88-102.

Wu, F. C., Tseng, R. L., & Juang, R. S. (2000). Comparative adsorption of metal and dye on flake-and bead-types of chitosans prepared from fishery wastes. *Journal of Hazardous Materials*, *73*(1), 63–75. doi:10.1016/S0304-3894(99)00168-5 PMID:10686379

Wu, F. C., Tseng, R. L., & Juang, R. S. (2001). Kinetic modeling of liquid-phase adsorption of reactive dyes and metal ions on chitosan. *Water Research*, *35*(3), 613–618. doi:10.1016/S0043-1354(00)00307-9 PMID:11228956

Xie, W., Xu, P., Wang, W., & Liu, Q. (2002). Preparation and antibacterial activity of a water-soluble chitosan derivative. *Carbohydrate Polymers*, *50*(1), 35–40. doi:10.1016/S0144-8617(01)00370-8

Yang, R., Li, H., Huang, M., Yang, H., & Li, A. (2016). A review on chitosan-based flocculants and their applications in water treatment. *Water Research*, *95*, 59–89. doi:10.1016/j.watres.2016.02.068 PMID:26986497

Yao, R., Meng, F., Zhang, L., Ma, D., & Wang, M. (2009). Defluoridation of water using neodymium-modified chitosan. *Journal of Hazardous Materials*, *165*(1), 454–460. doi:10.1016/j.jhazmat.2008.10.052 PMID:19046805

Yoshizuka, K., Lou, Z., & Inoue, K. (2000). Silver-complexed chitosan microparticles for pesticide removal. *Reactive & Functional Polymers*, *44*(1), 47–54. doi:10.1016/S1381-5148(99)00076-0

Younes, I., & Rinaudo, M. (2015). Chitin and chitosan preparation from marine sources. Structure, properties and applications. *Marine Drugs*, *13*(3), 1133–1174. doi:10.3390/md13031133 PMID:25738328

Zeng, D., Wu, J., & Kennedy, J. F. (2008). Application of a chitosan flocculant to water treatment. *Carbohydrate Polymers*, *71*(1), 135–139. doi:10.1016/j.carbpol.2007.07.039

Zhang, X., & Bai, R. (2003). Mechanisms and kinetics of humic acid adsorption onto chitosan-coated granules. *Journal of Colloid and Interface Science*, *264*(1), 30–38. doi:10.1016/S0021-9797(03)00393-X PMID:12885516

Zhang, Y., & Zhou, J. L. (2005). Removal of estrone and 17β-estradiol from water by adsorption. *Water Research*, *39*(16), 3991–4003. doi:10.1016/j.watres.2005.07.019 PMID:16126247

Zheng, S. K., Yang, Z. F., Jo, D. H., Park, Y. H., & Ho, Y. S. (2005). Comment on Removal of chlorophenols from groundwater by chitosan sorption. *Water Research*, *39*(1), 264–268. doi:10.1016/j.watres.2004.07.018 PMID:15607185

Zhou, C., Gao, X. L., Li, S. S., & Gao, C. J. (2013). Fabrication and characterization of novel composite nanofiltration membranes based on zwitterionic O-carboxymethyl chitosan. *Desalination*, *317*, 67–76. doi:10.1016/j.desal.2013.02.026

Chapter 10
Performance of Chitosan Micro/Nanoparticles to Remove Hexavalent Chromium From Residual Water

Jimena Bernadette Dima
CONICET-CENPAT, Argentina

Noemí Zaritzky
CIDCA-UNLP-CONICET, Argentina

ABSTRACT

Hexavalent chromium Cr(VI) is toxic to living systems and must be removed from wastewater. Chitosan is a cationic, biocompatible, biodegradable, biopolymer obtained from marine wastes. The performance of chitosan particles (CH) and chitosan nanoparticles (CHN) to remove Cr(VI) from aqueous solutions is discussed in the present chapter. CHN were obtained by reticulation with tripolyphosphate (TPP), and physico-chemically characterized. The performance of CHN decreased at higher pH due to the cross-linking process with TPP. Langmuir isotherm described the equilibrium adsorption values and pseudo-second order rate provided the best fitting to the kinetic data. Chemical analysis to determine the oxidation state of the adsorbed Cr, showed that Cr(VI) was adsorbed on CH particles without further reduction; in contrast Cr(VI) removed from the solution was reduced and bound to the CHN as Cr(III). Chitosan crosslinking was essential to adsorb Cr(VI) at pH<3 due to the dissolution of CH in acid media.

DOI: 10.4018/978-1-5225-2136-5.ch010

1. INTRODUCTION

Industrial and mining wastewaters are important sources of pollution of heavy metals. The use of chromate and dichromate has many industrial applications such as in textile, electroplating, leather tanning, cement preservations, paints, pigments and metallurgy industries. Chromium is a metal potentially toxic to humans as it is considered carcinogenic. Cr (VI) is toxic to living systems and must be removed from wastewater before it can be discharged. Cr(II), Cr(III) and Cr(VI) are the three oxidation states for chromium in nature, but only the last two are stable. Cr(VI) is 500 times more toxic, mutagenic and carcinogenic than Cr(III). The maximum permissible limit of for Cr(VI) in wastewater has been recommended as 0.005mg L^{-1} by World Health Organization.

For the removal of toxic pollutants from water and wastewater differences methods, such as oxidation/reduction, filtration, coagulation, membrane process, adsorption, osmosis, biological methods, etc, have generally been used. However, adsorption process is often considered the most appropriate technique owing to its advantages such as a variety of adsorbent materials and high efficiency.

Over the years, a wide range of clean-up technologies has been developed to remove toxic metals from water. The removal of metals, compounds, and particulates from solution by the biological material is recognized as an extension to adsorption and is named as biosorption.

Biosorbents have been given increasing attention as they can significantly reduce the cost of an adsorption system. Chitosan is an excellent biopolymer obtained cost-effectively by the derivation of chitin, which is a natural, abundance and non-toxicity, the material found widely in crustacean shells. Chitosan is the *N*-deacetylated derivative of chitin (2-acetamido-2-deoxy-β-D-glucose through a β (1→4) linkage).

Figure 1. Chitin and Chitosan

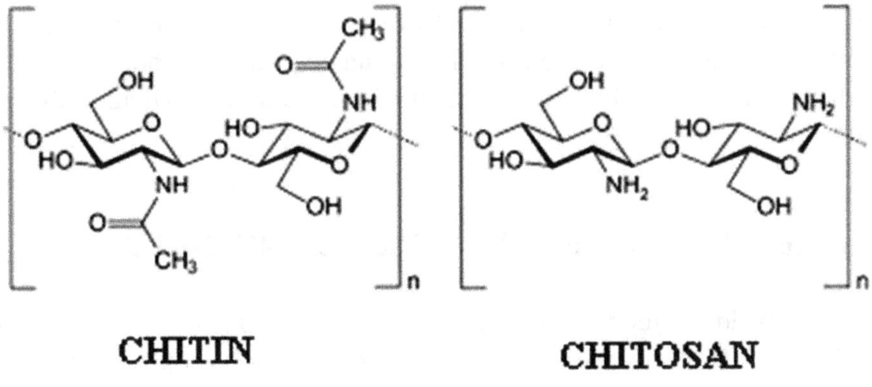

Chitin and chitosan are of commercial interest due to their high percentage of nitrogen (6.89%) compared to synthetically substituted cellulose (1.25%); this makes chitin a useful chelating agent. As most of the present-day polymers are synthetic materials, their biocompatibility and biodegradability are much more limited than those of natural polymers such as cellulose, chitosan, and their derivations. This biopolymer represents an attractive alternative to other biomaterials because of its physico-chemical characteristics, chemical stability, high reactivity, excellent chelation behavior and high selectivity toward pollutants.

Chitosan has been shown to effectively remove metals such as chromium, copper, mercury and lead from aqueous solutions (Schmuhl et al.,2001; Ngah et al., 2002; Ng et al.,2003; Qi &Xu, 2004;Laus et al., 2010; Hena, 2010; Yu et al., 2013,Sivakami, et al,2013).The adsorption of a metal ion on chitosan depends strongly on the pH of the solution. Chemical and physical modifications of chitosan have been commonly used to increase the stability of the polymer in acid solution and improve its adsorption capacity. In recent years, much active research has been performed on novel adsorbents and has explored various configurations of adsorbents to maximize their adsorptive capacity, particularly incorporating micro-nanoparticles

The general objective of this chapter was to analyze the use of contaminating crustaceans shellfish wastes for water purification, determining the simultaneous effects of pH and ionic crosslinking of chitosan particles on the removal of hexavalent chromium.

In this chapter the following topics are discussed:

1. Production of chitosan from seafood processing wastes (shrimp shells);
2. Synthesis of nanoparticles combining chitosan dissolved in acid solution with a nontoxic crosslinking agent;
3. Characterization of the obtained micro/nanoparticles by Scanning Electron Microscopy, particle size distribution, zeta potential and Infrared spectroscopy (FTIR);
4. Analysis of the performance of chitosan (CH) and chitosan nanoparticles (CHN) in the adsorption process of Cr from aqueous solutions at different pH values, initial chromium concentrations and contact times; and
5. Mathematical modeling of the equilibrium isotherms and the adsorption kinetics.

2. THE REMOVAL OF HEAVY METALS USING CHITOSAN

Around the world, seafood processing industry discards large amounts of crustaceans shellfish wastes; exoskeletons are converted in a solid residue, which accumulates in landfills becoming an environmental pollutant. The exoskeleton of the crustaceans represents approximately 50-60% of the total weight in crabs and more than 35% in shrimps. However, these residues contain chitin, which is a natural, abundant and non-toxic material. Chitosan (CH) is obtained cost-effectively from chitin (2-acetamido-2-deoxy-β-D-glucose through β(1-4) linkage); CH is the *N*-de-acetylated derivative of chitin and represents an attractive alternative to other biomaterials because of its physico-chemical characteristics, chemical stability, high reactivity, and excellent chelation behavior (Laus et al., 2010; Wu et al., 2010). Chitosan has three types of reactive functional groups, an amino group as well as both primary and secondary hydroxyl groups at the C-2, C-3 and C-6 positions, respectively (Schmuhl et al., 2001; Qi and Xu, 2004).Deacetylation degree and molecular weight are important properties to characterize this biopolymer when it is obtained from natural sources. Chitosan is a cationic, biocompatible, biodegradable, non-toxic biopolymer with numerous applications in agriculture, medicine, pharmacy, and waste -water treatments.

Water contamination with heavy metals is a critical problem because metals tend to persist and accumulate in the environment. Industrial and mining wastewaters are important sources of pollution of heavy metals. One of the heavy metals that have been a major focus in wastewater treatment is hexavalent chromium (Aydın &Aksoy, 2009; Gupta, 2012). The use of chromate and dichromate has many industrial applications such as in textile, electroplating, leather tanning, cement preservations, paints, pigments and metallurgy industries (Aydın &Aksoy, 2009; Bhatnagar&Sillanpää, 2009).

Cr(II), Cr(III) and Cr(VI) are the three oxidation states for chromium in nature, but only the last two are stable (Aydın & Aksoy, 2009).Cr(VI) is 500 times more toxic, mutagenic and carcinogenic than Cr(III) (Sarin et al, 2006; Dubey & Gopal, 2007) and must be removed from wastewater before it can be discharged. The permissible limit for hexavalent chromium in industrial wastewaters is 0.1 mg/L (Hena, 2010).

The removal of toxic metal ions from wastewater is an important research area. Different techniques have been used to remove metal ions from aqueous solutions, such as oxidation, reduction, precipitation, membrane filtration, ion exchange, and adsorption. Among these methods, adsorption using biopolymers, is one of the most recommended processes, due to economic and technical advantages (Bhatnagar & Sillanpää, 2009)

Chitosan, is a natural polyelectrolyte that has been used for remove cationic and anionic metals such as copper (Ngah et al., 2002; Laus et al., 2010; Yu et al., 2013), mercury and lead (Ng et al.,2003; Qi &Xu, 2004), chromium (Schmuhl et al.,2001; Hena, 2010,Sivakami, et al,2013), from aqueous solutions ;it combines with metal ions by three forms: ion exchange, adsorption, and chelation (Qi &Xu, 2004). The adsorption of a metal ion on chitosan depends strongly on the pH of the solution. (Schmuhl et al., 2001).

In recent years, research has been performed on novel adsorbents to maximize their adsorptive capacity (Yu et al., 2013); chemical and physical modifications of chitosan have been used to increase the stability of the polymer and to improve its functionality.

Chitosan is soluble in most dilute mineral acids, (except in sulfuric acid solutions) and in dilute organic acids, such as acetic, propionic, formic and lactic acids (Laus et al., 2010), consequently, its chemical stability needs to be reinforced through treatments using crosslinking agents for its application in acidic media. Chemical and physical cross-linking techniques are usually employed to modify chitosan; these treatments induce new linkages between the chitosan chains allowing the polymer to be highly resistant to dissolution, even in solutions, such as hydrochloric acid (Guibal et al., 2002; Beppu et al., 2004; Laus et al. 2010). The crosslinking procedure may be performed by reaction of CH with different agents such as glutaraldehyde, ethylene glycol diglycidyl ether, and epichlorohydrin. Tripolyphosphate (TPP) has also been proposed as a possible crosslinking agent (Laus et al., 2010, Dima et al, 2015).

Chitosan chains can be chemically cross-linked with glutaraldehyde leading to quite stable matrixes, with strength covalent bonds, however, glutaraldehyde is toxic and a strong irritant (López-León et al., 2005; Sano et al., 2005). Chitosan hydrogels can be obtained by ionic gelation, where micro or nanoparticles are formed by means of electrostatic interactions between the positively charged chitosan chains and polyanions employed as physical cross-linkers. Tripolyphosphate can be used as the polyanion for the cross-linking process. It is a condensed non-toxic phosphate suitable for water purification and also used as a food additive (Calvo et al., 1997; Laus et al., 2010). The protonated amine groups in chitosan interact with the negatively charged counter ions, through electrostatic interactions creating ionic cross-linked networks. It is important to remark the need to use non-toxic reagents such as TPP for water treatments. The correct interaction between chitosan and TPP leads to the formation of micro and nanoparticles (Calvo et al., 1997); therefore defined proportions of both components must be established in order to get adequate particle properties, related to size distribution and Z potential. The analysis of the pH effect on the performance of CH and CHN particles to remove Cr(VI) from contaminated water is complex because different factors, such as the proportion of Cr(VI) stable complexes (anions) formed, the presence of protonated groups in chitosan particles and the addition of TPP as reticulating agent are interacting.

3. CHITOSAN PRODUCTION AND CHARACTERIZATION

3. 1 Chitin Extraction and Chitosan Production

In the coastline of Patagonia-Argentina (42–43°S, 64–65°W), seafood processing industry discards large amounts of crustaceans shellfish wastes; exoskeletons are converted in a solid residue, which accumulates in landfills becoming an environmental pollutant. Shrimps shells (*Pleoticus muelleri*) provided by the industry were used for the extraction of chitin and chitosan (Dima et al. 2015).

Shrimp shells were ground using a mill to obtain a coarse powder (between 0.5 y 1mm mesh). The powder was treated with 1.5 N HCl solutions1:10 (w/v) at room temperature for 3h to remove minerals, then treated with aqueous sodium hydroxide solution (4.5%) at a ratio of 1:15 (w/v) at 80°C for 3 h to remove proteins. The mixture was filtered and washed with deionized water to reach neutrality. The solid thus obtained was discolorated using a solution water/acetone (75/25) at room temperature. Following discoloration, the solid was washed with deionized water to reach neutral pH. Finally, the wet mass was dried at 65°C for 24 h; the product obtained was designated as chitin. Chitin was deacetylated by treatment with 50% NaOH solution at 120°C for 120 min. After filtration, washing to neutral pH with deionized water and drying at 65°C for 24 h, the product obtained was designated as chitosan particles (CH). The yield of purified chitin was 24.8%± 0.4%. This value was within the range reported by different authors (Abdou& Nagy, 2008; Cocoletzi et al., 2009).After N-deacetylation, the yield of shrimp chitosan represented the 76.9%± 1.1% of the initial crude chitin.

Table 1 shows the yields of the different stages in the process of chitosan production. This values can be changed depending on the shell source (crab, shirmp, etc), methodology conditions, etc.

Table 1: Yields of the different stages in the process of chitosan production

Percentage of chitosan from chitin	77%
Percentage of chitin from dry shell	24.9%
Percentage of chitin from wet shell	42%
Percentage of wet shell from total shrimp	30%
Global yield	2.42%

3.2 Characterization of Chitosan

3.2.1. *Measurement of the Degree of N-Deacetylation by Potentiometric Titration*

The N-deacetylation degree (DD%) of shrimp chitosan samples was determined using the potentiometric technique proposed by Broussignac (1968). A sample of 0.5 g of chitosan was mixed with 20 ml of 0.3M HCl. The potentiometric evaluation was carried out with a 0.1M NaOH solution, using a pH-meter (Dima et al. 2015). The potentiometric curves were obtained by measuring the change in pH after each 2mL addition of base solution; additions were made slowly with continuous stirring. The percentage of N-deacetylation was carried out according to the following expression:

$$DD(\%) = \left(\frac{203M_{eq}}{1 + 42M_{eq}} \right) \tag{1}$$

where $M_{eq} = \frac{N\Delta V}{w}$; ΔV is the volume difference between higher inflection point and lower inflection point, N is the normality of NaOH solution, w is the chitosan mass used; 203 is the molar mass of glucosamine and 42 is the molar mass of acetyl group.

Chitosan performance depends on strongly on his deacetylation degree and molecular weight (Rinaudo, 2006).The titration curve (Fig.2) shows two inflection points; the difference between these points corresponds to the amount of acid required to protonate the amino groups of chitosan. The N-deacetylation calculated according to Eq 1 was 90.2%; this value falls into the range of commercial chitosan. The degree of deacetylation necessary to obtain a chitosan product must be 65% or higher (Kumar, 2000; Yen et al., 2008).

3.2.2 *Determination of the Degree of N-Deacetylation by Infrared Spectroscopy*

The degree of N-deacetylation (DD%) of shrimp chitosan, was also determined using Fourier-transform infrared spectra (FTIR, Bruker IFS 66). Samples of 2mg chitosan were mixed with 100mg of potassium bromide and compressed into pellets. The absorbances at 1320 cm^{-1} and 1420 cm^{-1} were used to calculate the DD% according to the following equation proposed by Brugnerotto et al. (2001):

$$DD(\%) = 100\text{-}[31.92*(A_{(1320cm^{-1})}/A_{(1420cm^{-1})}) - 12.20] \tag{2}$$

Figure 2. Potentiometric titration curve for shrimp chitosan. Peaks corresponding to the inflection points of the titration curve

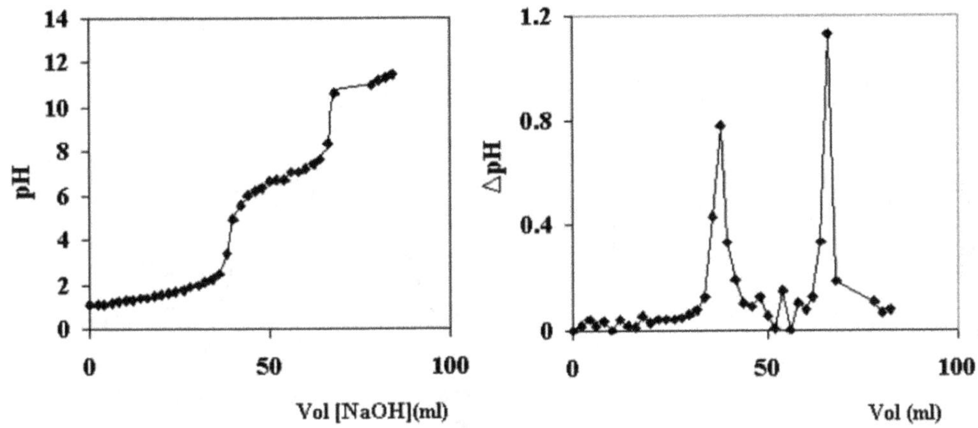

Fig. 3 shows the spectrum of shrimp chitosan; different characteristic bands can be observed: at 3450 cm^{-1} (N-H and O-H stretching vibrations), 2918 and 2852cm^{-1} (tension group C-H), 1721 cm^{-1} (C=O carbonyl group), 1618 cm^{-1} (Amine group), 1326 cm^{-1} (Amide III), 1382 cm^{-1} (–CH3 symmetrical deformation mode (scissoring) in amide group. The bands assigned to the stretching vibrations of C O C linkages in the polysaccharide structure are observed at 1151 cm^{-1}, 1098 cm^{-1} and 1021 cm^{-1} (skeletal stretching vibrations of C- O, characteristic peaks of polysaccharide structure in chitosan). According to Eq 2, the degree of deacetylation resulted in 86.9%, this value is similar to that obtained by potentiometric titration (Dima et al. 2015).

3.2.3 *Molecular Weight*

The molecular weight of CH was determined by a viscometric method. The intrinsic viscosity [η] of shrimp chitosan was measured by using an Ostwald capillary viscometer(Dima et al. 2015).

The chitosan sample was dissolved in 0.3 M acetic acid- 0.2M sodium acetate buffer and the measurements were performed at 25°C. The viscosity average molecular weight of chitosan (Mv) was calculated by measuring the intrinsic viscosity [η] according to Mark-Houwink-Kuhn-Sakurada (MHKS) equation:

$$[\eta] = k\, Mv^{a} \quad (3)$$

where k and a are viscometric constants, which depend on the polymer, the solvent system (buffer) used and the temperature.

Figure 3. Infrared spectra of CH obtained from shrimp

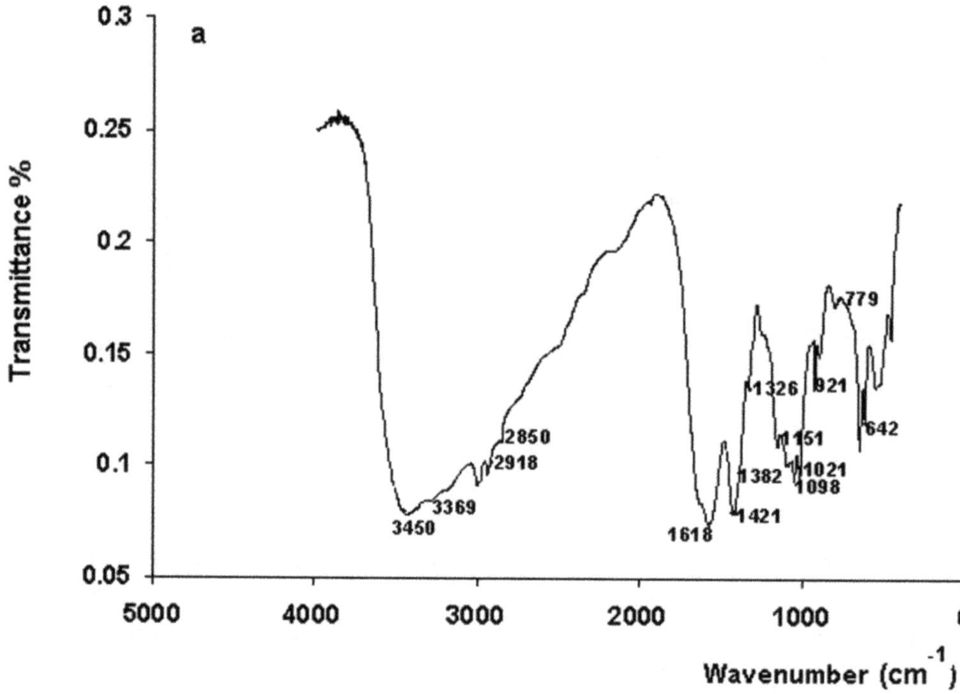

The specific viscosity (η_{sp}) is the fractional change in viscosity upon addition of polymer

$$\eta_{sp} = \frac{\eta_{solution} - \eta_{solvent}}{\eta_{solvent}} \tag{4}$$

The specific viscosity η_{sp} depends on the polymer concentration, thus to obtain the "intrinsic" properties of the polymer chain itself, it is necessary to extrapolate to zero concentration. Intrinsic viscosity $[\eta]$ was determined from $\frac{\eta_{sp}}{c}$ vs. concentration plot extrapolating at $c = 0$ and it has reciprocal concentration units.

$$[\eta] = \lim_{c \to 0} \frac{\eta_{sp}}{c} \tag{5}$$

According to Huggins equation:

$$\frac{\eta_{sp}}{c} = [\eta] + K_H [\eta]^2 c \qquad (6)$$

Figure 4 shows the Huggins equation to obtain the intrinsic viscosity [η]. The average molecular weight of chitosan (Mv) was determined by using Eq. 3.

Introducing the parameters proposed by Rinaudo et al (1993) k = 0.074 ml/g and a = 0.80 an average molecular weight of 1.46×10^5 Da was obtained by Dima et al.(2015); this value falls within the range reported by other authors, for different chitosan sources (1×10^5 - 5×10^5 Da) (Kumar, 2000; Rinaudo, 2006; Yen et al., 2008).

4. SYNTHESIS AND CHARACTERIZATION OF CHITOSAN RETICULATED NANOPARTICLES

Chitosan reticulated nanoparticles (CHN) were prepared by ionic gelation with TPP according to the technique described by Calvo et al (1997). Preliminary

Figure 4. Huggins equation applied to chitosan solutions from shrimps to obtain the intrinsic viscosity

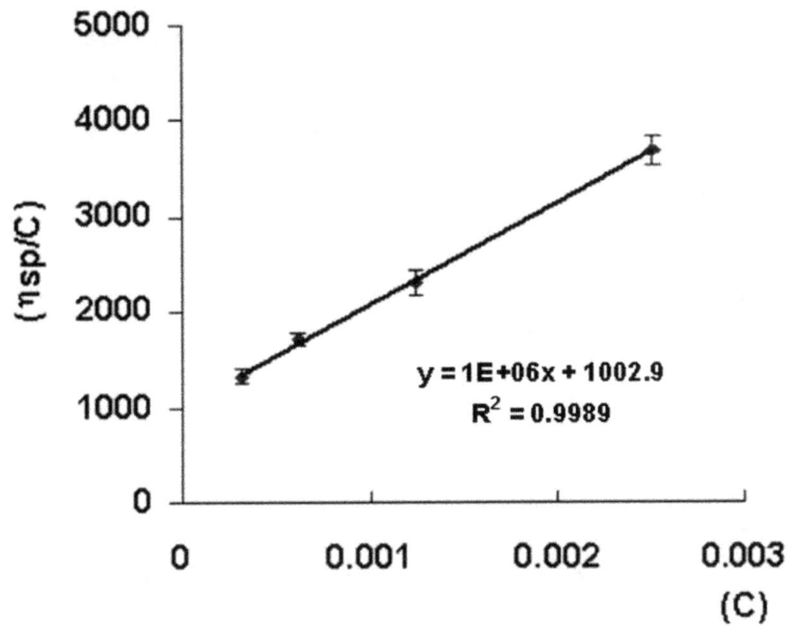

experiments were carried out in order to determine the adequate conditions for micro and nanoparticles formation (Fig 5). CH was dissolved in acetic acid aqueous solutions at various concentrations: 0.5, 0.75, 1.25, 2.5 and 5 g/L (w/v). Then, TPP was dissolved in purified water at concentrations of 0.25, 0.5, 1.0, 1.5 and 2.5 g/L (w/v). Finally, 1mL of the TPP solution was added to 3 mL of the CH solution under magnetic stirring at room temperature. Samples were visually analyzed and three different systems were identified: clear solution, opalescent suspension, and aggregates (Figure 5).

Figure 5. a) Scheme of nanoparticles formation, b) Photograph of the grade of opalescent suspension

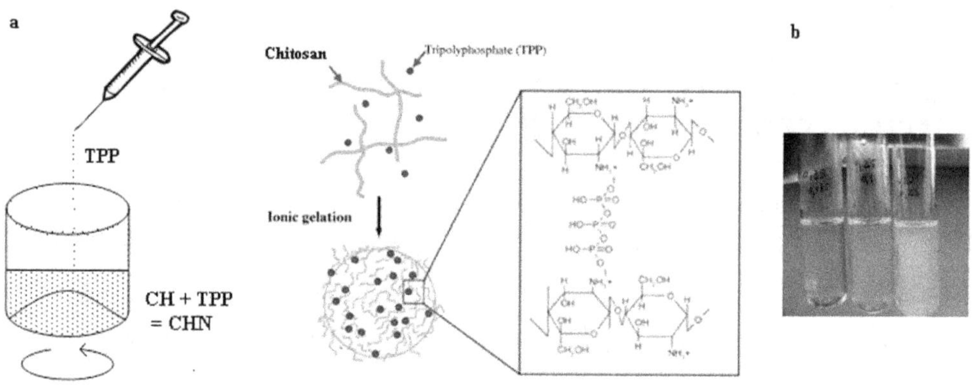

Figure 6. Zones of opalescent solutions: (♦) aggregates, (▲) clear solution, (○) opalescent suspension. Chitosan (g/L)- tripolyphosphate (TPP g/L)

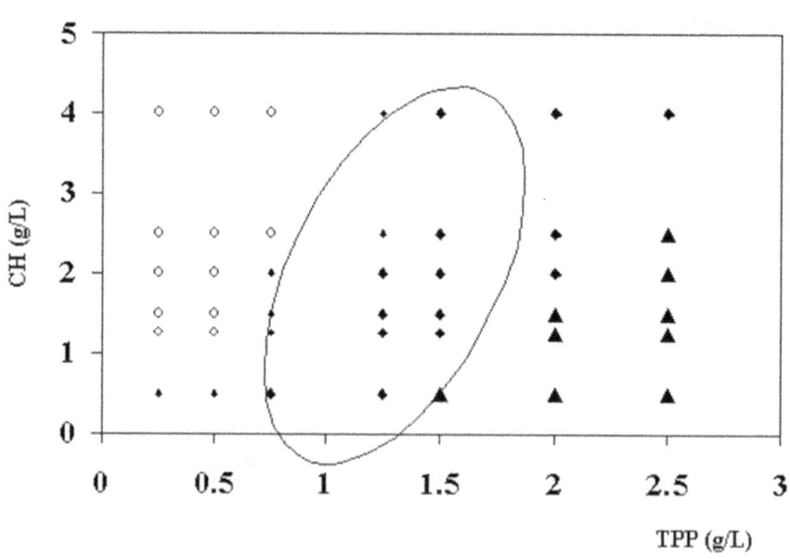

The opalescent suspension should correspond to the presence of very small particles (Figure 6). Micro-nanoparticles were purified by centrifugation at 10000xg for 30 min. Supernatants were discarded, and the CHN were extensively rinsed with distilled water to remove any excess of TPP residues or CH. The micro- nanoparticle suspensions were centrifuged and the sediment (wet mass of nanoparticles) dried in vacuum at 55°C for 24h to obtain the dried mass of nanoparticles production yield. CHN were characterized by the measurement of the size distribution and zeta potential as a function of the amount of TPP added to the dissolved CH and the pH of the system, using a Delsa™Nano C Instrument; Scanning electronic microscopy (SEM) was used for visualization of the CHN (Dima et al.2015).

The production mass yield for the ratio 3mL Chitosan: 1mL TPP at the proposed concentrations was 1.68 ± 0.09 mg of micro/nanoparticles (dry weight). The mass yield of these particles referred to the initial mass of chitosan (3ml CH 0.125%) was 44.8%.

For a final concentration of 1-1.5g/L TPP and 1.25g/L chitosan, at pH=4, the mean size of the micro/nanoparticles was 101 nm (P_{10}=88nm; P_{90}=145nm) with a narrow size distribution (polydispersity index <1) in a stable nano-suspension. The size of the reticulated micro/nanoparticles determined by the Beckman Coulter Equipment was in agreement with that observed by SEM microscopy (Figure 7). It is interesting to remark that being the average diameter of the reticulated particles close to 100 nm these particles could be considered as nanoparticles (Qu et al., 2013).

Figure 8 shows particle size, particle size distribution and zeta potential as a function of pH and TPP concentration. The particle size increased with the amount of added TPP (Figure 8a,b) and with increasing pH (Figure 8d,e).The CH–TPP

Figure 7. Particle size distribution determined by the Beckman Coulter Equipment and SEM micrograph of chitosan micro and nanoparticles for a ratio of CH:TPP of 1.25:1.50 (g/L)

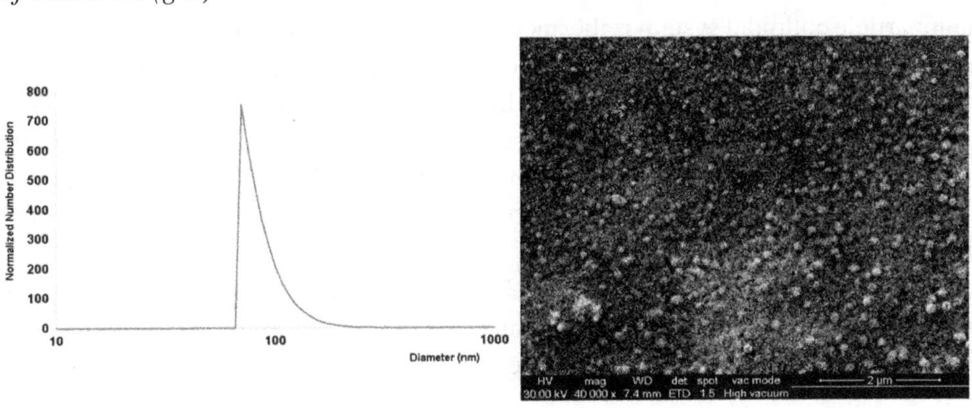

Figure 8. Characterization of chitosan-TPP nanoparticles. Effect of TPP concentration on: a)particle size distribution; b) average particle size; c) z-potential (pH =4): CHN in the original suspension (full line), CHN re-suspension in water (dotted line). Effect of pH on: d)particle size distribution; e) average particle size; f) z-potential (System of 1.25mg/ml Chitosan and 1.5mg/ml TPP)

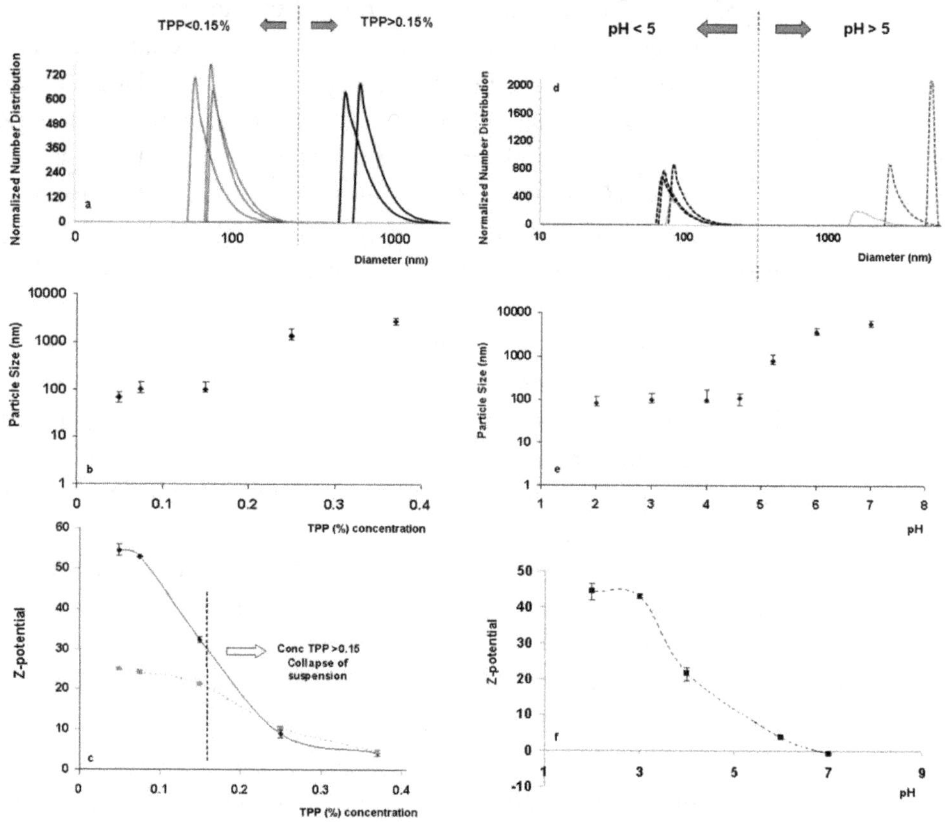

nanoparticle colloidal system is thermodynamically unstable, especially at unfavorable pH conditions and at high particle concentrations, because of the high surface energy associated with the nano scale dimensions; therefore when the initial TPP or chitosan concentrations were increased over and above a critical concentration, large aggregates are formed instantaneously (Figure 8,ab). pH has a marked effect on the electrokinetic behavior (z-potential), particle size and particle size distribution (Figure 8d, e,f).

At pH =5 the nano-suspension started to be unstable and a bimodal distribution was observed (full line); at pH> 5 the suspension was collapsed with a significant increase in particle size. This is because the reduction of the protonation degree at the surface of the particles decreased the electrostatic repulsion between the particles increasing the probability of particle aggregation (Gan et al, 2005).

When chitosan and TPP were mixed in diluted acetic acid, they spontaneously formed compact nano-complexes with an overall positive surface charge, and the density of the surface charge is reflected by the measured zeta potential values. The zeta potential of the particles decreased with the amount of TPP added (Figure 8c); a high magnitude of zeta potential denotes stability of particles in suspension. In the case of chitosan the positive values of zeta potential are attributed to the presence of protonated amine groups; when negatively charged TPP reacts with positively charged chitosan, the net z- potential decreases. Therefore, the higher the zeta potential, the lower the fraction of TPP present in the microparticles, in agreement with results reported by Qi & Xu (2004). For the concentration ratio proposed (1g/L-TPP: 1.25g/L- CH), the Z potential of chitosan nanoparticles decreased with the increase of pH (Figure 8f). The isoelectric point of CH–TPP reticulated system is around 7.

5. HEXAVALENT CHROMIUM ADSORPTION USING CHITOSAN (CH) AND CHITOSAN NANOPARTICLES (CHN)

Adsorption experiments were performed using different values of Cr(VI) initial concentration, contact times, and pH values. 80mg of CH particles or 40mg of CHN in 50 ml of Cr (VI) solutions of different initial concentrations ranging from 50 to 400 (mg/L)were tested. All the experiments to investigate the adsorption of Cr(VI) ions onto CH or on CHN were carried out in batch tests at 25°C, under constant stirring (200 rpm). Final concentrations of Cr (VI) were determined by flame atomic absorption spectrometry (FAAS). The equilibrium adsorption capacity of CH or CHN (Qe) is the amount of metal adsorbed per unit weight of adsorbent at equilibrium (mg/g) and was calculated as follows:

$$Qe = \frac{(Ci - Ceq)}{w} V \tag{7}$$

where: Ci and Ceq are the metal concentrations (mg/L) in the solution initially (time zero) and after equilibrium respectively, V (L) is the volume of the solution and w is the mass (g) of adsorbent used.

At each time the adsorption capacity Q (mg/g) was calculated as:

$$Q = \frac{(Ci - C)}{w} V \tag{8}$$

where C is the metal concentration in the solution after time t

The percentage of Cr (VI) removal was calculated according to:

$$(\%)\,\mathrm{Re}\,moval = \left(\frac{Ci - Ceq}{Ci}\right) x100 \tag{9}$$

The time required for the system to reach equilibrium conditions was established using CH or CHN in 50 mL of Cr (VI) solutions with an initial concentration of 100 mg/L, stirred for 3 h. After pre-established time periods (30min, 1, 1.5, 2 and 3h), the stirring was halted and the supernatant was removed and diluted to an adequate volume. Concentration was monitored with respect to time for kinetic analysis.

In order to analyze the effect of pH, samples of CH or CHN were placed in a series of flasks containing 50 mL of Cr (VI) metal ion at an initial concentration of 200 mg/L. The solution was buffered within a pH range of 2-6. After 3h, the sample was filtered from the solution, and quantified by FAAS.

Additional experiments were carried out using different initial concentrations of Cr (VI) ions in a range between 50 to 400 mg/L in contact with 80mg of CH or 40 mg of CHN. At the end of each mixing period, the liquid phase was separated from the solution by filtration and quantified by FAAS.

5.1. Effect of Contact Time

Contact time is an important parameter because this factor determines the adsorption kinetics of an adsorbent at a given initial concentration of the adsorbate.

The adsorption kinetic curves of Cr(VI) (initial concentration of 100mg/L) onto CH and CHN are observed in Figure 9. The curves show that equilibrium was reached after approximately 1h for CH (Figure 9a) and after 2h for CHN (Figure 9b). The maximum amounts of adsorbed chromium were produced after 3 h contact time obtaining equilibrium values of 66.9 mg/g for CH and 38.8mg/g for CHN (Dima et al., 2015).After these periods, both systems remained almost unchanged until the end of the experiment (3h).

5.2 Effect of pH

The effect of pH on the adsorption of Cr(VI) ions onto CH and CHN was studied in the pH range of 2-6 (Figure 10a, b). It can be observed that the optimum pH for the adsorption of Cr(VI) onto CH and CHN was pH=4 and pH=2 respectively (Figure 10a,b). When pH decreased from 4 to 2 the amount of Cr(VI) adsorbed onto CH decreased 29.0% (Figure 10a), while in the case of CHN increased 23.2% (Figure

Figure 9. Effect of contact time on Adsorption capacity (Q) of Chromium (VI) on: a) CH; b) CHN, at Cr (VI) initial concentration = 100 ppm and pH = 4

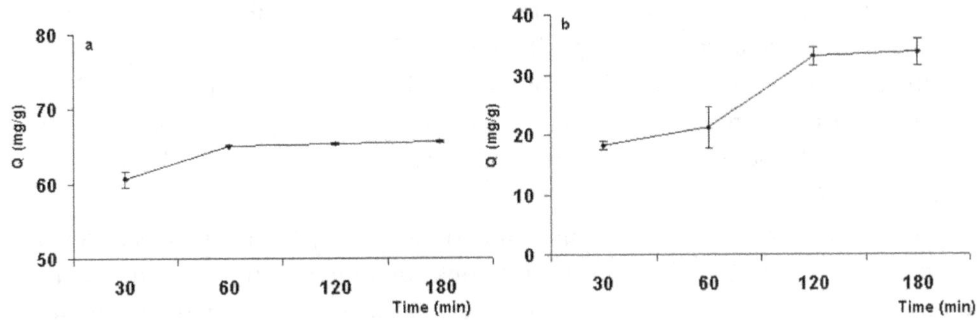

Figure 10. Effect of pH on Equilibrium adsorption capacity (Qe) of Chromium (VI) on: a) CH; b) CHN, at Cr(VI) initial concentration of 100 ppm

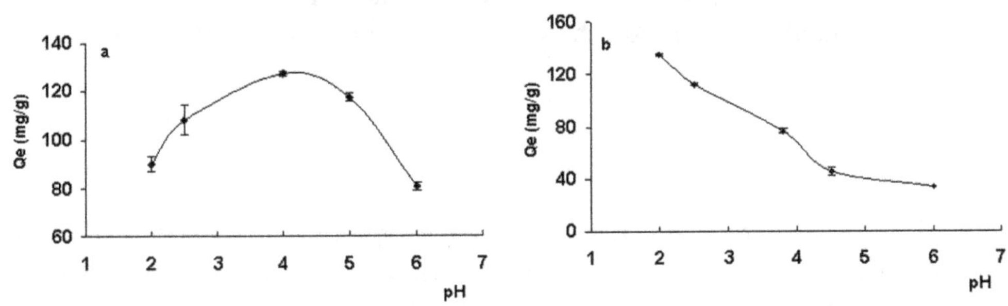

10b). The highest value of Qe, was 127.1mg/g for CH (pH=4), and in the case of CHN, a higher value of Qe=135.2mg/g was observed at pH=2.

The acid dissociation constant (pKa) of chitosan is in the range between 6.2 to 6.8, below the pKa value, chitosan is positively charged, while the metal is negatively charged; this leads to an electrostatic interaction. According to Hamadi et al. (2001) and Schmuhl et al. (2001),the decrease of Cr(VI) removal at a higher pH can be explained by the changes in protonated and unprotonated forms of CH. The decrease of the equilibrium adsorption capacity at pH= 2 may be attributed to the fact that chitosan is unstable at a low pH and dissolved in acidic medium (Qi &Xu, 2004). However, reticulated nanoparticles (CHN) improved the stability of chitosan at a low pH.

5.3. Influence of Initial Cr (VI) Concentration

Figure 11 shows the adsorption capacities of CH and CHN at different initial Cr(VI) concentrations. The adsorption capacity increased from 31.4 to 230.2 mg/g for CH (Figure 11a) and from 35.5 to 71.4 mg/g for CHN, when the initial Cr(VI) concentration varied from 50 to 400 mg/L, both at pH =4. At pH =2, the adsorption capacity of CHN increased (Figure 11b) and adsorption capacity of CH decreased. (Dima et al, 2015).

A concentration of 1.6g/L of chitosan particles (CH),at pH = 4, was necessary to remove 99% and 90% of Cr(VI) in solutions containing initial concentrations of 50 mg/L and 400mg/L of Cr (VI) respectively showing a very good performance. At pH=2, 0.8 g/L of CHN were necessary to remove 98% of Cr(VI) in solutions containing initial concentrations of 50 mg/L.

Figure 11 c, d shows the simultaneous effect of initial Cr(VI) concentration and pH on the percentage of Cr (VI) removal (defined in Eq.9) for CH and CHN respectively. Initial concentration showed a marked effect on Cr(VI) removal, especially for CHN.

Figure 11. Effect of initial Cr(VI) concentration on Equilibrium adsorption capacity (Qe): a) CH at pH =4; b) CHN at pH =2. Simultaneous effect of Chromium (VI) initial concentration and pH on the percentage of Cr (VI) removal using: c) CH ; d) CHN

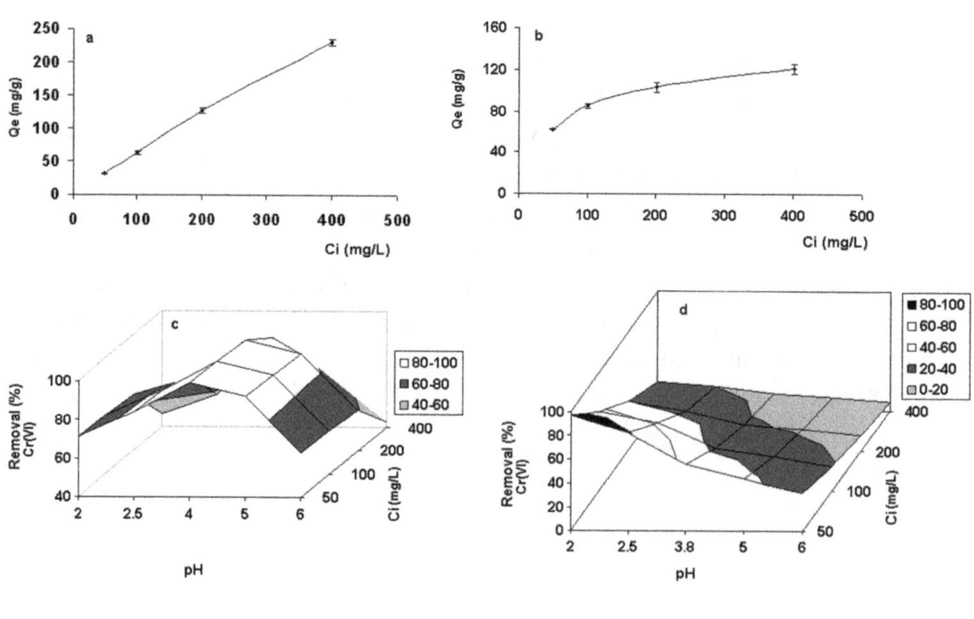

5.4 Equilibrium Adsorption Isotherms

Langmuir, Freundlich and Temkin isotherms were used for the mathematical description of the adsorption equilibrium of Cr(VI) ions on CH or CHN adsorbents(Sivakami, et al,2013). The operating parameters were: T= 25°C, pH= 4, time= 3 hours.

Adsorption was evaluated by the determination of Cr(VI) concentration in the solution under equilibrium conditions reached at the end of the experiments.

The Langmuir isotherm is given by the equation:

$$\frac{Ceq}{Qe} = \frac{Ceq}{Qm} + \frac{1}{K_L Qm} \qquad (10)$$

WhereQe is the amount adsorbed per unit weight of adsorbent at equilibrium (mg/g); Ceqis the equilibrium concentration of adsorbate in solution after adsorption (mg/L). K_L is the Langmuir constant (g/L) related to the affinity of the binding sites; Qm is the maximum monolayer adsorption capacity (mg/g).

The Freundlich isotherm is an empirical equation that assumes that the adsorption process takes place on heterogeneous surfaces and adsorption capacity is related to the concentration of Cr (VI) at equilibrium. It is defined as follows:

$$Qe = K_f Ceq^{1/n} \qquad (11)$$

Where K_f is the Freundlich constant or capacity factor (mg/g) and 1/n is the Freundlich exponent; n is the heterogeneity factor related to adsorption intensity.

Temkin isotherm assumes that the heat of adsorption of all the molecules in the layer decreases linearly with coverage due to adsorbent-adsorbate interactions.

The Temkin isotherm in its linear form is given by the equation:

$$Qe = B_t \ln(K_t) + B_t \ln(Ceq) \qquad (12)$$

The adsorption is characterized by a uniform distribution of binding energies, up to some maximum binding energy value. The Temkin parameter $B_t$is related to the heat of adsorption and $K_T$is the equilibrium binding constant.

The obtained parameters from Freundlich, Langmuir and Temkin adsorption equations and the corresponding R^2 values (coefficients of determination) are shown in Table 2. The regression values indicate that the adsorption data for chromium

Table 2. Parameters of the equilibrium isotherms for Cr(VI) adsorption upon chitosan (CH) and chitosan micro/nanoparticles (CHN)

	Langmuir			Freundlich			Temkin		
	Qm (mg/g)	K_L (L/mg)	R^2	1/n	K_f (mg/g)	R^2	B_t	Kt (L/mg)	R^2
Chitosan pH=4	250	0.018	0.999	0.43	44.70	0.941	45.01	2.76	0.997
Micro/nanoparticles (CHN) pH=4	68.9	0.014	0.990	0.36	7.02	0.983	13.95	1.76	0.989
Micro/nanoparticles (CHN) pH=2	124	0.086	0.990	0.12	60.42	0.977	22.55	1.46	0.939

removal fitted well to Langmuir and Freundlich isotherms, however, the Langmuir isotherm has higher R^2 when compared to the other models. Langmuir isotherm (Eq. 10) is representative of monolayer adsorption occurring on an energetically uniform surface on which the adsorbed molecules are not interactive. For nanoparticles, adsorption at pH=2 was more effective than at pH=4 (Table 2).

The obtained values of maximum Cr(VI) adsorption capacity were 250mg/g for CH and 68.8mg/g for CHN at pH =4 and T=25.5°C. This decrease of the maximum adsorption capacity is consistent with the fact that when chitosan is cross-linked, some of the binding sites are used for the cross-linking process instead of adsorbing the metal anions.The extent of capacity reduction strongly depends on the relative amount of complexing agent used. This effect should be compensated with the required properties of chitosan such as the stability of the crosslinked particles in acidic solutions (pH<2).

5.5 Kinetic Analysis

Kinetic analysis is required to get an insight of the rate of adsorption and the limiting step of the transport mechanism, which are primarily used in the modeling, and design of the process.Kinetic data were analyzed using the pseudo first order, pseudo second order kinetic models and Elovich equation.

The pseudo first order kinetic model of Lagergren has been widely used to predict the metal adsorption kinetics and is given by (Sivaraj et al 2001, Aydın &Aksoy, 2009):

$$dQ / dt = k_1 (Qe - Q) \qquad (13)$$

WhereQ is the amount of metal adsorbed at any time (mg/g), Qe is the amount of metal adsorbed at equilibrium time (mg/g) and k_1 is the pseudo first order rate constant (min^{-1}).Integrating Equation (13) becomes:

$$\ln\left(\frac{Qe}{Qe-Q}\right)=k_1 t \tag{14}$$

The adsorption kinetic data can be further analyzed using the pseudo second order kinetics, which is represented by (Sivaraj et al 2001, Aydın &Aksoy, 2009):

$$dQ / dt = k_2(Qe-Q)^2 \tag{15}$$

where k_2 is the second order rate constant. Integrating Equation (15) gives:

$$\frac{t}{Q}=\frac{1}{k_2 Qe^2}+\frac{1}{Qe}t \tag{16}$$

The intercept of the linearized pseudo-second order rate equation gives the second order rate constant, k_2.

The Elovich or Roginsky–Zeldovich equation is satisfied in chemical adsorption processes and is suitable for systems with heterogeneous adsorbing surfaces. It is generally expressed as follows (Sivaraj et al 2001, Aydın & Aksoy, 2009):

$$dQ / dt = \alpha \exp(-\beta\, Q) \tag{17}$$

where Q is the amount of metal adsorbed by chitosan at a time t, α is the initial adsorption rate ($mg\ g^{-1}\ h^{-1}$), and β is the Elovich constant.To simplify the Elovich equation, Chien& Clayton (1980) assumed $\alpha\beta t >> 1$, and on applying the initial condition Q = 0 at t = 0 the equation becomes Sparks (1999):

$$Q=\beta\ln(\alpha\beta)+\beta\ln t \tag{18}$$

The coefficients were obtained from the slope and intercept of a straight line plot of Qversus lnt.

The parameters obtained by the different tested kinetic equations with the corresponding correlation coefficients are given in Table 3.

Table 3. Kinetic parameters for Cr(VI) adsorption on chitosan flakes (CH) and chitosan micro-nanoparticles (CHN)

	Pseudo-First order		Pseudo-second order		Elovich		
	$k_1(h^{-1})$	R^2	k_2 (g/mgh)	R^2	β(g/mg)	α(mg/gh)	R^2
Chitosan (CH)	2.71	0.871	0.76	0.999	0.37	$6.0x10^{10}$	0.750
Chitosan Micro/nanoparticles (CHN)	1.94	0.932	0.078	0.946	0.10	116.8	0.877

For CH, the kinetic studies indicated a rapid removal of chromium from aqueous solutions. The large value of α in Elovich equation for chitosan particles indicates a very high initial adsorption rate in comparison to the micro/ nanoparticles.

The kinetic analysis of chromium adsorption showed that pseudo second order kinetic model fitted successfully the experimental data. According to Sağ & Aktay (2002) the pseudo-second order kinetic model is based on the assumption that the rate-limiting step may be chemisorption involving valency forces through sharing or exchange of electrons between sorbent and sorbate.

6. SOLUTIONS AND RECOMMENDATIONS

As mentioned previously, around the world, seafood processing industry discards large amounts of crustaceans shellfish wastes; exoskeletons are converted in a solid residue, which accumulates in landfills becoming an environmental pollutant. Particularly, in Patagonia Argentina hundreds of tons of exoskeletons are discarded. However, these residues contain up to 30% of chitin,from which chitosan is obtained. Chitosan is a natural, abundant and non-toxic material and is obtained cost-effectively from chitin. Chitosan has many applications and the local production of chitin and chitosan from crustaceans represents a good opportunity for the regional industry contributing also to improve environmental conditions.

One of the most interesting uses of chitosan is as a biadsorbent for water treatment. In this chapter a study about the use of chitosan and reticulated chitosan nanoparticles to remove a toxic residue such as hexavalent chromium is described. Chitosan reticulated with a nontoxic agent such as tripolyphosphate provides an alternative to adsorb Cr (VI) at low pH, due to the instability of CH particles in acid media. Chromium metal is added to alloy steel to increase hardenability and corrosion resistance. A high number of workers in a variety of occupations are potentially exposed to Cr(VI). This contaminant targets the respiratory system, kidneys, liver, skin, and eyes. Workplace exposures occur mainly in the following areas: tanning industry, welding and other types of "hot work" on stainless steel and other metals that contain chromium, use of pigments, spray paints and coatings, operating chrome plating baths. Chemical precipitation methods are commonly employed for the removal of chromium but this leads to the formation of chrome-bearing solid waste. Ion exchange and membrane separation methods are relatively expensive. The development of innovative metal clean-up technologies at industrial scale remains a challenge.

7. FUTURE RESEARCH DIRECTIONS

In the study carried out by Dima et al. (2015), chemical analysis to determine the oxidation state of the adsorbed Cr, showed that Cr(VI) was adsorbed on CH particles without further reduction. In contrast, Cr(VI) removed from the solution was reduced and bound to the CHN as Cr(III).The reduction of toxic Cr(VI) to the less or nontoxic Cr(III) by the reticulated chitosan micro/nanoparticles can be considered a very efficient detoxification technique for the treatment of Cr(VI) contaminated water.

The measurement of the oxidation state of the adsorbed chromium species indicated that when unmodified chitosan particles (CH) were used as the adsorbent material chromium is adsorbed as Cr(VI); in this case, when Cr was desorbed from the samples Cr(III) could not be detected.

In contrast, in the case of Cr (VI) removal, using reticulated chitosan micro and nanoparticles (CHN), more than 60% of the adsorbed chromium was reduced to Cr(III). Then CHN reduced Cr (VI) to Cr (III) thus favoring non-toxicity. One of the mechanisms proposed for removal of Cr (VI) by reticulated CHN could be adsorption which proceeds due to the electrostatic attraction between two oppositely charged ions since the chromium(VI) ions in the solution are present in the form of anions and chitosan micro/nanoparticles have an overall positive surface charge. Reticulated CHN used as adsorbents not only act as electron donors for the Cr(VI) reduction to Cr(III), but also act as the active sites for Cr(III) adsorption. The Cr(III) is then adsorbed on the surface of CHN by the attraction of amine groups and the precipitation of Cr(III); a light green color was observed in the CHN. The combination of the reduction of Cr(VI) to Cr(III) and the adsorption of Cr(III) on the adsorbent is expected as an effective approach to remove the Cr(VI) from contaminated wastewater. The reduction of toxic Cr(VI) to the less or nontoxic Cr(III) by chitosan nanoparticles reticulated with TPP, can be considered a very efficient detoxification technique for the treatment of Cr(VI) contaminated water. The study of the mechanism of action of CHN in reducing Cr (VI) is an innovative work in the area of environment and these results are very important from a toxicological point of view.

8. CONCLUSION

Chitosan particles (CH) were obtained from shrimp shells. The degree of deacetylation obtained by potentiometric titration and FTIR, were into de range of commercial chitosan (86%-90.1%). The molecular weight of chitosan was determined by intrinsic viscosity measurements. Chitosan micro and nanoparticles (CHN) were synthesized by inducing the gelation of a chitosan solution with tripolyphosphate (TPP), a non-toxic polyanion. Different mean sizes of nanoparticles were obtained by adjusting the ratio of chitosan to tripolyphosphate. Suspension pH and relative proportions of CH and TPP were modified obtaining particles with different properties, characterized by size distribution and Z-potential. For a given combination of TPP and CH the average particle size was 101 nm, obtaining nanoparticles.

The adsorption capacity of CH and reticulated CHN toward hexavalent chromium Cr(VI), under different conditions such as pH, contact time, and initial chromium concentration were analyzed. The amount of chromium adsorbed varied with the initial Cr (VI) concentration, solution pH and contact time.

CH showed a better performance to adsorb Cr(VI) than CHN at high pH values; this behavior can be explained considering that the cross-linking process with TPP reduced the number of sites for the adsorption of the metal ions onto reticulated chitosan; however, at low pH values the adsorption capacities were higher for the chitosan reticulated nanoparticles than for chitosan particles. At lower pH the functional groups of chitosan were protonated, resulting in a stronger attraction for a negatively charged chromium ion in the solution, leading to electrostatic interactions between the sorbent and the sorbate. The decrease of the equilibrium adsorption capacity of chitosan particles at pH= 2 was attributed to the fact that chitosan is unstable at a low pH. In contrast, cross-linking process improved the stability of CHN at $pH<3$.

Langmuir, Freundlich, and Temkin isotherms were applied for the mathematical description of the adsorption equilibrium curves. Langmuir isotherm which is valid for monolayer adsorption on a surface with a limited number of sites, predicting a homogeneous distribution of adsorption energies, provided the best fit to the experimental data.

The kinetic analysis was conducted using the pseudo first-order, pseudo second order and Elovich models. The pseudo-second order kinetic model provided the best correlation of the experimental data for both, chitosan particles and reticulated chitosan particles.

Reticulated chitosan nanoparticles and chitosan particles can be effectively applied for the removal of Cr(VI) from wastewater. However, the present work demonstrated that chitosan crosslinking is necessary to adsorb Cr(VI) at $pH<2.5$ due to the dissolution of the non-modified CH particles in acid media.

REFERENCES

Abdou, E. S., Nagy, K. S., & Elsabee, M. Z. (2008). Extraction and characterization of chitin and chitosan from local sources. *Bioresource Technology*, *99*(5), 1359–1367. doi:10.1016/j.biortech.2007.01.051 PMID:17383869

Aydýn, Y. A., & Aksoy, N. D. (2009). Adsorption of chromium on chitosan: Optimization, kinetics and thermodynamics. *Chemical Engineering Journal*, *151*(1), 188–194. doi:10.1016/j.cej.2009.02.010

Beppu, M. M., Arruda, E. J., Vieira, R. S., & Santos, N. N. (2004). Adsorption of Cu (II) on porous chitosan membranes functionalized with histidine. *Journal of Membrane Science*, *240*(1), 227–235. doi:10.1016/j.memsci.2004.04.025

Bhatnagar, A., & Sillanpää, M. (2009). Applications of chitin-and chitosan-derivatives for the detoxification of water and wastewater—a short review. *Advances in Colloid and Interface Science*, *152*(1), 26–38. doi:10.1016/j.cis.2009.09.003 PMID:19833317

Broussignac, P. (1968). Chitosan: A natural polymer not well known by the industry. *Chim. Ind. Genie Chim*, *99*(9), 1241–1247.

Brugnerotto, J., Lizardi, J., Goycoolea, F. M., Argüelles-Monal, W., Desbrieres, J., & Rinaudo, M. (2001). An infrared investigation in relation with chitin and chitosan characterization. *Polymer*, *42*(8), 3569–3580. doi:10.1016/S0032-3861(00)00713-8

Calvo, P., Remuñan-López, C., Vila-Jato, J. L., & Alonso, M. J. (1997). Chitosan and chitosan/ethylene oxide-propylene oxide block copolymer nanoparticles as novel carriers for proteins and vaccines. *Pharmaceutical Research*, *14*(10), 1431–1436. doi:10.1023/A:1012128907225 PMID:9358557

Chien, S. H., & Clayton, W. R. (1980). Application of Elovich equation to the kinetics of phosphate release and sorption in soils. *Soil Science Society of America Journal*, *44*(2), 265–268. doi:10.2136/sssaj1980.03615995004400020013x

Cocoletzi, H. H., Almanza, E. Á., Agustin, O. F., Nava, E. V., &cCassellis, E. R. (2009). Obtención y caracterización de quitosano a partir de exoesqueletos de camarón. *Superficies y vacío, 22*(3), 57-60.

Dima, J. B., Sequeiros, C., & Zaritzky, N. E. (2015). Hexavalent chromium removal in contaminated water using reticulated chitosan micro/nanoparticles from seafood processing wastes. *Chemosphere*, *141*, 100–111. doi:10.1016/j.chemosphere.2015.06.030 PMID:26151484

Dubey, S. P., & Gopal, K. (2007). Adsorption of chromium (VI) on low cost adsorbents derived from agricultural waste material: A comparative study. *Journal of Hazardous Materials*, *145*(3), 465–470. doi:10.1016/j.jhazmat.2006.11.041 PMID:17175098

Gan, Q., Wang, T., Cochrane, C., & McCarron, P. (2005). Modulation of surface charge, particle size and morphological properties of chitosan–TPP nanoparticles intended for gene delivery. *Colloids and Surfaces. B, Biointerfaces*, *44*(2), 65–73. doi:10.1016/j.colsurfb.2005.06.001 PMID:16024239

Guibal, E., Sweeney, N. V. O., Vincent, T., & Tobin, J. M. (2002). Sulfur derivatives of chitosan for palladium sorption. *Reactive & Functional Polymers*, *50*(2), 149–163. doi:10.1016/S1381-5148(01)00110-9

Gupta, V. K., Agarwal, S., & Saleh, T. A. (2011). Chromium removal by combining the magnetic properties of iron oxide with adsorption properties of carbon nanotubes. *Water Research*, *45*(6), 2207–2212. doi:10.1016/j.watres.2011.01.012 PMID:21303713

Hamadi, N. K., Chen, X. D., Farid, M. M., & Lu, M. G. (2001). Adsorption kinetics for the removal of chromium (VI) from aqueous solution by adsorbents derived from used tyres and sawdust. *Chemical Engineering Journal*, *84*(2), 95–105. doi:10.1016/S1385-8947(01)00194-2

Kumar, M. N. R. (2000). A review of chitin and chitosan applications. *Reactive & Functional Polymers*, *46*(1), 1–27. doi:10.1016/S1381-5148(00)00038-9

Laus, R., Costa, T. G., Szpoganicz, B., & Fávere, V. T. (2010). Adsorption and desorption of Cu (II), Cd (II) and Pb (II) ions using chitosan crosslinked with epichlorohydrin-triphosphate as the adsorbent. *Journal of Hazardous Materials*, *183*(1), 233–241. doi:10.1016/j.jhazmat.2010.07.016 PMID:20674156

López-León, T., Carvalho, E. L. S., Seijo, B., Ortega-Vinuesa, J. L., & Bastos-González, D. (2005). Physicochemical characterization of chitosan nanoparticles: Electrokinetic and stability behavior. *Journal of Colloid and Interface Science*, *283*(2), 344–351. doi:10.1016/j.jcis.2004.08.186 PMID:15721903

Ng, J. C. Y., Cheung, W. H., & McKay, G. (2003). Equilibrium studies for the sorption of lead from effluents using chitosan. *Chemosphere*, *52*(6), 1021–1030. doi:10.1016/S0045-6535(03)00223-6 PMID:12781235

Ngah, W. W., Endud, C. S., & Mayanar, R. (2002). Removal of copper (II) ions from aqueous solution onto chitosan and cross-linked chitosan beads. *Reactive & Functional Polymers*, *50*(2), 181–190. doi:10.1016/S1381-5148(01)00113-4

Qi, L., & Xu, Z. (2004). Lead sorption from aqueous solutions on chitosan nanoparticles. *Colloids and Surfaces. A, Physicochemical and Engineering Aspects*, *251*(1), 183–190. doi:10.1016/j.colsurfa.2004.10.010

Qu, X., Alvarez, P. J., & Li, Q. (2013). Applications of nanotechnology in water and wastewater treatment. *Water Research*, *47*(12), 3931–3946. doi:10.1016/j.watres.2012.09.058 PMID:23571110

Rinaudo, M. (2006). Chitin and chitosan: Properties and applications. *Progress in Polymer Science*, *31*(7), 603–632. doi:10.1016/j.progpolymsci.2006.06.001

Rinaudo, M., Milas, M., & Le Dung, P. (1993). Characterization of chitosan. Influence of ionic strength and degree of acetylation on chain expansion. *International Journal of Biological Macromolecules*, *15*(5), 281–285. doi:10.1016/0141-8130(93)90027-J PMID:8251442

Saǒ, Y., & Aktay, Y. (2002). Kinetic studies on sorption of Cr (VI) and Cu (II) ions by chitin, chitosan and Rhizopusarrhizus. *Biochemical Engineering Journal*, *12*(2), 143–153. doi:10.1016/S1369-703X(02)00068-2

Sano, L. L., Krueger, A. M., & Landrum, P. F. (2005). Chronic toxicity of glutaraldehyde: Differential sensitivity of three freshwater organisms. *Aquatic Toxicology (Amsterdam, Netherlands)*, *71*(3), 283–296. doi:10.1016/j.aquatox.2004.12.001 PMID:15670634

Sarin, V., Singh, T. S., & Pant, K. K. (2006). Thermodynamic and breakthrough column studies for the selective sorption of chromium from industrial effluent on activated eucalyptus bark. *Bioresource Technology*, *97*(16), 1986–1993. doi:10.1016/j.biortech.2005.10.001 PMID:16311033

Schmuhl, R., Krieg, H. M., & Keizer, K. (2004). Adsorption of Cu (II) and Cr (VI) ions by chitosan: Kinetics and equilibrium studies. *Water S.A.*, *27*(1), 1–8. doi:10.4314/wsa.v27i1.5002

Sivakami, M. S., Gomathi, T., Venkatesan, J., Jeong, H. S., Kim, S. K., & Sudha, P. N. (2013). Preparation and characterization of nano chitosan for treatment wastewaters. *International Journal of Biological Macromolecules*, *57*, 204–212. doi:10.1016/j.ijbiomac.2013.03.005 PMID:23500442

Sivaraj, R., Namasivayam, C., & Kadirvelu, K. (2001). Orange peel as an adsorbent in the removal of acid violet 17 (acid dye) from aqueous solutions. *Waste Management (New York, N.Y.)*, *21*(1), 105–110. doi:10.1016/S0956-053X(00)00076-3 PMID:11150127

Sparks, D. L. (Ed.). (1998). *Soil physical chemistry*. Boca Raton, FL: CRC Press.

Wu, F. C., Tseng, R. L., & Juang, R. S. (2010). A review and experimental verification of using chitosan and its derivatives as adsorbents for selected heavy metals. *Journal of Environmental Management*, *91*(4), 798–806. doi:10.1016/j.jenvman.2009.10.018 PMID:19917518

Yen, M. T., Yang, J. H., & Mau, J. L. (2008). Antioxidant properties of chitosan from crab shells. *Carbohydrate Polymers*, *74*(4), 840–844. doi:10.1016/j.carbpol.2008.05.003

Yu, K., Ho, J., McCandlish, E., Buckley, B., Patel, R., Li, Z., & Shapley, N. C. (2013). Copper ion adsorption by chitosan nanoparticles and alginate microparticles for water purification applications. *Colloids and Surfaces. A, Physicochemical and Engineering Aspects*, *425*, 31–41. doi:10.1016/j.colsurfa.2012.12.043

KEY TERMS AND DEFINITIONS

Crustaceans: A very large group of arthropods, which includes such familiar animals as crabs, lobsters, crayfish, shrimp.

Exoskeleton: The external skeleton that supports and protects an animal's body, in contrast to the internal skeleton. In crustaceans the exoskeleton is form by chitin.

Patagonia: Is a sparsely populated region located at the southern end of South America, shared by Argentina and Chile.

Shrimp: Free-swimming crustacean with an elongated body, typically marine and frequently of commercial importance as food.

Sodium Tripolyphosphate or Tripolyphosphate (TPP): Is an inorganic compound with formula Na5P3O10. It is the sodium salt of the polyphosphatepenta-anion, which is the conjugate base of triphosphoric acid.

Chapter 11
Applications of Nanomaterials for Water Treatment:
A Future Avenue

Anupreet Kaur
Punjabi University – Patiala, India

ABSTRACT

Development of new technologies, progressive urbanization, increasing consumerism and industrial boom in developing countries has led to elevated pollution of the environment. The spectrum of pollutants produced and released to the environment has increased in the last few decades including the agricultural, industrial, pharmaceutical and plastic industries. In the developed and underdeveloped countries where environmental pollution goes on increasing day by day, the concern of mankind to the threat of humanity increases which comes from anthropogenic degradation of the environment. An analytical chemist is always in search of cheaper, quicker, more sensitive, more reliable, precise methods of analysis. To achieve such a goal many properties of the materials are studied. Nanotechnology meets many of the conditions mentioned above and is very economic. So, analytical nanotechnology is an important tool for preconcentration and separation of pollutants at low levels.

INTRODUCTION

Rapid urbanization and industrialization of human civilization leads to the discharge of several hazardous contaminants and pollutants into aquatic streams, these effluents contains toxic wastes such as metals, organic impurities, etc., which directly produces

DOI: 10.4018/978-1-5225-2136-5.ch011

a detrimental impact on the biotic resources (Mara 2003; Moore et al. 2003). The unmanageable discharge of pollutants and contaminants due to their severe toxicity and more carcinogenic nature as comparable to other water contaminants, metal ions can cause brutal health troubles for aquatic fauna as well as flora but it also troubles human health through the prevailing ecological food chain (Hutton et al. 2007). Thus, the instantaneous removal of these toxic pollutants from wastewater is a noticeable issue in the aerobic and aquatic world.

The severity of pure drinking water scarcity has to be looked at from two aspects: First, quantity of available water and second, the quality of drinking water. With the rapid development of industries, such as batteries, metal plating facilities, mining operations, fertilizers industries, tanneries, and pesticides (Lvovich 1979).

As nanotechnology continues to proceed steadily, the primary focus of the research agenda has progressively advanced from the fundamental discovery and characterizations to look into the opportunities of presenting nanotechnology and the resultant nanomaterials as the major key to address the challenges of improving global sustainability, particularly in a wide spectrum of pivotal areas involving energy and environment (Seckler et al. 2003).

The properties of nanotechnology innovations are overpowering. Nanoscience and nanotechnology inspired images of "science fiction" with the word invoking visions of technologies wonders such as atoms and molecules joining to create materials with extraordinary physical and chemical properties which can mimic all of the things developed by nature in millions of years. This great deal of optimism and global enthusiasm about"small" encourages developing a lot of nanosensors which are used for detection of various pollutants. Nanotechnology is the creation and utilization of materials and devices at the levels of atoms, molecules, clusters supramolecular structures and exploitation of their unique properties at the nanoscale. It has introduced these novel Nanomaterials including Nanoparticles into our environment. Environmental protection, water pollution, food safety and human health are major issues which need to be addressed as soon as possible. Nanotechnology endeavored to provide" smatter" materials used in environmental sensing and remediation of water pollution problem as well as protecting human health from various toxic materials. The advent of nanotechnology has given immeasurable opportunities to purify water even at ionic state. The various nanostructures materials have been fabricated with features such as high aspect ratio, reactivity, and tunable pore volume, electrostatic, hydrophilic and hydrophobic interactions which are useful in adsorption, catalysis, sensing, and optoelectronics. Nanotechnology, which literally refers to the manipulations of materials and processes that are engineered to the molecular scale of 1–100 nm, is an emerging field that poses immense ability to hierarchically construct materials from the bottom up approach to significantly improve the performance as well as add new functionality to the existing products. The evolutions and advancements explored in this field have urged seawater desalination progress as a more economical and sustainable option (Shannon et al. 2008). In this context, the paradigm shift made by the new approaches and technological innovation based on the new frontier in enabling nanoscience and nanotechnology, although not one size fits all, has advocated the progress of maintaining the sustainability in terms of reduced membrane materials, chemical and energy usage that eventually leads to minimized sources of environmental issues (Cloete 2010). Nanomaterials are the drivers of the nanotechnology revolution and a key bottleneck to the applications of nanotechnology to solve this world water crisis. Nanomaterials have a number of key physicochemical properties that make them particularly attractive as separation media for water purification. On a mass basis, they have much large surface areas than bulk particles. Nanomaterials can also be functionalized with various chemical groups to increase their affinity toward a given compound. They can also serve as high capacity/ selectivity and recyclable legends for toxic metal ions, radionuclides, organic and inorganic solutes/ anions in aqueous solutions. Nanomaterials also provide unprecedented opportunities to develop more efficient water-purification catalysts due to their large surface areas and their size and shape-dependent optical, electronic and catalytic properties.

Water pollutants have huge impacts on the entire living systems including terrestrial, aquatic, and aerial flora and fauna. In addition to conventional priority, and newly emerging micro/nano-pollutants, increasing global warming and consequent climate changes are posing major threats to the freshwater availability. Global warming and climate change are constantly increasing the salinity level of both land and sea water, dwindling the availability of existing fresh water for household, agriculture and industry. This has made it urgent to invent an appropriate water treatment technology that not only removes macro-, micro- and nano-pollutants but also desalinates water to a significant extension. Continuing population growth and urbanization as well as rapid industrialization are constantly putting pressure on global water demand and resulting in an alarming scenario where accessible freshwater supply is no longer matching the patterns of these developments. Desalination has long served as a feasible option to provide safe drinking water in many deserted areas, coastal regions or remote locations (Crittenden et al. 2005). Water, being a universal solvent dissolves all kinds of contaminants like metal ions, inorganic anions, organic dyes, pesticides and insecticides, radioactive elements, *etc.* Therefore, great challenges lie in the removal of contaminants from drinking water at ultra-trace levels at an affordable cost.A common problem in developing countries in drinking water that is contaminated with bacteria and viruses, which are the main cause of water-borane diseases (Patzke et al. 2002). Due to change in climatic conditions, growing pollution, water will become an even scarer, especially

Table 1. The interesting facts are shown in Table 1

Figures	Facts
3.4 millions	In developing countries people dies from water-related diseases every year
63 millions	In developing countries like Bangladesh, India and Nepal people suffers from arsenic pollution
6Km	Women's from Asia and Asian continent walk to fetch water
80%	Water-related deaths in children between age 0-14 years
40%	Water-related deaths are due to diarrhoea

in developing countries. Moreover, in these countries, available water is often unsafe to drink. The interesting facts are shown in Table 1.

This chapter mainly concerned with providing a review of the interests of scientists and researchers in the application of nanomaterials in the improvement of water treatment, and application of nanotech in water treatments has several advantages than the traditional water treatments and this review also identifying the grand challenges and directions for future avenues in the field of Nanotechnology applications in water treatment.

2. NANOPARTICLES IN WATER TREATMENT

Nanoparticles have two key properties that make them particularly attractive as sorbents. First, on a mass basis, they have much larger surface areas than bulk particles. Second, they can also be functionalized with various chemical groups to increase their affinity towards target compounds.

Applications of nanoparticles as adsorbents for high efficient removal of heavy metal ions from wastewater must satisfy the following criterions:

1. The nanosorbents themselves should be nontoxic
2. The sorbents must show relatively high sorption capacities and selectivity even to the low
3. The concentration of pollutants.
4. The adsorbed pollutant could be removed from the surface of the nanoadsorbent easily,
5. The sorbents could be infinitely recycled
6. The process should be reversible so that we can get our adsorbent back; the Nanoparticles
7. Studied in the removal of heavy metal ions from wastewater should show high adsorption
8. Capacity.

2.1 Alumina Nanoparticles for Water Remediation

Alumina is a traditional adsorbent for pollutants and Nanoparticles of alumina is anticipated to be more active than alumina in bulk state. Nanosized Al_2O_3 can be prepared by sol-gel method and has been employed as solid phase extraction materials for separation/ preconcentration of trace pollutants. Physical /Chemical modification of Al_2O_3 with various functional groups containing some donors atoms such as nitrogen, oxygen, and phosphorus have improved their separation/preconcentration toward emerging trace pollutants. Chitosan-grafted polyacrylamide nano-alumina Nanoparticles had been applied for the removal of arsenic from drinking water. Sorption of As(V) on aluminosilicate treated with iron nanocomposites and also alumina-silica nanocomposites had been applied for the adsorption of cadmium, lead from waste water (Medina et al. 2010). Figure 1 shows the structures of some alumina samples.

Figure 1. The structures of some alumina samples

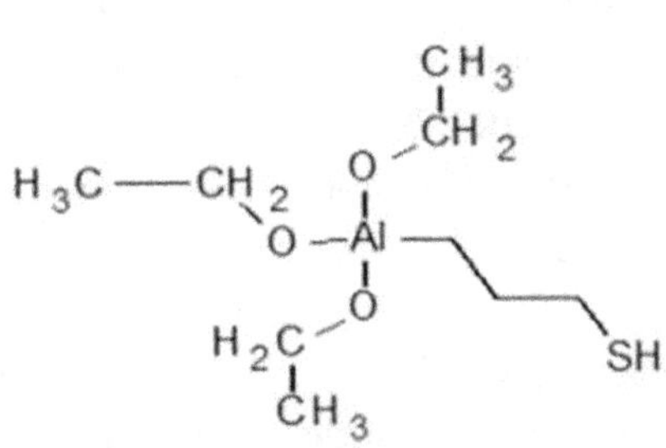

Al_2O_3-3-mercaptopropyltriethoxysilane Nanoparticles

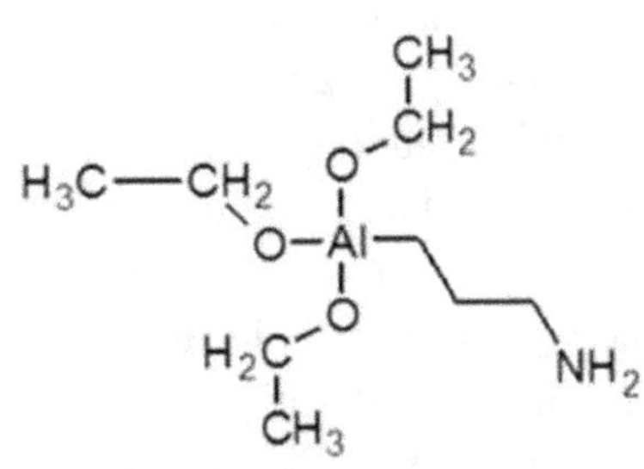

Al_2O_3-3-aminoropyltriethoxysilane Nanoparticles

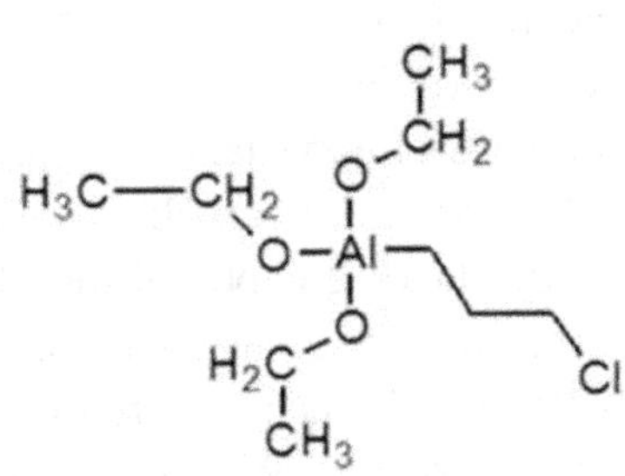

Al_2O_3-3-cholropropyltriethoxysilane nanoparticles

2.2 Silica Nanoparticles for Water Remediation

Silica nanoparticles is a promising material as a solid phase extractant because of its large surface area, high adsorption capacity, low-temperature modification, less degree of unsaturation and low electrophilicity (Kaur et al. 2009). The sequence of reactivity is expressed as follows:

$$Zr(OR)_4, Al(OR)_4 > Ti(OR)_4 > Sn(OR)_4 >> Si(OR)_4$$

Some pollutants are poorly adsorbed on nanoparticles. To overcome this problem, physical or chemical modification of the surface of these nanoparticles with certain functional groups containing some donor atoms such as oxygen, nitrogen, sulfur and phosphorus are necessary. The most often used method is to load a kind of specific chelating reagent by the physical or chemical procedure. The formal method is simple but the loaded reagent is prone to leaking out from the sorbent, while the chemically bonded material is more stable and can be used repeatedly. The modification of nanometer sized materials is usually required in order to prevent a conglomeration of particles and to improve its consistency in relation to other materials. Also, the modification of nanometer-sized material can improve the selectivity of nanometer–sized materials toward pollutants. Selectivity of suitable specific functional groups towards metal ions depends on certain factors such as (1) size of the modifiers (2) activity of loaded group and also (3) on the basis of the concept of hard-soft acids and bases. Chemisorption of nanoparticles provides immobility, mechanical stability, and water insolubility, thereby increases the efficiency, sensitivity and selectivity.

Chemical modification is a process that leads to change in chemical characteristics of the surface of nanoparticles. By the modification, the adsorption properties are significantly affected. Chemisorption of chelating molecules on nanoparticle surface provides immobility, mechanical stability, and water insolubility, thereby increases the efficiency, sensitivity and selectivity of nanoparticles for the analytical application. Chemical modification of nanoparticles by silylation procedure using different silylating agents such as 3-aminopropyltriethoxysilane, 3-chloropropyltriethoxysilane and 3-mercaptopropyltriethoxysilane provides immobility, mechanical stability and water insolubility, Figure 2. 1-(2-pydriylazo)-2-naphthol-modified SiO_2 nanoparticles have been used for the preconcentration of some toxic heavy metal ions such as Cd(II),Zn(II), Mn(II), Ni(II), Sb(III), Pb(II) and also ferbam pesticides from various samples (Kaur and Gupta 2010). Resacetophenone-modified silica nanoparticles have also been used for the preconcentration of Pb(II) and Cd(II) from water samples.

Figure 2. Chemical modification of silica nanoparticles with various silanes and ligands

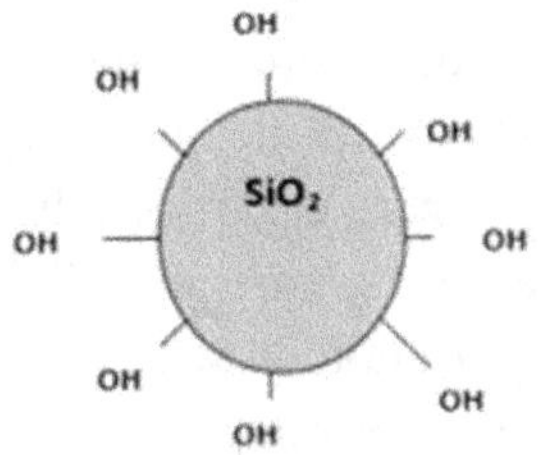

Silica nanoparticle

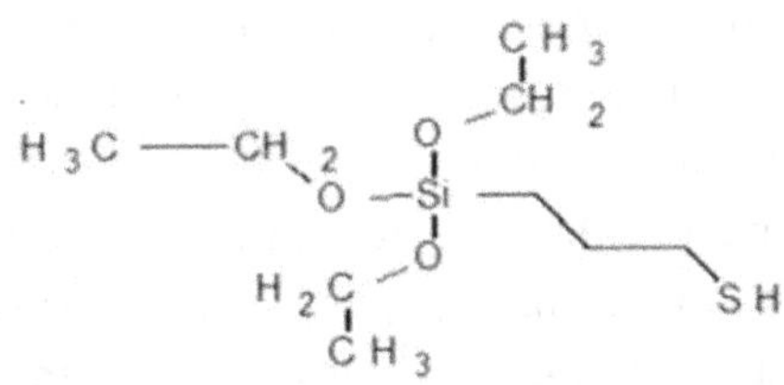

SiO_2-3-aminopropyltriethoxysilane nanoparticle

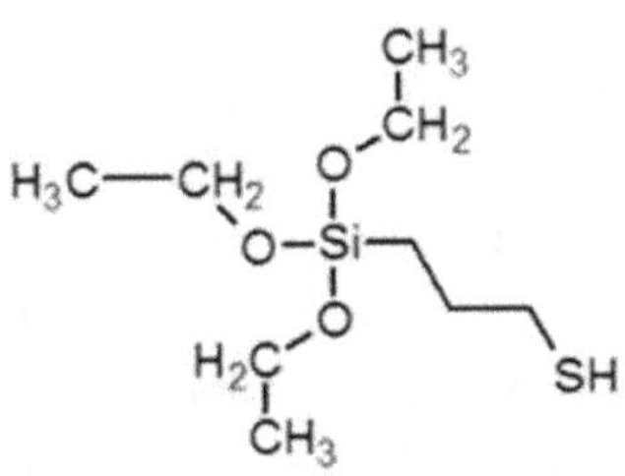

SiO_2-3-mercaptopropyltriethoxysilane nanoparticle

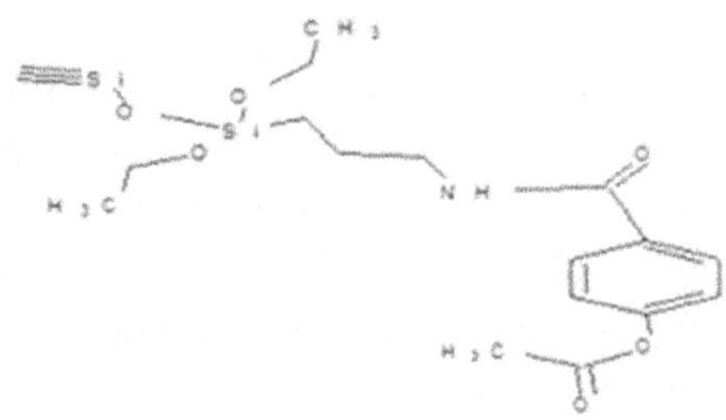

SiO_2-acetylsalicylic acid nanoparticles

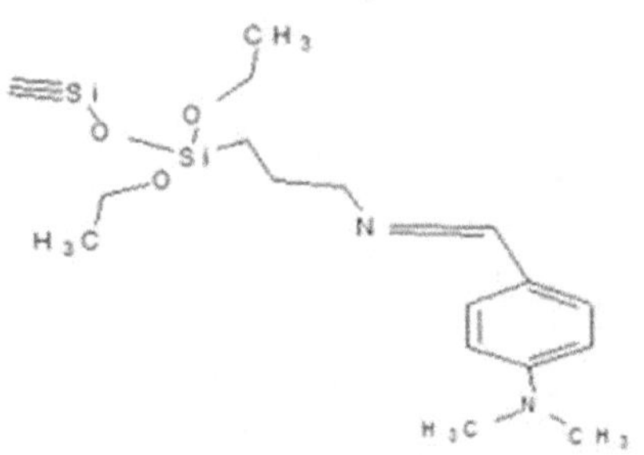

SiO_2-p-dimethylaminobenzaldehyde nanoparticle

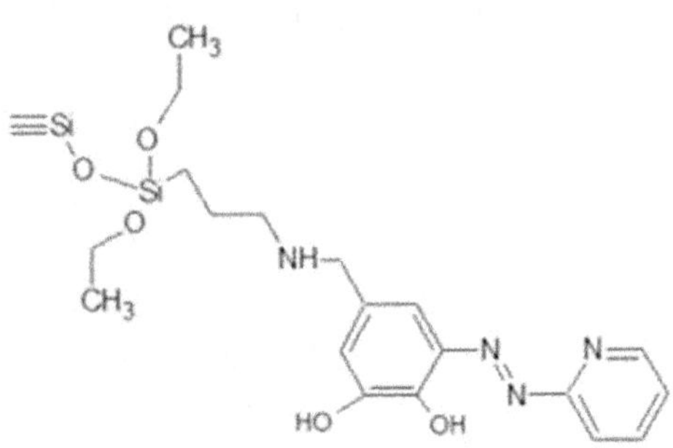

SiO_2-4-(2-pyridylazo)-resorcinol nanoparticle

2.3 Titania Nanoparticles for Water Remediation

Tiny particles of titanium dioxide are found as key ingredients in wall paints, sunscreens, and toothpaste; they act as reflectors of light or as abrasives. However, with decreasing particle size and a corresponding change in their surface-to-volume ratio, their properties change so that crystalline titanium dioxide nanoparticles acquire catalytic ability; activated by the UV component in sunlight, they break down toxins or catalyze other relevant reactions. Titanium oxide photocatalysts have been widely studied for solar energy conversion and environmental applications in the past several decades because of their high chemical stability, good photoactivity, relatively low cost and nontoxicity. However, the photocatalytic capability of TiO2 is limited to only ultraviolet light; to overcome this problem both chemical and physical modifi cation approaches were developed to extend the absorption band-edge of TiO2 into visible region. In the photocatalytic oxidation process, organic pollutants are destroyed in the presence of semiconductor photocatalysts, an energetic light source, and an oxidizing agent such as oxygen or air. Only photons with energies greater than the bandgap energy (ΔE) can result in the excitation of the valence band (VB) electrons which then promote the possible reactions. The absorption of photons with energy lower than ΔE or longer wavelength usually causes energy dissipation in the form of heat.

TiO_2 nanoparticles have emerged as promising photocatalysts for water purification, Figure 3. The removal of total organic carbon from waters contaminated with organic wastes was greatly enhanced by the addition of TiO_2 nanoparticles in the presence of ultraviolet light. It was successfully used to degrade organic compounds (e.g. chlorinated alkanes and benzenes, dioxins, furans) pentacholorophenol and aromatic chlorinated compounds had been photo-catalytically studied using titania

Figure 3. Photocatalytic studies by titania nanoparticles

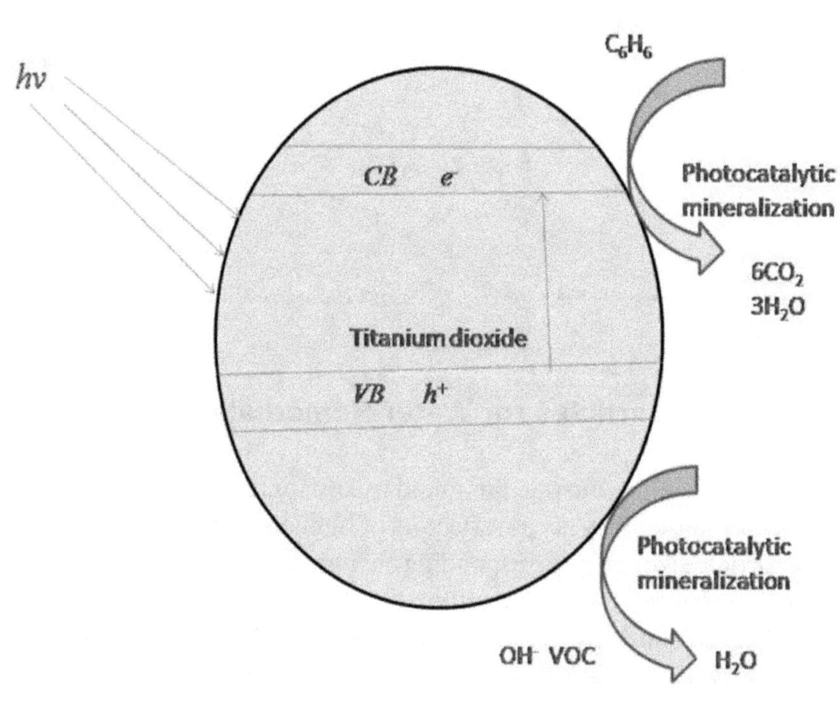

Nanoparticles (Mills Hoffmann 1993). Separation and microcolumn preconcentration of traces of rare earth elements on nanoscale TiO_2 and their determination in geological samples (Jardim et al. 1997).

2.5 Nanoscale Zero-Valent Metals

Iron is one of the most abundant elements on earth. Elemental iron has been used as an ideal candidate for remediation because it is inexpensive, abundant easy to prepare and apply to a variety of systems, and devoid of any known toxicity induced by its usage. The concept of using metals, such as iron, as remediation agents are based on reduction–oxidation or "redox" reactions, in which a neutral electron donor (a metal) chemically reduces an electron acceptor (a contaminant). Nanoscale iron particles have surface areas significantly greater than larger-sized powders or granular iron, which leads to enhanced reactivity for the redox process. As a result, iron nanoparticles have been extensively investigated for the decomposition of halogenated hydrocarbons to benign hydrocarbons and the remediation of many other contaminants including anions and heavy metals. Zero-valent iron nanoparticles are highly reactive and react rapidly with surrounding media in the subsurface (Zhang Elliot, 2006). A significant loss of reactivity can occur before the particles are able to reach the target contaminant. In addition, zero-valent iron nanoparticles tend to flocculate when added to water, resulting in a reduction in the effective surface area of the metal. Therefore, the effectiveness of a remediation depends on the accessibility of the contaminants to the nanoparticles; and the maximum efficiency of remediation will be achieved only if the metal nanoparticles can effectively migrate without oxidation to the contaminant or the water/ contaminant interface. To overcome such difficulties, a commonly used strategy is to incorporate iron nanoparticles within support materials, such as polymers, porous carbon, and polyelectrolytes. Dissolved As(III) had been removed by iron Nanoparticles-embedded macrporous composites. Amine-functionalized Nanoparticles used for the rapid capture and removal of bacterial pathogens. Nanoscale zero-valent iron and bimetallic particles for a contaminated site for water remediation. Zerovalent iron Nanoparticles were applied for reductive decholrination of octacholorodibenzo-p-dioxin. Bentonite supported Fe/Pd Nanoparticles for the dechlorination of p-cholorophenol and also for PCB (Zhao et al. 2011), Figure 4.

3.Other Materials for Nanoremediation

Graphene is the allotrope of carbon. Especially, graphene–metal oxide composites have attracted extensive attention for their potential applications in environmental

Figure 4. Dechlorination by Zerovalent iron Nanoparticles

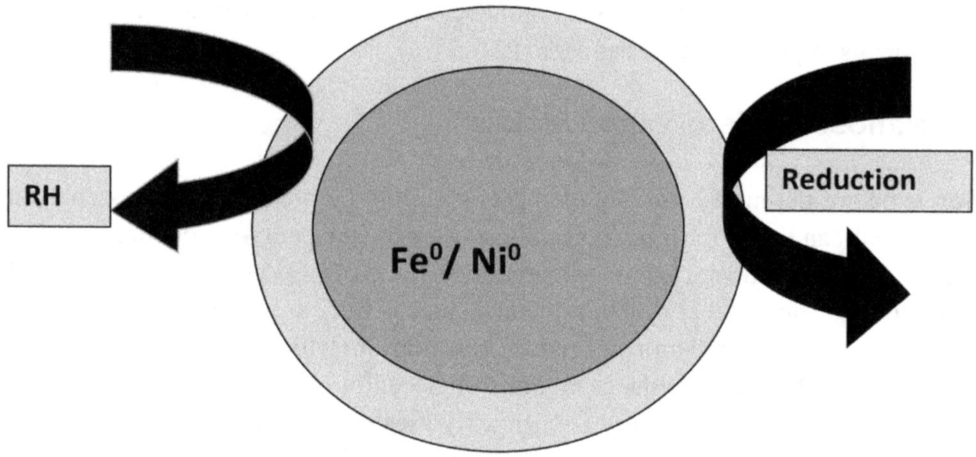

and energy-related areas. In this brief review, we will summarize recent achievements of effective strategies for synthesizing high-quality graphene–metal oxide composites and their photocatalytic applications. Photo-catalytically degradation of volatile aromatic pollutant had done by TiO_2-graphene nanocomposites. Removal of Pb(II) ions from aqueous solutions on few-layered graphene oxide nanosheets (Zhao et al. 2011). Preconcentration of U(VI) ions on few-layered graphene oxide nanosheets from aqueous solutions. Folding/aggregation of graphene oxide and its application in Cu^{2+} removal. Adsorptive remediation of environmental pollutants using novel graphene-based Nanomaterials.Graphene-based PANI nanocomposites have also been used as sensing platforms due to their good electrocatalytic activity, high specific surface area, excellent reliability, and low-cost properties. These composites are successfully used for DNA, 4-aminophenol, dopamine, artesunate, hydrogen peroxide and hydrazine detection, as well as hydrogen (H_2), methane (CH_4) and ammonia (NH_3) gas sensing.

Innovative sensors with high sensitivity and selectivity, and fast response are in great need, and also we know the fast response is in great need. Biosensors and affinity sensor devices have been shown to have the ability to provide rapid, cost-effective, specific and reliable quantitative and qualitative analysis. To date, the developments in nanomaterials and biosensor fabrications technology are moving rapidly with new and novel nano biorecognition materials being developed which can be applied as the sensing receptors for mycotoxins analysis. Biosensors, as tools have proofed to be able to provide rapid, sensitive, robust and cost effective quantitative methods for on-site testing. Developing biosensor devices for different mycotoxins are attracting much research interest in recent years with a range of devices are being developed and reported in the scientific literature. However, with the advent of nanotechnology and its impact on developing ultrasensitive devices, mycotoxins analysis is benefiting also from the advances taking place in applying nanomaterials in sensors development (Saini and Kaur, 2012).

It was established that gold supported on various metal oxides is a useful candidate for alkenes hydrogenation. A few interesting revelations arose from this research. Thiol–modified gold-coated polystyrene particles had been used for adsorption of aromatic compounds. Dansyl-norvaline stereospecfic recognition had been done by albumin gold nanoparticles. Gold nanoparticles were also employed for an immunoassay for the detection of aflatoxin B1 (AFB1) in foods (Liu et al. 2003).

Recently, MOFs, obtained by linking metal cations (or cationic metal clusters) with organic linkers, have attracted significant interest in the last years mainly due to the advantage of showing a large variety of structural types and chemical compositions, high surface area and permanent nanoscale porosity (Abhijith and Thakur 2012). MOFs have been widely studied as materials for catalysis gas storage and separation sensing and drug delivery and, more recently, the analytical applications of MOFs have emerged (Sumida et al. 2011). In this field, MOFs have shown to be promising materials as sorbents for sample preparation as chromatographic stationary phases, as well as for the development of improved detection systems and sensors. However, MOFs crystalline powders generally possess a random crystal size and shape, which makes troublesome their direct application and have led to engineer hybrid materials containing them, such as through supports, magnetic beads, beads coated with a MOF shell, or MOF crystals entrapped porous monolith. Porous materials are defined as solids containing empty voids which can host other molecules. The fundamental features of these materials are their porosity, the ratio between total occupied and empty space, the (average) size of the pores and the surface area (Meek et al. 2011). Typical surface area values for the porous materials applied in technological processes range between 2,000 and 8,000 $m^2 g^{-1}$. The most important applications of such materials are the storage of small molecules and filtering .The metal organic frameworks are defined as a nanocomposite material which can be consisted of either inorganic or organic materials. MOFs have shown high potential in gas storage, separation, chemical sensing, drug delivery, and heterogeneous catalysis applications. In general, the flexible and highly porous structure of MOFs allows guest species such as metal ions to diffuse into their bulk structure. The shape and size of the pores lead to selectivity over the guests that may be adsorbed. These features make MOFs an ideal sorbent in solid phase extraction of heavy metals. However, there is little information about MOFs as an adsorbent.

Nanomaterials are well known to possess excellent electrical, optical, thermal, catalytic properties and strong mechanical strength, which offer great opportunities to construct nanomaterials-based sensors or devices for monitoring environmental contaminations in air, water, and soil. Various nanomaterials, such as carbon nanotubes, gold nanoparticles, silicon nanowires and quantum dots, have been extensively explored in detecting and measuring toxic metal ions, toxic gasses, pesticides, and hazardous industrial chemicals with high sensitivity, selectivity, and simplicity.

CNTs and graphene have been extensively studied for nanodevices or nanosensors (Bagheri et al. 2012). CNTs and graphene ideal candidates for environmental pollution detection, such as heavy metal ions (Hg^{2+}, Pb^{2+}, As^{3+}, Ag^{+}), organophosphate pesticides,toxic gasses,industrial wastewater. Kim *et al.* constructed a selective, sensitive, and fast single-walled carbon nanotube (SWCNT)- based field effect transistor nanosensor for Hg^{2+} ion detection (Kim et al. 2009; Zhang et al. 2010).

Metallic nanoparticles, especially gold nanoparticles (AuNPs) and silver nanoparticles (AgNPs) have received much attention due to their potential applications in chemical and biological sensors, medical diagnostics and therapeutics and biological imaging (Rosi et al. 2006). AuNPs or AgNPs exhibit high extinction coefficients and strong size- and distance-dependent optical properties, and are considered as a new class of reporters and have been applied in colorimetric sensing (Lee et al. 2007). Mirkin's group developed a series of colorimetric sensors based on the color of the AuNPs or AgNPs from the red to blue, which allow in situ, real-time qualitative or semiquantitative detection without complicated analytical instruments (Mirkin et al. 1997). On this basis, Lu et al. developed colorimetric sensors for environmental monitoring, such as heavy metal ions ($Pb^{2+,}$ $Hg^{2+),}$ radioactive metal detection (UO_2^{2+}) (Liu and Lu, 2003).

Chen et al. synthesized b-cyclodextrin (CD)-functionalized silver nanoparticles as probes for highly robust, quantitative, sensitive, and observable by the naked eye colorimetric detection of different isomers of aromatic compounds (Chen et al. 2010). He *et al.* developed an oligo-T-based gold nanoprobe for rapid and portable detection of mercury with the help of a power-free poly(dimethylsiloxane) (PDMS) microfluidic device (He et al. 2008).

Silicon-based nanomaterials are an important type of nanomaterial, which has been extensively studied due to their with single-molecule sensitivity. Therefore, metallic nanoparticles are also excellent candidate substrates for environmental monitoring by surface-enhanced Raman scattering techniques, such as phenol, pesticides, TNT and heavy metal ions.

Dasary *et al.* 2009, used a cysteine modified AuNPs-based miniaturized, inexpensive, ultrasensitive SERS probe for highly sensitive and selective screening of 2,4,6-trinitrotoluene (TNT) from aqueous solution, which can induce aplastic anemia and abnormal liver functions.

Other nanomaterials also have been used for environmental monitoring, such as Fe_3O_4, SiO_2, ZrO_2*etc*. For example, Zuo et al. 2009, used a nanocomplex made of

MWNTs and magnetic nanoparticles (MNPs, Fe3O4) for highly efficient catalytic oxidation of phenols and their removal from wastewater.

More importantly, several portable nanomaterials-based devices have been designed in this direction. The unique chemical and physical properties of nanomaterials provide a versatile toolbox for developing various detection techniques with high sensitivity, specificity, speed, simplicity and cost-effectiveness for environmental monitoring. However, there is still an urgent need to further improve the sensitivity, and probably more importantly, the operation convenience from a practical point of view. The possibility for mass production is also a critical point for widespread application of nanomaterials-based sensors. In this regard, the design simplicity and compatibility with industry standards should be considered. Overall, we expect that the exploration of new nanomaterials and/or elaborate nanostructures will become an innovative source for developing sensors for environmental monitoring, and such sensors may eventually find widespread applications with the efforts from both academics and industry.

CONCLUSION

Decontamination is the reduction or removal of chemical and biological agents by means of physical, chemical neutralization or detoxication techniques. Nanotechnology revolution will play an essential role in solving the problem of rising demands of clean water and decreasing of the available supplies of fresh water. Nanomaterials have a number of physicochemical properties that make them particularly attractive as separation media for water purification.Nanotechnology has shown huge potential in areas as diverse as drug development, water decontamination, information and communication technologies, and the production of stronger, lighter materials and human health care. Water and air are two vital components of life on earth; the existence of life on earth is made possible largely because of their importance to metabolic processes within the body. Clean and fresh water and air are essential for the existence of life. The recent development of nanotechnology has raised the possibility of environmental decontamination through several nanomaterials, processes, and tools. This chapter summarizes the expertise of various approaches of decontamination for the successful realization of remediation in the environment.

REFERENCES

Abhijith, K. S., & Thakur, M. S. (2012). Application of green synthesis of gold nanoparticles for sensitive detection of aflatoxin B1 based on metal enhanced fluorescence Anal. *Methods (San Diego, Calif.)*, *4*, 4250–4256.

Chen, X., Parker, S. G., Zou, G., Su, W., & Zhang, Q. (2010). β-cyclodextrin-functionalized silver nanoparticles for the naked eye detection of aromatic isomers. *ACS Nano*, *4*(11), 6387–6394. doi:10.1021/nn1016605 PMID:20973513

Cloete, T.E. (2010). Nanotechnology in Water Treatment Applications. New York: Horizon Scientific Press.

Crittenden, J. C., Trussell, R. R., David, W., Hand, P., Howe, K. J., & Tchobanoglous, G. (2005). Water Treatment: Principles and Design. John Wiley and Sons.

Dasary, S. S. R., Singh, A. K., Senapati, D., Yu, H., & Ray, P. C. (2009). Gold nanoparticle based label-free SERS probe for ultrasensitive and selective detection of trinitrotoluene. *Journal of the American Chemical Society*, *131*(38), 13806–13812. doi:10.1021/ja905134d PMID:19736926

He, S., Li, D., Zhu, C., Song, S., Wang, L., Long, Y., & Fan, C. (2008). A graphene-based fluorescent nanoprobe for silver(I) ions detection by using graphene oxide and a silver-specific oligonucleotide. *Chemical Communications (Cambridge)*, *46*, 4885–4887. doi:10.1039/b811528a

Hutton, Haller, & Bartram. (2007). *Economic and Health Effects of Increasing Coverage of Low Cost Household Drinking Water Supply and Sanitation Interventions*. World Health Organization.

Jardim, W. F., Moraes, S. G., & Takiyama, M. M. K. (1997). Photocatalytic degradation of aromatic chlorinated compounds using TiO2: Toxicity of intermediates. *Water Research*, *31*(7), 1728–1732. doi:10.1016/S0043-1354(96)00349-1

Kaur, A., & Gupta, U. (2009). A review on applications of nanoparticles for the preconcentration of environmental pollutants. *J. Mater. Chem. A*, *19*(44), 8279–8289. doi:10.1039/b901933b

Kaur, A., & Gupta, U. (2010). Solid-Phase Extraction of Antimony Using Chemically Modified SiO2-PAN Nanoparticles. *Journal of AOAC International*, *93*, 1302–1307. PMID:20922965

Kim, T. H., Lee, J., & Hong, S. (2009). Highly Selective Environmental Nanosensors Based on Anomalous Response of Carbon Nanotube Conductance to Mercury Ions. *The Journal of Physical Chemistry C, 113*(45), 19393–19396. doi:10.1021/jp908902k

Lee, J. S., Han, M., & Mirkin, C. A. (2007). Colorimetric detection of mercuric ion (Hg2+) in aqueous media using DNA-functionalized gold nanoparticles. *Angewandte Chemie*, *119*(22), 4171–4174. doi:10.1002/ange.200700269 PMID:17461429

Liu, F. K., Wei, G. T., & Cheng, F. C. (2003). Immobilization of a Monolayer of Bovine Serum Albumin on Gold Nanoparticles for Stereo-specified Recognition of Dansyl-norvaline J. *Chin. Chem. Soc.*, *50*(4), 931–937. doi:10.1002/jccs.200300131

Liu, J., & Lu, Y. (2003). A Colorimetric Lead Biosensor Using DNAzyme-Directed Assembly of Gold Nanoparticles. *Journal of the American Chemical Society*, *125*(22), 6642–6643. doi:10.1021/ja034775u PMID:12769568

Lvovich, M. I. (1979). *World Water Resources and Their Future*. Washington, DC: American Geophysical Union. doi:10.1029/SP013

Mara, D. D. (2003). Water, sanitation and hygiene for the health of developing nations. *Public Health, 117*(6), 452–456. doi:10.1016/S0033-3506(03)00143-4 PMID:14522162

Medina, M., Tapia, J., Pacheco, S., Espinosa, M., & Rodriguez, R. (2010). Adsorption of lead ions in aqueous solution using silica–alumina Nanoparticles. *Journal of Non-Crystalline Solids*, *356*(6-8), 383–387. doi:10.1016/j.jnoncrysol.2009.11.032

Meek, S. T., Greathouse, J. A., & Allendorf, M. D. (2011). Metal-organic frameworks: A rapidly growing class of versatile nanoporous materials. *Advanced Materials*, *23*(2), 249–267. doi:10.1002/adma.201002854 PMID:20972981

Mills, G., & Hoffmann, M. R. (1993). Photocatalytic degradation of pentachlorophenol on titanium dioxide particles: Identification of intermediates and mechanism of reaction. *Environmental Science & Technology*, *27*(8), 1681–1689. doi:10.1021/es00045a027

Mirkin, C. A., Letsinger, R. L., Mucic, R. C., & Storhoff, J. J. (1996). A DNA-based method for rationally assembling nanoparticle into macroscopic materials. *Nature*, *382*(6592), 607–609. doi:10.1038/382607a0 PMID:8757129

Moore, M., Gould, P., & Keary, B. S. (2003). Global urbanization and impact on health. *International Journal of Hygiene and Environmental Health*, *206*(4-5), 269–278. doi:10.1078/1438-4639-00223 PMID:12971682

Patzke, G. R., Krumeich, F., & Nesper, R. (2002). Oxidic nanotubes and nanorods-anisotropic modules for a future nanotechnology. *Angewandte Chemie International Edition*, *41*(14), 2446–2461. doi:10.1002/1521-3773(20020715)41:14<2446::AID-ANIE2446>3.0.CO;2-K PMID:12203509

Rosi, N. L., Giljohann, D. A., Thaxton, C. S., Lytton-Jean, A. K. R., Han, S., & Mirkin, C. A. (2006). Oligonucleotide-modified gold nanoparticles for intracellular gene regulation. *Science*, *312*(5776), 1027–1030. doi:10.1126/science.1125559 PMID:16709779

Saini & Kaur. (2012). The Analysis of Aflatoxin M1 in Dairy Products. *Separation Science*, *4*, 13-17.

Seckler, Molden, & Sakthivadivel. (2003). *The Concept of Efficiency in Water Resources Management and Policy*. Academic Press.

Shannon, M. A., Bohn, P. W., Elimelech, M., Georgiadis, J. G., Mariñas, B. J., & Mayes, A. M. (2008). Science and technology for water purification in the coming decades. *Nature*, *452*(7185), 301–310. doi:10.1038/nature06599 PMID:18354474

Sumida, K., Rogow, D. L., Mason, J. A., McDonald, T. M., Bloch, E. D., Herm, Z. R., & Long, J. R. et al. (2011). Carbon dioxide capture in metal-organic frameworks. *Chemical Reviews*, *112*(2), 724–781. doi:10.1021/cr2003272 PMID:22204561

Zhang, L., Li, T., Li, B., Li, J., & Wang, E. (2010). Carbon nanotube–DNA hybrid fluorescent sensor for sensitive and selective detection of mercury (II) ion. *Chemical Communications (Cambridge)*, *46*(9), 1476–1478. doi:10.1039/b921191h PMID:20162153

Zhang, W. X., & Elliot, D. W. (2006). Applications of iron nanoparticles for groundwater remediation. *Remediation Journal*, *16*(2), 7–21. doi:10.1002/rem.20078

Zhao, G., Ren, X., Gao, X., Tan, X., Li, J., Chen, C., & Wang, X. et al. (2011). Removal of Pb(II) ions from aqueous solutions on few-layered graphene oxide nanosheets. *Dalton Transactions (Cambridge, England)*, *40*(41), 10945–10952. doi:10.1039/c1dt11005e PMID:21918761

Zuo, X., Peng, C., Huang, Q., Song, S., Wang, L., Li, D., & Fan, C. (2009). Design of a carbon nanotube/magnetic nanoparticle-based peroxidase-like nanocomplex and its application for highly efficient catalytic oxidation of phenols. *Nano Res.*, *2*(8), 617–623. doi:10.1007/s12274-009-9062-3

Compilation of References

Abbas, A. O. M. (2010). *Chitosan for biomedical applications*. Academic Press.

Abbasi, M., Saeed, F., & Rafique, U. (2014). Preparation of silver nanoparticles from synthetic and natural sources: Remediation model for PAHs. *Materials Science and Engineering*, *60*, 012061.

Abdou, E. S., Nagy, K. S., & Elsabee, M. Z. (2008). Extraction and characterization of chitin and chitosan from local sources. *Bioresource Technology*, *99*(5), 1359–1367. doi:10.1016/j.biortech.2007.01.051 PMID:17383869

Abebe, L. S., Chen, X., & Sobsey, M. D. (2016). Chitosan Coagulation to Improve Microbial and Turbidity Removal by Ceramic Water Filtration for Household Drinking Water Treatment. *International Journal of Environmental Research and Public Health*, *13*(3), 269. doi:10.3390/ijerph13030269 PMID:26927152

Abhijith, K. S., & Thakur, M. S. (2012). Application of green synthesis of gold nanoparticles for sensitive detection of aflatoxin B1 based on metal enhanced fluorescence Anal. *Methods (San Diego, Calif.)*, *4*, 4250–4256.

Abou-Gamra, Z. M., & Ahmed, M. A. (2015). TiO_2 nanoparticles for removal of malachite green dye from waste water. *Advances in Chemical Engineering and Science*, *5*(03), 373–388. doi:10.4236/aces.2015.53039

Abu, M. R., & Gedanken, S. (2005). Sonochemical synthesis of stable hydrosol of Fe_3O_4 nanoparticles. *Journal of Colloid and Interface Science*, *284*(2), 489–494. doi:10.1016/j.jcis.2004.10.073 PMID:15780286

Adams, C., Wang, Y., Loftin, K., & Meyer, M. (2002). Removal of antibiotics from surface and distilled water in conventional water treatment processes. *Journal of Environmental Engineering*, *128*(3), 253–260. doi:10.1061/(ASCE)0733-9372(2002)128:3(253)

Ademe, M., & Mekonnen, Z. (2014). Repeated reuse of insulin injection syringes and incidence of bacterial contamination among diabetic patients in Jimma University Specialized Hospital, Jimma, Ethiopia. *Asian Pac J Trop Dis*, *4*(Suppl 2), S712–S716. doi:10.1016/S2222-1808(14)60712-X

Adhikari, R., Gyawali, G., kim, T. H., Sekino, T., & Lee, S. W. (2014). EDTA mediated microwave hydrothermal synthesis of WO 3 hierarchical structure and its photoactivity under simulated solar light. *J. Env. Chem. Eng.*, *2*(3), 1365–1370. doi:10.1016/j.jece.2014.02.019

Adjdir, M., Bendeddouche, C. K., Benhaoua, H., Kaid, M., Karmaoui, M., Boudinar, M., & Weidler, P. G. (2015). Increasing the efficiency of silicon and aluminum extraction from Volclay by a water iteration treatment for the synthesis of MCM-41 nanomaterials. *Comptes Rendus. Chimie*, *18*(4), 385–390. doi:10.1016/j.crci.2014.07.003

Agency for Toxic Substances and Disease Registry (ATSDR) Toxicological Profile for Arsenic TP-92. (2000). *Georgia*. Atlanta, GA: Center for Disease Control.

Agency for Toxic Substances and Disease Registry (ATSDR). (1993). *Toxicological Profile for Chromium*. Atlanta, GA: US Public Health Service, US Public Health Service, US Department of Health and Human Services.

Agency for Toxic Substances and Disease Registry (ATSDR). (1999). Public Health Service. Atlanta, GA: U.S. Department of Health and Human Services.

Agency for Toxic Substances and Disease Registry (ATSDR). (2008). Atlanta, GA: Draft Toxicological Profile for Cadmium.

Agnihotri, S., Mukherji, S., & Mukherji, S. (2014). Size-controlled silver nanoparticles synthesized over the range 5–100 nm using the same protocol and their antibacterial efficacy. *RSC Advances*, *4*(8), 3974–3983. doi:10.1039/C3RA44507K

Aguayo-Villarreal, I. A., & Bonilla-Petriciolet, A. (2011). Batch and column studies of Zn2þ removal from aqueous solution using chicken feathers as sorbents. *Chemical Engineering Journal*, *167*, 67–76. doi:10.1016/j.cej.2010.11.107

Ahandani, P. Z., & Habibi-Yangjeh, A. (2011). Microwave-assisted preparation of Zn1−xCuxS nanoparticles by a fast, green, and template-free method and photocatalytic. *Desal.*, *271*(1-3), 273–278. doi:10.1016/j.desal.2010.12.042

Ahmad, A. L., Sumathi, S., & Hameed, B. H. (2006). Coagulation of residue oil and suspended solid in palm oil mill effluent by chitosan, alum and PAC. *Chemical Engineering Journal*, *118*(1), 99–105. doi:10.1016/j.cej.2006.02.001

Ahmad, M., Ahmed, E., Hong, Z. L., Ahmed, W., Elhissi, A., & Khalid, N. R. (2014). Photocatalytic, sonocatalytic and sonophotocatalytic degradation of Rhodamine B using ZnO/CNTs composites photocatalysts. *Ultrasonics Sonochemistry*, *21*(2), 761–773. doi:10.1016/j.ultsonch.2013.08.014 PMID:24055646

Ahmad, S., Sumathi, , & Hameed, . (2004). Chitosan: A natural biopolymer for the adsorption of residue oil from oily wastewater. *Adsorption Science and Technology*, *22*(1), 75–88. doi:10.1260/026361704323151015

Ai, L., Zhang, C., & Chen, Z. (2011). Removal of methylene blue from aqueous solution by a solvothermal-synthesized graphene/magnetite composite. *Journal of Hazardous Materials*, *192*(3), 1515–1524. doi:10.1016/j.jhazmat.2011.06.068 PMID:21782325

Ai, L., Zhang, C., Liao, F., Wang, Y., Li, M., Meng, L., & Jiang, J. (2011). Removal of methylene blue from aqueous solution with magnetite loaded multi-wall carbon nanotube: Kinetic, isotherm and mechanism analysis. *Journal of Hazardous Materials*, *198*, 282–290. doi:10.1016/j.jhazmat.2011.10.041 PMID:22040800

Akbarzadeh, A., Samiei, M., & Davaran, S. (2012). Magnetic nanoparticles: Preparation, physical properties, and applications in biomedicine. *Nanoscale Research Letters*, *7*(1), 144. doi:10.1186/1556-276X-7-144 PMID:22348683

Akhavan, O., Azimirad, R., Safa, S., & Larijani, M. (2010). Visible light photo-induced antibacterial activity of CNT–doped TiO_2 thin films with various CNT contents. *Journal of Materials Chemistry*, *20*(35), 7386–7392. doi:10.1039/c0jm00543f

Akkaya, G., Uzun, İ., & Güzel, F. (2007). Kinetics of the adsorption of reactive dyes by chitin. *Dyes and Pigments*, *73*(2), 168–177. doi:10.1016/j.dyepig.2005.11.005

Alam, J., Dass, L. A., Ghasemi, M., & Alhoshan, M. (2013). Synthesis and optimization of PES-Fe_3O_4 mixed matrix nanocomposite membrane: Application studies in water purification. *Polymer Composites*, *34*(11), 1870–1877. doi:10.1002/pc.22593

Alansi, A. M., Alkayali, W. Z., Al-qunaibit, M. H., Qahtan, T. F., & Saleh, T. A. (2015). Synthesis of exfoliated polystyrene/anionic clay MgAl-layered double hydroxide: Structural and thermal properties. *RSC Advances*, *5*(87), 71441–71448. doi:10.1039/C5RA10545E

Alansi, A. M., Alkayali, W. Z., Al-Qunaibit, M. H., Qahtan, T. F., & Saleh, T. A. (2015). Synthesis, Surface Morphology and Properties of Polystyrene Modified Synthetic Clay Nanocomposites. *Asian Journal of Chemistry*, *27*(10), 3900–3906. doi:10.14233/ajchem.2015.19070

Al-Asheh, S., Banat, F., & Al-Rousan, D. (2002). Adsorption of copper, zinc and nickel ions from single and binary metal ion mixtures on to chicken feathers. *Adsorption Science and Technology*, *20*(9), 849–864. doi:10.1260/02636170260555778

Al-Asheh, S., Banat, F., & Al-Rousan, D. (2003). Beneficial use of chicken feathers in removal of heavy metals from wastewater. *Journal of Cleaner Production*, *11*(3), 321–326. doi:10.1016/S0959-6526(02)00045-8

Al-Qodah, A. (2006). Biosorption of heavy metal ions from aqueous solutions by activated sludge. *Desalination*, *196*(1-3), 1–3, 164–176. doi:10.1016/j.desal.2005.12.012

Al-Saadi, A. A., Saleh, A. A., & Gupta, V. K. (2013). Spectroscopic and computational evaluation of cadmium adsorption using activated carbon produced from rubber tires. *Journal of Molecular Liquids*, *188*, 136–142. doi:10.1016/j.molliq.2013.09.036

Al-Salem, S. M., Lettieri, P., & Baeyens, J. (2009). Recycling and recovery routes of plastic solid waste (PSW): A review. *Waste Management (New York, N.Y.)*, *29*(10), 2625–2643. doi:10.1016/j.wasman.2009.06.004 PMID:19577459

Al-Shalalfeh, M. M., Saleh, T. A., & Al-Saadi, A. A. (2016). Silver colloid and film substrates in surface-enhanced Raman scattering for 2-thiouracil detection. *RSC Advances*, *6*(79), 75282–75292. doi:10.1039/C6RA14832H

Al-sheh, S., Banat, F., & Mohai, F. (1999). Sorption of copper and nickel by spent animal bones. *Chemosphere*, *39*(12), 2087–2096. doi:10.1016/S0045-6535(99)00098-3 PMID:10576108

Alswat, A. A., Ahmad, M. B., & Saleh, T. A. (2016). Zeolite modified with copper oxide and iron oxide for lead and arsenic adsorption from aqueous solutions. *Journal of Water Supply: Research & Technology - Aqua*, *65*(6), 465–479. doi:10.2166/aqua.2016.014

Altaher, H. (2012). The use of chitosan as a coagulant in the pre-treatment of turbid sea water. *Journal of Hazardous Materials*, *233*, 97–102. doi:10.1016/j.jhazmat.2012.06.061 PMID:22819482

Amarasekara, A., & Owereh, O. S. (2009). Homogenous phase synthesis of cellulose carbamate silica hybrid materials using 1-n-butyl-3-methylimidazolium chloride ionic liquid. *Carbohydrate Polymers*, *78*(3), 635–636. doi:10.1016/j.carbpol.2009.05.032

Ambashta, R. D., & Sillanpää, M. (2010). Water purification using magnetic assistance: A review. *Journal of Hazardous Materials*, *180*(1-3), 38–49. doi:10.1016/j.jhazmat.2010.04.105 PMID:20488616

Amin, M. T., Alazba, A. A., & Manzoor, U. (2014). A Review of Removal of Pollutants from Water/Wastewater Using Different Types of Nanomaterials. *Advances in Materials Science and Engineering*, 825910–825934.

An, B., & Zhao, D. (2012). Immobilization of As(III) in soil and groundwater using a new class of polysaccharide stabilized Fe–Mn oxide nanoparticles. *Journal of Hazardous Materials*, *211–212*, 332–341. doi:10.1016/j.jhazmat.2011.10.062 PMID:22119304

Anderson, C. G., De Pablo, N., & Romo, C. R. (1978). Antarctic krill (Euphausia superba) as a source of chitin and chitosan. *Proceedings of the First International Conference on Chitin/Chitosan. MIT Sea Grant Program,* 54.

Andrea, K., Axel, F., & Titirici, M.-M. (2013). Hydrothermal conversion of biomass to fuels and energetic materials. *Current Opinion in Chemical Biology*, *17*(3), 515–521. doi:10.1016/j.cbpa.2013.05.004 PMID:23707262

Andres, M. V., Costo, R., Roca, A. G., Marco, J. F., Goya, G. F., Serna, C. J., & Morales, M. P. (2008).. . *Journal of Physics. D, Applied Physics*, *41*(13).

Ang, P. K., Chen, W., Wee, A. T. S., & Loh, K. P. (2008). Solution-gated epitaxial graphene as pH sensor. *Journal of the American Chemical Society*, *130*(44), 14392–14393. doi:10.1021/ja805090z PMID:18850701

Anju, S., Yesodharan, S., & Yesodharan, E. (2012). Zinc oxide mediated sonophotocatalytic degradation of phenol in water. *Chemical Engineering Journal, 189*, 84–93. doi:10.1016/j.cej.2012.02.032

Annadurai, G., & Krishnan, M. R. V. (1997). Adsorption of acid dye from aqueous solution by chitin: Equilibrium studies. *Indian Journal of Chemical Technology, 4*, 217-222.

Annadurai, G., Chellapandian, M., & Krishnan, M. R. V. (1999). Adsorption of reactive dye on chitin. *Environmental Monitoring and Assessment, 59*(1), 111–119. doi:10.1023/A:1006072119624

Ao, Y., Xu, J., Fu, D., Ba, L., & Yuan, C. (2008). Deposition of anatase titania onto carbon encapsulated magentic nanoparticles. *Nanotechnology, 19*(40), 405604. doi:10.1088/0957-4484/19/40/405604

Aranaz, I., Mengíbar, M., Harris, R., Paños, I., Miralles, B., Acosta, N., & Heras, Á. et al. (2009). Functional characterization of chitin and chitosan. *Current Chemical Biology, 3*(2), 203–230.

Aravid, P. V., & Jong, W. (2012). Evaluation of high temperature gas cleaning options for biomass gasification product gas for solid oxide fuel cells. *Progress in Energy and Combustion Science, 38*(6), 737–764. doi:10.1016/j.pecs.2012.03.006

Arvanitoyannis, I., & Bosnea, L. (2001). Recycling of polymeric materials used for food packaging: Current status and perspectives. *Food Reviews International, 17*(3), 291–346. doi:10.1081/FRI-100104703

Ashbolt, N. J. (2004). Microbial contamination of drinking water and disease outcomes in developing regions. *Toxicology, 198*(1-3), 229–238. doi:10.1016/j.tox.2004.01.030 PMID:15138046

Ashraf, M. H. M., & Ali, S. A. (2015). Reusing waste plastic bottles as an alternative sustainable building material. *Energy for Sustainable Development, 24*, 79–85. doi:10.1016/j.esd.2014.11.001

Asmaly, H. A., Abussaud, B., Ihsanullah, , Saleh, T. A., Gupta, V. K., & Atieh, M. A. (2015). Ferric oxide nanoparticles decorated carbon nanotubes and carbon nanofibers: From synthesis to enhanced removal of phenol. *J. Saudi Chem. Soc, 19*(5), 511–520. doi:10.1016/j.jscs.2015.06.002

Awual, M., & Hasan, M. M. (2014). Novel conjugate adsorbent for visual detection and removal of toxic lead(II) ions from water. *Microporous and Mesoporous Materials, 196*, 261–269. doi:10.1016/j.micromeso.2014.05.021

Awual, M., Yaita, T., Suzuki, S., & Shiwaku, H. (2015). Ultimate selenium(IV) monitoring and removal from water using a new class of organic ligand based composite adsorbent. *Journal of Hazardous Materials, 91*, 111–119. doi:10.1016/j.jhazmat.2015.02.066 PMID:25771216

Ayati, A., Ahmadpour, A., Bamoharram, F. F., Tanhaei, B., Manttari, M., & Sillanpa, M. (2014). A review on catalytic applications of Au/TiO_2 nanoparticles in the removal of water pollutant. *Chemosphere, 107*, 163–174. doi:10.1016/j.chemosphere.2014.01.040 PMID:24560285

Aydýn, Y. A., & Aksoy, N. D. (2009). Adsorption of chromium on chitosan: Optimization, kinetics and thermodynamics. *Chemical Engineering Journal*, *151*(1), 188–194. doi:10.1016/j.cej.2009.02.010

Ayoub, A., Venditti, R. A., Pawlak, J. J., Salam, A., & Hubbe, M. A. (2013). Novel hemicellulose–chitosan biosorbent for water desalination and heavy metal removal. *ACS Sustainable Chemistry & Engineering*, *1*(9), 1102–1109. doi:10.1021/sc300166m

Azam, A., Ahmed, A. S., Oves, M., Khan, M. S., Habib, S. S., & Memic, A. (2012). Antimicrobial activity of metal oxide nanoparticles against Gram-positive and Gram-negative bacteria: A comparative study. *International Journal of Nanomedicine*, *7*, 6003–6009. doi:10.2147/IJN.S35347 PMID:23233805

Azari, A., Kalantary, R. R., Ghanizadeh, G., Kakavandi, B., Farzadkia, M., & Ahmadi, E. (2015). Iron–silver oxide nanoadsorbent synthesized by co-precipitation process for fluoride removal from aqueous solution and its adsorption mechanism. *RSC Advances*, *5*(106), 87377–87391. doi:10.1039/C5RA17595J

Azlan, K., Wan Saime, W. N., & Lai Ken, L. (2009). Chitosan and chemically modified chitosan beads for acid dyes sorption. *Journal of Environmental Sciences (China)*, *21*(3), 296–302. doi:10.1016/S1001-0742(08)62267-6 PMID:19634439

Ba-Abbad, M. M., Kadhum, A. A. H., Mohamad, A. B., Takriff, M. S., & Sopian, K. (2012). Synthesis and catalytic activity of TiO_2 nanoparticles for photochemical oxidation of concentrated chlorophenols under direct solar radiation. *International Journal of Electrochemical Science*, *7*, 4871–4888.

Babu, C. M., Vinodh, R., Sundaravel, B., Abidov, A., Peng, M. M., Cha, W. S., & Jang, H.-T. (2016). Characterization of reduced graphene oxide supported mesoporous Fe_2O_3/TiO_2 nanoparticles and adsorption of As(III) and As(V) from potable water. *Journal of the Taiwan Institute of Chemical Engineers*, *62*, 199–208. doi:10.1016/j.jtice.2016.02.005

Badruddoza, A. Z. M., Shawon, Z. B. Z., Tay, W. J. D., Hidajat, K., & Uddin, M. S. (2013). Fe_3O_4/cyclodextrin polymer nanocomposites for selective heavy metals removal from industrial wastewater. *Carbohydrate Polymers*, *91*(1), 322–332. doi:10.1016/j.carbpol.2012.08.030 PMID:23044139

Baek, I. G., You, S. J., & Park, E. D. (2012). Direct conversion of cellulose into polyols over Ni/W/SiO2. *Bioresource Technology*, *114*, 684–690. doi:10.1016/j.biortech.2012.03.059 PMID:22497706

Bae, M. J., Shin, H. S., Kim, E. K., Kim, J., & Shon, D. H. (2013). Oral administration of chitin and chitosan prevents peanut-induced anaphylaxis in a murine food allergy model. *International Journal of Biological Macromolecules*, *61*, 164–168. doi:10.1016/j.ijbiomac.2013.06.017 PMID:23806320

Bagabas, A., Alshammari, A., Aboud, M., & Kosslick, H. (2013). Room-temperature synthesis of zinc oxide nanoparticles in different media and their application in cyanide photodegradation. *Nanoscale Research Letters*, *8*(1), 516. doi:10.1186/1556-276X-8-516 PMID:24314056

Baghbanzadeh, M., Carbone, L., Cozzoli, P. D., & Kappe, C. O. (2011). Microwave-assisted synthesis of colloidal inorganic nanocrystals. *Chem. Int. Ed. Engl.*, *50*(48), 11312–11359. doi:10.1002/anie.201101274 PMID:22058070

Baghbanzadeh, M., Rana, D., Lan, C. Q., & Matsuura, T. (2016). Effects of inorganic nano-additives on properties and performance of polymeric membranes in water treatment. *Separation & Purification Reviews*, *45*(2), 141–167. doi:10.1080/15422119.2015.1068806

Bai, A., Zhao, X., Jin, Y., Yang, G., & Feng, Y. (2013). A novel thermophilic-B-glucosidase from Caldicellulosiruptorbescir: Characterization and its synhergistic catalysis with other cellulose. *Journal of Molecular Catalysis*, *85*, 248–256. doi:10.1016/j.molcatb.2012.09.016

Baleizao, C., Gigante, B., Garcia, H., & Corma, A. (2002). Ionic liquids as green solvents for the asymmetric synthesis of cyanohydrins catalyzed by VOC(salen) complexes. *Green Chemistry*, *4*, 272–274. doi:10.1039/b201497c

Banat, F., Al-Asheh, S., & Al-Rousan, D. (2002). Comparison between different keratin-composed biosorbents for the removal of heavy metal ions from aqueous solutions. *Adsorption Science and Technology*, *20*(4), 393–416. doi:10.1260/02636170260295579

Banat, F., Al-Asheh, S., & Mohai, F. (2002). Multi-metal sorption by spent animal bones. *Separation Science and Technology*, *37*(2), 311–327. doi:10.1081/SS-120000790

Bardajee, G., & Hooshyar, Z. (2013). Degradation of 2-chlorophenol from wastewater Using γ-Fe_2O_3 nanoparticles. *International Journal of Nanoscience and Nanotechnology*, *9*(1), 3–6.

Baributsa, D., Djibo, K., Lowenberg-DeBoer, J., Moussa, B., & Baoua, I. (2014). J. Lowenberg-DeBoer, Bokar Moussa, Ibrahim Baoua, "The fate of triple-layer plastic bags used for cowpea storage. *Journal of Stored Products Research*, *58*, 97–102. doi:10.1016/j.jspr.2014.02.011

Baselt, R. C., & Cravey, R. H. (1995). *Disposition of Toxic Drugs and Chemicals in Man* (4th ed.). Chicago, IL: Year Book Medical Publishers.

Basu, D., Blackburn, K., Harris, B., Neal, M., & Stoss, F. (1983). *Health assessment document for chromium*. Springfield, VA: Environmental Protection Agency.

Becher, G. (1999). Drinking water chlorination and health, *Acta Hydrochim. EtHydrobiol.*, *27*, 100–102.

Behnam, R., Morshed, M., Tavanai, H., & Ghiaci, M. (2013). Destructive adsorption of diazinon pesticide by activated carbon nanofibers containing Al_2O_3 and MgO nanoparticles. *Bulletin of Environmental Contamination and Toxicology*, *91*(4), 475–480. doi:10.1007/s00128-013-1064-x PMID:23912227

Behrens, S. (2011). Preparation of functionalized magnetic nanocomposites and hybrid materials: Recent progress and future directions. *Nanoscale*, *3*(3), 877–892. doi:10.1039/C0NR00634C PMID:21165500

BeMILLER, J. N., & Whistler, R. L. (1962). Alkaline Degradation of Amino Sugars1. *The Journal of Organic Chemistry*, *27*(4), 1161–1164. doi:10.1021/jo01051a009

Beppu, M. M., Arruda, E. J., Vieira, R. S., & Santos, N. N. (2004). Adsorption of Cu (II) on porous chitosan membranes functionalized with histidine. *Journal of Membrane Science*, *240*(1), 227–235. doi:10.1016/j.memsci.2004.04.025

Beyersmann, D., & Hartwig, A. (2008). Carcinogenic metal compounds: Recent insight into molecular and cellular mechanisms. *Archives of Toxicology*, *82*(8), 493–512. doi:10.1007/s00204-008-0313-y PMID:18496671

Bhatnagar, A., & Sillanpää, M. (2009). Applications of chitin-and chitosan-derivatives for the detoxification of water and wastewater—a short review. *Advances in Colloid and Interface Science*, *152*(1), 26–38. doi:10.1016/j.cis.2009.09.003 PMID:19833317

Bhattacharyya, K., & Gupta, S. S. (2008). Adsorption of a few heavy metals on natural and modified kaolinite and montmorillonite: A review. *Advances in Colloid and Interface Science*, *140*(2), 114–131. doi:10.1016/j.cis.2007.12.008 PMID:18319190

Bhaumik, M., Maity, A., Srinivasu, V. V., & Onyango, M. S. (2011). Enhanced removal of Cr(VI) from aqueous solution using polypyrrole/Fe_3O_4 magnetic nanocomposite. *Journal of Hazardous Materials*, *190*(1-3), 381–390. doi:10.1016/j.jhazmat.2011.03.062 PMID:21497438

Bilecka, I., Djerdj, I., & Niederberger, M. (2008). One-minute synthesis of crystalline binary and ternary metal oxide nanoparticles. *Chemical Communications (Cambridge)*, (7): 886–888. doi:10.1039/B717334B PMID:18253537

Bilecka, I., Kubli, M., Amstad, E., & Niederberger, M. (2011). Simultaneous formation of ferrite nanocrystals and deposition of thin films via a microwave-assisted nonaqueous sol–gel process. *Journal of Sol-Gel Science and Technology*, *57*(3), 313–322. doi:10.1007/s10971-010-2165-1

Bissen, M., & Frimmel, F. H. (2000). Speciation of As(III), As(V), MMA and DMA in contaminated soil extracts by HPLC-ICP/MS. *Journal of Analytical Chemistry*, *367*(1), 51–55. doi:10.1007/s002160051597 PMID:11227433

Blumberg, R., Southall, C. L., Van Rensburg, N. J., & Volckman, O. B. (1951). South african fish products. XXXII.—The rock lobster: A study of chitin production from processing wastes. *Journal of the Science of Food and Agriculture*, *2*(12), 571–576. doi:10.1002/jsfa.2740021210

Bootharaju, M. S., & Pradeep, T. (2010). Uptake of Toxic Metal Ions from Water by Naked and Monolayer Protected Silver Nanoparticles: An X-ray Photoelectron Spectroscopic Investigation. *The Journal of Physical Chemistry C*, *114*(18), 8328–8336. doi:10.1021/jp101988h

Boparai, H. K., Joseph, M., & OCaroll, D. M. (2011). Kinetics and thermodynamics of cadmium ion removal by adsorption onto nano zerovalent iron particles. *Journal of Hazardous Materials*, *186*(1), 458–465. doi:10.1016/j.jhazmat.2010.11.029 PMID:21130566

Bora, T., & Dutta, J. (2014). Applications of nanotechnology in wastewater treatment—a review. *Journal of Nanoscience and Nanotechnology*, *14*(1), 613–626. doi:10.1166/jnn.2014.8898 PMID:24730286

Bottero, J.-Y., Rose, J., & Wiesner, M. R. (2006). Nanotechnologies: Tools for sustainability in a new wave of water treatment processes. *Integrated Environmental Assessment and Management*, *2*(4), 391–395. doi:10.1002/ieam.5630020411 PMID:17069181

Bough, W. A., Salter, W. L., Wu, A. C. M., & Perkins, B. E. (1978). Influence of manufacturing variables on the characteristics and effectiveness of chitosan products. I. Chemical composition, viscosity, and molecular-weight distribution of chitosan products. *Biotechnology and Bioengineering*, *20*(12), 1931–1943. doi:10.1002/bit.260201208

Bough, W. A., Shewfelt, A. L., & Salter, W. L. (1975). Use of chitosan for the reduction and recovery of solids in poultry processing waste effluents. *Poultry Science*, *54*(4), 992–1000. doi:10.3382/ps.0540992

Bratskaya, S., Avramenko, V., Schwarz, S., & Philippova, I. (2006). Enhanced flocculation of oil-in-water emulsions by hydrophobically modified chitosan derivatives. *Colloids and Surfaces. A, Physicochemical and Engineering Aspects*, *275*(1), 168–176. doi:10.1016/j.colsurfa.2005.09.036

Bratskaya, S., Schwarz, S., & Chervonetsky, D. (2004). Comparative study of humic acids flocculation with chitosan hydrochloride and chitosan glutamate. *Water Research*, *38*(12), 2955–2961. doi:10.1016/j.watres.2004.03.033 PMID:15223291

Brine, C. J., & Austin, P. R. (1981). Chitin variability with species and method of preparation. *Comparative Biochemistry and Physiology Part B: Comparative Biochemistry*, *69*(2), 283–286. doi:10.1016/0305-0491(81)90242-X

British Plastic Federation (BPF). (2012). *UK Plastics Industry Capability Guide*. Retrieved from http://www.bpf.co.uk/Industry/Default.aspx

British Plastics Federation. (2008). *Oil consumption*. Retrieved from http://www.bpf.co.uk/Oil_Consumption.aspx

Broussignac, P. (1968). Chitosan: A natural polymer not well known by the industry. *Chim. Ind. Genie Chim*, *99*(9), 1241–1247.

Broussignac, P. (1968). Un haut polymère naturel peu connu dans l'industrie, Le chitosane. *Chim. Ind. Genie Chim*, *99*, 1241–1247.

Brugnerotto, J., Lizardi, J., Goycoolea, F. M., Argüelles-Monal, W., Desbrieres, J., & Rinaudo, M. (2001). An infrared investigation in relation with chitin and chitosan characterization. *Polymer*, *42*(8), 3569–3580. doi:10.1016/S0032-3861(00)00713-8

Brzeski, M. M. (1982). Concept of chitin/chitosan isolation from Antarctic Krill (Euphausia superba) shells on a technique scale.*Proceedings of the 2nd international conference on chitin and chitosan,* 15-29.

Bykkam, S., Narsingam, S., Ahmadipour, M., Dayakar, T., Rao, K. V., Chakra, C. S., & Kalakotla, S. (2015). Few layered graphene Sheet decorated by ZnO Nanoparticles for anti-bacterial application. *Superlattices and Microstructures*, *83*, 776–784. doi:10.1016/j.spmi.2015.03.063

Bystrzejewski, M., & Pyrzyńska, K. (2011). Kinetics of copper ions sorption onto activated carbon, carbon nanotubes and carbon-encapsulated magnetic nanoparticles. *Colloids and Surfaces. A, Physicochemical and Engineering Aspects*, *377*(1), 402–408. doi:10.1016/j.colsurfa.2011.01.041

Bystrzejewski, M., Pyrzyńska, K., Huczko, A., & Lange, H. (2009). Carbon-encapsulated magnetic nanoparticles as separable and mobile sorbents of heavy metal ions from aqueous solutions. *Carbon*, *47*(4), 1201–1204. doi:10.1016/j.carbon.2009.01.007

Cai, H., Li, C., Wang, A., Xu, G., & Zhang, T. (2012). Zeolite-promoted hydrolusis of cellulose in ionic liquid, insight into the mutual behavior of zeolite, cellulose and ionic liquid. *Applied Catalysis B: Environmental*, *123*, 333–338. doi:10.1016/j.apcatb.2012.04.041

Calvo, P., Remuñan-López, C., Vila-Jato, J. L., & Alonso, M. J. (1997). Chitosan and chitosan/ethylene oxide-propylene oxide block copolymer nanoparticles as novel carriers for proteins and vaccines. *Pharmaceutical Research*, *14*(10), 1431–1436. doi:10.1023/A:1012128907225 PMID:9358557

Cao, B., Cao, S., Dong, P., Gao, J., & Wang, J. (2013). High antibacterial activity of ultrafine TiO_2/graphene sheets nanocomposites under visible light irradiation. *Materials Letters*, *93*, 349–352. doi:10.1016/j.matlet.2012.11.136

Cao, C., Xiao, L., Chen, C., Shi, X., Cao, Q., & Gao, L. (2014). In situ preparation of magnetic Fe_3O_4/chitosan nanoparticles via a novel reduction–precipitation method and their application in adsorption of reactive azo dye. *Powder Technology*, *260*, 90–97. doi:10.1016/j.powtec.2014.03.025

Cao, J., Li, J., Liu, L., Xie, A., Li, S., Qiu, L., & Shen, Y. et al. (2014). One-pot synthesis of novel Fe_3O_4/Cu_2O/PANI nanocomposites as absorbents in water treatment. *Journal of Materials Chemistry A*, *2*(21), 7953–7959. doi:10.1039/c4ta00303a

Cao, W., Chen, L., & Qi, Z. (2015). Microwave-assisted synthesis of Ag/Ag2SO4/ZnO nanostructures for efficient visible-light-driven photocatalysis. *Journal of Molecular Catalysis A Chemical*, *401*, 81–89. doi:10.1016/j.molcata.2015.02.023

Cao, Y., & Li, X. (2014). Adsorption of graphene for the removal of inorganic pollutants in water purification: A review. *Adsorption*, *20*(5-6), 713–727. doi:10.1007/s10450-014-9615-y

Carlson, T. R., Vispute, T. P., & Huber, G. W. (2008). Green gasoline by catalytic fast pyrolysis of solid biomass derived compounds. *ChemSusChem*, *1*(5), 397–400. doi:10.1002/cssc.200800018 PMID:18702131

Carniti, P., Gervasini, A., Biella, S., & Auroux, A. (2005). Instrinsic and effective acidity study of niobic acid and niobium phosphate by multitechnique approach. *Chemistry of Materials*, *17*(24), 6128–6135. doi:10.1021/cm0512070

Centeno, J. A., Tchounwou, P. B., Patlolla, A. K., Mullick, F. G., Murakat, L., Meza, E., & Yedjou, C. G. et al. (2005). Environmental pathology and health effects of arsenic poisoning: a critical review. In R. Naidu, E. Smith, J. Smith, & P. Bhattacharya (Eds.), *Managing Arsenic in the environment: from soil to human health*. Adelaide, Australia: CSIRO Publishing Corporation.

Chai, L., Wang, Y., Zhao, N., Yang, W., & You, X. (2013). Sulfate-doped Fe_3O_4/Al_2O_3 nanoparticles as a novel adsorbent for fluoride removal from drinking water. *Water Research*, *47*(12), 4040–4049. doi:10.1016/j.watres.2013.02.057 PMID:23602616

Chanda, M., & Roy, S. (2007). *Plastics technology handbook* (4th ed.). Boca Raton, FL: CRC Press.

Chang, J., Jianchao, M., Qingliang, M., Duoduo, Z., Nannan, Q., Mengxiao, H., & Hongzhu, M. (2016). Adsorption of methylene blue onto Fe_3O_4/activated montmorillonite nanocomposite. *Applied Clay Science*, *119*, 132–140. doi:10.1016/j.clay.2015.06.038

Chang, Y.-N., Ou, X.-M., Zeng, G.-M., Gong, J.-L., Deng, C.-H., Jiang, Y., & He, X. et al. (2015). Synthesis of magnetic graphene oxide–TiO_2 and their antibacterial properties under solar irradiation. *Applied Surface Science*, *343*, 1–10. doi:10.1016/j.apsusc.2015.03.082

Chatterjee, S., Chatterjee, S., Chatterjee, B. P., & Guha, A. K. (2007). Adsorptive removal of congo red, a carcinogenic textile dye by chitosan hydrobeads: Binding mechanism, equilibrium and kinetics. *Colloids and Surfaces. A, Physicochemical and Engineering Aspects*, *299*(1), 146–152. doi:10.1016/j.colsurfa.2006.11.036

Chatterjee, S., Lee, D. S., Lee, M. W., & Woo, S. H. (2009). Enhanced adsorption of congo red from aqueous solutions by chitosan hydrogel beads impregnated with cetyl trimethyl ammonium bromide. *Bioresource Technology*, *100*(11), 2803–2809. doi:10.1016/j.biortech.2008.12.035 PMID:19208471

Chatterjee, S., Lee, D. S., Lee, M. W., & Woo, S. H. (2009). Nitrate removal from aqueous solutions by cross-linked chitosan beads conditioned with sodium bisulfate. *Journal of Hazardous Materials*, *166*(1), 508–513. doi:10.1016/j.jhazmat.2008.11.045 PMID:19111975

Chatterjee, S., & Woo, S. H. (2009). The removal of nitrate from aqueous solutions by chitosan hydrogel beads. *Journal of Hazardous Materials*, *164*(2), 1012–1018. doi:10.1016/j.jhazmat.2008.09.001 PMID:18977085

Chen, J., Liao, W., Jiang, Y., Yu, D., Zou, M., Zhu, H., . . . Du, M. (2016). Facile fabrication of ZnO/TiO_2 heterogeneous nanofibres and their photocatalytic behaviour and mechanism towards rhodamine B. *Nanomaterials and Nanotechnology*, *6*(9).

Chen, C. J., Hsueh, Y. M., Lai, M. S., Shyu, P., Chen, S. Y., Wu, M. M., & Tai, T. Y. et al. (1995). Increased prevalence of hypertension and long term arsenic exposure. *Hypertension*, *25*(1), 53–60. doi:10.1161/01.HYP.25.1.53 PMID:7843753

Chen, C., Hu, J., Shao, D., Li, J., & Wang, X. (2009). Adsorption behavior of multiwall carbon nanotube/iron oxide magnetic composites for Ni(II) and Sr(II). *Journal of Hazardous Materials*, *164*(2), 923–928. doi:10.1016/j.jhazmat.2008.08.089 PMID:18842337

Chen, D., Zeng, Z., Zeng, Y., Zhang, F., & Wang, M. (2016). Removal of methylene blue and mechanism on magnetic γ-Fe_2O_3/SiO_2 nanocomposite from aqueous solution. *Water Resources and Industry*, *15*, 1–13. doi:10.1016/j.wri.2016.05.003

Cheng, W. P., Chi, F. H., Yu, R. F., & Lee, Y. C. (2005). Using chitosan as a coagulant in recovery of organic matters from the mash and lauter wastewater of brewery. *Journal of Polymers and the Environment*, *13*(4), 383–388. doi:10.1007/s10924-005-5533-0

Cheng, W., Zhang, W., Hu, L., Ding, W., Wu, F., & Li, J. (2016). Etching synthesis of iron oxide nanoparticles for adsorption of arsenic from water. *RSC Advances*, *6*(19), 15900–15910. doi:10.1039/C5RA26143K

Chen, X., Parker, S. G., Zou, G., Su, W., & Zhang, Q. (2010). β-cyclodextrin-functionalized silver nanoparticles for the naked eye detection of aromatic isomers. *ACS Nano*, *4*(11), 6387–6394. doi:10.1021/nn1016605 PMID:20973513

Chien, S. H., & Clayton, W. R. (1980). Application of Elovich equation to the kinetics of phosphate release and sorption in soils. *Soil Science Society of America Journal*, *44*(2), 265–268. doi:10.2136/sssaj1980.03615995004400020013x

Chi, F. H., & Cheng, W. P. (2006). Use of chitosan as coagulant to treat wastewater from milk processing plant. *Journal of Polymers and the Environment*, *14*(4), 411–417. doi:10.1007/s10924-006-0027-2

Chiou, M. S., Ho, P. Y., & Li, H. Y. (2004). Adsorption of anionic dyes in acid solutions using chemically cross-linked chitosan beads. *Dyes and Pigments*, *60*(1), 69–84. doi:10.1016/S0143-7208(03)00140-2

Chiou, M. S., & Li, H. Y. (2002). Equilibrium and kinetic modeling of adsorption of reactive dye on cross-linked chitosan beads. *Journal of Hazardous Materials*, *93*(2), 233–248. doi:10.1016/S0304-3894(02)00030-4 PMID:12117469

Chiou, M. S., & Li, H. Y. (2003). Adsorption behavior of reactive dye in aqueous solution on chemical cross-linked chitosan beads. *Chemosphere*, *50*(8), 1095–1105. doi:10.1016/S0045-6535(02)00636-7 PMID:12531717

Chong, Y., Chen, M., Xu, T., Zhang, Y., & Wang, Q. (2014). Tantalum and aluminum co-doped iron oxide as a robust photocatalyst for water oxidation. *App Catal E*, *147*, 733–740. doi:10.1016/j.apcatb.2013.10.009

Chung, J., Chun, J., Lee, J., Lee, S. H., Lee, Y. J., & Hong, S. W. (2012). Sorption of Pb(II) and Cu(II) onto multi-amine grafted mesoporous silica embedded with nano-magnetite: Effects of steric factors. *Journal of Hazardous Materials*, *183*, 239–240. PMID:22985819

Chung, Y. C. (2006). Improvement of aquaculture wastewater using chitosan of different degrees of deacetylation. *Environmental Technology*, *27*(11), 1199–1208. doi:10.1080/09593332708618734 PMID:17203601

Chung, Y. C., Li, Y. H., & Chen, C. C. (2005). Pollutant removal from aquaculture wastewater using the biopolymer chitosan at different molecular weights. *Journal of Environmental Science and Health*, *40*(9), 1775–1790. doi:10.1081/ESE-200068058 PMID:16134368

Clausen, L., Fabricius, I., & Madsen, L. (2001). Adsorption of pesticides onto quartz, calcite, kaolinite and α-alumina. *Journal of Environmental Quality*, *30*(3), 846–857. doi:10.2134/jeq2001.303846x PMID:11401273

Cloete, T.E. (2010). Nanotechnology in Water Treatment Applications. New York: Horizon Scientific Press.

Cocoletzi, H. H., Almanza, E. Á., Agustin, O. F., Nava, E. V., &cCassellis, E. R. (2009). Obtención y caracterización de quitosano a partir de exoesqueletos de camarón. *Superficies y vacío*, *22*(3), 57-60.

Coey, J. M. D. (2009). *Magnetism and magnetic materials*. New York, NY: Cambridge University Press.

Cohen-Tanugi, D., & Grossman, J. C. (2012). Water desalination across nanoporous graphene. *Nano Letters*, *12*(7), 3602–3608. doi:10.1021/nl3012853 PMID:22668008

Corsini, E., Liesivuori, J., Vergieva, T., Loveren, H. V., & Colosio, C. (2008). Effects of pesticide exposure on the human immune system. *Human and Experimental Toxicology*, *27*(9), 671–680. doi:10.1177/0960327108094509 PMID:19042949

Crini, G., Gimbert, F., Robert, C., Martel, B., Adam, O., Morin-Crini, N., & Badot, P. M. et al. (2008). The removal of Basic Blue 3 from aqueous solutions by chitosan-based adsorbent: Batch studies. *Journal of Hazardous Materials*, *153*(1), 96–106. doi:10.1016/j.jhazmat.2007.08.025 PMID:17888569

Crini, G., Martel, B., & Torri, G. (2008). Adsorption of CI Basic Blue 9 on chitosan-based materials. *International Journal of Environment and Pollution*, *34*(1-4), 451–465. doi:10.1504/IJEP.2008.020809

Cristina, Y. M., Rodriguez, J., Freer, J., Duran, N., & Mansilla, H. D. (2000). Photocatalytic degradation of cellulose bleaching effluent by supported TiO2 and ZnO. *Chemosphere*, *41*, 1193–1197. doi:10.1016/S0045-6535(99)00551-2 PMID:10901246

Crittenden, J. C., Trussell, R. R., David, W., Hand, P., Howe, K. J., & Tchobanoglous, G. (2005). Water Treatment: Principles and Design. John Wiley and Sons.

Cun, W., Jincai, Z., Xinming, W., Bixian, M., Guoying, S., Pingan, P., & Jiamo, F. (2002). Preparation, characterization and photocatalytic activity of nano-sized ZnO/SnO_2 coupled photocatalysts. *Applied Catalysis B: Environmental*, *39*(3), 269–279. doi:10.1016/S0926-3373(02)00115-7

Curlee, T. R., & Das, S. (1991). *Plastic Wastes (Management, Control, Recycling and Disposal)*. Noyes Data Corporation.

Czaja, A. U., Trkhan, N., & Muller, U. (2009). Industrial applications of metal-organic frameworks. *Chemical Society Reviews*, *38*(5), 1284–1293. doi:10.1039/b804680h PMID:19384438

Czechowska-Biskup, R., Jarosińska, D., Rokita, B., Ulański, P., & Rosiak, J. M. (2012). Determination of degree of deacetylation of chitosan-comparision of methods. *Prog. Chem. Appl. Chitin Its Deriv*, *17*, 5–20.

Dadfarnia, S., Shabani, A. M. H., Moradi, S. E., & Emami, S. (2015). Methyl red removal from water by iron based metal-organic frameworks loaded onto iron oxide nanoparticle adsorbent. *Applied Surface Science*, *330*, 85–93. doi:10.1016/j.apsusc.2014.12.196

Daels, N., Radoicic, M., Radetic, M., & Hulle, S. W. H. V. (2015). Electrospunnnanofibre membranes functionalised with TiO_2 nanoparticles: Evaluation of humic acid and bacterial removal from polluted water. *Separation and Purification Technology*, *149*, 488–49. doi:10.1016/j.seppur.2015.06.016

Dahal, N., García, S., Zhou, J., & Humphrey, S. M. (2012). Beneficial effects of microwave-assisted heating versus conventional heating in noble metal nanoparticle synthesis. *ACS Nano*, *6*(11), 9433–9446. doi:10.1021/nn3038918 PMID:23033897

Danmaliki, G. I., & Saleh, T. A. (2016). Influence of conversion parameters of waste tires to activated carbon on adsorption of dibenzothiophene from model fuels. *Journal of Cleaner Production*, *117*, 50–55. doi:10.1016/j.jclepro.2016.01.026

Danmaliki, G. I., & Saleh, T. A. (2017). Effects of bimetallic Ce/Fe nanoparticles on the desulfurization of thiophenes using activated carbon. *Chemical Engineering Journal*, *307*, 914–927. doi:10.1016/j.cej.2016.08.143

Danmaliki, G. I., Saleh, T. A., & Shamsuddeen, A. A. (2017). Response Surface Methodology Optimization of Adsorptive Desulfurization on Nickel/Activated Carbon. *Chemical Engineering Journal.*

Darwish, M., Mohammadi, A., & Assi, N. (2016). Microwave-assisted polyol synthesis and characterization of pvp-capped cds nanoparticles for the photocatalytic degradation of tartrazine. *Research Bulletin (Sun Chiwawitthaya Thang Thale Phuket)*, *74*, 387–396. doi:10.1016/j.materresbull.2015.11.002

Dasary, S. S. R., Singh, A. K., Senapati, D., Yu, H., & Ray, P. C. (2009). Gold nanoparticle based label-free SERS probe for ultrasensitive and selective detection of trinitrotoluene. *Journal of the American Chemical Society*, *131*(38), 13806–13812. doi:10.1021/ja905134d PMID:19736926

Das, N., Vimala, R., & Karthika, P. (2008). Bio-sorption of heavy metals- An overview. *Indian Journal of Biotechnology*, *7*, 159–169.

Davis, T. A., Volesky, B., & Mucci, A. (2003). A review of the biochemistry of heavy metal biosorption by brown algae. *Water Research*, *37*(18), 4311–4330. doi:10.1016/S0043-1354(03)00293-8 PMID:14511701

De la Hoz, A., Díaz-Ortiz, A., & Moreno, A. (2005). Microwaves in organic synthesis. Thermal and non-thermal microwave effects. *Chemical Society Reviews*, *34*(2), 164–178. doi:10.1039/B411438H PMID:15672180

de la Rosa, G., Reynel-Avila, H. E., Bonillia-Petriciolet, A., Cano-Rodriguez, I., Velasco-Santos, C., & Martinez-Hernandez, A. L. (2008). Recycling poultry feathers for Pb removal from wastewater: Kinetic and equilibrium studies. *Proceedings of World Academy of Science, Engineering and Technology*, *30*, 1011–1019.

De, Isayev, & Khait. (Eds.). (2005). *Rubber Recycling*. Taylor and Francis.

Debrassi, A., Baccarin, T., Demarchi, C. A., Nedelko, N., Slawska-Waniewska, A., Dhuzewski, P., & Rodrigues, C. A. et al. (2012). Adsorption of Remazol Red198 onto magnetic N-lauryl chitosan particles: Equilibrium, kinetics, reuse and factorial design. *Environ. Sci. Pollut. Res*, *19*(5), 1594–1604. doi:10.1007/s11356-011-0662-6 PMID:22134860

Dědková, K., Lang, J., Matějová, K., Peikertová, P., Holešinský, J., Vodárek, V., & Kukutschová, J. (2015). Nanostructured composite material graphite/TiO_2 and its antibacterial activity under visible light irradiation. *Journal of Photochemistry and Photobiology. B, Biology*, *149*, 265–271. doi:10.1016/j.jphotobiol.2015.06.018 PMID:26114221

Demarchi, C. A., Debrassi, A., de Campos Buzzi, F., Nedelko, N., Slawska-Waniewska, A., Dłuzewski, P., & Rodrigues, C. A. (2015). *Adsorption of the dye Remazol Red 198 (RR198) by O-carboxymethylchitosan-N-lauryl/γ-Fe_2O_3 magnetic nanoparticles*. Arabian Journal of Chemistry.

Demirbas, A. (2001). Carbonization ranking of selected biomass for charcoal, liquid and gaseous products. *Energy Conv & Mang*, *42*(10), 1229–1238. doi:10.1016/S0196-8904(00)00110-2

Demirbas, A. (2008). Heavy metal adsorption onto agro-based waste materials: A review. *Journal of Hazardous Materials*, *157*(2-3), 220–229. doi:10.1016/j.jhazmat.2008.01.024 PMID:18291580

Deng, C., & Tian, X. (2013). Facile microwave-assisted aqueous synthesis of CdS nanocrystals with their photocatalytic activities under visible lighting. *Materials Research Bulletin*, *48*(10), 4344–4350. doi:10.1016/j.materresbull.2013.07.019

Deng, J.-H., Zhang, X.-R., Zeng, G.-M., Gong, J.-L., Niu, Q.-Y., & Liang, J. (2013). Simultaneous removal of Cd(II) and ionic dyes from aqueous solution using magnetic graphene oxide nanocomposite as an adsorbent. *Chemical Engineering Journal*, *226*, 189–200. doi:10.1016/j.cej.2013.04.045

Desalination industry enjoys growth spurt as scarcity starts to bite. (n.d.). Retrieved from globalwaterintel.com

Dhruv, M., Siddharth, M., & Singha, S. K. (2015). Magnetic adsorbents for the treatment of water/wastewater—A review. *Journal of Water Process Engineering*, *7*, 244–265. doi:10.1016/j.jwpe.2015.07.001

Diale, P. P., & Muzenda, E. (2011). A Study of South African Natural Zeolites Properties and Applications Properties and Applications. *Proceedings of the World Congress on Engineering and Computer Science*, 2.

Dima, J. B., Sequeiros, C., & Zaritzky, N. E. (2015). Hexavalent chromium removal in contaminated water using reticulated chitosan micro/nanoparticles from seafood processing wastes. *Chemosphere*, *141*, 100–111. doi:10.1016/j.chemosphere.2015.06.030 PMID:26151484

Dimple, L. (2014). Adsorption of heavy metals: A Review. *International Journal of Environmental Research and Development*, *4*, 41–44.

Divakaran, R., & Pillai, V. S. (2002). Flocculation of river silt using chitosan. *Water Research*, *36*(9), 2414–2418. doi:10.1016/S0043-1354(01)00436-5 PMID:12108734

Divakaran, R., & Pillai, V. S. (2004). Mechanism of kaolinite and titanium dioxide flocculation using chitosan assistance by fulvic acids? *Water Research*, *38*(8), 2135–2143. PubMed

Dizaj, S. M., Lotfipour, F., Barzegar-Jalali, M., Zarrintan, M. H., & Adibkia, K. (2014). Antimicrobial activity of the metals and metal oxide nanoparticles. *Materials Science and Engineering C*, *44*, 278–284. doi:10.1016/j.msec.2014.08.031 PMID:25280707

Dmitri, A. B., Sergey, B., Pavel, E. P., Yurii, V. S., Valerii, I. B., Sergey, V. K., & Julian, R. H. R. (2013). Vapour phase formic acid decomposition over PdAu/c-Al2O3 catalysts: Effect of composition of metallic particle. *Journal of Catalysis*, *299*, 171–180. doi:10.1016/j.jcat.2012.12.009

Dobrucka, R., & Długaszewska, J. (2016). Biosynthesis and antibacterial activity of ZnO nanoparticles using *Trifolium pratense* flower extract. *Saudi Journal of Biological Sciences*, *23*(4), 517–523. doi:10.1016/j.sjbs.2015.05.016 PMID:27298586

Dolphen, R., Sakkayawong, N., Thiravetyan, P., & Nakbanpote, W. (2007). Adsorption of Reactive Red 141 from wastewater onto modified chitin. *Journal of Hazardous Materials*, *145*(1), 250–255. doi:10.1016/j.jhazmat.2006.11.026 PMID:17184913

Domard, A. (2011). A perspective on 30 years research on chitin and chitosan. *Carbohydrate Polymers*, *84*(2), 696–703. doi:10.1016/j.carbpol.2010.04.083

Domingo, C., Clemente, R. R., & Blesa, M. (1994). Morphological properties of a-FeOOH, g-FeOOH and Fe_3O_4 obtained by oxidation of aqueous Fe (II) solutions. *Journal of Colloid and Interface Science*, *165*(1), 244–252. doi:10.1006/jcis.1994.1225

Dong, L., Yang, H., Liu, S., Wang, X., & Xie, Y. F. (2015). Fabrication and anti-biofouling properties of alumina and zeolite nanoparticle embedded ultrafiltration membranes. *Desalination*, *365*, 70–78. doi:10.1016/j.desal.2015.02.023

Dopp, E., Hartmann, L. M., Florea, A. M., Rettenmier, A. W., & Hirner, A. V. (2004). Environmental distribution, analysis, and toxicity of organometal (loid) compounds. *Critical Reviews in Toxicology*, *34*(3), 301–333. doi:10.1080/10408440490270160 PMID:15239389

Dora, S., Bhasker, T., Singh, R., Naik, D. V., & Adhikari, D. K. (2012). Effective catalytic conversion of cellulose into high yields of methyl glycosides over sulfonated carbon based catalyst. *Bioresource Technology*, *120*, 318–321. doi:10.1016/j.biortech.2012.06.036 PMID:22776237

Dreher, M., Jonson, B., Peterson, A. A., Nachtegaal, M., Wambach, J., & Vogel, F. (2013). Catalysis in supercritical water: Pathway of the methanation reaction and sulfur poisoning over a Ru/C catalyst during the reforming of biomolecules. *Journal of Catalysis*, *301*, 38–45. doi:10.1016/j.jcat.2013.01.018

Dreyer, D. R., Park, S., Bielawski, C. W., & Ruoff, R. S. (2010). The chemistry of graphene oxide. *Chemical Society Reviews*, *39*(1), 228–240. doi:10.1039/B917103G PMID:20023850

Dubey, S. P., & Gopal, K. (2007). Adsorption of chromium (VI) on low cost adsorbents derived from agricultural waste material: A comparative study. *Journal of Hazardous Materials*, *145*(3), 465–470. doi:10.1016/j.jhazmat.2006.11.041 PMID:17175098

Dursun, A. Y., & Kalayci, C. S. (2005). Equilibrium, kinetic and thermodynamic studies on the adsorption of phenol onto chitin. *Journal of Hazardous Materials*, *123*(1), 151–157. doi:10.1016/j.jhazmat.2005.03.034 PMID:15993297

Dutta, S., De, S., & Saha, B. (2013). Advances in biomass transformation to 5-hydroxymethylfurfural and mechanic aspects. *Biomass & Biores.*, *55*, 355–369. doi:10.1016/j.biombioe.2013.02.008

Edward, L. K. (2010). Tyre char preparation from waste tyre rubber for dye removal from effluents. *Journal of Hazardous Materials*, *175*(1-3), 151–158. doi:10.1016/j.jhazmat.2009.09.142 PMID:19854570

Edward, W., Yong, T. K., Geoffrey, A. T., Alejandro, F., Kee, S. H., Edward, W. H., & Hubber, G. W. et al. (2013). Conversion of glucose into levulinic acid with solid metal (IV) phosphate catalyts. *Journal of Catalysis*, *304*, 123–134. doi:10.1016/j.jcat.2013.03.023

Elkady, M. F., Shokry Hassan, H., & Salama, E. (2016). Sorption profile of phosphorus ions onto ZnO nanorods synthesized via Sonic technique. *Journal of Engineering*, *9*. doi:10.1155/2016/2308560

Elson Santiago de Alvarenga. (2011). *Characterization and Properties of Chitosan, Biotechnology of Biopolymers*. InTech. Retrieved from http://www.intechopen.com/books/biotechnology-of-biopolymers/characterization-and-properties-of-chitosan

Elwakeel, K. Z., & Guibal, E. (2015). Arsenic(V) sorption using chitosan/$Cu(OH)_2$ and chitosan/CuO composite sorbents. *Carbohydrate Polymers*, *134*, 190–204. doi:10.1016/j.carbpol.2015.07.012 PMID:26428116

Emadi, M., Shams, E., & Amini, M. K. (2013). Removal of zinc from aqueous solutions by magnetite silica core-shell nanoparticles. *J. Chem.*, 1–10.

Eoff, L. S., & Reddy, B. R. (2004). *U.S. Patent No. 6,764,981*. Washington, DC: U.S. Patent and Trademark Office.

Fan, G., Liao, C., Fang, T., Wang, M., & Song, G. (2013). Hydrolysis of cellulose catalyzed by sulfonated poly(styrene-co-divinylbenzene) in the ionic liquid 1-n-butyl-3-methylimidazolium bromide. *Fuel Processing Technology*, *116*, 142–148. doi:10.1016/j.fuproc.2013.05.009

Fan, L., Luo, C., Lv, Z., Lu, F., & Qiu, H. (2011). Preparation of magnetic modified chitosan and adsorption of Zn^{2+} from aqueous solutions. *Colloids and Surfaces. B, Biointerfaces*, *88*(2), 574–581. doi:10.1016/j.colsurfb.2011.07.038 PMID:21868204

Fan, L., Luo, C., Sun, M., & Qiu, H. (2012). Synthesis of graphene oxide decorated with magnetic cyclodextrin for fast chromium removal. *Journal of Materials Chemistry*, *22*(47), 24577–24583. doi:10.1039/c2jm35378d

Fan, L., Zhang, Y., Li, X., Luo, C., Lu, F., & Qiu, H. (2012). Removal of alizarin red from water environment using magnetic chitosan with Alizarin Red as imprinted molecules. *Colloids and Surfaces. B, Biointerfaces*, *91*, 250–257. doi:10.1016/j.colsurfb.2011.11.014 PMID:22119217

Fan, L., Zhang, Y., Luo, C., Lu, F., Qiu, H., & Sun, M. (2012). Synthesis and characterization of magnetic β-cyclodextrin–chitosan nanoparticles as nano-adsorbents for removal of methyl blue. *International Journal of Biological Macromolecules*, *50*(2), 444–450. doi:10.1016/j.ijbiomac.2011.12.016 PMID:22227306

Farnad, N., Farhadi, K., & Voelcker, N. H. (2012). Polydopamine nanoparticles as a new and highly selective biosorbent for the removal of copper (ii) ions from aqueous solution. *Water, Air, and Soil Pollution*, *223*(6), 3535–3544. doi:10.1007/s11270-012-1131-7

Farooq, M., Raja, I. A., & Pervez, A. (2009). Photocatalytic degradation of TCE in water using TiO_2 catalyst. *Solar Energy*, *83*(9), 1527–1533. doi:10.1016/j.solener.2009.04.009

Fast, S. A., & Gude, V. G. (2015). Ultrasound-chitosan enhanced flocculation of low algal turbid waters. *Journal of Industrial and Engineering Chemistry*, *24*, 153–160. doi:10.1016/j.jiec.2014.09.023

Fast, S. A., Kokabian, B., & Gude, V. G. (2014). Chitosan enhanced coagulation of algal turbid waters–Comparison between rapid mix and ultrasound coagulation methods. *Chemical Engineering Journal*, *244*, 403–410. doi:10.1016/j.cej.2014.01.081

Faulconer, E. K., Hoogesteijn von Reitzenstein, N. V., & Mazyck, D. W. (2012). Optimization of magnetic powdered activated carbon for aqueous Hg(II) removal and magnetic recovery. *Journal of Hazardous Materials*, *199-200*, 9–14. doi:10.1016/j.jhazmat.2011.10.023 PMID:22104766

Fawell, J., & Nieuwenhuijsen, M. J. (2003). Contaminants in drinking water. *British Medical Bulletin*, *68*(1), 199–208. doi:10.1093/bmb/ldg027 PMID:14757718

Fayazi, M., Ghanei-Motlagh, M., & Taher, M. A. (2015). The adsorption of basic dye (Alizarin red S) from aqueous solution onto activated carbon/γ-Fe_2O_3 nano-composite: Kinetic and equilibrium studies. *Materials Science in Semiconductor Processing*, *40*, 35–43. doi:10.1016/j.mssp.2015.06.044

Feldmann, C. (2003). Polyol-Mediated Synthesis of Nanoscale Functional Materials. *Advanced Functional Materials*, *13*(2), 101–107. doi:10.1002/adfm.200390014

Felica, L.-D., Courson, C., Foscoto, P. U., & Kiennemann, A. (2011). Iron and nickel doped alkaline earth catalysis for biomass gasification with simultenous for reformation and CO2 capture. *International Journal of Hydrogen Energy*, *36*(9), 5296–5310. doi:10.1016/j.ijhydene.2011.02.008

Fellenz, N. A., Marchetti, S. G., Bengoa, J. F., Mercader, R. C., & Stewart, S. J. (2006). Synthesis and magnetic characterization of magnetite particles embedded in mesoporous MCM-41. *J. Magn. Mater.*, *306*(1), 30–34. doi:10.1016/j.jmmm.2006.02.101

Feng, N., Guo, X., Liang, S., Zhu, Y., & Liu, J. (2011). Biosorption of heavy metals from aqueous solutions by chemically modified orange peel. *Journal of Hazardous Materials*, *185*(1), 49–54. doi:10.1016/j.jhazmat.2010.08.114 PMID:20965652

Figueiredo, S. A., Loureiro, J. M., & Boaventura, R. A. (2005). Natural waste materials containing chitin as adsorbents for textile dyestuffs: Batch and continuous studies. *Water Research*, *39*(17), 4142–4152. doi:10.1016/j.watres.2005.07.036 PMID:16140355

Fisher, M. (2003). Plastics recycling. In A. Andrady (Ed.), *Plastics and the environment* (pp. 563–627). Hoboken, NJ: Wiley Interscience.

Förstner, U., & Wittmann, G. T. W. (1981). *Metal Pollution in the Aquatic Environment*. Berlin: Springer-Verlag. doi:10.1007/978-3-642-69385-4

Fry, A. (2005). Water: Facts and trends: WBCSD.

Fungaro, A., Yamaura, M., & Carvalho, T. E. M. (2011). Adsorption of anionic dyes from aqueous solution on zeolite from fly ash-iron oxide magnetic nanocomposite. *J. At. Mol. Sci.*, *2*(4), 305–316.

Furst, A., Lederer, E. W. H., Fensterheim, R. J., & Reinhold, V. N. (1983). Arsenic: Industrial, Biomedical, Environmental Prospectives. New York: Academic Press.

Gabriele, C., Paola, L., & Siglinda, P. (2011). Analysis of the alternative routes in the catalytic transformational lignocellulosic materials. *Catalysis Today*, *167*(1), 14–30. doi:10.1016/j.cattod.2010.10.099

Gamage, A., & Shahidi, F. (2007). Use of chitosan for the removal of metal ion contaminants and proteins from water. *Food Chemistry*, *104*(3), 989–996. doi:10.1016/j.foodchem.2007.01.004

Gandhi, M. R., & Meenakshi, S. (2012). Preparation and characterization of silica gel/chitosan composite for the removal of Cu(II) and Pb(II). *International Journal of Biological Macromolecules*, *50*(3), 650–657. doi:10.1016/j.ijbiomac.2012.01.012 PMID:22281115

Gandhi, M. R., Viswanathan, N., & Meenakshi, S. (2010). Preparation and application of alumina/chitosan biocomposite. *International Journal of Biological Macromolecules*, *47*(2), 146–154. doi:10.1016/j.ijbiomac.2010.05.008 PMID:20471415

Gan, Q., Wang, T., Cochrane, C., & McCarron, P. (2005). Modulation of surface charge, particle size and morphological properties of chitosan–TPP nanoparticles intended for gene delivery. *Colloids and Surfaces. B, Biointerfaces*, *44*(2), 65–73. doi:10.1016/j.colsurfb.2005.06.001 PMID:16024239

Gao, J., Yu, W., & Pan, N. (2007). Structures and properties of the goose down as a material for thermal insulation. *Textile Research Journal*, *77*(8), 617–626. doi:10.1177/0040517507079408

Gao, Q., Luo, J., Wang, X., Gao, C., & Ge, M. (2015). Novel hollow α-Fe_2O_3 nanofibers via electrospinning for dye adsorption. *Nanoscale Research Letters*, *10*(1), 1–8. doi:10.1186/s11671-015-0874-7 PMID:25918495

Gao, W., Majumder, M., Alemany, L. B., Narayanan, T. N., Ibarra, M. A., Pradhan, B. K., & Ajayan, P. M. (2011). Engineered graphite oxide materials for application in water purification. *ACS Applied Materials & Interfaces*, *3*(6), 1821–1826. doi:10.1021/am200300u PMID:21568266

Gao, Y., & Liu, H. (2005). Preparation and catalytic property study of a novel kind of suspended photocatalyst of TiO2-activated carbon immobilized on silicon ribber film. *Materials Chemistry and Physics*, *92*(2-3), 604–608. doi:10.1016/j.matchemphys.2005.02.018

Gao, Z., Li, W., Liu, B., Liang, F., He, H., Yang, S., & Sun, C. (2011). Nano-structured polyaniline-ionic liquid composite film coated steel wire for headspace solid-phase microextraction of organochlorine pesticides in water. *Journal of Chromatography. A*, *1218*(37), 6285–6291. doi:10.1016/j.chroma.2011.07.041 PMID:21821255

Gautam, A., Kshirsagar, A., Biswas, R., Banerjee, S., & Khanna, P. K. (2016). Photodegradation of organic dyes based on anatase and rutile TiO_2 nanoparticles. *RSC Advances*, *6*(4), 2746–2759. doi:10.1039/C5RA20861K

Gautam, R. K., Mudhoo, A., Lofrano, G., & Chattopadhyaya, M. C. (2014). Biomass-derived biosorbents for metal ions sequestration: Adsorbent modification and activation methods and adsorbent regeneration. *Journal of Environmental Chemical Engineering*, *2*(1), 239–259. doi:10.1016/j.jece.2013.12.019

Gautam, R. K., Rawat, V., Banerjee, S., Sanroman, M. A., Soni, S., Singh, S. K., & Chattopadhyaya, M. C. (2015). Synthesis of bimetallic Fe–Zn nanoparticles and its application towards adsorptive removal of carcinogenic dye malachite green and Congo red in water. *Journal of Molecular Liquids*, *212*, 227–236. doi:10.1016/j.molliq.2015.09.006

Gayathri, S., Jayabal, P., Kottaisamy, M., & Ramakrishnan, V. (2015). Synthesis of the graphene-$ZnTiO_3$ nanocomposite for solar light assisted photodegradation of methylene blue. *Journal of Physics. D, Applied Physics*, *48*(41), 415305. doi:10.1088/0022-3727/48/41/415305

Geboers, J. A., Vyver, S. V., Ooms, R., Beeck, B. O., Jacobs, P. A., & Sels, B. F. (2011). Chemocatalytic conversion of cellulose: Opportunities, advances and pitfalls. *Catal Sci Technol.*, *1*(5), 714–720. doi:10.1039/c1cy00093d

Ge, F., Li, M., Ye, H., & Zhao, B. (2012). Effective removal of heavy metal ions Cd^{2+}, Zn^{2+}, Pb^{2+}, Cu^{2+} from aqueous solution by polymer-modified magnetic nanoparticles. *Journal of Hazardous Materials, 211–212*, 366–372. doi:10.1016/j.jhazmat.2011.12.013 PMID:22209322

Ghasemi, Z., Seif, A., Ahmadi, T. S., Zargar, B., Rashidi, F., & Rouzbahani, G. M. (2012). Thermodynamic and kinetic studies for the adsorption of Hg (II) by nano-TiO 2 from aqueous solution. *Advanced Powder Technology, 23*(2), 148–156. doi:10.1016/j.apt.2011.01.004

Ghasemzadeh, G., Momenpour, M., Omidi, F., Hosseini, M. R., Ahani, M., & Barzegari, A. (2014). Applications of nanomaterials in water treatment and environmental remediation. *Frontiers of Environmental Science & Engineering, 8*(4), 471–482. doi:10.1007/s11783-014-0654-0

Ghosh, S., Badruddoza, A.Z., Hidajat, K. & Uddin, M.S. (2013). Adsorptive removal ofemerging contaminants from water using superparamagnetic Fe_3O_4 nanoparticles bearing aminated-cyclodextrin. *J. Environ. Chem. Eng.*, 122–130.

Gilpin, R., Wagel, D., & Solch, J. (2003). Production, distribution, and fate of polycholorinated dibenzo-p-dioxins, dibenzofurans, and related organohalogens in the environment. In A. Schecter & T. Gasiewicz (Eds.), *Dioxins and health* (2nd ed.). Hoboken, NJ: John Wiley & Sons Inc.

Giraldo, L., Erto, A., & Moreno-Piraján, J. C. (2013). Magnetite nanoparticles for removal of heavy metals from aqueous solutions: Synthesis and characterization. *Adsorption, 19*(2-4), 465–474. doi:10.1007/s10450-012-9468-1

Girginova, P. I., Daniel-da-Silva, A. L., Lopes, C. B., Figueira, P., Otero, M., Amaral, V. S., & Trindade, T. et al. (2010). Silica coated magnetite particles for magnetic removal of Hg^{2+} from water. *Journal of Colloid and Interface Science, 345*(2), 234–240. doi:10.1016/j.jcis.2010.01.087 PMID:20202646

Girisuta, B., Dussan, K., Haverty, D., Leahy, J. J., & Hayes, M. H. B. (2013). A kinetic study of acid catalysed hydrolysis of sugar cane bagasse to levulinic acid. *Chemical Engineering Journal, 217*, 61–70. doi:10.1016/j.cej.2012.11.094

Gkementzoglou, C., Kotrotsiou, O., Koronaiou, M., & Kiparissides, C. (2016). Development of a sandwich-type filtration unit packed with MIP nanoparticles for removal of atrazine from water source. *Chemical Engineering Journal, 287*, 233–240. doi:10.1016/j.cej.2015.11.018

Gohari, R. J., Lau, W. J., Matsuura, T., Halakoo, E., & Ismail, A. F. (2013). Adsorptive removal of Pb(II) from aqueous solution by novel PES/HMO ultrafiltration mixed matrix membrane. *Separation and Purification Technology, 120*, 59–68. doi:10.1016/j.seppur.2013.09.024

Gondal, M. A., Sayeed, M. N., Yamani, Z. H., & Al-Arfaj, A. R. (2009). Efficient removal of phenol from water using Fe_2O_3 semiconductor catalyst under UV laser irradiation. *Journal of Environmental Science and Health. Part A, 44*(5), 515–521. doi:10.1080/10934520902720033

Gonzalez, C. M., Hernandez, J., Parsons, J. G., & Gardea-Torresdey, J. L. (2010). A study of the removal of selenite and selenate from aqueous solutions using a magnetic iron/manganese oxide nanomaterial and ICP-MS. *Microchemical Journal*, *96*(2), 324–329. doi:10.1016/j.microc.2010.05.005

Goon, I. Y., Zhang, C., Lim, M., Gooding, J. J., & Amal, R. (2010). Controlled fabrication of polyethylenimine-functionalized magnetic nanoparticles for the sequestration and quantification of free Cu2+. *Langmuir*, *26*(14), 12247–12252. doi:10.1021/la101196r PMID:20527930

Guan, G., Chen, G., Kasai, Y., Lim, E. W. C., Hao, X., Kaewpanha, M., & Tsutmi, A. et al. (2012). Catalytic stream reforming of biomass tar over iron and nickel-based catalyst supported on calcinaed scallop shell. *Applied Catalysis B: Environmental*, *115*, 159–168. doi:10.1016/j.apcatb.2011.12.009

Guibal, E., & Roussy, J. (2007). Coagulation and flocculation of dye-containing solutions using a biopolymer (Chitosan). *Reactive & Functional Polymers*, *67*(1), 33–42. doi:10.1016/j.reactfunctpolym.2006.08.008

Guibal, E., Sweeney, N. V. O., Vincent, T., & Tobin, J. M. (2002). Sulfur derivatives of chitosan for palladium sorption. *Reactive & Functional Polymers*, *50*(2), 149–163. doi:10.1016/S1381-5148(01)00110-9

Guibal, E., Van Vooren, M., Dempsey, B. A., & Roussy, J. (2006). A review of the use of chitosan for the removal of particulate and dissolved contaminants. *Separation Science and Technology*, *41*(11), 2487–2514. doi:10.1080/01496390600742807

Guisnet, M. (2013). Ideal bifunctional catalysis over Pt acid zeolites. *Catalysis Today*, *218*, 123–134. doi:10.1016/j.cattod.2013.04.028

Gunawan, P., Guan, C., Song, X., Zhang, Q., Leong, S. S. J., Tang, C., & Xu, R. et al. (2011). Hollow fiber membrane decorated with Ag/MWNTs: Toward effective water disinfection and biofouling control. *ACS Nano*, *5*(12), 10033–10040. doi:10.1021/nn2038725 PMID:22077241

Gupta, V. K., & Imran, A. M. (2004). Adsorbents for water treatment: development of low-cost alternatives to carbon. Encyclopedia of Surface and Colloid Science, 2004 Update Supplement, 5, 1.

Gupta, V., Jain, R., Mittal, A., Saleh, T. A., & Nayak, A. (2012). Photo-catalytic degradation of toxic dye amaranth on TiO 2/UV in aqueous suspensions. *Materials Sci. Engineer.*, *C 32*, 12-17.

Gupta, V. K., Agarwal, S., & Saleh, T. A. (2011). Synthesis and characterization of alumina-coated carbon nanotubes and their application for lead removal. *Journal of Hazardous Materials*, *185*(1), 17–23. doi:10.1016/j.jhazmat.2010.08.053 PMID:20888691

Gupta, V. K., Ali, I., Saleh, T. A., Nayak, A., & Agarwal, S. (2012). Chemical treatment technologies for waste-water recycling—an overview. *RSC Advances*, *2*(16), 6380–6388. doi:10.1039/c2ra20340e

Gupta, V. K., & Nayak, A. (2012). Cadmium removal and recovery from aqueous solutions by novel adsorbents prepared from orange peel and Fe_2O_3 nanoparticles. *Chemical Engineering Journal, 180*, 81–90. doi:10.1016/j.cej.2011.11.006

Gupta, V. K., & Saleh, T. A. (2013). Sorption of pollutants by porous carbon, carbon nanotubes and fullerene—an overview. *Environmental Science and Pollution Research International, 20*(5), 2828–2843. doi:10.1007/s11356-013-1524-1 PMID:23430732

Gupta, V., Agarwal, S., & Saleh, T. A. (2011). Chromium removal by combining the magnetic properties of iron oxide with adsorption properties of carbon nanotubes. *Water Research, 45*(6), 2207–2212. doi:10.1016/j.watres.2011.01.012 PMID:21303713

Gu, Q., Long, J., Fan, L., Chen, L., Zhao, L., Lin, H., & Wang, X. (2013). Single site Sn-grafted Ru/TiO2 photocatalysis for biomass reforming synergetic effect of dual co-catalyst and molecular mechanism. *Journal of Catalysis, 30*, 141–155. doi:10.1016/j.jcat.2013.03.014

Gurbuz, E., Bond, J. Q., Domesic, J. A., & Roman-Leshkov, Y. (2013). Role of acid catalysis in the conversion of lignocellulosic biomass to fuels and chemicals. In The role of catalysis for the sustainable production of bio-fuels and bio-chemicals, (pp. 261-288). Academic Press.

Hackman, R. (1954). Studies on chitin I. Enzymic degradation of chitin and chitin esters. *Australian Journal of Biological Sciences, 7*(2), 168–178. doi:10.1071/BI9540168 PMID:13189822

Hackman, R. H., & Goldberg, M. (1974). Light-scattering and infrared-spectrophotometric studies of chitin and chitin derivatives. *Carbohydrate Research, 38*, 35–45. doi:10.1016/S0008-6215(00)82336-8

Haixin, G., Xinhua, Q., Luyang, L., & Richard, L. S. J. (2012). Hydrolysis of cellulose over functionalized glucose-derived carbon catalyst in ionic liquid. *Bioresource Technology, 116*, 355–359. doi:10.1016/j.biortech.2012.03.098 PMID:22522013

Haladu, S., Muhammad, A. M., Saleh, T. A., & Ali, S. A. (2016). A., A. M. Muhammad, T. A. Saleh and S. A. Ali, "Synthesis of novel cross-linked cyclopolymer bearing polyzwitterion-dianionic moieties and its sorption efficiency for Ni(II) removal from waters. *Chemical Engineering Research & Design, 106*, 337–346. doi:10.1016/j.cherd.2015.12.021

Haldorai, Y., & Shim, J.-J. (2013). Multifunctional chitosan-copper oxide hybrid material: Photocatalytic and antibacterial activities. *International Journal of Photoenergy.*

Haldorai, Y., & Shim, J.-J. (2014). An efficient removal of methyl orange dye from aqueous solution by adsorption onto chitosan/MgO composite: A novel reusable adsorbent. *Applied Surface Science, 292*, 447–453. doi:10.1016/j.apsusc.2013.11.158

Hamadi, N. K., Chen, X. D., Farid, M. M., & Lu, M. G. (2001). Adsorption kinetics for the removal of chromium (VI) from aqueous solution by adsorbents derived from used tyres and sawdust. *Chemical Engineering Journal, 84*(2), 95–105. doi:10.1016/S1385-8947(01)00194-2

Hameed, A., Montini, T., Gombac, V., & Fornasiero, P. (2009). Photocatalytic decolourization of dyes on NiO–ZnO nano-composites. *Photochemical & Photobiological Sciences*, *8*(5), 677–682. doi:10.1039/b817396f PMID:19424542

Hamer, U., Marschner, B., Brodowski, S., & Amelong, W. (2004). Interactive priming of black carbon and glucose mineralisation. *Organic Geochemistry*, *36*(7), 823–830. doi:10.1016/j.orggeochem.2004.03.003

Han, J., & Li, C. J. (2013). Characterization of TiO2/ACF prparation in caprolactom tetrabutyl ammonium bromide ionic liquid. *Adv Mater Res.*, *750*, 1864–1867. doi:10.4028/www.scientific.net/AMR.750-752.1864

Han, J., Liu, Y., Singhal, N., Wang, L., & Gao, W. (2012). Comparative photocatalytic degradation of estrone in water by ZnO and TiO_2 under artificial UVA and solar irradiation. *Chemical Engineering Journal*, *213*, 150–162. doi:10.1016/j.cej.2012.09.066

Hara, M. (2010). Biomass conversion by a solid acid catalyst. *Energy Envi Sci.*, *3*(5), 601–607. doi:10.1039/b922917e

Hasan, S., Ghosh, A., Race, K., Schreiber, R. Jr, & Prelas, M. (2014). Dispersion of FeOOH on Chitosan Matrix for Simultaneous Removal of As(III) and As(V) from Drinking Water. *Separation Science and Technology*, *49*(18), 2863–2877. doi:10.1080/01496395.2014.949774

Hasany, S.F., Ahmed, I., Rajin, I. & Rehman, A. (2012). Systematic Review of the preparation techniques of iron oxide magnetic nanoparticles. *Nanoscience Nanotechnology*, *2*, 148-158

Hashem, F. (2012). Adsorption of methylene blue from aqueous solutions using Fe_3O_4/bentonite nanocomposite. *Hydrol Current Res*, *3*(5), 1–6.

Hashemian, S., Saffari, H., & Ragabion, S. (2015). Adsorption of cobalt(II) from aqueous solutions by Fe_3O_4/bentonite nanocomposite. *Water, Air, and Soil Pollution*, *226*(1), 1–10. doi:10.1007/s11270-014-2212-6

Hay. (1873). *Patent No. 140*. Washington, DC: US Patent Office.

Hayat, K., Gondal, M., Khaled, M. M., Ahmed, S., & Shemsi, A. M. (2011). Nano ZnO synthesis by modified sol gel method and its application in heterogeneous photocatalytic removal of phenol from water. *Applied Catalysis A, General*, *393*(1), 122–129. doi:10.1016/j.apcata.2010.11.032

Hayes, W. J., Hayes, W., & Lawes, E. R. (1991). *Handbook of Pesticides Toxicology*. San Diego, CA: Academic Press.

Heinze, T., & Schaller, J. (2000). New water soluble cellulose esters synthesized by on effective acylation procedure. *Macromolecular Chemistry and Physics*, *201*(12), 1214–1218. doi:10.1002/1521-3935(20000801)201:12<1214::AID-MACP1214>3.0.CO;2-9

He, L., Liu, Y., Mustapha, A., & Lin, M. (2011). Antifungal activity of zinc oxide nanoparticles against *Botrytis cinerea* and *Penicillium expansum*. *Microbiological Research*, *166*(3), 207–215. doi:10.1016/j.micres.2010.03.003 PMID:20630731

He, M., & Xia, H. (2012). Photocatalytic performance of activated carbon fibers (ACFs) supported La-S/TiO2 photocatalyst. *Adv Mater Res.*, *518*(523), 764–767. doi:10.4028/www.scientific.net/AMR.518-523.764

Heredia, A., Jimenez, A., & Guillen, R. (1995). Composition of plant cell walls. *Zeitschrift fur Lebensmittel-Untersuchung und -Forschung*, *200*(1), 24–31. doi:10.1007/BF01192903 PMID:7732730

Hertz-Picciotto, I. (2000). The evidence that lead increases the risk for spontaneous abortion. *American Journal of Industrial Medicine*, *38*(3), 300–309. doi:10.1002/1097-0274(200009)38:3<300::AID-AJIM9>3.0.CO;2-C PMID:10940968

He, S., Li, D., Zhu, C., Song, S., Wang, L., Long, Y., & Fan, C. (2008). A graphene-based fluorescent nanoprobe for silver(I) ions detection by using graphene oxide and a silver-specific oligonucleotide. *Chemical Communications (Cambridge)*, *46*, 4885–4887. doi:10.1039/b811528a

Hokkanen, S., Bhatnagar, A., & Sillanpää, M. A. (2016). A review on modification methods to cellulose-based adsorbents to improve adsorption capacity. *Water Research*, *91*, 156–173. doi:10.1016/j.watres.2016.01.008 PMID:26789698

Homaeigohar, S., & Elbahri, M. (2014). Nanocomposite electrospun nanofiber membranes for environmental remediation. *Materials (Basel)*, *7*(2), 1017–1045. doi:10.3390/ma7021017

Hopewell, J., Dvorak, R., & Kosior, E. (2009). Plastics recycling: Challenges and opportunities. *Phil. Trans. R. Soc. B*, *364*(1526), 2115–2126. doi:10.1098/rstb.2008.0311 PMID:19528059

Horowitz, S. T., Roseman, S., & Blumenthal, H. J. (1957). The Preparation of Glucosamine Oligosaccharides. I. Separation1, 2. *Journal of the American Chemical Society*, *79*(18), 5046–5049. doi:10.1021/ja01575a059

Howard, G. T. (2002). Biodegredation of polyurethane: A review. *International Biodeterioration & Biodegradation*, *49*(1), 245–252. doi:10.1016/S0964-8305(02)00051-3

Hristodor, C. M., Cochecl, L., Pode, R., Copcia, V. E., & Popovici, E. (2011). The Removal of p-cloro-phenol from synthetic waste water on Fe_2O_3-Clays nanocomposites ystems. *REVISTA DE CHIMIE*, *62*(11), 1119–1122.

Hsi, H.-C., & Tsai, C.-Y. (2012). Preparation of oxygen-vacant TiO2-x and activated carbon fiber composite using a single step thermal plasma method for low-concentration elemental mercury removal. *Chemical Engineering Journal*, *200*, 18–24. doi:10.1016/j.cej.2012.06.021

Hsu, W.-H., Lee, Y.-Y., Peng, W.-H., Kelven, C., & Wu, C.-W. (2011). Cellulosic conversion in ionic liquids (Ils): Effects of H2O/ cellulose molr ratios, temperature, times and different Ils on the production of monosaccharides and 5-hydroxymethylfurfural (HMF). *Catalysis Today*, *174*(1), 65–69. doi:10.1016/j.cattod.2011.03.020

Hua, M., Zhang, S., Pan, B., Zhang, W., Lv, L., & Zhang, Q. (2012). Heavy metal removal from water/wastewater by nanosized metal oxides: A review. *Journal of Hazardous Materials*, *211*, 317–331. doi:10.1016/j.jhazmat.2011.10.016 PMID:22018872

Huang, C., & Huang, C. P. (1996). Application of Aspergillus oryzae and Rhizopus oryzae for Cu(II) removal. *Water Research*, *30*(9), 1985–1990. doi:10.1016/0043-1354(96)00020-6

Huang, D., Miyamoto, Y., Ding, J., Gu, J., Zhu, S., Liu, Q., & Zhang, D. et al. (2011). A new method prepared high surface-area N-TiO2/activated carbon. *Materials Letters*, *65*(2), 326–328. doi:10.1016/j.matlet.2010.10.025

Huang, D., Miyamoto, Y., Matsumoto, T., Tojo, T., Fan, T., Ding, J., & Zhang, D. et al. (2011). Preparation and characterization of high-surface-area TiO2/activated carbon by low-temperature impregnation. *Separation and Purification Technology*, *78*(1), 9–15. doi:10.1016/j.seppur.2011.01.010

Huang, J., Xia, C., Cao, L., & Zeng, X. (2008). Facile microwave hydrothermal synthesis of zinc oxide one-dimensional nanostructure with three-dimensional morphology. *Materials Science and Engineering B*, *150*(3), 187–193. doi:10.1016/j.mseb.2008.05.014

Huang, S.-H., & Chen, D.-H. (2009). Rapid removal of heavy metal cations and anions from aqueous solutions by an amino-functionalized magnetic nano-adsorbent. *Journal of Hazardous Materials*, *163*(1), 174–179. doi:10.1016/j.jhazmat.2008.06.075 PMID:18657903

Huang, X., Qi, X., Boey, F., & Zhang, H. (2012). Graphene-based composites. *Chemical Society Reviews*, *41*(2), 666–686. doi:10.1039/C1CS15078B PMID:21796314

Hu, C., Lan, Y., Qu, J., Hu, X., & Wang, A. (2006). Ag/AgBr/TiO_2 visible light photocatalyst for destruction of azodyes and bacteria. *The Journal of Physical Chemistry B*, *110*(9), 4066–4072. doi:10.1021/jp0564400 PMID:16509698

Hui, K. S., Chao, C. Y. H., & Kot, S. C. (2005). Removal of mixed heavy metal ions in wastewater by zeolite 4A and residual products from recycled coal fly ash. *Journal of Hazardous Materials*, *127*(1-3), 89–101. doi:10.1016/j.jhazmat.2005.06.027 PMID:16076523

Hu, J., Chen, G. H., & Lo, I. M. C. (2005). Removal and recovery of Cr(VI) from wastewater by maghemite nanoparticles. *Water Research*, *39*(18), 4528–4536. doi:10.1016/j.watres.2005.05.051 PMID:16146639

Hu, J., Chen, G., & Lo, I. M. (2006). Selective removal of heavy metals from industrial wastewater using maghemite nanoparticle: Performance and mechanisms. *Journal of Environmental Engineering*, *132*(7), 709–715. doi:10.1061/(ASCE)0733-9372(2006)132:7(709)

Hummers, W. Jr, & Offeman, R. (1958). Preparation of graphitic oxide. *Journal of the American Chemical Society*, *80*(6), 1339. doi:10.1021/ja01539a017

Hur, J., Shin, J., Yoo, J., & Seo, Y.-S. (2015). Competitive adsorption of metals onto magnetic graphene oxide: Comparison with other carbonaceous adsorbents. *The Scientific World Journal*. PMID:25861683

Hussein, A.-Y., El Barbary, H., & Philip, S. (2013). Rapid conversion of cellulose to 5-hydroxymethylfurfural using single and combined metal chloride catalyst in ionic liquid. *J Fuel Chem & Tech.*, *41*(2), 214–222. doi:10.1016/S1872-5813(13)60013-4

Hutton, G., Haller, L., & Bartram, J. (2007). *Economic and Health Effects of Increasing Coverage of Low Cost Household Drinking Water Supply and Sanitation Intervention*. World Health Organization.

Hutton, Haller, & Bartram. (2007). *Economic and Health Effects of Increasing Coverage of Low Cost Household Drinking Water Supply and Sanitation Interventions*. World Health Organization.

Hu, X. D., Deng, H., & Du, L. (2011). Preparation and characterization of Fe^{3+}/TiO_2 thin films loaded activated carbon and degradation of methyl orange. *Advanced Materials Research*, *332-334*, 134–137. doi:10.4028/www.scientific.net/AMR.332-334.134

Idris, A., Ismail, N. S., Hassan, N., Misran, E., & Ngomsik, A. F. (2012). Synthesis of magnetic alginate beads based on maghemite nanoparticles for Pb(II) removal in aqueous solution. *Journal of Industrial and Engineering Chemistry*, *18*(5), 1582–1589. doi:10.1016/j.jiec.2012.02.018

Iguchi, M., Aida, T. M., Watanabe, M., & Smith, R. L. J. Jr. (2013). Dissolution and recovery of cellulose from 1-butyl-3-methylimidazolium chloride in presence of water. *Carbohydrate Polymers*, *92*(1), 651–658. doi:10.1016/j.carbpol.2012.09.021 PMID:23218349

Igwe, J. C., & Abia, A. A. (2007). Adsorption kinetics and intraparticulate diffusivities for bioremediation of Co(II), Fe(II) and Cu(II) ions from waste water using modified and unmodified maize cob. *International Journal of Physical Sciences*, *2*(5), 119–127.

Ihsanullah, A. (2015). Heavy metal removal from aqueous solution by advanced carbon nanotubes: Critical review of adsorption applications. *Separation and Purification Technology*, *157*(8), 141–161.

Iijima, S. (1991). Helical microtubules of graphitic carbon. *Nature*, *354*(6348), 56–58. doi:10.1038/354056a0

Ilhan, S., Nurbas, M., & Nourbaksh, S. (2004). Removal of chromium, lead and copper ions from industrial waste water by *Staphylococcus Saprophyticus*. *Turkish Electronics Journal of Biotechnology*, *2*, 50–57.

Independent Statistics and Analysis, US Energy Information Administration. (2014). *Frequently Asked Questions: How much oil is used to make plastic?*. Retrieved from http://www.eia.gov/tools/faqs/faq.cfm?id=34&t=6

Indira, T. K., & Lakshmi, P. K. (2010). Magnetic nanoparticles. A review. *International Journal of Pharmaceutical Sciences and Nanotechnology*, *3*, 1035–1042.

Ishikawa, S., Sekine, S., Miura, N., Suyama, K., Arihara, K., & Itoh, M. (2004). Removal of selenium and arsenic by animal biopolymers. *Biological Trace Element Research*, *102*(1), 113–127. doi:10.1385/BTER:102:1-3:113 PMID:15621933

Ishikawa, T., Kataoka, S., & Kandori, K. (1993). The influence of carboxylated ions on the growth of β-FeOOH particles. *Journal of Materials Science*, *28*(10), 2693–2698. doi:10.1007/BF00356205

Ismail, R. A., Sulaiman, G. M., Abdulrahman, S. A., & Marzoog, T. R. (2015). Antibacterial activity of magnetic iron oxide nanoparticles synthesized by laser ablation in liquid. *Materials Science and Engineering C*, *53*, 286–297. doi:10.1016/j.msec.2015.04.047 PMID:26042717

Ito, A., Sato, T., & Goto, T. (2014). Transparent anatase and rutile TiO2 films grown by laser chemical vapor deposition. *Thin Solid Films*, *551*, 37–41. doi:10.1016/j.tsf.2013.11.089

Izakmehri, Z., Ardjmand, M., Ganji, M. D., Babanezhad, E., & Heydarinasab, A. (2015). Removal of dioxane pollutants from water by using Al-doped single walled carbon nanotubes. *RSC Advances*, *5*(59), 48124–48132. doi:10.1039/C5RA07611K

Jagtap, S., Thakre, D., Wanjari, S., Kamble, S., Labhsetwar, N., & Rayalu, S. (2009). New modified chitosan-based adsorbent for defluoridation of water. *Journal of Colloid and Interface Science*, *332*(2), 280–290. doi:10.1016/j.jcis.2008.11.080 PMID:19181328

Jamshidi, M., Ghaedi, M., Dashtian, K., Hajati, S., & Bazrafshan, A. (2016). Sonochemical assisted hydrothermal synthesis of ZnO: Cr nanoparticles loaded activated carbon for simultaneous ultrasound-assisted adsorption of ternary toxic organic dye: Derivative spectrophotometric, optimization, kinetic and isotherm study. *Ultrasonics Sonochemistry*, *32*, 119–131. doi:10.1016/j.ultsonch.2016.03.004 PMID:27150752

Jardim, W. F., Moraes, S. G., & Takiyama, M. M. K. (1997). Photocatalytic degradation of aromatic chlorinated compounds using TiO2: Toxicity of intermediates. *Water Research*, *31*(7), 1728–1732. doi:10.1016/S0043-1354(96)00349-1

Javadian, H. (2014). Application of kinetic, isotherm and thermodynamic models for the adsorption of Co(II) ions on polyaniline/polypyrrole copolymer nanofibers from aqueous solution. *Journal of Industrial and Engineering Chemistry*, *20*(6), 4233–4241. doi:10.1016/j.jiec.2014.01.026

Javanbakht, T., Laurent, S., Stanicki, D., & Wilkinson, K. J. (2016). Relating the surface properties of superparamagnetic iron oxide nanoparticles (SPIONs) to their bactericidal effect towards a biofilm of *Streptococcus mutans*. *PLoS ONE*, *11*(4), e0154445. doi:10.1371/journal.pone.0154445 PMID:27115356

Jawor, A., & Hoek, E. M. V. (2010). Removing cadmium ions from water via nanoparticle-enhanced ultrafiltration. *Environmental Science & Technology*, *44*(7), 2570–2576. doi:10.1021/es902310e PMID:20230021

Jayarathna, L., Bandara, A., Ng, W. J., & Weerasooriya, R. (2015). Fluoride adsorption on $\gamma - Fe_2O_s$ nanoparticles. *Journal of Environmental Health Science and Engineering*, *13*(1), 1–10. doi:10.1186/s40201-015-0210-2 PMID:25763187

Jegadeesan, G., Al-Abed, S. A., Sundaram, V., Choi, H., Scheckel, K. G., & Dionysiou, D. D. (2010). Arsenic sorption on TiO_2 nanoparticles: Size and crystallinity effects. *Water Research*, *44*(3), 965–973. doi:10.1016/j.watres.2009.10.047 PMID:20022353

Jeon, I.-Y., & Baek, J.-B. (2010). Nanocomposites derived from polymers and inorganic nanoparticles. *Materials (Basel)*, *3*(6), 3654–3674. doi:10.3390/ma3063654

Jiang, H., Chen, P., Luo, S., Luo, X., Tu, X., Cao, Q., & Zhang, W. et al. (2013). Synthesis of novel biocompatible composite Fe_3O_4/ZrO_2/chitosan and its application for dye removal. *Journal of Inorganic and Organometallic Polymers and Materials*, *23*(2), 393–400. doi:10.1007/s10904-012-9792-7

Jiang, W., Mashayekhi, H., & Xing, B. (2009). Bacterial toxicity comparison between nano- and micro-scaled oxide particles. *Environmental Pollution*, *157*(5), 1619–1625. doi:10.1016/j.envpol.2008.12.025 PMID:19185963

Jian, W., Kitanaka, A., Nishijima, W., Baes, A. U., & Okada, M. (1999). Removal of oil pollutants in seawater as pretreatment of reverse osmosis desalination process. *Water Research*, *33*(8), 1857–1863. doi:10.1016/S0043-1354(98)00404-7

Ji, L., Bai, X., Zhou, L., Shi, H., Chen, W., & Hua, Z. (2013). One-pot preparation of graphene oxide magnetic nanocomposites for the removal of tetrabromobisphenol A. *Front. Environ. Sci. Eng.*, *7*(3), 442–450. doi:10.1007/s11783-013-0515-2

Jimenez, I. O., Lopez, X., Arbiol, J., & Puntes, V. (2012). Citrate-coated gold nanoparticles as smart scavengers for mercury(II) removal from polluted waters. *ACS Nano*, *6*(3), 2253–2260. doi:10.1021/nn204313a PMID:22332645

Jimenez-Morales, I., Santamaria-Gonzalez, J., Maireles-Torres, P., & Jimenez-Lopez, A. (2011). Aluminium doped SBA-15 silica as acid catalyst for the methanolysis of sunflower oil. *Applied Catalysis B: Environmental*, *105*(1-2), 199–205. doi:10.1016/j.apcatb.2011.04.015

Jing-po, Y., Fei, L., Jing, C., & Jin-di, Y. (2015). Sensing and magnetic removal of Hg(II) using core–shell structured nanocomposite grafted with fluorescence Off–On probe Microporous. *Mesoporous Mater.*, *202*, 175–182. doi:10.1016/j.micromeso.2014.09.054

Jin, Y., Deng, J., Liang, J., Shan, C., & Tong, M. (2015). Efficient bacteria capture and inactivation by cetyltrimethylammonium bromide modified magnetic nanoparticles. *Colloids and Surfaces. B, Biointerfaces*, *136*, 659–665. doi:10.1016/j.colsurfb.2015.10.009 PMID:26496472

Jin, Z., Gao, H., & Hu, L. (2015). Removal of Pb (II) by nano-titanium oxide investigated by batch, XPS and model techniques. *RSC Advances*, *5*(107), 88520–88528. doi:10.1039/C5RA14004H

Jokar, M., Foroutani, R., Safaralizadeh, M. H., & Farhadi, K. (2014). Synthesis and characterization of polyaniline/Fe_3O_4 magnetic nanocomposite as practical approach for fluoride removal process. *Annual Research & Review in Biology*, *4*(21), 3262–3273. doi:10.9734/ARRB/2014/9108

José, A., Arsuaga, J. M., Arencibia, A., Lindo, M., & Gascón, V. (2009). Aqueous heavy metals removal by adsorption on amine-functionalized mesoporous silica. *Journal of Hazardous Materials*, *163*(1), 213–221. doi:10.1016/j.jhazmat.2008.06.080 PMID:18675509

Joshua, P., Erin, S., Jessica, H., Brenton, D. B., & Brett, L. L. (2011). Effect of NaCl on the conversion of cellulose of glucose and levulinic acid via solid supported acid cataylst. *Tetrahend Lett.*, *52*(44), 5891–5893. doi:10.1016/j.tetlet.2011.09.013

Jouhannaud, J., Rossignol, J., & Stuerga, D. (2008). Rapid synthesis of tin (IV) oxide nanoparticles by microwave induced thermohydrolysis. *Journal of Solid State Chemistry*, *181*(6), 1439–1444. doi:10.1016/j.jssc.2008.02.040

Juan, C. S.-R., Antonio, P., Alina, M. B., Rafeal, L., Juan, M. C., Antonio, A. R., & Jose, M. R.-F. (2012). Catalytic transformations of biomass-derived acids into advanced biofuels. *Catalysis Today*, *195*(1), 162–268. doi:10.1016/j.cattod.2012.01.009

Juang, R. S., Tseng, R. L., Wu, F. C., & Lee, S. H. (1997). Adsorption behavior of reactive dyes from aqueous solutions on chitosan. *Journal of Chemical Technology and Biotechnology (Oxford, Oxfordshire)*, *70*(4), 391–399. doi:10.1002/(SICI)1097-4660(199712)70:4<391::AID-JCTB792>3.0.CO;2-V

Juang, R. S., Tseng, R. L., Wu, F. C., & Lin, S. J. (1996). Use of chitin and chitosan in lobster shell wastes for color removal from aqueous solutions. *Journal of Environmental Science & Health Part A*, *31*(2), 325–338. doi:10.1080/10934529609376360

Juang, R. S., Wu, F. C., & Tseng, R. L. (2000). Mechanism of adsorption of dyes and phenols from water using activated carbons prepared from plum kernels. *Journal of Colloid and Interface Science*, *227*(2), 437–444. doi:10.1006/jcis.2000.6912 PMID:10873331

Jung, H.-J., Hong, J.-S., & Suh, J.-K. (2013). A comparision of fenton oxidation and photocatalyst reaction efficiency for humic acid degradation. *Journal of Industrial and Engineering Chemistry*, *19*(4), 1325–1330. doi:10.1016/j.jiec.2012.12.036

Kaegi, R., Voegelin, A., Sinnet, B., Zuleeg, S., Hagendorfer, H., Burkhardt, M., & Siegrist, H. (2011). Behavior of metallic silver nanoparticles in a pilot wastewater treatment Plant. *Environmental Science & Technology*, *45*(9), 3902–3908. doi:10.1021/es1041892 PMID:21466186

Kajiwara, N., Hirata, O., Takigami, H., Noma, Y., Tachifuji, A., & Matsufuji, Y. (2014). Leaching of brominated flame retardants from mixed wastes in lysimeters under conditions simulating landfills in developing countries. *Chemosphere*, *116*, 46–53. doi:10.1016/j.chemosphere.2014.01.025 PMID:24560282

Kalantari, K., Ahmad, M. B., Masoumi, H. R. F., Shameli, K., Basri, M., & Khandanlou, R. (2014). Rapid adsorption of heavy metals by Fe_3O_4/talc nanocomposite and optimization study using response surface methodology. *International Journal of Molecular Sciences*, *15*(7), 12913–12927. doi:10.3390/ijms150712913 PMID:25050784

Kamble, S. P., Jagtap, S., Labhsetwar, N. K., Thakare, D., Godfrey, S., Devotta, S., & Rayalu, S. S. (2007). Defluoridation of drinking water using chitin, chitosan and lanthanum-modified chitosan. *Chemical Engineering Journal*, *129*(1), 173–180. doi:10.1016/j.cej.2006.10.032

Kaminsky, W., & Sinn, H. (1996). *Book Recycling and Recovery of Plastics* (J. Brandrup, M. Bittner, W. Michaeli, & G. Menges, Eds.). Munich: Hanser Publishers.

Kangwanwatana, W., Salwan, C., & Tontiwachwuthikul, P. (2013). Study of CO2 adsorption using adsorbent modified with pipercizine. *Chem Eng Trans*, *35*, 403-408.

Kapoor, P. N., Bhagi, A. K., Mulukutla, R. S., & Klabunde, K. J. (2004). *Mixed metal oxide nanoparticle*. Dekker Encyclopedia of Nanoscience and Nanotechnology.

Kappe, C. O. (2004). Controlled microwave heating in modern organic synthesis. *Angewandte Chemie International Edition in English*, *43*(46), 6250–6284. doi:10.1002/anie.200400655 PMID:15558676

Kara, A., Demirbel, E., Tekin, N., Osman, B., & Beşirli, N. (2015). Magnetic vinylphenyl boronic acid microparticles for Cr(VI) adsorption: Kinetic, isotherm and thermodynamic studies. *Journal of Hazardous Materials*, *286*, 612–623. doi:10.1016/j.jhazmat.2014.12.011 PMID:25666882

Karimi, H., Mousavi, S., & Sadeghian, B. (2012). Silver nanoparticle loaded on activated carbon as efficient adsorbent for removal of methyl orange. *Ind. J. Sci. Technol.*, *5*, 2346–2356.

Kar, P., & Misra, M. (2004). Use of keratin fiber for separation of heavy metals from water. *Journal of Chemical Technology and Biotechnology (Oxford, Oxfordshire)*, *79*(11), 1313–1319. doi:10.1002/jctb.1132

Kashinath, L., Namratha, K., & Byrappa, K. (2015). Microwave assisted facile hydrothermal synthesis and characterization of zinc oxide flower grown on graphene oxide sheets for enhanced photodegradation of dyes. *App. Surface Sci.*, *357*, 1849–1856. doi:10.1016/j.apsusc.2015.09.072

Kaur, A., & Gupta, U. (2009). A review on applications of nanoparticles for the preconcentration of environmental pollutants. *J. Mater. Chem. A*, *19*(44), 8279–8289. doi:10.1039/b901933b

Kaur, A., & Gupta, U. (2010). Solid-Phase Extraction of Antimony Using Chemically Modified SiO2-PAN Nanoparticles. *Journal of AOAC International*, *93*, 1302–1307. PMID:20922965

Kaur, M., Kaur, N., Jeet, K., & Kaur, P. (2015). $MgFe_2O_4$ nanoparticles loaded on activated charcoal for effective removal of Cr(VI) – A novel approach. *Ceramics International*, *41*(10), 13739–13750. doi:10.1016/j.ceramint.2015.08.040

Kaur, M., Singh, M., Mukhopadhyay, S. S., Singh, D., & Gupta, M. (2015). Structural, magnetic and adsorptive properties of clay ferrite nanocomposite and its use for effective removal of Cr(VI) from water. *Journal of Alloys and Compounds*, *653*, 202–211. doi:10.1016/j.jallcom.2015.08.265

Kaur, T., Sraw, A., Wanchoo, R. K., & Toor, A. P. (2016). Visible –light induced photocatalytic degradation of fungicide with Fe and Si doped TiO_2 nanoparticles. *Materials Today: Proceedings*, *3*(2), 354–361. doi:10.1016/j.matpr.2016.01.020

Kemp, K. C., Seema, H., Saleh, M., Le, N. H., Mahesh, K., Chandra, V., & Kim, K. S. (2013). Environmental applications using graphene composites: Water remediation and gas adsorption. *Nanoscale*, *5*(8), 3149–3171. doi:10.1039/c3nr33708a PMID:23487161

Khan, M., & Lo, I. M. (2016). Removal of ionizable aromatic pollutants from contaminated water using nano γ-Fe_2O_3 based magnetic cationic hydrogel: Sorptive performance, magnetic separation and reusability. *Journal of Hazardous Materials*. doi:10.1016/j.jhazmat.2016.01.051 PMID:26857338

Khan, N. A., Hasan, Z., & Jhung, S. H. (2013). Adsorptive removal of hazardous materials using metal-organic frameworks (MOFs): A review. *Journal of Hazardous Materials, 244-245*, 444–456. doi:10.1016/j.jhazmat.2012.11.011 PMID:23195596

Khan, S., Suriyaprabha, R., Pathak, B., & Fulekar, M. (2015). Photocatalytic degradation of organophosphate pesticides (Chlorpyrifos) using synthesized zinc oxide nanoparticle by membrane filtration reactor under UV irradiation. *Frontier of Nanoscience and Nanotechnology*, *1*(1), 23–27. doi:10.15761/FNN.1000105

Khatamian, M., & Alaji, Z. (2012). Efficient adsorption-photodegradation of 4-nitrophenol in aqueous solution by using ZnO/HZSM-5 nanocomposites. *Desalination*, *286*, 248–253. doi:10.1016/j.desal.2011.11.031

Kim, S., Kim, J., & Seo, G. (2013). Iron oxide nanoparticle-impregnated powder-activated carbon (IPAC) for NOM removal in MF membrane water treatment system. *Desalination and Water Treatment*, *51*(31-33), 6392–6400. doi:10.1080/19443994.2013.781000

Kim, T. H., Lee, J., & Hong, S. (2009). Highly Selective Environmental Nanosensors Based on Anomalous Response of Carbon Nanotube Conductance to Mercury Ions. *The Journal of Physical Chemistry C*, *113*(45), 19393–19396. doi:10.1021/jp908902k

Kim, Y., Osako, M., & Lee, D. (2002). Removal of hydrophobic organic pollutants by coagulation-precipitation process with dissolved humic matter. *Waste Management & Research*, *20*(4), 341–349. doi:10.1177/0734247X0202000405 PMID:12363094

Klamrassamee, T., Champreda, V., Reunglek, V., & Laosiripojana, N. (2013). Comparison of homogeneous and heterogeneous acid promoters in single-step aqueous-organosolv fractionation of eucalyptus wood chips. *Bioresource Technology*, *147*, 276–284. doi:10.1016/j.biortech.2013.08.015 PMID:24001557

Klemm, D., Heublein, B., Fink, H. F., & Bohn, A. (2005). Cellulose: Fascinating biopolymer and sustainable raw material. *Angewandte Chemie International Edition*, *44*(22), 3358–3393. doi:10.1002/anie.200460587 PMID:15861454

Knop, A., & Pilato, L. A. (1985). *Phenolic Resins Chemistry, Applications and Performance*. Springer-Verlag. doi:10.1007/978-3-662-02429-4

Kohler, A., Hellweg, S., Escher, B. I., & Hungerbuhler, K. (2006). Organic pollutant removal versus toxicity reduction in industrial wastewater treatment: The example of wastewater from fluorescent whitening agent production. *Environmental Science & Technology*, *40*(10), 3395–3401. doi:10.1021/es060555f PMID:16749712

Kojima, N., Mitomo, A., Itaya, Y., Mori, S., & Yoshida, S. (2002). Adsorption removal of pollutants by active cokes produced from sludge in the energy recycle process of wastes. *Waste Management (New York, N.Y.)*, *22*(4), 399–404. doi:10.1016/S0956-053X(02)00022-3 PMID:12099497

Kolthoff, I. M., Elving, P. J., & Stross, F. H. (1971). Treatise on Analytical Chemistry. Wiley Interscience.

Kosynkin, D. V., Higginbotham, A. L., Sinitskii, A., Lomeda, J. R., Dimiev, A., Price, B. K., & Tour, J. M. (2009). Mechanically Assisted Exfoliation and Functionalization of Thermally Converted Graphene Sheets. *Nature*, *458*, 872–876. doi:10.1038/nature07872 PMID:19370030

Kotas, J., & Stasicka, Z. (2000). Chromium occurrence in the environment and methods of its speciation. *Environmental Pollution*, *107*(3), 263–283. doi:10.1016/S0269-7491(99)00168-2 PMID:15092973

Kremsner, J. M., & Kappe, C. O. (2006). Silicon carbide passive heating elements in microwave-assisted organic synthesis. *The Journal of Organic Chemistry*, *71*(12), 4651–4658. doi:10.1021/jo060692v PMID:16749800

Krishnan, D., Raidongia, R., Shao, J., & Huang, J. (2014). Graphene Oxide Assisted Hydrothermal Carbonization of Carbon Hydrates. *ACS Nano*, *8*(1), 449–457. doi:10.1021/nn404805p PMID:24298909

Kumar, G. V., Ali, I., Saleh, T. A., Siddiqui, M. N., & Agarwal, S. (2013). Chromium removal from water by activated carbon developed from waste rubber tires. *Environmental Science and Pollution Research International*, *20*(3), 1261–1268. doi:10.1007/s11356-012-0950-9 PMID:22573097

Kumar, M. N. R. (2000). A review of chitin and chitosan applications. *Reactive & Functional Polymers*, *46*(1), 1–27. doi:10.1016/S1381-5148(00)00038-9

Kumar, N. S., Suguna, M., Subbaiah, M. V., Reddy, A. S., Kumar, N. P., & Krishnaiah, A. (2010). Adsorption of phenolic compounds from aqueous solutions onto chitosan-coated perlite beads as biosorbent. *Industrial & Engineering Chemistry Research*, *49*(19), 9238–9247. doi:10.1021/ie901171b

Kumar, R., & Chawla, J. (2014). Removal of cadmium ion from water/wastewater by nano-metal oxides: A review. *Water Quality. Exposure and Health*, *5*(4), 215–226. doi:10.1007/s12403-013-0100-8

Kumar, S., Kumar, B., Surender, T., & Shanker, V. (2014). G-C3N4/NaTaO3 organic inorganic hybrid nanocomposite: High performance and recyclable visible light driven photocatalyst. *Materials Research Bulletin*, *49*, 310–318. doi:10.1016/j.materresbull.2013.09.013

Kumar, V., Talreja, N., Deva, D., Sankararamakrishnan, N., Sharma, A., & Verma, N. (2011). Development of bi-metal doped micro- and nano multifunctional polymeric adsorbents for the removal of fluoride and arsenic (V) from wastewater. *Desalination*, *282*, 27–38. doi:10.1016/j.desal.2011.05.013

Kuo, C. Y., Wu, C. H., & Chen, S. T. (2014). Decolorization of CI Reactive Red 2 by UV/TiO2/PAC and visible light/TiO2/PAC systems. *Desalination and Water Treatment*, *52*(4-6), 834–843. doi:10.1080/19443994.2013.826409

Kuriacose, J. C., & Rajaram, J. (1987). *Chemistry in Engineering and Technology*. New York: McGraw Hill, Publishing Company Ltd.

Kurita, K., Tomita, K., Tada, T., Ishii, S., Nishimura, S. I., & Shimoda, K. (1993). Squid chitin as a potential alternative chitin source: Deacetylation behavior and characteristic properties. *Journal of Polymer Science. Part A, Polymer Chemistry*, *31*(2), 485–491. doi:10.1002/pola.1993.080310220

Kuyucak, N., & Volesky, B. (1988). Biosorbents for recovery of metals from industrial solutions. *Biotechnology Letters*, *10*(2), 137–142. doi:10.1007/BF01024641

Kuzyakos, Y., Subbotino, I., Chen, H., Bagomolova, I., & Xu, X. (2009). Black carbon decomposition and incorporation into soil microbial biomass estimated by 14C labeling. *Soil Biology & Biochemistry*, *41*(2), 210–219. doi:10.1016/j.soilbio.2008.10.016

Kyzas, G. Z., & Lazaridis, N. K. (2009). Reactive and basic dyes removal by sorption onto chitosan derivatives. *Journal of Colloid and Interface Science*, *331*(1), 32–39. doi:10.1016/j.jcis.2008.11.003 PMID:19026426

Labus, K., Gryglewicz, S., & Machnikowski, J. (2014). Granular KOH-activated carbons from coal based cokes and their CO2 adsorption capacity. *Fuel*, *118*, 9–15. doi:10.1016/j.fuel.2013.10.042

Langa, F., de la Cruz, P., de la Hoz, A., Diaz-Ortiz, A., & Diez-Barra, E. (1997). Microwave irradiation: More than just a method for accelerating reactions. *Contemporary Organic Synthesis*, *4*(5), 373–386. doi:10.1039/CO9970400373

Larraza, I., López-Gónzalez, M., Corrales, T., & Marcelo, G. (2012). Hybrid materials: Magnetite–Polyethylenimine–Montmorillonite, as magnetic adsorbents for Cr(VI) water treatment. *Journal of Colloid and Interface Science*, *385*(1), 24–33. doi:10.1016/j.jcis.2012.06.050 PMID:22841705

Laus, R., Costa, T. G., Szpoganicz, B., & Fávere, V. T. (2010). Adsorption and desorption of Cu (II), Cd (II) and Pb (II) ions using chitosan crosslinked with epichlorohydrin-triphosphate as the adsorbent. *Journal of Hazardous Materials*, *183*(1), 233–241. doi:10.1016/j.jhazmat.2010.07.016 PMID:20674156

Lazaridis, N. K., Kyzas, G. Z., Vassiliou, A. A., & Bikiaris, D. N. (2007). Chitosan derivatives as biosorbents for basic dyes. *Langmuir*, *23*(14), 7634–7643. doi:10.1021/la700423j PMID:17530870

Leaner, V. D., Donninger, H., & Birrer, M. J. (2007). Transcription Factors as Targets for Cancer Therapy: AP-1 a Potential Therapeutic Target. *Current Cancer Therapy Reviews*, *3*(1), 1–6. doi:10.2174/157339407780126665

Lee, J. S., Han, M., & Mirkin, C. A. (2007). Colorimetric detection of mercuric ion (Hg2+) in aqueous media using DNA-functionalized gold nanoparticles. *Angewandte Chemie*, *119*(22), 4171–4174. doi:10.1002/ange.200700269 PMID:17461429

Lee, J.-Y., & Jo, W.-K. (2012). Control of methyl tertiary-butyl ether via carbon-doped photocatalysts under visible light irradiation. *Env Engn Res.*, *17*(4), 179–184. doi:10.4491/eer.2012.17.4.179

Lee, K. M., Lai, C. W., Ngai, K. S., & Juan, J. C. (2016). Recent developments of zinc oxide based photocatalyst in water treatment technology: A review. *Water Research*, *88*, 428–448. doi:10.1016/j.watres.2015.09.045 PMID:26519627

Lee, Y.-F., Nan, F.-H., Chen, M.-J., Wu, H.-Y., Ho, C.-W., Chen, Y.-Y., & Huan, C.-C. (2012). Detection and removal of mercury and lead ions by using gold nanoparticle-based gel membrane. *Anal. Methods*, *4*(6), 1709–1717. doi:10.1039/c2ay05872c

Lemine, O. M., Omri, K., Zhang, B., El Mir, L., Sajieddine, M., Alyamani, A., & Bououdina, M. (2012). Solgel synthesis of 8 nm magnetite (Fe_3O_4) nanoparticles and magnetic properties. *Superlattices and Microstructures*, *52*(4), 793–799. doi:10.1016/j.spmi.2012.07.009

Li, X. M., Xu, G., Liu, Y. & He, T. (2011). Magnetic Fe_3O_4 nanoparticles: Synthesis and application in water treatment. *Nanoscience and Nanotechnology-Asia*, *1*, 14-24.

Liang, J., Huang, Y., Zhang, L., Wang, Y., Ma, Y., Guo, T., & Chen, Y. (2009). Molecular-level dispersion of graphene into poly (vinyl alcohol) and effective reinforcement of their nanocomposites. *Advanced Functional Materials*, *19*(14), 2297–2302. doi:10.1002/adfm.200801776

Liang, P., Qin, Y., Hu, B., Li, C., Peng, T., & Jiang, Z. (2000). Study of the adsorption behavior of heavy metal ions on nanometer-size titanium dioxide with ICP-AES. *Fresenius Journal of Analytical Chemistry*, *368*(6), 638–640. doi:10.1007/s002160000546 PMID:11228717

Liang, P., Shi, T. Q., & Li, J. (2004). Nanometer-size titanium dioxide separation/preconcentration and FAAS determination of trace Zn and Cd in water sample. *Intern J Environ. Analytical Chemistry*.

Liang, Y., Wang, H., Casalongue, H. S., Chen, Z., & Dai, H. (2014). TiO2 nanocrystal grown on graphene as advanced photocatalytic hybrid materials. *Nano Res.*, *3*(10), 701–705. doi:10.1007/s12274-010-0033-5

Liao, W., Zheng, T., Wang, P., Tu, S., & Pan, W. (2010). Efficient microwave-assisted photocatalytic degradation of endocrine disruptor dimethyl phthalate over composite catalyst ZrO(x)/ZnO. *Journal of Environmental Sciences (China)*, *22*(11), 1800–1806. doi:10.1016/S1001-0742(09)60322-3 PMID:21235170

Liao, Z.-H., Chen, J.-J., Yao, K. F., Zhao, F. H., & Li, R.-X. (2004). Preparation and characterization of nanometer-sized magnetic photocatalyst. *J Inorg Mater.*, *04*, 1–10.

Lidström, P., Tierney, J., Wathey, B., & Westman, J. (2001). Microwave Reactors: A Brief Review on Its Fundamental Aspects and Applications. *Tetrahedron*, *57*, 9225–9283.

Li, J. M., Meng, X. G., Hu, C. W., & Du, J. (2009). Adsorption of phenol, p-chlorophenol and p-nitrophenol onto functional chitosan. *Bioresource Technology*, *100*(3), 1168–1173. doi:10.1016/j.biortech.2008.09.015 PMID:18930394

Li, J., Shi, Y., Cai, Y., Mou, S., & Jiang, G. (2008). Adsorption of di-ethyl-phthalate from aqueous solutions with surfactant-coated nano/microsized alumina. *Chemical Engineering Journal*, *140*(1), 214–220.

Li, M., Zhu, L., & Lin, D. (2011). Toxicity of ZnO nanoparticles to *Escherichia coli*: Mechanism and the influence of medium components. *Environmental Science & Technology*, *45*(5), 1977–1983. doi:10.1021/es102624t PMID:21280647

Lincai, P., Lu, L., Junhua, Z., Jianbin, S., & Shijie, L. (2011). Solid acid catalyzed glucose conversion to ethyl levulinate. *App Catal Gen*, *397*(1-2), 259–265. doi:10.1016/j.apcata.2011.03.008

Lin, S. H. (1993). Adsorption of disperses dye by powdered activated carbon. *Journal of Chemical Technology and Biotechnology (Oxford, Oxfordshire)*, *57*(4), 387–391. doi:10.1002/jctb.280570415

Lin, X.-X., Ji, X., Fu, D.-G., & Hao, L.-Y. (2013). Photocatalytic properties of magnetic activated carbon supported F-doped TiO2. *J Inorg Mater.*, *28*(9), 997–1002. doi:10.3724/SP.J.1077.2013.12671

Li, Q., Mahendra, S., Lyon, D. Y., Brunet, L., Liga, M. V., Li, D., & Alvarez, P. J. (2008). Antimicrobial nanomaterials for water disinfection and microbial control: Potential applications and implications. *Water Research*, *42*(18), 4591–4602. doi:10.1016/j.watres.2008.08.015 PMID:18804836

Li, Q., Su, H., & Tan, T. (2008). Synthesis of ion-imprinted chitosan-TiO_2 adsorbent and its multi-functional performances. *Biochemical Engineering Journal*, *38*(2), 212–218. doi:10.1016/j.bej.2007.07.007

Li, T., Ren, Y., & Wei, C. (2014). Study on preparation and properties of PVA-SA-PHB-AC composite carrier for microorganism immobilization. *Journal of Applied Polymer Science*, *131*(3), n/a. doi:10.1002/app.39837

Liu, F. K., Wei, G. T., & Cheng, F. C. (2003). Immobilization of a Monolayer of Bovine Serum Albumin on Gold Nanoparticles for Stereo-specified Recognition of Dansyl-norvaline J. *Chin. Chem. Soc.*, *50*(4), 931–937. doi:10.1002/jccs.200300131

Liu, F., Chung, S., Oh, G., & Seo, T. S. (2012). Three-dimensional graphene oxide nanostructure for fast and efficient water- soluble dye removal. *ACS Applied Materials & Interfaces*, *4*(2), 922–927. doi:10.1021/am201590z PMID:22206476

Liu, J., & Lu, Y. (2003). A Colorimetric Lead Biosensor Using DNAzyme-Directed Assembly of Gold Nanoparticles. *Journal of the American Chemical Society*, *125*(22), 6642–6643. doi:10.1021/ja034775u PMID:12769568

Liu, L. F., Zhang, P. H., & Yang, F. L. (2010). Adsorptive removal of 2,4-DCP from water by fresh or regenerated chitosan/ACF/TiO2 membrane. *Separation and Purification Technology*, *70*(3), 354–361. doi:10.1016/j.seppur.2009.10.022

Liu, M., Xi, B., Hou, L., & Yu, S. (2013). Magnetic multi-functional nano-fly ash-derived zeolite composites for environmental applications. *J. Mater. Chem. A*, *1*(40), 12617–12626. doi:10.1039/c3ta13180g

Liu, X., Hu, Q., Fang, Z., Zhang, X., & Zhang, B. (2008). Magnetic chitosan nanocomposites: A useful recyclable tool for heavy metal ion removal. *Langmuir*, *25*(1), 3–8. doi:10.1021/la802754t PMID:19032060

Liu, X., Zhang, H., Ma, Y., Wu, X., Meng, L., Guo, Y., & Liu, Y. et al. (2013). Graphene-coated silica as a highly efficient sorbent for residual organophosphorus pesticides in water. *J. Mater. Chem. A*, *1*(5), 1875–1884. doi:10.1039/C2TA00173J

Liu, Y., Chen, M., & Hao, Y. (2013). Study on the adsorption of Cu(II) by EDTAfunctionalized Fe_3O_4 magnetic nanoparticles. *Chemical Engineering Journal*, *218*, 46–54. doi:10.1016/j.cej.2012.12.027

Li, X., Qin, Y., Picraux, S. T., & Guo, Z.-X. (2011). Noncovalent assembly of carbon nanotube-inorganic hybrids. *Journal of Materials Chemistry*, *21*(21), 7527–7547. doi:10.1039/c1jm10516g

Li, X., Qi, Y., Li, Y., Zhang, Y., He, X., & Wang, Y. (2013). Novel magnetic beads based on sodium alginate gel crosslinked by zirconium(IV) and their effective removal for Pb^{2+} in aqueous solutions by using a batch and continuous systems. *Bioresource Technology*, *142*, 611–619. doi:10.1016/j.biortech.2013.05.081 PMID:23771001

Li, X., Wang, D., Cheng, G., Luo, Q., An, J., & Wang, Y. (2008). Preparation of polyaniline-modified TiO_2 nanoparticles and their photocatalytic activity under visible light illumination. *Applied Catalysis B: Environmental*, *81*(3), 267–273. doi:10.1016/j.apcatb.2007.12.022

Li, Y., Li, J., & Deng, C. (2014). Occurrence, characteristics and leakage of polybrominated diphenyl ethers in leachate from municipal solid waste landfills in China. *Environmental Pollution*, *184*, 94–100. doi:10.1016/j.envpol.2013.08.027 PMID:24041483

Lojkowski, W., Gedanken, A., Grzanka, E., Opalinska, A., Strachowski, T., Pielaszek, R., & Matysiak, H. et al. (2009). Solvothermal synthesis of nanocrystalline zinc oxide doped with Mn2+, Ni2+, Co2+ and Cr3+ ions. *Journal of Nanoparticle Research*, *11*(8), 1991–2002. doi:10.1007/s11051-008-9559-9

Long, M., & Cai, W. (2011). Visible light responsive TiO_2 modification with nonmetal elements. *Frontiers of Chemistry in China*, *6*(3), 190–199. doi:10.1007/s11458-011-0243-8

López-León, T., Carvalho, E. L. S., Seijo, B., Ortega-Vinuesa, J. L., & Bastos-González, D. (2005). Physicochemical characterization of chitosan nanoparticles: Electrokinetic and stability behavior. *Journal of Colloid and Interface Science*, *283*(2), 344–351. doi:10.1016/j.jcis.2004.08.186 PMID:15721903

Lotfi, M., Merabet, D., Bouzaza, A., & Belkhiri, L. (2011). Adsorption of Pb(II) from aqueous solutions using activated carbon developed from Apricot stone. *Desalination*, *276*(1-3), 148–153. doi:10.1016/j.desal.2011.03.038

Loyo, R. L. A., Nikitenko, S. I., Scheinost, A. C., & Simonoff, M. (2008). Immobilization of selenite on Fe_3O_4 and Fe/Fe_3C ultrasmall particles. *Environmental Science & Technology*, *42*(7), 2451–2456. doi:10.1021/es702579w PMID:18504980

Lu, A. H., Salabas, E. L., & Schuth, F. (2007). Magnetic Nanoparticles: Synthesis, protection, functionalization and application[Review]. *Angewandte Chemie International Edition in English*, *46*(8), 1222–1244. doi:10.1002/anie.200602866 PMID:17278160

Lu, J. W., Li, Q., Qi, Q. L., Guo, Z. X., & Yu, J. (2009). Surface engineering of poly(D,L-lactic acid) by entrapment of soluble eggshell membrane protein. *Journal of Biomedical Materials Research. Part A*, *91*(3), 701–707. doi:10.1002/jbm.a.32304 PMID:19048638

Luo, C., Wei, R., Guo, D., Zhang, S., & Yan, S. (2013). Adsorption behavior of MnO_2 functionalized multi-walled carbon nanotubes for the removal of cadmium from aqueous solutions. *Chemical Engineering Journal*, *225*, 406–415. doi:10.1016/j.cej.2013.03.128

Luo, X., Wang, C., Luo, S., Dong, R., Tu, X., & Zeng, G. (2012). Adsorption of As(III) and As(V) from water using magnetite Fe_3O_4-reduced graphite oxide–MnO_2 nanocomposites. *Chemical Engineering Journal*, *187*, 45–52. doi:10.1016/j.cej.2012.01.073

Lvovich, M. I. (1979). World water resources and their future. American Geophysical Union.

Lvovich, M. I. (1979). *World Water Resources and Their Future*. Washington, DC: American Geophysical Union. doi:10.1029/SP013

Madhavan, P., & Ramachandran Nair, K. G. (1974). Utilization of prawn waste: Isolation of chitin and its conversion to chitosan. *Fishery Technology*, *11*(1), 50–53.

Mahdavi, S., Jalali, M., & Afkhami, A. (2012). Removal of heavy metals from aqueous solutions using Fe_3O_4, ZnO and CuO nanoparticles. *Journal of Nanoparticle Research*, *14*(8), 1–18. doi:10.1007/s11051-012-0846-0 PMID:22448125

Mahdavi, S., Jalali, M., & Afkhami, A. (2015). Heavy metals removal from aqueous solutions by Al_2O_3 nanoparticles modified with natural and chemical modifiers. *Clean Technologies and Environmental Policy*, *17*(1), 85–102. doi:10.1007/s10098-014-0764-1

Mahlambi, M. M., Ngila, C. J., & Mamba, B. B. (2015). Recent developments in environmental photocatalytic degradation of organic pollutants: The case of titanium dioxide nanoparticles–A review. *Journal of Nanomaterials*, *29*, 790173.

Mahmoodi, N. M. (2013). Magnetic ferrite nanoparticle–alginate composite: Synthesis, characterization and binary system dye removal. *J. Taiwan Inst. Chem. Eng*, *44*(2), 322–330. doi:10.1016/j.jtice.2012.11.014

Majidi, S., Sehrig, F. Z., Farkhani, S. M., Goloujeh, M. S., Akbazadeh, A. (2016). Current methods of synthesis of magnetic nanoparticles. *Artificial Cells, Nanomedicine and Biotechnology*, *4*, 722-734

Manchon-Vizuete, E., Macias-Garcia, A., Gisbert, A. N., & Fernandez-Gonzalez, C. (2000). Use of froth flotation to separate PVC/PET mixtures. *Waste Management (New York, N.Y.)*, *20*(4), 265–269. doi:10.1016/S0956-053X(99)00333-5

Mani, A. D., Reddy, P. M. K., Srinivaas, M., Ghosal, P., Xanthopoulos, N., & Subrahmanyam, C. (2015). Facile synthesis of efficient visible active C-doped TiO_2 nanomaterials with high surface area for the simultaneous removal of phenol and Cr(VI). *Materials Research Bulletin*, *61*, 391–399. doi:10.1016/j.materresbull.2014.10.051

Manya, J. J., Velo, E., & Puigjaner, L. (2003). Kinetics of biomass pyrolysis: A reformulated three-parallel-reaction model. *Industrial & Engineering Chemistry Research*, *42*(3), 433–441. doi:10.1021/ie020218p

Mara, D. D. (2003). Water, sanitation and hygiene for the health of developing nations. *Public Health*, *117*(6), 452–456. doi:10.1016/S0033-3506(03)00143-4 PMID:14522162

Mario, S., Robin, J. W., Tobias, Z., Titirici, M.-M., Markus, A., Regina, R., & Ferdi, S. (2013). Direct methane oxidation over Pt modified nitrogen doped carbons. *Chemical Communications (Cambridge)*, *49*(3), 240–242. doi:10.1039/C2CC36232E PMID:23172370

Martinson, C. A., & Reddy, K. J. (2009). Adsorption of arsenic (III) and arsenic (V) by cupric oxide nanoparticles. *Journal of Colloid and Interface Science*, *336*(2), 406–411. doi:10.1016/j.jcis.2009.04.075 PMID:19477461

Masahiro, T., Takashi, Y., Tsumura, T., Yutaka, A., & Michio, I. (2006). Effect of carbon coating on TinO2n-1 for decomposition of imioctadine triacetate in aqeous solution under visible light. *J Adv Oxidation Tech.*, *9*, 1–10.

Mascolo, M. C., Pei, Y., & Ring, T. A. (2013). Room temperature co-precipitation synthesis of magnetite nanoparticles in a large pH window with different bases. *Materials (Basel)*, *6*(12), 5549–5567. doi:10.3390/ma6125549

Masoumi, A., Hemmati, K., & Ghaemy, M. (2016). Low-cost nanoparticles sorbent from modified rice husk and a copolymer for efficient removal of Pb(II) and crystal violet from water. *Chemosphere*, *146*, 253–262. doi:10.1016/j.chemosphere.2015.12.017 PMID:26735725

Mathieu-Denoncourt, Wallace, de Solla, & Langlois. (2014). Plasticizer endocrine disruption: Highlighting developmental and reproductive effects in mammals and non-mammalian aquatic species. *General and Comparative Endocrinology*.

Matos, J., Garcia-Lopez, E., Palmisano, L., Garcia, A., & Marci, G. (2010). Influence of activated carbon in TiO2 and ZnO mediated photo-assisted degradation of 2-propanol in gas solid regime. *Applied Catalysis B: Environmental*, *99*(1-2), 170–180. doi:10.1016/j.apcatb.2010.06.014

Maximous, N. N., Nakhla, G. F., & Wan, W. (2010). Removal of heavy metals from wastewater by adsorption and membrane processes: A comparative study. *International Journal of Environmental, Chemical, Ecological. Geological and Geophysical Engineering*, *4*(4), 125–130.

Maximous, N., Nakhla, G., Wan, W., & Wong, K. (2009). Preparation, characterization and performance of Al_2O_3/PES membrane for wastewater filtration. *Journal of Membrane Science*, *341*(1), 67–75. doi:10.1016/j.memsci.2009.05.040

Maximous, N., Nakhla, G., Wan, W., & Wong, K. (2010). Effect of the metal oxide particle distributions on modified PES membranes characteristics and performance. *Journal of Membrane Science*, *361*(1), 213–222. doi:10.1016/j.memsci.2010.05.051

Maya-Treviño, M. L., Guzmán-Mar, J. L., Hinojosa-Reyes, L., Ramos-Delgado, N. A., Maldonado, M. I., & Hernández-Ramírez, A. (2014). Activity of the ZnO–Fe_2O_3 catalyst on the degradation of Dicamba and 2,4-D herbicides using simulated solar light. *Ceramics International*, *40*(6), 8701–8708. doi:10.1016/j.ceramint.2014.01.088

Mayo, J. T., Yavuz, C., Yean, S., Cong, L., Shipley, H., Yu, W., & Colvin, V. L. et al. (2007). The effect of nanocrystalline magnetite size on arsenic removal. *Science and Technology of Advanced Materials*, *8*(1-2), 71–75. doi:10.1016/j.stam.2006.10.005

Medina, M., Tapia, J., Pacheco, S., Espinosa, M., & Rodriguez, R. (2010). Adsorption of lead ions in aqueous solution using silica–alumina Nanoparticles. *Journal of Non-Crystalline Solids*, *356*(6-8), 383–387. doi:10.1016/j.jnoncrysol.2009.11.032

Meek, S. T., Greathouse, J. A., & Allendorf, M. D. (2011). Metal-organic frameworks: A rapidly growing class of versatile nanoporous materials. *Advanced Materials*, *23*(2), 249–267. doi:10.1002/adma.201002854 PMID:20972981

Mehdizadeh, S., Sadjadi, S., Ahmadi, S. J., & Outokesh, M. (2014). Removal of heavy metals from aqueous solution using platinum nanopartcles/Zeolite-4A. *J. Environ. Health Sci. Eng.*, *12*(1), 7. doi:10.1186/2052-336X-12-7 PMID:24397886

Meng, A., Xing, J., Li, Z., & Li, Q. (2015). Cr-doped ZnO nanoparticles: Synthesis, characterization, adsorption property and recyclability. *ACS Applied Materials & Interfaces*, *7*(49), 27449–27457. doi:10.1021/acsami.5b09366 PMID:26600320

Meral, K., & Metin, Ö. (2014). Graphene oxide-magnetite nanocomposite as an efficient and magnetically separable adsorbent for methylene blue removal from aqueous solution. *Turkish Journal of Chemistry*, *38*(5), 775–782. doi:10.3906/kim-1312-28

Merian, E. (1984). Introduction on environmental chemistry and global cycles of chromium, nickel, cobalt, beryllium, arsenic, cadmium and selenium, and their derivatives. *Toxicological and Environmental Chemistry*, *8*(1), 9–38. doi:10.1080/02772248409357038

Miao, J., Chen, G. H., & Gao, C. J. (2005). A novel kind of amphoteric composite nanofiltration membrane prepared from sulfated chitosan (SCS). *Desalination*, *181*(1), 173–183. doi:10.1016/j.desal.2005.03.001

Milbrandt, A. R., Heimiler, D. M., Perry, A. D., & Field, C. B. (2014). Renewable energy potential on marginal lands in the United States. *Renewable & Sustainable Energy Reviews*, *29*, 473–481. doi:10.1016/j.rser.2013.08.079

Miller, S. M., & Zimmerman, J. B. (2010). Novel, bio-based, photoactive arsenic sorbent: TiO_2-impregnated chitosan bead. *Water Research*, *44*(19), 5722–5729. doi:10.1016/j.watres.2010.05.045 PMID:20594571

Mills, G., & Hoffmann, M. R. (1993). Photocatalytic degradation of pentachlorophenol on titanium dioxide particles: Identification of intermediates and mechanism of reaction. *Environmental Science & Technology*, *27*(8), 1681–1689. doi:10.1021/es00045a027

Mima, S., Miya, M., Iwamoto, R., & Yoshikawa, S. (1983). Highly deacetylated chitosan and its properties. *Journal of Applied Polymer Science*, *28*(6), 1909–1917. doi:10.1002/app.1983.070280607

Ming, H., Zhang, S., Pan, B., Zhang, W., Lu, L., & Zhang, Q. (2012). Heavy metal removal from water/wastewater by nanosized metal oxides: A review. *Journal of Hazardous Materials*, 211–212, 317–331.

Mingxing, C., Tian, S., Hongyu, G., Shengtian, W., Xiaohong, W., & Zijiang, J. (2011). Cean production of glucose from polysaccharides using a micellar heteropolyacid as a heterogeous catalyst. *Applied Catalysis B: Environmental*, *107*(1-2), 104–109. doi:10.1016/j.apcatb.2011.07.002

Miranda, S. M., Romanos, G. E., Likodimos, V., Marques, R. R., Favvas, E. P., Katsaros, F. K., & Falaras, P. et al. (2014). Pore structure, interface properties and photocatalytic efficiency of hydration/dehydration derived TiO_2/CNT composites. *Applied Catalysis B: Environmental*, *147*, 65–81. doi:10.1016/j.apcatb.2013.08.013

Mirkin, C. A., Letsinger, R. L., Mucic, R. C., & Storhoff, J. J. (1996). A DNA-based method for rationally assembling nanoparticle into macroscopic materials. *Nature*, *382*(6592), 607–609. doi:10.1038/382607a0 PMID:8757129

Mishra, D. K., Tripathy, J., Srivastava, A., Mishra, M. M., & Behari, K. (2008). Graft copolymer (chitosan-g-N- vinyl formamide): Synthesis and study of its properties like swelling, metal ion uptake and flocculation. *Carbohydrate Polymers*, *74*(3), 632–639. doi:10.1016/j.carbpol.2008.04.015

Mittal, A., Kurup, L., & Mittal, J. (2007). Freundlich and Langmuir adsorption isotherms and kinetics for the removal of Tartrazine from aqueous solutions using hen feathers. *Journal of Hazardous Materials*, *146*(1-2), 243–248. doi:10.1016/j.jhazmat.2006.12.012 PMID:17222509

Mkhoyan, K. A., Contryman, A. W., Silcox, J., Stewart, D. A., Eda, G., Mattevi, C., & Chhowalla, M. et al. (2009). Atomic and electronic structure of graphene-oxide. *Nano Letters*, *9*(3), 1058–1063. doi:10.1021/nl8034256 PMID:19199476

Mohanta, O., Singhbabu, Y. N., Giri, S. K., Dadhich, D., Das, N. N., & Sahu, R. K. (2013). Degradation of Congo red pollutants using microwave derived SrFe12O19: An efficient magnetic photocatalyst under visible light. *Journal of Alloys and Compounds*, *564*, 78–83. doi:10.1016/j.jallcom.2013.02.074

Mohapatra, M., Rout, K., Singh, P., Anand, S., Layek, S., Verma, H., & Mishra, B. (2011). Fluoride adsorption studies on mixed-phase nano iron oxides prepared by surfactant mediation-precipitation technique. *Journal of Hazardous Materials*, *186*(2), 1751–1757. doi:10.1016/j.jhazmat.2010.12.076 PMID:21232851

Monvisade, P., & Siriphannon, P. (2009). Chitosan intercalated montmorillonite: Preparation, characterization and cationic dye adsorption. *Applied Clay Science*, *42*(3), 427–431. doi:10.1016/j.clay.2008.04.013

Moore, M., Gould, P., & Keary, B. S. (2003). Global urbanization and impact on health. *International Journal of Hygiene and Environmental Health, 206*(4-5), 69–278. doi:10.1078/1438-4639-00223 PMID:12971682

Moorjani, M. N., Achutha, V., & Khasim, D. I. (1975). Parameters affecting the viscosity of chitosan from prawn waste. *Journal of Food Science and Technology.*

Moreira, M. L., Paris, E. C., do Nascimento, G. S., Longo, V. M., Sambrano, J. R., Mastelaro, V. R., & Longo, E. et al. (2009). Structural and optical properties of CaTiO 3 perovskite-based materials obtained by microwave-assisted hydrothermal synthesis: An experimental and theoretical insight. *Acta Materialia, 57*(17), 5174–5185. doi:10.1016/j.actamat.2009.07.019

Mostafa, M. S., Bakr, A.-S. A., Naggar, A. M. A. E., & Sultan, E.-S. A. (2016). Water decontamination via the removal of Pb (II) using a new generation of highly energetic surface nano-material: Co^{+2}, Mo^{+6} LDH. *Journal of Colloid and Interface Science, 461*, 261–272. doi:10.1016/j.jcis.2015.08.060 PMID:26402785

Motlagh, M. M., Hassanzadeh-Tabrizi, S., & Saffar-Teluri, A. (2015). Sol–gel synthesis of Mn_2O_3/ Al_2O_3/ SiO_2 hybrid nanocomposite and application for removal of organic dye. *Journal of Sol-Gel Science and Technology, 73*(1), 9–13. doi:10.1007/s10971-014-3487-1

Motshekga, S. C., Ray, S. S., Onyango, M. S., & Momba, M. N. (2013). Microwave-assisted synthesis, characterization and antibacterial activity of Ag/ZnO nanoparticles supported bentonite clay. *Journal of Hazardous Materials, 262*, 439–446. doi:10.1016/j.jhazmat.2013.08.074 PMID:24076479

Motshekga, S. C., Ray, S. S., Onyango, M. S., & Momba, M. N. B. (2015). Preparation and antibacterial activity of chitosan-based nanocomposites containing bentonite-supported silver and zinc oxide nanoparticles for water disinfection. *Applied Clay Science, 114*, 330–339. doi:10.1016/j.clay.2015.06.010

Motsi, T., Rowson, N. A., & Simmons, M. J. H. (2009). Adsorption of heavy metals from acid mine drainage by natural zeolite. *International Journal of Mineral Processing, 92*(1-2), 42–48. doi:10.1016/j.minpro.2009.02.005

Mou, F., Guan, J., Ma, H., Xu, L., & Shi, W. (2012). Magnetic iron oxide chestnut like hierarchical nanostructures: Preparation and their excellent arsenic removal capabilities. *ACS Application Materials Interfaces, 4*(8), 3987-3993.

Mui, E. L. K., Cheung, W. H., Valix, M., & McKay, G. (2010a). Dye adsorption onto activated carbons from tyre rubber waste using surface coverage analysis. *J. Coll. Interf. Sci., 347*(2), 290–300. doi:10.1016/j.jcis.2010.03.061

Muzzarelli, R. A. A., Tanfani, F., Emanuelli, M., & Gentile, S. (1980). The chelation of cupric ions by chitosan membranes[Callinectes sapidus, blue crab shell]. *Journal of Applied Biochemistry.*

Nabi, D., Aslam, I., & Qazi, I. A. (2009). Evaluation of the adsorption potential of titanium dioxide nanoparticles for arsenic removal. *Journal of Environmental Sciences (China)*, *21*(3), 402–408. doi:10.1016/S1001-0742(08)62283-4 PMID:19634455

Nadavala, S. K., Swayampakula, K., Boddu, V. M., & Abburi, K. (2009). Biosorption of phenol and o-chlorophenol from aqueous solutions on to chitosan–calcium alginate blended beads. *Journal of Hazardous Materials*, *162*(1), 482–489. doi:10.1016/j.jhazmat.2008.05.070 PMID:18573601

Nalbandian, M. J., Zhang, M., Sanchez, J., Choa, Y.-H., Nam, J., Cwiertny, D. M., & Myung, N. V. (2016). Synthesis and optimization of Fe_2O_3 nanofibers for chromate adsorption from contaminated water sources. *Chemosphere*, *144*, 975–981. doi:10.1016/j.chemosphere.2015.08.056 PMID:26433935

Namchot, W., Panyacharang, N., Jonglertjunya, W., & Sakdarannarong, C. (2014). Hydrolysis of delignified sugarcane bagasse using hydrothermal technique catalyzed by carbonaceous acid catalysts. *Fuel*, *116*, 608–616. doi:10.1016/j.fuel.2013.08.062

Nawaz, H., Pires, P. A. R., & El Seoud, O. A. (2013). Kinetics and mechanism of imidozole-catalyzed acylation of cellulose in LiCl/ N,N-dimethylacetamide. *Carbohydrate Polymers*, *92*(2), 997–1005. doi:10.1016/j.carbpol.2012.10.009 PMID:23399120

Nawi, M., Sabar, S., & Sheilatina, . (2012). Photocatalytic decolourisation of Reactive Red 4 dye by an immobilised TiO_2/chitosan layer by layer system. *Journal of Colloid and Interface Science*, *372*(1), 80–87. doi:10.1016/j.jcis.2012.01.024 PMID:22321991

Nazari, M., & Halladj, R. (2015). Optimization of fluoride adsorption onto a sonochemically synthesized nano-MgO/γ-Al_2O_3 composite adsorbent through applying the L16 Taguchi orthogonal design. *Desalination and Water Treatment*, *56*(9), 2464–2476. doi:10.1080/19443994.2014.961558

Ngah, W. W., Endud, C. S., & Mayanar, R. (2002). Removal of copper (II) ions from aqueous solution onto chitosan and cross-linked chitosan beads. *Reactive & Functional Polymers*, *50*(2), 181–190. doi:10.1016/S1381-5148(01)00113-4

Ngah, W. W., Kamari, A., & Koay, Y. J. (2004). Equilibrium and kinetics studies of adsorption of copper (II) on chitosan and chitosan/PVA beads. *International Journal of Biological Macromolecules*, *34*(3), 155–161. doi:10.1016/j.ijbiomac.2004.03.001 PMID:15225987

Ngah, W. W., Teong, L., & Hanafiah, M. (2011). Adsorption of dyes and heavy metal ions by chitosan composites: A review. *Carbohydrate Polymers*, *83*(4), 1446–1456. doi:10.1016/j.carbpol.2010.11.004

Ngeno, E., Orata, F., Lilechi, D. B., Shikuku, V. O., & Kimosop, S. (2016). Adsorption of caffeine and ciprofloxacin onto pyrolytically derived water hyacinth biochar: Isothermal, kinetics and thermodynamics. *Journal of Chemistry and Chemical Engineering*, *10*, 185–194.

Ng, J. C. Y., Cheung, W. H., & McKay, G. (2003). Equilibrium studies for the sorption of lead from effluents using chitosan. *Chemosphere*, *52*(6), 1021–1030. doi:10.1016/S0045-6535(03)00223-6 PMID:12781235

Nguyen, K. C., Ngoc, M. P., & Van Nguyen, M. (2016). Enhanced photocatalytic activity of nanohybrids TiO_2/CNTs materials. *Materials Letters*, *165*, 247–251. doi:10.1016/j.matlet.2015.12.004

Nguyen-Phan, T.-D., Pham, V. H., Kim, E. J., Oh, E.-S., Hur, S. H., Chung, J. S., & Shin, E. W. et al. (2012). Reduced graphene oxide–titanate hybrids: Morphologic evolution by alkali-solvothermal treatment and applications in water purification. *Applied Surface Science*, *258*(10), 4551–4557. doi:10.1016/j.apsusc.2012.01.026

Niu, H., Wang, Y., Zhang, X., Meng, Z., & Cai, Y. (2011). Easy synthesis of surface-tunable carbon-encapsulated magnetic nanoparticles: Adsorbents for selective isolation and preconcentration of organic pollutants. *ACS Applied Materials & Interfaces*, *4*(1), 286–295. doi:10.1021/am201336n PMID:22148634

Niu, S.-F., Liu, Y., Xu, X.-H., & Lou, Z.-H. (2005). Removal of hexavalent chromium from aqueous solution by iron nanoparticles. *Journal of Zhejiang University. Science. B.*, *6*(10), 1022–1027. doi:10.1631/jzus.2005.B1022 PMID:16187417

No, H. K., & Meyers, S. P. (1989). Crawfish chitosan as a coagulant in recovery of organic compounds from seafood processing streams. *Journal of Agricultural and Food Chemistry*, *37*(3), 580–583. doi:10.1021/jf00087a002

No, H. K., Meyers, S. P., & Lee, K. S. (1989). Isolation and characterization of chitin from crawfish shell waste. *Journal of Agricultural and Food Chemistry*, *37*(3), 575–579. doi:10.1021/jf00087a001

Nouri, J., Nabizadeh, R., Jahangiri-rad, M., Yunesian, M., & Moattar, F. (2014). Fluoride removal from aqueous solution by functionalized-polyacrylonitrile coated with iron oxide nano particles: Characterization and sorption studies. *Desalination and Water Treatment*, *52*(22-24), 4369–4375. doi:10.1080/19443994.2013.801792

Nriagu, J. O., & Pacyna, J. M. (1988). Quantitative assessment of worldwide contamination of air, water and soils by trace metals. *Nature*, *333*(6169), 134–139. doi:10.1038/333134a0 PMID:3285219

Nunes, M. R., Perez, G. M., Loguercio, L. F., Alves, E. W., Carreno, N. L. V., Martins, J. L., & Garcia, I. T. S. (2011). Active carbon preparation from treads of tire waste for dye removal in waste water. *Journal of the Brazilian Chemical Society*, *22*(11), 2027–2035. doi:10.1590/S0103-50532011001100002

Nyutu, E. K., Chen, C.-H., Dutta, P. K., & Suib, S. L. (2008). Effect of Microwave Frequency on Hydrothermal Synthesis of Nanocrystalline Tetragonal Barium Titanate. *The Journal of Physical Chemistry C*, *112*(26), 9659–9667. doi:10.1021/jp7112818

Obare, S. O., & Meyer, G. J. (2004). Nanostructured materials for environmental remediation of organic contaminants in water, *J. Environ. Sci. Health—Part A*, *39*, 2549–2582.

Obeid, L., Bée, A., Talbot, D., Jaafar, S. B., Dupuis, V., Abramson, S., & Welschbillig, M. et al. (2013). Chitosan/maghemite composite: A magsorbent for the adsorption of methyl orange. *Journal of Colloid and Interface Science*, *410*, 52–58. doi:10.1016/j.jcis.2013.07.057 PMID:23998368

OConnell, D. W., Birkinshaw, C., & ODwyer, T. F. (2008). Heavy metal adsorbents prepared from the 1123 modification of cellulose: A review. *Bioresource Technology*, *99*(15), 6709–6724. doi:10.1016/j.biortech.2008.01.036 PMID:18334292

Oehlmann, J., Oetken, M., & Schulte-Oehlmann, U. (2008). A critical evaluation of the environmental risk assessment for plasticizers in the freshwater environment in Europe, with special emphasis on bisphenol A and endocrine disruption. *Environmental Research*, *108*(2), 140–149. doi:10.1016/j.envres.2008.07.016 PMID:18949832

Ojedokun, A. T., & Bello, O. S. (2016). Sequestering heavy metals from wastewater using cow dung. *Water Resources and Industry*, *13*, 7–13. doi:10.1016/j.wri.2016.02.002

Olushola, S., Olalekan, S., Folahan, A., Leslie, F., & Ximba, B. J. (2014). Application of nano zinc oxide (nZnO) for the removal of triphenyltin chloride (TPT) from dockyard wastewater. *Water S.A.*, *40*(4), 659–664. doi:10.4314/wsa.v40i4.11

Onal, Y., Schimpt, S., & Claus, P. (2004). Structure sensitivity and kinetics of e-glucose oxidation to D-gluconic acid over carbon supported gold catalyst. *Journal of Catalysis*, *223*(1), 122–133. doi:10.1016/j.jcat.2004.01.010

Ong, Y. T., Ahmad, A. L., Zein, S. H. S., & Tan, S. H. (2010). A review on carbon nanotubes in an environmental protection and green engineering perspective. *Brazilian Journal of Chemical Engineering*, *27*(2), 227–242.

Otomo, R., Yokoi, T., Kondo, J.N., & Tatsumi, T. (2014). *Dealuminated beta zeolite as effective bifunctional catalyst for direct transformation of glucose to 5-hydroxymethylfurfural*. Academic Press.

Palza, H. (2015). Antimicrobial polymers with metal nanoparticles. *International Journal of Molecular Sciences*, *16*(1), 2099–2116. doi:10.3390/ijms16012099 PMID:25607734

Pambi, R. L. L., & Musonge, P. (2014). *Influence of Effluent Type on the Performance of Chitosan as a Coagulant*. Academic Press.

Pambi, R. L., & Musonge, P. (2015). The Efficiency of Chitosan as a Coagulant in the Treatment of the Effluents from the Sugar Industry. *Journal of Polymer Materials*, *32*(1), 57.

Pang, K. M., Ng, S., Chung, W. K., & Wong, P. K. (2007). Removal of pentachlorophenol by adsorption on magnetite-immobilized chitin. *Water, Air, and Soil Pollution*, *183*(1-4), 355–365. doi:10.1007/s11270-007-9384-2

Parham, H., Zargar, B., & Shiralipour, R. (2012). Fast and efficient removal of mercury from water samples using magnetic iron oxide nanoparticles modified with 2-mercaptobenzothiazole. *Journal of Hazardous Materials*, *205-206*, 94–100. doi:10.1016/j.jhazmat.2011.12.026 PMID:22244341

Park, S., & Jung, W. Y. (2001). Removal of chromium by activated carbon fibers plated with copper metal. *Carbon Science*, *2*(1), 15–21.

Park, S., Lee, J., Lee, D., & Lee, J. (2003). Photocatalytic ZnO nanopowders prepared by solution combustion method for noble metal recovery. *Journal of Materials Science*, *38*(22), 4493–4497. doi:10.1023/A:1027329501367

Park, S., & Ruoff, R. S. (2009). Chemical methods for the production of graphenes. *Nature Nanotechnology*, *4*(4), 217–224. doi:10.1038/nnano.2009.58 PMID:19350030

Patel, K. S., Shrivas, K., Brandt, R., Jakubowski, N., Corns, W., & Hoffman, P. (2005). Arsenic contamination in water, soil, sediment and rice of central India. *Environmental Geochemistry and Health*, *27*(2), 131–145. doi:10.1007/s10653-005-0120-9 PMID:16003581

Patel, K. S., Shrivas, K., Hoffmann, P., & Jakubowski, N. (2006). A survey of lead pollution in Chhattisgarh State, central India. *Environmental Geochemistry and Health*, *28*(1-2), 11–17. doi:10.1007/s10653-005-9006-0 PMID:16528599

Patel, S., & Hota, G. (2016). Iron oxide nanoparticle-immobilized PAN nanofibers: Synthesis and adsorption studies. *RSC Advances*, *6*(19), 15402–15414. doi:10.1039/C5RA20345G

Patlolla, A., Barnes, C., Yedjou, C., Velma, V., & Tchounwou, P. B. (2009). Oxidative stress, DNA damage and antioxidant enzyme activity induced by hexavalent chromium in Sprague Dawley rats. *Environmental Toxicology*, *24*(1), 66–73. doi:10.1002/tox.20395 PMID:18508361

Patra, A. K., Dutta, A., & Bhaumik, A. (2012). Self-assembled mesoporous γ- Al_2O_3 spherical nanoparticles and their efficiency for the removal of arsenic from water. *Journal of Hazardous Materials*, *201–202*, 170–177. doi:10.1016/j.jhazmat.2011.11.056 PMID:22169241

Patra, C. R., & Gedanken, A. (2004). Rapid synthesis of nanoparticles of hexagonal type In2O3 and spherical type Tl2O3 by microwave irradiation. *New Journal of Chemistry*, *28*(8), 1060–1065. doi:10.1039/b400206g

Patra, S., Roy, E., Madhuri, R., & Sharma, P. K. (2016). Agar based bimetallic nanoparticles as high-performance renewable adsorbent for removal and degradation of cationic organic dyes. *Journal of Industrial and Engineering Chemistry*, *33*, 226–238. doi:10.1016/j.jiec.2015.10.008

Patzke, G. R., Krumeich, F., & Nesper, R. (2002). Oxidic nanotubes and nanorods-anisotropic modules for a future nanotechnology. *Angewandte Chemie International Edition*, *41*(14), 2446–2461. doi:10.1002/1521-3773(20020715)41:14<2446::AID-ANIE2446>3.0.CO;2-K PMID:12203509

Peng, X., Xu, F., Zhang, W., Wang, J., Zeng, C. & Niu, M. (2014). Magnetic Fe_3O_4@silica xanthan gum composites for aqueous removal and recovery of Pb^{2+}. *Colloids Surf. A Physicochem. Eng. Aspects,* 27–36.

Peng, L.-M., Zhang, Z., & Wang, S. (2014). Carbon nanotube electronics: Recent advances. *Materials Today*, *17*(9), 433–442. doi:10.1016/j.mattod.2014.07.008

Peng, X., Luan, Z., Ding, J., Di, Z., Li, Y., & Tian, B. (2005). Ceria nanoparticles supported on carbon nanotubes for the removal of arsenate from water. *Materials Letters*, *59*(4), 399–403. doi:10.1016/j.matlet.2004.05.090

Peng, X., Luan, Z., Di, Z., Zhang, Z., & Zhu, C. (2005). Carbon nanotubes-iron oxides magnetic composites as adsorbent for removal of Pb(II) and Cu(II) from water. *Carbon*, *43*(4), 880–883. doi:10.1016/j.carbon.2004.11.009

Perez-Urquiza, M., Ferrer, R., & Beltran, J. L. (2000). Determination of sulfonatedazodyes in river water samples by capillary zone electrophoresis. *Journal of Chromatography. A*, *883*(1-2), 277–283. doi:10.1016/S0021-9673(00)00431-3 PMID:10910220

Periasamy, K., & Namasivayam, C. (1996). Removal of copper (II) by adsorption onto peanut hull carbon from water and copper plating industry water. *Chemosphere*, *32*(4), 769–789. doi:10.1016/0045-6535(95)00332-0

Pinna, N. (2007). The benzyl alcohol route: An elegant approach towards organic–inorganic hybrid nanomaterials. *Journal of Materials Chemistry*, *17*(27), 2769–2774. doi:10.1039/B702854G

Pino, L., Vita, A., Lagana, M., & Recupero, V. (2014). Hydrogen from biogas: Catalytic tri-reforming process with Ni/L---Ce---O mixed oxides. *Applied Catalysis B: Environmental*, *148*, 91–105. doi:10.1016/j.apcatb.2013.10.043

PlasticsEurope. (2013). *Plastics- The facts 2013: an analysis of European latest plastics production, demand and waste*. PlasticsEurope.

Poletto, M., & Zattera, A. J. (2013). Materials produced from plant biomass. part III: Degradation kinetics and hydrogen bonding in lignin. *Materials Research*, *16*(5), 1065–1070. doi:10.1590/S1516-14392013005000112

Postel, S. L., Daily, G. C., & Ehrlich, P. R. (1996). Human appropriation of renewable fresh water. *Science*, *271*(5250), 785–788. doi:10.1126/science.271.5250.785

Prabhakaran, S. K. (2006). *A comparative study of low-cost biomaterials for the removal of chromium (VI/III) from aqueous solutions* (Master of Engineering Thesis). Department of Chemical and Biomolecular Engineering, National University of Singapore, Singapore.

Puntes, V. F., Krishan, K. M., & Alivisatos, A. P. (2001). Colloidal nanocrystal shape and size control: The case of cobalt. *Science*, *291*(5511), 2115–2117. doi:10.1126/science.1057553 PMID:11251109

Pyrzyńska, K., & Bystrzejewski, M. (2010). Comparative study of heavy metal ions sorption onto activated carbon, carbon nanotubes, and carbon-encapsulated magnetic nanoparticles. *Colloids and Surfaces. A, Physicochemical and Engineering Aspects*, *362*(1), 102–109. doi:10.1016/j.colsurfa.2010.03.047

Qi, J., Zhang, G., & Li, H. (2015). Efficient removal of arsenic from water using a granular adsorbent: Fe–Mn binary oxide impregnated chitosan bead. *Bioresource Technology*, *193*, 243–249. doi:10.1016/j.biortech.2015.06.102 PMID:26141284

Qi, L., & Xu, Z. (2004). Lead sorption from aqueous solutions on chitosan nanoparticles. *Colloids and Surfaces. A, Physicochemical and Engineering Aspects*, *251*(1-3), 183–190. doi:10.1016/j.colsurfa.2004.10.010

Qi, L., Xu, Z., Jiang, X., Hu, C., & Zou, X. (2004). Preparation and antibacterial activity of chitosan nanoparticles. *Carbohydrate Research*, *339*(16), 2693–2700. doi:10.1016/j.carres.2004.09.007 PMID:15519328

Quek, A., & Balasubramanian, R. (2009). Article. *Journal of the Air & Waste Management Association*, *59*, 747. doi:10.3155/1047-3289.59.6.747 PMID:19603742

Quinones, D. H., Rey, A., Alvarez, P. M., Beltran, F. J., & Plucinski, P. K. (2014). Enchanced activity and reusability of TiO2 loaded magnetic activated carbon for solar photocatalytic ozonation. *Applied Catalysis B: Environmental*, *144*, 96–106. doi:10.1016/j.apcatb.2013.07.005

Rabea, E. I., Badawy, M. E.-T., Stevens, C. V., Smagghe, G., & Steurbaut, W. (2003). Chitosan as antimicrobial agent: Applications and mode of action. *Biomacromolecules*, *4*(6), 1457–1465. doi:10.1021/bm034130m PMID:14606868

Radzig, M. A., Nadtochenko, V. A., Koksharova, O. A., Kiwi, J., Lipasova, V. A., & Khmel, A. (2013). Antibacterial effects of silver nanoparticles on gram-negative bacteria: Influence on the growth and biofilms formation, mechanisms of action. *Colloids and Surfaces. B, Biointerfaces*, *102*, 300–306. doi:10.1016/j.colsurfb.2012.07.039 PMID:23006569

Rafiq, Z., Nazir, R., Shah, M. R., & Ali, S. (2014). Utilization of magnesium and zinc oxide nano-adsorbents as potential materials for treatment of copper electroplating industry wastewater. *Journal of Environmental Chemical Engineering*, *2*(1), 642–651. doi:10.1016/j.jece.2013.11.004

Rahamani, A. R., & Mahvi, A. H. (2006). Use of ion exchange for removal of ammonium: A biological regeneration of zeolite. *Global NEST Journal*, *8*, 146–150.

Rangel-Mendez, J. R., Davila-Rodriguez, J. L., & Barrios, V. A. E. (2010). Chitin based biocomposites for removal of contaminants from water: A case study of fluoride adsorption. In M. Elnashar (Ed.), *Biopolymers*. INTECH Open Access Publisher.

Rao, C. N., Müller, A., & Cheetham, A. K. (2006). *The Chemistry of nanomaterials: Synthesis, properties and applications*. Wiley.

Rao, G. P., Lu, C., & Su, F. (2007). Sorption of divalent metal ions from aqueous solution by carbon nanotubes: A review. *Separation and Purification Technology*, *58*(1), 224–231. doi:10.1016/j.seppur.2006.12.006

Rao, S. S., Yashodha, K. P., Mahendrakar, N. S., & Puttarajappa, P. (1987). deacetylation of chitin at low-temperature by a novel alkali impregnation technique. *Indian Journal of Technology*, *25*(4), 194–196.

Rashidi, F., Sarabi, R. S., Ghasemi, Z., & Seif, A. (2010). Kinetic, equilibrium and thermodynamic studies for the removal of lead (II) and copper (II) ions from aqueous solutions by nanocrystalline. *Superlattices and Microstructures*, *48*(6), 577–591. doi:10.1016/j.spmi.2010.09.011

Rauf, M., Meetani, M., & Hisaindee, S. (2011). An overview on the photocatalytic degradation of azo dyes in the presence of TiO_2 doped with selective transition metals. *Desalination, 276*(1), 13–27. doi:10.1016/j.desal.2011.03.071

Ray, P. Z., & Shipley, H. J. (2015). Inorganic nano-adsorbents for the removal of heavy metals and arsenic: A review. *RSC Advances, 5*(38), 29885–29907. doi:10.1039/C5RA02714D

Reddad, Z., Gerente, C., Andres, Y., Ralet, M. C., Thibault, J. F., & Le Cloirec, P. (2002). Ni(II) and Cu(II) binding properties of native and modified sugar beet pulp. *Carbohydrate Polymers, 49*(1), 23–31. doi:10.1016/S0144-8617(01)00301-0

Reddy, D. H. K., & Lee, S.-M. (2013). Application of magnetic chitosan composites for the removal of toxic metal and dyes from aqueous solutions. *Advances in Colloid and Interface Science, 201*, 68–93. doi:10.1016/j.cis.2013.10.002 PMID:24182685

Remya, N., & Lin, J. G. (2011). Current status of microwave application in wastewater treatment—A review. *Chemical Engineering Journal, 166*(3), 797–813. doi:10.1016/j.cej.2010.11.100

Renault, F., Sancey, B., Badot, P. M., & Crini, G. (2009). Chitosan for coagulation/flocculation processes–an eco- friendly approach. *European Polymer Journal, 45*(5).

Rezan, D.-C., Niki, B., Markus, A., & Maria-Magdalena, T. (2009). Carboxylate-rich carbonaceous materials via open-step hydrothermal carbonization of glucose in the presence of arcylic acid. *Chemistry of Materials, 21*(3), 484–490. doi:10.1021/cm802141h

Riaz, S., Faima, S., & Naseem, S. (2016). *Bacteria Removal from Water by using Al_2O_3 and FeO Nanoparticles*. Paper presented at the The 2016 World Congress on Advances in Civil, Environmental, and Materials Research (ACEM16), Jeju Island, Korea.

Richard, S. (2003). *Polymers, Recycling* (3rd ed.). Reference Module in Chemistry, Molecular Sciences and Chemical Engineering, from Encyclopedia of Physical Science and Technology.

Rickerby, D. G., & Morrison, M. (2007). Nanotechnology and the environment: A European perspective. *Science and Technology of Advanced Materials, 8*(1-2), 19–24. doi:10.1016/j.stam.2006.10.002

Rick, O., James, R. K., & Joby, M. (2012). Hemicellulose hydrolysis using solid acid catalysts generated from biochar. *Catalysis Today, 190*(1), 89–97. doi:10.1016/j.cattod.2012.02.050

Ricordel, S., Taha, S., Cisse, I., & Dorange, G. (2001). Heavy metals removal by adsorption onto peanut husks carbon: Characterization, kinetic study and modelling. *Separation and Purification Technology, 24*(3), 389–401. doi:10.1016/S1383-5866(01)00139-3

Rinaudo, M. (2006). Chitin and chitosan: Properties and applications. *Progress in Polymer Science, 31*(7), 603–632. doi:10.1016/j.progpolymsci.2006.06.001

Rinaudo, M., Milas, M., & Le Dung, P. (1993). Characterization of chitosan. Influence of ionic strength and degree of acetylation on chain expansion. *International Journal of Biological Macromolecules, 15*(5), 281–285. doi:10.1016/0141-8130(93)90027-J PMID:8251442

Ritter, L., Solomon, K., Sibley, P., Hall, K., Keen, P., Mattu, G., & Linton, B. (2002). Sources, pathways, and relative risks of contaminants in surface water and groundwater: A perspective prepared for the Walkerton inquiry. *Journal of Toxicology and Environmental Health. Part A., 65*, 1–142. doi:10.1080/152873902753338572 PMID:11809004

Rivera-Utrilla, J., Sánchez-Polo, M., Gómez-Serrano, V., Álvarez, P. M., Alvim-Ferraz, M. C. M., & Dias, J. M. (2011). Activated carbon modifications to enhance its water treatment applications. An overview. *Journal of Hazardous Materials, 187*(1-3), 1–23. doi:10.1016/j.jhazmat.2011.01.033 PMID:21306824

Rizzo, L., Di Gennaro, A., Gallo, M., & Belgiorno, V. (2008). Coagulation/chlorination of surface water: A comparison between chitosan and metal salts. *Separation and Purification Technology, 62*(1), 79–85. doi:10.1016/j.seppur.2007.12.020

Roberts, G. A. (1992). Preparation of chitin and chitosan. In *Chitin chemistry* (pp. 54–84). Macmillan Education UK. doi:10.1007/978-1-349-11545-7_2

Rodrigues, A. C., Boroski, M., Shimada, N. S., Garcia, J. C., Nozaki, J., & Hioka, N. (2008). Treatment of paper pulp and paper mill wastewater by coagulation–flocculation followed by heterogeneous photocatalysis. *Journal of Photochemistry and Photobiology A Chemistry, 194*(1), 1–10. doi:10.1016/j.jphotochem.2007.07.007

Ronald, G.-B., Xie, Z.-L., Mike, N., Cristina, G., Ruben, L. B., Shiori, K., & Andreas, T. et al. (2012). Synthesis of mesoporous carbon/iron carbide hydrids with unusually high surface areas from ionic liquid precusor[BMIM] [FeCl4]. *CrystEngComm, 14*(15), 4946–4951. doi:10.1039/c2ce25064k

Ronen, W., Geoffrey, A. T., Conner, W. C. J., & George, W. H. (2011). Design of solid acid catalysts for aqueous-phase dehydration of carbohydrates: The role of Lewis and Bronsted acid sites. *Journal of Catalysis, 279*(1), 174–182. doi:10.1016/j.jcat.2011.01.013

Rong, M., Zhang, M., & Ruan, W. (2006). Surface modification of nanoscale fillers for improving properties of polymer nanocomposites: A review. *Materials Science and Technology, 22*(7), 787–796. doi:10.1179/174328406X101247

Roquerol, F., Roquerol, J., & Sing, K. (1999). *Adsorption by powders and porous solids: Principles, Methodology and Applications* (2nd ed.). Elsevier.

Rosal, R., Rodriguez, A., Perdigon-Melon, J. A., Petre, A., Garcia-Calvo, E., Gomez, M. J., & Fernandez-Alba, A. R. et al. (2010). Occurrence of emerging pollutants in urban wastewater and their removal through biological treatment followed by ozonation. *Water Research, 44*(2), 578–588. doi:10.1016/j.watres.2009.07.004 PMID:19628245

Rosa, S., Laranjeira, M. C., Riela, H. G., & Fávere, V. T. (2008). Cross-linked quaternary chitosan as an adsorbent for the removal of the reactive dye from aqueous solutions. *Journal of Hazardous Materials, 155*(1), 253–260. doi:10.1016/j.jhazmat.2007.11.059 PMID:18180101

Rosi, N. L., Giljohann, D. A., Thaxton, C. S., Lytton-Jean, A. K. R., Han, S., & Mirkin, C. A. (2006). Oligonucleotide-modified gold nanoparticles for intracellular gene regulation. *Science*, *312*(5776), 1027–1030. doi:10.1126/science.1125559 PMID:16709779

Rossi, L. M., Costa, N. J. S., Silva, F. P., & Wojcieszak, R. (2014). Magnetic nanomaterials in catalysis: Advanced catalysts for magnetic separation and beyond. *Green Chemistry*, *16*(6), 2906–2933. doi:10.1039/c4gc00164h

Roussy, J., Van Vooren, M., Dempsey, B. A., & Guibal, E. (2005). Influence of chitosan characteristics on the coagulation and the flocculation of bentonite suspensions. *Water Research*, *39*(14), 3247–3258. doi:10.1016/j.watres.2005.05.039 PMID:16019052

Roussy, J., Van Vooren, M., & Guibal, E. (2005). Influence of chitosan characteristics on coagulation and flocculation of organic suspensions. *Journal of Applied Polymer Science*, *98*(5), 2070–2079. doi:10.1002/app.22411

Rubi, H., Fall, C., & Ortega, R. E. (2009). Pollutant removal from oily wastewater discharged from car washes through sedimentation-coagulation. *Water Science and Technology*, *59*(12), 2359–2369. doi:10.2166/wst.2009.307 PMID:19542641

Sadiq, I. M., Pakrashi, S., Chandrasekaran, N., & Mukherjee, A. (2011). Studies on toxicity of aluminum oxide (Al_2O_3) nanoparticles to microalgae species: *Scenedesmus sp.* and *Chlorella sp. Journal of Nanoparticle Research*, *13*(8), 3287–3299. doi:10.1007/s11051-011-0243-0

Sað, Y., & Aktay, Y. (2002). Kinetic studies on sorption of Cr (VI) and Cu (II) ions by chitin, chitosan and Rhizopusarrhizus. *Biochemical Engineering Journal*, *12*(2), 143–153. doi:10.1016/S1369-703X(02)00068-2

Saha, B., Das, S., Saikia, J., & Das, G. (2011). Preferential and enhanced adsorption of different dyes on iron oxide nanoparticles: A comparative study. *The Journal of Physical Chemistry C*, *115*(16), 8024–8033. doi:10.1021/jp109258f

Saha, S., & Sarkar, P. (2012). Arsenic remediation from drinking water by synthesized nano-alumina dispersed in chitosan-grafted polyacrylamide. *Journal of Hazardous Materials*, *227–228*, 68–78. doi:10.1016/j.jhazmat.2012.05.001 PMID:22647233

Sahiner, N., Ozay, O., Aktas, N., Blake, D. A., & John, V. T. (2011). Arsenic (V) removal with modifiable bulk and nano p(4-vinylpyridine)-based hydrogels: The effect of hydrogel sizes and quarternization agents. *Desalination*, *79*(1-3), 344–352. doi:10.1016/j.desal.2011.06.028

Sai Subhashini, S., Kaliappan, S., & Manickam, V. (2011). Removal of heavy metal from aqueous solution using Schizosaccharomyces Pombo in free and alginate immobilized cells. *2nd International Conference on Environmental Science and Technology*, *6*, 108–111.

Saini & Kaur. (2012). The Analysis of Aflatoxin M1 in Dairy Products. *Separation Science, 4*, 13-17.

Salam, O. E. A., Reiad, N. A., & ElShafei, M. M. (2011). A study of the removal characteristics of heavy metals from wastewater by low-cost adsorbents. *Journal of Advanced Research*, *2*(4), 297–303. doi:10.1016/j.jare.2011.01.008

Saleh T.A. (2013). *The role of carbon nanotubes in enhancement of photocatalysis*. doi: 10.5772/51050

Saleh, T. A. (2014) *Activated carbon/aluminum oxide/polyethylenimine composites and methods thereof*. US Patent App. 14/580,311.

Saleh, T. A., & Mohammed, A. (2016). *Graphene Science Handbook Fabrication Methods*. CRC Press, Taylor & Francis Inc. doi: 10.1201/b19606-39

Salehi, R., Arami, M., Mahmoodi, N. M., Bahrami, H., & Khorramfar, S. (2010). Novel biocompatible composite (chitosan–zinc oxide nanoparticle): Preparation, characterization and dye adsorption properties. *Colloids and Surfaces. B, Biointerfaces*, *80*(1), 86–93. doi:10.1016/j.colsurfb.2010.05.039 PMID:20566273

Saleh, T. A. (2011). The influence of treatment temperature on the acidity of MWCNT oxidized by HNO3 or a mixture of HNO3/H2SO4. *Applied Surface Science*, *257*(17), 7746–7751. doi:10.1016/j.apsusc.2011.04.020

Saleh, T. A. (2013). A strategy for integrating basic concepts of nanotechnology to enhance undergraduate nano-education: Statistical evaluation of an application study. *Journal of Nano Education*, *4*(1-2), 1–7.

Saleh, T. A. (2015a). Mercury sorption by silica/carbon nanotubes and silica/activated carbon: A comparison study. *Journal of Water Supply: Research & Technology - Aqua*, *64*(8), 892–903. doi:10.2166/aqua.2015.050

Saleh, T. A. (2015b). Isotherm, kinetic, and thermodynamic studies on Hg (II) adsorption from aqueous solution by silica-multiwall carbon nanotubes. *Environmental Science and Pollution Research International*, *22*(21), 16721–16731. doi:10.1007/s11356-015-4866-z PMID:26087931

Saleh, T. A. (2016). Nanocomposite of carbon nanotubes/silica nanoparticles and their use for adsorption of Pb (II): From surface properties to sorption mechanism. *Desalination and Water Treatment*, *57*(23), 10730–10744. doi:10.1080/19443994.2015.1036784

Saleh, T. A., Agarwal, S., & Gupta, V. K. (2011). Synthesis of MWCNT/MnO 2 and their application for simultaneous oxidation of arsenite and sorption of arsenate. *Applied Catalysis B: Environmental*, *106*(1), 46–53.

Saleh, T. A., Agarwal, S., & Gupta, V. K. (2011). Synthesis of MWCNT/MnO2 and their application for simultaneous oxidation of arsenite and sorption of arsenate. *Applied Catalysis B: Environmental*, *106*(1), 46–53.

Saleh, T. A., & Al-Saadi, A. A. (2013). Adsorption of lead ions from aqueous solution using porous carbon derived from rubber tires: Experimental and computational study. *Journal of Colloid and Interface Science*, *396*, 264–269. doi:10.1016/j.jcis.2013.01.037 PMID:23433519

Saleh, T. A., & Al-Saadi, A. A. (2015). Surface characterization and sorption efficacy of tire-obtained carbon: Experimental and semiempirical study of rhodamine B adsorption. *Surface and Interface Analysis*, *47*(7), 785–792. doi:10.1002/sia.5775

Saleh, T. A., Al-Saadi, A. A., & Gupta, V. K. (2014). Carbonaceous adsorbent prepared from waste tires: Experimental and computational evaluations of organic dye methyl orange. *Journal of Molecular Liquids*, *191*, 85–91. doi:10.1016/j.molliq.2013.11.028

Saleh, T. A., Al-Shalalfeh, M. M., & Al-Saadi, A. A. (2016). Graphene Dendrimer-stabilized silver nanoparticles for detection of methimazole using Surface-enhanced Raman scattering with computational assignment. *Scientific Reports*, *6*, 32185. doi:10.1038/srep32185 PMID:27572919

Saleh, T. A., & Danmaliki, G. I. (2016). Adsorptive desulfurization of dibenzothiophene from fuels by rubber tyres-derived carbons: Kinetics and isotherms evaluation. *Process Safety and Environmental Protection*, *102*, 9–19. doi:10.1016/j.psep.2016.02.005

Saleh, T. A., & Danmaliki, G. I. (2016). Influence of acidic and basic treatments of activated carbon derived from waste rubber tires on adsorptive desulfurization of thiophenes. *Journal of the Taiwan Institute of Chemical Engineers*, *60*, 460–468. doi:10.1016/j.jtice.2015.11.008

Saleh, T. A., & Danmaliki, G. I. (2016). Influence of acidic and basic treatments of activated carbon derived from waste rubber tires on adsorptive desulfurization of thiophenes. *JTICE*, *60*, 460–468.

Saleh, T. A., Gondal, M. A., & Drmosh, Q. A. (2010). Preparation of a MWCNT/ZnO nanocomposite and its photocatalytic activity for the removal of cyanide from water using a laser. *Nanotechnology*, *21*(49), 495705. doi:10.1088/0957-4484/21/49/495705 PMID:21079289

Saleh, T. A., Gondal, M. A., Drmosh, Q. A., Yamani, Z. Y., & Al-Yamani, A. (2011). Enhancement in photocatalytic activity for acetaldehyde removal by embedding ZnO nano particles on multiwall carbon nanotubes. *Chemical Engineering Journal*, *166*(1), 407–412. doi:10.1016/j.cej.2010.10.070

Saleh, T. A., & Gupta, V. (2011). Functionalization of tungsten oxide into MWCNT and its application for sunlight-induced degradation of rhodamine B. *Journal of Colloid and Interface Science*, *362*(2), 337–344. doi:10.1016/j.jcis.2011.06.081 PMID:21788030

Saleh, T. A., & Gupta, V. (2012). Synthesis and characterization of alumina nano-particles polyamide membrane with enhanced flux rejection performance. *Separation and Purification Technology*, *89*, 245–251. doi:10.1016/j.seppur.2012.01.039

Saleh, T. A., & Gupta, V. (2016). *Nanomaterial and Polymer Membranes: Synthesis, Characterization, and Applications*. Elsevier.

Saleh, T. A., & Gupta, V. K. (2012). Column with CNT/magnesium oxide composite for lead(II) removal from water. *Environmental Science and Pollution Research International*, *19*(4), 1224–1228. doi:10.1007/s11356-011-0670-6 PMID:22124802

Saleh, T. A., & Gupta, V. K. (2012). Photo-catalyzed degradation of hazardous dye methyl orange by use of a composite catalyst consisting of multi-walled carbon nanotubes and titanium dioxide. *Journal of Colloid and Interface Science*, *371*(1), 101–106. doi:10.1016/j.jcis.2011.12.038 PMID:22297021

Saleh, T. A., & Gupta, V. K. (2014). Processing methods, characteristics and adsorption behavior of tire derived carbons: A review. *Advances in Colloid and Interface Science*, *211*, 93–101. doi:10.1016/j.cis.2014.06.006 PMID:25001042

Saleh, T. A., & Gupta, V. K. (2016). Nanomaterial and Polymer Membranes. *Synthesis, Characterization, and Applications*.

Saleh, T. A., Muhammad, A. M., & Ali, S. A. (2016). Synthesis of hydrophobic cross-linked polyzwitterionic acid for simultaneous sorption of Eriochrome black T and chromium ions from binary hazardous waters. *Journal of Colloid and Interface Science*, *468*, 324–333. doi:10.1016/j.jcis.2016.01.057 PMID:26854746

Saleh, T. A., Muhammad, A. M., Tawabini, B., & Ali, S. A. (2016). Aminomethylphosphonate Chelating Ligand and Octadecyl Alkyl Chain in a Resin for Simultaneous Removal of Co (II) Ions and Organic Contaminants. *Journal of Chemical & Engineering Data*, *61*(9), 3377–3385. doi:10.1021/acs.jced.6b00475

Saleh, T. A., Sarı, A., & Tuzen, M. (2016b). Chitosan-modified vermiculite for As (III) adsorption from aqueous solution: Equilibrium, thermodynamic and kinetic studies. *Journal of Molecular Liquids*, *219*, 937–945. doi:10.1016/j.molliq.2016.03.060

Saleh, T. A., Sarı, A., & Tuzen, M. (2017). Effective adsorption of antimony (III) from aqueous solutions by polyamide-graphene composite as a novel adsorbent. *Chemical Engineering Journal*, *307*, 230–238. doi:10.1016/j.cej.2016.08.070

Saleh, T. A., Tuzen, M., & Sarı, A. (2017). Polyethylenimine modified activated carbon as novel magnetic adsorbent for the removal of uranium from aqueous solution. *Chemical Engineering Research & Design*, *117*, 218–227. doi:10.1016/j.cherd.2016.10.030

Samus, T., Lang, B., & Rohn, H. (2013). Assesing the natural resource use and the resource efficiency potential of the Desertec Concept. *Solar Eng*, *87*, 176–183. doi:10.1016/j.solener.2012.10.011

Sánchez-Duarte, R. G., Sánchez-Machado, D. I., López-Cervantes, J., & Correa-Murrieta, A. (2012). Adsorption of allura red dye by cross-linked chitosan from shrimp waste. *Water Science and Technology*, *65*(4), 618–623. doi:10.2166/wst.2012.900 PMID:22277220

Sani, H., Ahmad, A., & Saleh, T. A. (2016). 2016, Synthesis of Zinc Oxide/Talc Nanocomposite for Enhanced Lead Adsorption from Aqueous Solutions. *RSC Advances*, *6*(110), 108819–108827. doi:10.1039/C6RA24615J

Sano, L. L., Krueger, A. M., & Landrum, P. F. (2005). Chronic toxicity of glutaraldehyde: Differential sensitivity of three freshwater organisms. *Aquatic Toxicology (Amsterdam, Netherlands)*, *71*(3), 283–296. doi:10.1016/j.aquatox.2004.12.001 PMID:15670634

Santhosh, C., Kollu, P., Doshi, S., Sharma, M., Bahadur, D., Vanchinathan, M. T., & Grace, A. N. et al. (2014). Adsorption, photodegradation and antibacterial study of graphene-Fe_3O_4 nanocomposite for multipurpose water purification application. *RSC Advances*, *4*(54), 28300–28308. doi:10.1039/c4ra02913e

Sarin, V., Singh, T. S., & Pant, K. K. (2006). Thermodynamic and breakthrough column studies for the selective sorption of chromium from industrial effluent on activated eucalyptus bark. *Bioresource Technology*, *97*(16), 1986–1993. doi:10.1016/j.biortech.2005.10.001 PMID:16311033

Savina, I. N., English, C. J., Whitby, R. L. D., Zheng, Y., Leistner, A., Mikhalovsky, S. V., & Cundy, A. B. (2011). High efficiency removal of dissolved As(III) using iron nanoparticle-embeddedmacroporous polymer composites. *Journal of Hazardous Materials*, *192*(3), 1002–1008. doi:10.1016/j.jhazmat.2011.06.003 PMID:21715089

Schamberger, R. J., & Frieden, E. (1984). Selenium. In *Biochemistry of the essential ultratrace elements* (pp. 201–237). New York, NY: Plenum Press. doi:10.1007/978-1-4684-4775-0_9

Schmuhl, R., Krieg, H. M., & Keizer, K. (2004). Adsorption of Cu (II) and Cr (VI) ions by chitosan: Kinetics and equilibrium studies. *Water S.A.*, *27*(1), 1–8. doi:10.4314/wsa.v27i1.5002

Seckler, Molden, & Sakthivadivel. (2003). *The Concept of Efficiency in Water Resources Management and Policy*. Academic Press.

Seiler, H. G., Sigel, H., & Dekker, M. (1988). *Handbook on Toxicity of Inorganic Compounds*. New York.

Selvakumar, R., Jothi, N. A., Jayavignesh, V., Karthikaiselvi, K., Immanual Antony, G., Sharmila, P. R., & Swaminathan, K. et al. (2011). As (V) removal using carbonized yeast cells containing silver nanoparticles. *Water Research*, *45*(2), 583–592. doi:10.1016/j.watres.2010.09.034 PMID:20947119

Serrano-Ruiz, J. C., Luque, R., & Clark, J. H. (2013). The role of heterogenous catalysis in the biorefinering of the future. In The role of catal for sust poduct biofuel, (pp. 557-576). Academic Press.

Shang, H.-J., Zhou, Y.-L., Zhao, Y., Zhang, G., Zhao, D., Zhang, S.-H., & Zheng, X.-M. (2012). Study on heterogenous photocatalytic degradation of NH4+ N with immobilized AC/TiO2 photocatalyst. *Modern Chem Indust.*, *32*, 59–62.

Shannon, M. A., Bohn, P. W., Elimelech, M., Georgiadis, J. G., Mariñas, B. J., & Mayes, A. M. (2008). Science and technology for water purification in the coming decades. *Nature*, *452*(7185), 301–310. doi:10.1038/nature06599 PMID:18354474

Sharma, Y. C. (2003). Chromium removal from industrial effluents by adsorption on anindigenous low-cost material. *Colloid Surf. A*, *215*(1-3), 155–162. doi:10.1016/S0927-7757(02)00485-5

Sharma, Y. C., Srivastava, V., Singh, V. K., Kaul, S. N., & Weng, C. H. (2009). Nano-adsorbents for the removal of metallic pollutants from water and wastewater. *Environmental Technology*, *30*(6), 583–609. doi:10.1080/09593330902838080 PMID:19603705

Shawabkeh, R. A., Al-Harthi, M., & Al-Ghamdi, S. M. (2014). The synthesis and characterization of microporous high surface area activated carbon. *Energy Sources A.*, *36*(1), 93–103. doi: 10.1080/15567036.2010.535103

Sheela, T., Nayaka, Y. A., Viswanatha, R., Basavanna, S., & Venkatesha, T. (2012). Kinetics and thermodynamics studies on the adsorption of Zn (II), Cd (II) and Hg (II) from aqueous solution using zinc oxide nanoparticles. *Powder Technology*, *217*, 163–170. doi:10.1016/j.powtec.2011.10.023

Shen, L., & Worrell, E. (2014). Plastic Recycling. In Handbook of Recycling (pp. 179-190). Academic Press.

Shen, C., Shen, Y., Wen, Y., Wang, H., & Liu, W. (2011). Fast and highly efficient removal of dyes under alkaline conditions using magnetic chitosan-Fe (III) hydrogel. *Water Research*, *45*(16), 5200–5210. doi:10.1016/j.watres.2011.07.018 PMID:21839488

Shengsen, W., Bin, G., Yuncong, L., Anne, E., & Feng, H. (2016). Adsorptive removal of arsenate from aqueous solutions by biochar supported zero-valent iron nanocomposite: Batch and continuous flow tests. *Journal of Hazardous Materials*. doi:10.1016/j.jhazmat.2016.01.052

Shen, J., Ruan, H., Wu, L., & Gao, C. (2011). Preparation and characterization of PES–SiO_2 organic–inorganic composite ultrafiltration membrane for raw water pretreatment. *Chemical Engineering Journal*, *168*(3), 1272–1278. doi:10.1016/j.cej.2011.02.039

Shenvi, S. S., Rashid, S. A., Ismail, A. F., Kassim, M. A., & Isloor, A. M. (2013). Preparation and characterization of PPEES/chitosan composite nanofiltration membrane. *Desalination*, *315*, 135–141. doi:10.1016/j.desal.2012.09.009

Shen, X., Wang, Q., Chen, W. L., & Pang, Y. (2014). One-step synthesis of water-dispersible cysteine functionalized magnetic Fe_3O_4 nanoparticles for mercury (II) removal from aqueous solutions. *Applied Surface Science*, *317*, 1028–1034. doi:10.1016/j.apsusc.2014.09.033

Shikuku, V. O. (2016). *Adsorption of sulfonamides by γ-Fe_2O_3 modified natural clay: Equilibrium, kinetics and thermodynamics studies* (Unpublished Doctoral Dissertation). Maseno University, Kenya.

Shinde, S., Shinde, P., Bhosale, C., & Rajpure, K. (2011). Zinc oxide mediated heterogeneous photocatalytic degradation of organic species under solar radiation. *Journal of Photochemistry and Photobiology. B, Biology*, *104*(3), 425–433. doi:10.1016/j.jphotobiol.2011.04.010 PMID:21592815

Shipley, H. J., Yean, S., Kan, A. T., & Tomson, M. B. (2009). Adsorption of arsenic to magnetite nanoparticles: Effect of particle concentration, pH, ionic strength, and temperature. *Environmental Toxicology and Chemistry*, *28*(3), 509–515. doi:10.1897/08-155.1 PMID:18939890

Shrifian-Esfahni, A., Salehi, M. T., Esfahani, M., & Ekramian, E. (2015). Chitosan-modified superparamgnetic iron oxide nanoparticles: Design, fabrication, characterization and antibacterial activity. *CHEMIK*, *69*(1), 19–32.

Shrivastava, S., Bera, T., Roy, A., Singh, G., Ramachandrarao, P., & Dash, D. (2007). Characterization of enhanced antibacterial effects of novel silver nanoparticles. *Nanotechnology*, *18*(22), 225103. doi:10.1088/0957-4484/18/22/225103

Sievers, D. M., Jenner, M. W., & Hanna, M. (1994). Treatment of dilute manure wastewaters by chemical coagulation. *Transactions of the ASAE. American Society of Agricultural Engineers*, *37*(2), 597–601. doi:10.13031/2013.28118

Singh, K. P., Gupta, S., Singh, A. K., & Sinha, S. (2011). Optimizing adsorption of crystal violet dye from water by magnetic nanocomposite using response surface modeling approach. *Journal of Hazardous Materials*, *186*(2-3), 1462–1473. doi:10.1016/j.jhazmat.2010.12.032 PMID:21211903

Singh, S., Barick, K. C., & Bahadu, D. (2011). Surface engineered magnetic nanoparticles for removal of toxic metal ions and bacterial pathogens. *Journal of Hazardous Materials*, *192*(3), 1539–1547. doi:10.1016/j.jhazmat.2011.06.074 PMID:21784580

Singh, S., Barick, K. C., & Bahadur, D. (2013). Functional oxide nanomaterials and nanocomposites for the removal of heavy metals and dyes. *Nanomater. Nanotechnology*, *3*, 3–20.

Singh, V., Sharma, A. K., Tripathi, D. N., & Sanghi, R. (2009). Poly (methylmethacrylate) grafted chitosan: An efficient adsorbent for anionic azo dyes. *Journal of Hazardous Materials*, *161*(2), 955–966. doi:10.1016/j.jhazmat.2008.04.096 PMID:18547715

Singla, P., Goel, N., Kumar, V., & Singhal, S. (2015). Boron nitride nanomaterials with different morphologies: Synthesis, characterization and efficient application in dye adsorption. *Ceramics International*, *41*(9), 10565–10577. doi:10.1016/j.ceramint.2015.04.151

Sirelkhatim, A., Mahmud, S., Seeni, A., Kaus, N. H. M., Ann, L. C., Bakhori, S. K. M., & Mohamad, D. et al. (2015). Review on zinc oxide nanoparticles: Antibacterial activity and toxicity mechanism. *Nano-Micro Letters*, *7*(3), 219–242. doi:10.1007/s40820-015-0040-x

Sismanoglu, T., & Pura, S. (2001). Adsorption of aqueous nitrophenols on clinoptilolite. *Colloids and Surfaces. A, Physicochemical and Engineering Aspects*, *180*(1), 1–6. doi:10.1016/S0927-7757(00)00751-2

Sivakami, M. S., Gomathi, T., Venkatesan, J., Jeong, H. S., Kim, S. K., & Sudha, P. N. (2013). Preparation and characterization of nano chitosan for treatment wastewaters. *International Journal of Biological Macromolecules*, *57*, 204–212. doi:10.1016/j.ijbiomac.2013.03.005 PMID:23500442

Sivaraj, R., Namasivayam, C., & Kadirvelu, K. (2001). Orange peel as an adsorbent in the removal of acid violet 17 (acid dye) from aqueous solutions. *Waste Management (New York, N.Y.)*, *21*(1), 105–110. doi:10.1016/S0956-053X(00)00076-3 PMID:11150127

Sivashankar, R., Sathya, A. B., Vasantharaj, K., & Sivasubramanian, V. (2014). Magnetic composite an environmental super adsorbent for dye sequestration- A review, *Environ. Nanotechnol. Monitor. Manag.*, *1–2*, 36–49.

Skoulou, V., & Zabaniotou, A. (2012). Fe catalysis for lignocellulosic biomass conversion to fuels and materials via thermochemical process. *Catal Today, 196*, 56-66.

Skubal, L. R., Meshkov, N. K., Rajh, T., & Thurnauer, M. (2002). Cadmium removal from water using thiolactic acid-modified titanium dioxide nanoparticles. *Journal of Photochemistry and Photobiology A Chemistry*, *148*(1-3), 393–397. doi:10.1016/S1010-6030(02)00069-2

Slimen, H., Houas, A., & Nogier, J. P. (2011). Elaboration of stable anatase TiO2 through activated carbon addition with light photocatalytic activity under visible light. *J Photochem Photobio A.*, *221*(1), 13–21. doi:10.1016/j.jphotochem.2011.04.013

Smith, A. G., & Gangolli, S. D. (2002). Organochlorine chemicals in seafood: Occurrence and health concerns. *Food and Chemical Toxicology*, *40*(6), 767–779. doi:10.1016/S0278-6915(02)00046-7 PMID:11983271

Somasani, S. L. (2012). Removal of heavy metals from drinking water by adsorption onto limestone with a focus on copper and aluminum applications (Masters Theses & Specialist Projects). The Faculty of the Department of Chemistry, Western Kentucky University, Kentucky.

Sondi, I., & Salopek-Sondi, B. (2004). Silver nanoparticles as antimicrobial agent: A case study on *E. coli*as a model for Gram-negative bacteria. *Journal of Colloid and Interface Science*, *275*(1), 177–182. doi:10.1016/j.jcis.2004.02.012 PMID:15158396

Sparks, D. L. (Ed.). (1998). *Soil physical chemistry*. Boca Raton, FL: CRC Press.

Stein, R. S. (1992, February). Polymer recycling: Opportunities and limitations[Colloquium Paper]. *Proceedings of the National Academy of Sciences of the United States of America*, *89*(3), 835–838. doi:10.1073/pnas.89.3.835 PMID:11607263

Stephanie, G. W., David, M. A., Elif, I. G., & James, A. D. (2012). A roadmap for conversion of lignocellulosic biomass to chemicals and fuels. *Current Opinion Chem Eng.*, *1*(3), 218–224. doi:10.1016/j.coche.2012.04.002

Stern, A. H., Gochefld, M., Weisel, C., & Burger, J. (2001). Mercury and methyl mercury exposure in the New Jersey Pregnant population. *Archives of Environmental Health*, *56*(1), 4–10. doi:10.1080/00039890109604048 PMID:11256856

Stietiya, M. H., & Wang, J. J. (2014). Zinc and cadmium adsorption to aluminum oxide nanoparticles affected by naturally occurring ligands. *Journal of Environmental Quality*, *43*(2), 498–506. doi:10.2134/jeq2013.07.0263 PMID:25602651

Stocker, M. (2005). Gas phase catalysis by zeolites. *Microporous and Mesoporous Materials*, *82*(3), 257–292. doi:10.1016/j.micromeso.2005.01.039

Subramanian, P. M. (2000). Plastics recycling and waste management in the US. *Resources, Conservation and Recycling*, *28*(3-4), 253–263. doi:10.1016/S0921-3449(99)00049-X

Suchithra, P. S., Shadiya, C. P., Peer Mohamed, A., Velusamy, P., & Ananthakumar, S. (2013). One-pot microwave mediated growth of heterostructured ZnO@AlSi as a potential dual-function eco-catalyst for treating hazardous pollutants in water resources. *Applied Catalysis B: Environmental*, *130*, 44–53. doi:10.1016/j.apcatb.2012.10.010

Sumesh, E., Bootharaju, M. S., Anshup, , & Pradeep, T. (2011). A practical silver nanoparticle-based adsorbent for the removal of Hg^{2+} from water. *Journal of Hazardous Materials, 189*(1-2), 450–457. doi:10.1016/j.jhazmat.2011.02.061 PMID:21398028

Sumida, K., Rogow, D. L., Mason, J. A., McDonald, T. M., Bloch, E. D., Herm, Z. R., & Long, J. R. et al. (2011). Carbon dioxide capture in metal-organic frameworks. *Chemical Reviews, 112*(2), 724–781. doi:10.1021/cr2003272 PMID:22204561

Sundaram, C. S., Viswanathan, N., & Meenakshi, S. (2009). Defluoridation of water using magnesia/chitosan composite. *Journal of Hazardous Materials, 163*(2), 618–624. doi:10.1016/j.jhazmat.2008.07.009 PMID:18723280

Sun, H., Cao, L., & Lu, L. (2011). Magnetite/reduced graphene oxide nanocomposites: One step solvothermal synthesis and use as a novel platform for removal of dye pollutants. *Nano Research, 4*(6), 550–562. doi:10.1007/s12274-011-0111-3

Sun, J. Z., Liao, Z. H., Si, R. W., Kingori, G. P., Chang, F. X., Gao, L., & Yong, Y. C. et al. (2014). Adsorption and removal of triphenylmethane dyes from water by magnetic reduced graphene oxide. *Water Science and Technology, 70*(10), 1663–1669. doi:10.2166/wst.2014.427 PMID:25429455

Sun, M., & Zhang, Y. (2013). Study on the preparation of activated carbons with Baiju Vinasse. *App Mech & Mater., 448*, 669–693.

Sun, X. F., Wang, S. G., Liu, X. W., Gong, W. X., Bao, N., Gao, B. Y., & Zhang, H. Y. (2008). Biosorption of Malachite Green from aqueous solutions onto aerobic granules: Kinetic and equilibrium studies. *Bioresource Technology, 99*(9), 3475–3483. doi:10.1016/j.biortech.2007.07.055 PMID:17855080

Suresh, P., Vijaya, J. J., & Kennedy, L. J. (2015). Photocatalytic degradation of textile dyeing wastewater through microwave synthesized Zr−AC, Ni−AC and Zn−AC. *Trans. Nonferrous Met. Soc. China, 25*(12), 4216–4225. doi:10.1016/S1003-6326(15)64072-9

Suriyaraj, S., Vijayaraghavan, T., Biji, P., & Selvakumar, R. (2014). Adsorption of fluoride from aqueous solution using different phases of microbially synthesized TiO_2 nanoparticles. *Journal of Environmental Chemical Engineering, 2*(1), 444–454. doi:10.1016/j.jece.2014.01.013

Suslick, K. S., & Price, G. J. (1999). Application of ultrasound to materials chemistry. *Annual Review of Materials Science, 29*(1), 295–326. doi:10.1146/annurev.matsci.29.1.295

Suyama, K., Fukazawa, Y., & Suzumura, H. (1996). Biosorption of precious metal ions by chicken feather. *Applied Biochemistry and Biotechnology, 57/67*(1), 67–74. doi:10.1007/BF02941689

Szekeres, M., Tombácz, E., Ferencz, K., & Dékány, I. (1998). Adsorption of salicylate on alumina surfaces. *Colloids and Surfaces. A, Physicochemical and Engineering Aspects, 141*(3), 319–325. doi:10.1016/S0927-7757(98)00242-8

Tang, S. C. N., & Lo, I. M. C. (2013). Magnetic nanoparticles: Essential factors for sustainable environmental applications. *Water Research, 47*(8), 2613–2632. doi:10.1016/j.watres.2013.02.039 PMID:23515106

Tanhaei, B., Ayati, A., Lahtinen, M., & Sillanpää, M. (2015). Preparation and characterization of a novel chitosan/Al_2O_3/magnetite nanoparticles composite adsorbent for kinetic, thermodynamic and isotherm studies of methyl orange adsorption. *Chemical Engineering Journal*, *259*, 1–10. doi:10.1016/j.cej.2014.07.109

Tania, D. (Ed.). (2012). *Nanotechnology for Water Purification*. Brown Walker Press.

Tawabini, B., Al-Khaldi, S., Atieh, M., & Khaled, M. (2010). Removal of mercury from water by multi-walled carbon nanotubes. *Water Science and Technology*, *61*(3), 591–598. doi:10.2166/wst.2010.897 PMID:20150694

Tayyebi, A., Outokesh, M., Moradi, S., & Doram, A. (2015). Synthesis and characterization of ultrasound assisted graphene oxide–magnetite hybrid, and investigation of its adsorption properties for Sr(II) and Co(II) ions. *Applied Surface Science*, *353*, 350–362. doi:10.1016/j.apsusc.2015.06.087

Tchounwou, P. B., Ayensu, W. K., Ninashvilli, N., & Sutton, D. (2003). Environmental exposures to mercury and its toxicopathologic implications for public health. *Environmental Toxicology*, *18*(3), 149–175. doi:10.1002/tox.10116 PMID:12740802

Tchounwou, P. B., Patlolla, A. K., & Centeno, J. A. (2003). Carcinogenic and systemic health effects associated with arsenic exposure-a critical review'. *Toxicologic Pathology*, *31*(6), 575–588. PMID:14585726

Tchounwou, P. B., Yedjou, C. G., Patlolla, A. K., & Sutton, D. J. (2012). Heavy metals toxicity and the Environment. *EXS*, *101*, 133–164. doi:10.1007/978-3-7643-8340-4_6 PMID:22945569

Teja, A. S., & Koh, P.-Y. (2009). Synthesis, properties, and applications of magnetic iron oxide nanoparticles. *Progress in Crystal Growth and Characterization of Materials*, *55*(1), 22–45. doi:10.1016/j.pcrysgrow.2008.08.003

Tessonnier, J. P., Villa, A., Majoulet, O., Su, D. S., & Sclogl, R. (2009). Defect-mediated functionalization of carbon nanotubes as a route to design single site basic heterogenous catalysts for biomass conversion. *Ange Chemic Int*, *48*(35), 6543–6546. doi:10.1002/anie.200901658 PMID:19630046

Teuten, E. L., Saquing, J. M., Knappe, D. R. U., Barlaz, M. A., Jonsson, S., Bjorn, A., & Takada, H. et al. (2009). Transport and release of chemicals from plastics to the environment and to wildlife. *Phil. Trans. R. Soc. B*, *364*(1526), 2027–2045. doi:10.1098/rstb.2008.0284 PMID:19528054

Teymourian, H., Salimi, A., & Khezrian, S. (2013). Fe_3O_4 magnetic nanoparticles/reduced graphene oxide nanosheets as a novel electrochemical and bioeletrochemical sensing platform. *Biosensors & Bioelectronics*, *49*, 1–8. doi:10.1016/j.bios.2013.04.034 PMID:23708810

Thanh, D. N., Singh, M., Ulbrich, P., Strnadova, N., & Stepanek, F. (2011). Perlite incorporating γ-Fe_2O_3 and α-MnO_2 nanomaterials: Preparation and evaluation of a new adsorbent for As (V) removal. *Separation and Purification Technology*, *82*, 93–101. doi:10.1016/j.seppur.2011.08.030

Theepharaksapan, S., Chiemchaisri, C., Chiemchaisri, W., & Yamamoto, K. (2011). Removal of pollutants and reduction of bio-toxicity in a full scale chemical coagulation and reverse osmosis leachate treatment system. *Bioresource Technology*, *102*(9), 5381–5388. doi:10.1016/j.biortech.2010.11.091 PMID:21147528

Thirugnanasambandham, K., Sivakumar, V., Maran, J. P., & Kandasamy, S. (2014). Chitosan based grey wastewater treatment—A statistical design approach. *Carbohydrate Polymers*, *99*, 593–600. doi:10.1016/j.carbpol.2013.08.058 PMID:24274548

Thirugnanasambandham, K., Sivakumar, V., & Prakash, M. J. (2014). Treatment of egg processing industry effluent using chitosan as an adsorbent. *Journal of the Serbian Chemical Society*, *79*(6), 743–757. doi:10.2298/JSC130201053T

Thompson, R. C., Moore, C. J., vom Saal, F. S., & Swan, S. H. (2009). Plastics, the environment and human health: Current consensus and future trends. *Phil. Trans. R. Soc. B*, *364*(1526), 2153–2166. doi:10.1098/rstb.2009.0053 PMID:19528062

Tian, F., Wu, Z., Chen, Q., Yan, Y., Cravotto, G., & Wu, Z. (2015). Microwave-induced crystallization of AC/TiO2 for improving the performance of rhodamine B dye degradation. *App. Surface Sci.*, *351*, 104–112. doi:10.1016/j.apsusc.2015.05.133

Tian, N., Tian, X., Ma, L., Yang, C., Wang, Y., Wang, Z., & Zhang, L. (2015). Well-dispersed magnetic iron oxide nanocrystals on sepiolite nanofibers for arsenic removal. *RSC Advances*, *5*(32), 25236–25243. doi:10.1039/C5RA01592H

Tolaimate, A., Desbrieres, J., Rhazi, M., & Alagui, A. (2003). Contribution to the preparation of chitins and chitosans with controlled physico-chemical properties. *Polymer*, *44*(26), 7939–7952. doi:10.1016/j.polymer.2003.10.025

Tran, N., Mir, A., Mallik, D., Sinha, A., Nayar, S., & Webster, T. J. (2010). Bactericidal effect of iron oxide nanoparticles on *Staphylococcus aureus*. *International Journal of Nanomedicine*, *5*, 277–283. PMID:20463943

Tratnyek, P. G., & Johnson, R. L. (2006). Nanotechnologies for environmental cleanup. *Nano Today*, *1*(2), 44–48. doi:10.1016/S1748-0132(06)70048-2

Tryba, B., Morawski, A. W., & Inagaki, M. (2003). Application of TiO2-mounted activated carbon to the removal of phenol from water. *Applied Catalysis B: Environmental*, *41*(4), 427–433. doi:10.1016/S0926-3373(02)00173-X

Tuutijärvi, T., Repo, E., Vahala, R., Sillanpää, M., & Chen, G. (2012). Effect of competing anions on arsenate adsorption onto maghemite nanoparticles. *Chinese Journal of Chemical Engineering*, *20*(3), 505–514. doi:10.1016/S1004-9541(11)60212-7

U.S. Environmental Protection Agency. (1980). *Ambient water quality criteria for selenium.* Report No. EPA-440/5-80-070, Criteria and Standards Division, Office of Water Regulations and Standards.

U.S. EPA Environmental Criteria and Assessment Office. (1992). Cincinnati, OH: United States Environmental Protection Agency. Available online at https://cfpub.epa.gov/ncea/iris/iris_documents/documents/subst/0373_summary.pdf

U.S. EPA. (1993). Silver: Reregistration eligibility decision fact sheet. prepared by the office of prevention, pesticides and toxic substances. Environmental Protection Agency.

Umar, M., & Aziz, H. A. (2013). Photocatalytic degradation of organic pollutants in water, organic pollutants - Monitoring, risk and treatment. In Organic Pollutants - Monitoring, risk and treatment. InTech Open.

UNEP Report. (2008). *Vital water graphics: An overview of the state of the world's fresh and marine waters* (2nd ed.). Available online at: http://www.unep.org/dewa/vitalwater/article141.html

UNwater. (n.d.). Retrieved from UNwater Statistics: http://www.unwater.org/statistics/en/

Upadhyay, R. K., Soin, N., & Roy, S. S. (2014). Role of graphene/metal oxide composites as photocatalysts, adsorbents and disinfectants in water treatment: A review. *RSC Advances*, *4*(8), 3823–3851. doi:10.1039/C3RA45013A

Upadhyayula, V. K. K., Deng, S., Mitchell, M. C., & Smith, G. B. (2009). Application of carbon nanotube technology for removal of contaminants in drinking water: A review. *The Science of the Total Environment*, *408*(1), 1–13. doi:10.1016/j.scitotenv.2009.09.027 PMID:19819525

Urase, T., & Kikuta, T. (2005). Separate estimation of adsorption and degradation of pharmaceutical substances and estrogens in the activated sludge process. *Water Research*, *39*(7), 1289–1300. doi:10.1016/j.watres.2005.01.015 PMID:15862328

US Environmental Protection Agency. (n.d.). *Mercury and Compounds*. Retrieved from http://www.epa.gov/opptintr/pbt/mercury.html

USEPA. (2008). *Municipal Solid Waste in the United States: 2007 Facts and Figures. Executive summary.* Office of solid waste management and emergency response (5306P), EPA530-R-08-010, November. Available at: www.epa.gov

USEPA. (2012). *Municipal Solid Waste Generation, Recycling, and Disposal in the United States: Facts and Figures for2012*. Available at: www.epa.gov

Vachoud, L., Zydowicz, N., & Domard, A. (2001). Sorption and desorption studies on chitin gels. *International Journal of Biological Macromolecules*, *28*(2), 93–101. doi:10.1016/S0141-8130(00)00151-3 PMID:11164225

Velizarov, S., Rodrigues, C. M., Reis, M. A., & Crespo, J. G. (2000-2001). Mechanism of charged pollutants removal in an ion exchange membrane bioreactor: Drinking water denitrification. *Biotechnology and Bioengineering*, *71*(4), 245–254. doi:10.1002/1097-0290(2000)71:4<245::AID-BIT1014>3.0.CO;2-F PMID:11291034

Velo-Gala, I., Lopez-Penalver, J. J., Sanchez-Polo, M., & Rivera-Utrilla, J. (2013). Activated carbon as photocatalyst of reaction in aqeuous phase. *Applied Catalysis B: Environmental, 142*, 694–704. doi:10.1016/j.apcatb.2013.06.003

Venkatesham, V., Madhu, G., Satyanarayana, S., & Preetham, H. (2013). Adsorption of lead on gel combustion derived nano ZnO. *Procedia Engineering, 51*, 308–313. doi:10.1016/j.proeng.2013.01.041

Venkateswarlu, S., & Yoon, M. (2015). Core–Shell ferromagnetic nanorod based on amine polymer composite (Fe_3O_4@ DAPF) for fast removal of Pb(II) from aqueous solutions. *ACS Applied Materials & Interfaces, 7*(45), 25362–25372. doi:10.1021/acsami.5b07723 PMID:26496966

Vieno, N., Tuhkanen, T., & Kronberg, L. (2006). Removal of pharmaceuticals in drinking water treatment: Effect of chemical coagulation. *Environmental Technology, 27*(2), 183–192. doi:10.1080/09593332708618632 PMID:16506514

Viswanathan, N., & Meenakshi, S. (2010). Enriched fluoride sorption using alumina/chitosan composite. *Journal of Hazardous Materials, 178*(1), 226–232. doi:10.1016/j.jhazmat.2010.01.067 PMID:20144851

Viswanathan, N., Sundaram, C. S., & Meenakshi, S. (2009). Removal of fluoride from aqueous solution using protonated chitosan beads. *Journal of Hazardous Materials, 161*(1), 423–430. doi:10.1016/j.jhazmat.2008.03.115 PMID:18479817

Vitali, L., Justi, K. C., Laranjeira, M., & Fávere, V. T. (2006). Impregnation of chelating agent 3, 3-bis-N, N bis- (carboxymethyl) aminomethyl-o-cresolsulfonephthalein in biopolymer chitosan: Adsorption equilibrium of Cu (II) in aqueous medium. *Polímeros, 16*(2), 116–122. doi:10.1590/S0104-14282006000200011

Vukcevic, M., Pejic, B., Kalijadis, A., Pajic-Lijakovic, I., Kostic, M., Lausevic, Z., & Lausevic, M. (2014). Carbon materials from waste short hemp fibers as a sorbent for heavy metal ions matermatical modelling of sorbent structure and ions transport. *Chemical Engineering Journal, 235*, 284–292. doi:10.1016/j.cej.2013.09.047

Vukovic, G. D., Marinkovic, A. D., Colic, M., Ristic, M. D., Aleksic, R., Peric-Grujic, A. A., & Uskokovic, P. S. (2010). Removal of cadmium from aqueous solutions by oxidized and ethylenediamine-functionalized multi-walled carbon nanotubes. *Chemical Engineering Journal, 157*(1), 238–248. doi:10.1016/j.cej.2009.11.026

Vyver, S. V., Peng, L., Geboers, J., Schepers, H., Clippel, F., Gommes, C. J., & Seis, B. F. et al. (2010). Sulfonated silica/carbon nanocomposites as novel catalyst for hydrolysis of cellulose to glucose. *Green Chemistry, 12*(9), 1560–1563. doi:10.1039/c0gc00235f

Waalkes, M. P., Misra, R. R., & Chang, L. W. (Eds.). (1996). *Toxicology of Metals*. Boca Raton, FL: CRC Press.

Wada, S., Ichikawa, H., & Tastsumi, K. (1995). Removal of phenols and aromatic amines from wastewater by a combination treatment with tyrosinase and a coagulant. *Biotechnology and Bioengineering*, *45*(4), 304–309. doi:10.1002/bit.260450404 PMID:18623183

Walcarius, A., & Mercier, L. (2010). Mesoporousorganosilica adsorbents: Nanoengineered materials for removal of organic and inorganic pollutants. *Journal of Materials Chemistry*, *20*(22), 4478–4511. doi:10.1039/b924316j

Walha, K., Amar, R. B., Firdaous, L., Quemeneur, F., & Jaouen, P. (2007). Brackish groundwater treatment by nanofiltration, reverse osmosis and electrodialysis in Tunisia: Performance and cost comparison. *Desalination*, *207*(1-3), 95–106. doi:10.1016/j.desal.2006.03.583

Wan Ngah, W. S., & Hanafiah, M. A. K. M. (2008). Removal of heavy metal ions from wastewater by chemically modified plant wastes as adsorbents: A review. *Bioresource Technology*, *99*(10), 3935–3948. doi:10.1016/j.biortech.2007.06.011 PMID:17681755

Wang, C.-C., Lee, C.-K., Lyu, M.-D., & Juang, L.-C. (2008). Photocatalytic degradation of CI Basic Violet 10 using TiO_2 catalysts supported by Y zeolite: An investigation of the effects of operational parameters. *Dyes and Pigments*, *76*(3), 817–824. doi:10.1016/j.dyepig.2007.02.004

Wang, H., Xie, C., Zhang, W., Cai, S., Yang, Z., & Gui, Y. (2007). Comparison of dye degradation efficiency using ZnO powders with various size scales. *Journal of Hazardous Materials*, *141*(3), 645–652. doi:10.1016/j.jhazmat.2006.07.021 PMID:16930825

Wang, H., Zhu, L., Peng, S., Peng, F., Yu, H., & Yang, J. (2012). High efficient conversion of cellulose to polyols with Ru/CNTs as catalyst. *Renewable Energy*, *37*(1), 192–196. doi:10.1016/j.renene.2011.06.020

Wang, J. L., & Chen, C. (2006). Biosorption of heavy metals by Saccharomyces cerevisiae: A review. *Biotechnology Advances*, *24*(5), 427–451. doi:10.1016/j.biotechadv.2006.03.001 PMID:16737792

Wang, J., Jiang, Z., Zhang, L., Kang, P., Xie, Y., Lv, Y., & Zhang, X. et al. (2009). Sonocatalytic degradation of some dyestuffs and comparison of catalytic activities of nano-sized TiO_2, nano-sized ZnO and composite TiO_2/ZnO powders under ultrasonic irradiation. *Ultrasonics Sonochemistry*, *16*(2), 225–231. doi:10.1016/j.ultsonch.2008.08.005 PMID:18842448

Wang, J., Xu, W., Chen, L., Huang, X., & Liu, J. (2014). Preparation and evaluation of magnetic nanoparticles impregnated chitosan beads for arsenic removal from water. *Chemical Engineering Journal*, *251*, 25–34. doi:10.1016/j.cej.2014.04.061

Wang, K., Huang, T., Liu, H., Zhao, Y., Liu, H., & Sun, C. (2008). Size control synthesis of palladium oxide nanoparticles by microwave irradiation, Colloids Surfaces A Physicochem. *Eng. Asp.*, *325*(1-2), 21–25. doi:10.1016/j.colsurfa.2008.04.027

Wang, L., Zhang, J., & Wang, A. (2008). Removal of methylene blue from aqueous solution using chitosan-g-poly (acrylic acid)/montmorillonite superadsorbent nanocomposite. *Colloids and Surfaces. A, Physicochemical and Engineering Aspects*, *322*(1), 47–53. doi:10.1016/j.colsurfa.2008.02.019

Wang, S., & Shi, X. (2001). Molecular mechanisms of metal toxicity and carcinogenesis. *Molecular and Cellular Biochemistry*, *222*(1/2), 1, 3–9. doi:10.1023/A:1017918013293 PMID:11678608

Wang, T., Jin, X., Chen, Z., Megharaj, M., & Naidu, R. (2014). Simultaneous removal of Pb (II) and Cr (III) by magnetite nanoparticles using various synthesis conditions. *Journal of Industrial and Engineering Chemistry*, *20*(5), 3543–3549. doi:10.1016/j.jiec.2013.12.047

Wang, W., Mao, Q., He, H., & Zhou, M. (2013). Fe_3O_4 nanoparticles as an efficient heterogeneous Fenton catalyst for phenol removal at relatively wide pH values. *Water Science and Technology*, *68*(11), 2367–2373. doi:10.2166/wst.2013.497 PMID:24334884

Wang, X., Yu, J. C., Chen, Y., Wu, L., & Fu, X. (2006). ZrO_2-modified mesoporous nanocrystalline $TiO_{2-x}N_x$ as efficient visible light photocatalysts. *Environmental Science & Technology*, *40*(7), 2369–2374. doi:10.1021/es052000a PMID:16646476

Wang, Y., Ren, P., Feng, C., Zheng, X., Wang, Z., & Li, D. (2014). Photocatalytic behavior and photo-corrosion of visible light active silver carbonate/titanium oxide. *Materials Letters*, *115*, 85–88. doi:10.1016/j.matlet.2013.10.025

Wang, Y.-W., Cao, A., Jiang, Y., Zhang, X., Liu, J.-H., Liu, Y., & Wang, H. (2014). Superior antibacterial activity of zinc oxide/graphene oxide composites originating from high zinc concentration localized around bacteria. *ACS Applied Materials & Interfaces*, *6*(4), 2791–2798. doi:10.1021/am4053317 PMID:24495147

Wang, Z., Wu, D., Wu, G., Yang, N., & Wu, A. (2013). Modifying Fe_3O_4 microspheres with rhodamine hydrazide for selective detection and removal of Hg^{2+} ion in water. *Journal of Hazardous Materials*, *244-245*, 621–627. doi:10.1016/j.jhazmat.2012.10.050 PMID:23177242

Wei, Z., Li, Y., Thurshara, D., Liu, Y., & Ren, Q. (2011). Novel dehydration of carbohydrates to 5-hydroxymethylfurfural catalyzed by Ir and Au chlorides in ionic liquids. *J Taiwan Inst Chem Eng.*, *42*(2), 363–370. doi:10.1016/j.jtice.2010.10.004

Westervelt, R. M. (2008). Graphene nanoelectronics. *Science*, *320*(5874), 324–325. doi:10.1126/science.1156936 PMID:18420920

WHO. (1989). *Environmental health criteria 85: lead.* Geneva: WHO.

WHO. (1999). *Arsenic in drinking water: 210.* Geneva: WHO.

Wibowo, S., Velazquez, G., Savant, V., & Torres, J. A. (2007). Effect of chitosan type on protein and water recovery efficiency from surimi wash water treated with chitosan–alginate complexes. *Bioresource Technology*, *98*(3), 539–545. doi:10.1016/j.biortech.2006.02.014 PMID:16580193

Woan, K., Pyrgiotakis, G., & Sigmund, W. (2009). Photocatalytic carbon-nanotube-TiO2 composites. *Advanced Materials*, *21*(21), 2233–2239. doi:10.1002/adma.200802738

Wong, Y. C., Szeto, Y. S., Cheung, W., & McKay, G. (2004). Adsorption of acid dyes on chitosan—equilibrium isotherm analyses. *Process Biochemistry*, *39*(6), 695–704. doi:10.1016/S0032-9592(03)00152-3

Woo, K., Hung, S., Choi, S., Lee, H. W., Ahn, J. P., Kim, C. S., & Lee, S. W. (2004). Easy synthesis and magnetic properties of iron oxide magnetic nanoparticles. *Chemistry of Materials*, *16*(14), 2814–2818. doi:10.1021/cm049552x

World Health Organization and UNICEF. (2013). *Progress on Sanitation and Drinking-Water, World*. Geneva, Switzerland: Health Organization.

Wu, A. C. M., & Bough, W. A. (1978). A study of variables in the chitosan manufacturing process in relation to molecular-weight distribution, chemical characteristics and waste-treatment effectiveness.*Proceedings of the First International Conference on Chitin/Chitosan,*88-102.

Wu, D., Zheng, P., Chang, P. R., & Ma, X. (2011). Preparation and characterization of magnetic rectorite/iron oxide nanocomposites and its application for the removal of the dyes. *Chemical Engineering Journal*, *174*(1), 489–494. doi:10.1016/j.cej.2011.09.029

Wu, F. C., Tseng, R. L., & Juang, R. S. (2000). Comparative adsorption of metal and dye on flake-and bead-types of chitosans prepared from fishery wastes. *Journal of Hazardous Materials*, *73*(1), 63–75. doi:10.1016/S0304-3894(99)00168-5 PMID:10686379

Wu, F. C., Tseng, R. L., & Juang, R. S. (2001). Kinetic modeling of liquid-phase adsorption of reactive dyes and metal ions on chitosan. *Water Research*, *35*(3), 613–618. doi:10.1016/S0043-1354(00)00307-9 PMID:11228956

Wu, F. C., Tseng, R. L., & Juang, R. S. (2010). A review and experimental verification of using chitosan and its derivatives as adsorbents for selected heavy metals. *Journal of Environmental Management*, *91*(4), 798–806. doi:10.1016/j.jenvman.2009.10.018 PMID:19917518

Wu, Q., Feng, C., Wang, C., & Wang, Z. (2013). A facile one-pot solvothermal method to produce superparamagnetic graphene–Fe_3O_4 nanocomposite and its application in the removal of dye from aqueous solution. *Colloids and Surfaces. B, Biointerfaces*, *101*, 210–214. doi:10.1016/j.colsurfb.2012.05.036 PMID:23010021

Wu, Q., Zhao, G., Feng, C., Wang, C., & Wang, Z. (2011). Preparation of a graphene-based magnetic nanocomposite for the extraction of carbamate pesticides from environmental water samples. *Journal of Chromatography. A*, *1218*(44), 7936–7942. doi:10.1016/j.chroma.2011.09.027 PMID:21962496

WWDR. (2014). *Water and Enegy: Facts and Figures*. Division of Water Sciences, UNESCO.

Xiang,, Y-Z., Luo,, X., & Han,, W.Y., Yang, J-J., Ma, C.P., Xin, L., & Wang, L.B. (2013). Preparation and catalytic activities of hydrolysis catalyst for carbonyl sulfide in light hydrocarbon. *Modern Chem.*, *33*, 48–51.

Xiao, J., Xie, Y., & Cao, H. (2015). Organic pollutants removal in wastewater by heterogeneous photocatalyticozonation. *Chemosphere*, *121*, 1–17. doi:10.1016/j.chemosphere.2014.10.072 PMID:25479808

Xiaolei, Q., Pedro, J. J., & Qilin, L. (2013). Applications of nanotechnology in water and wastewater treatment. *Water Research*, *47*(12), 3931–3946. doi:10.1016/j.watres.2012.09.058 PMID:23571110

Xiao, Q., Sun, Y., Zhang, J., & Li, Q. (2015). Size-dependent of chromium (VI) adsorption on nano α-Fe_2O_3 surface. *Applied Surface Science*, *356*, 18–23. doi:10.1016/j.apsusc.2015.08.005

Xiao, Z., Yuan, M., Yang, B., Liu, Z., Huang, J., & Sun, D. (2016). Plant-mediated synthesis of highly active iron nanoparticles for Cr (VI) removal: Investigation of the leading biomolecules. *Chemosphere*, *150*, 357–364. doi:10.1016/j.chemosphere.2016.02.056 PMID:26921588

Xie, W., Xu, P., Wang, W., & Liu, Q. (2002). Preparation and antibacterial activity of a water-soluble chitosan derivative. *Carbohydrate Polymers*, *50*(1), 35–40. doi:10.1016/S0144-8617(01)00370-8

Xing, M., Lejin, X., & Jianlong, W. (2016). Mechanism of Co(II) adsorption by zero valent iron/graphene nanocomposite. *Journal of Hazardous Materials*, *301*, 286–296. doi:10.1016/j.jhazmat.2015.09.004 PMID:26368802

Xu, C., Cui, A., Xu, Y., & Fu, X. (2013). Graphene oxide–TiO_2 composite filtration membranes and their potential application for water purification. *Carbon*, *62*, 465–471. doi:10.1016/j.carbon.2013.06.035

Xu, P., Zeng, G. M., Huang, D. L., Feng, C. L., Hu, S., Zhao, M. H., & Xie, G. X. et al. (2012). Use of iron oxide nanomaterials in wastewater treatment: A review. *The Science of the Total Environment*, *424*, 1–10. doi:10.1016/j.scitotenv.2012.02.023 PMID:22391097

Yabushita, M., Kobayashi, H., & Fukuoka, A. (2014). Catalytic transformation of cellulose into platform chemical. *Applied Catalysis*, *145*, 1–9. doi:10.1016/j.apcatb.2013.01.052

Yallapu, M. M., Othman, S. F., Curtis, E. T., Gupta, B. K., Jaggi, M., & Chauhan, S. C. (2011). Multi-functional magnetic nanoparticles for magnetic resonance imaging and cancer therapy. *Biomaterials*, *32*(7), 1890–1905. doi:10.1016/j.biomaterials.2010.11.028 PMID:21167595

Yamani, J. S., Miller, S. M., Spaulding, M. L., & Zimmerman, J. B. (2012). Enhanced arsenic removal using mixed metal oxide impregnated chitosan beads. *Water Research*, *46*(14), 4427–4434. doi:10.1016/j.watres.2012.06.004 PMID:22743162

Yang, D. J., Paul, B., Xu, W. J., Yuan, Y., Liu, E. M., Ke, X. B., & Zhu, H. Y. et al. (2010). Alumina nanofibers grafted with functional groups: A new design in efficient sorbents for removal of toxic contaminants from water. *Water Research*, *44*(3), 741–750. doi:10.1016/j.watres.2009.10.014 PMID:19889440

Yang, J. L., An, S. J., Park, W. I., Yi, G. C., & Choi, W. (2004). Photocatalysis using ZnO thin films and nanoneedles grown by metal–organic chemical vapor deposition. *Advanced Materials*, *16*(18), 1661–1664. doi:10.1002/adma.200306673

Yang, K., Wu, W., Jing, Q. F., & Zhu, L. (2008). Aqueous adsorption of aniline, phenol, and their substitutes by multi-walled carbon nanotubes. *Environmental Science & Technology*, *42*(21), 7931–7936. doi:10.1021/es801463v PMID:19031883

Yang, L.-Y., Dong, S.-Y., Sun, J.-H., Feng, J.-L., Wu, Q.-H., & Sun, S.-P. (2010). Microwave-assisted preparation, characterization and photocatalytic properties of a dumbbell-shaped ZnO photocatalyst. *Journal of Hazardous Materials*, *179*(1-3), 438–443. doi:10.1016/j.jhazmat.2010.03.023 PMID:20403660

Yang, P., Kobayashi, H., & Fukuoka, A. (2011). Recent developments in the catalytic conversion of cellulose into valuable chemicals. *Chinese Journal of Catalysis*, *32*, 1–5. doi:10.1016/j.jcat.2011.08.005

Yang, Q., Liao, Y., & Mao, L. (2012). Kinetics of photocatalytic degradation of gaseous organic compounds on modified TiO2/AC composite photocatalyst. *Chinese Journal of Chemical Engineering*, *20*(3), 572–576. doi:10.1016/S1004-9541(11)60221-8

Yang, R., Li, H., Huang, M., Yang, H., & Li, A. (2016). A review on chitosan-based flocculants and their applications in water treatment. *Water Research*, *95*, 59–89. doi:10.1016/j.watres.2016.02.068 PMID:26986497

Yang, S.-T., Chang, Y., Wang, H., Liu, G., Chen, S., Wang, Y., & Cao, A. et al. (2010). Folding/aggregation of graphene oxide and its application in Cu^{2+} removal. *Journal of Colloid and Interface Science*, *351*(1), 122–127. doi:10.1016/j.jcis.2010.07.042 PMID:20705296

Yang, W., Ding, P., Zhou, L., Yu, J., Chen, X., & Jiao, F. (2013). Preparation of diamine modified mesoporous silica on multi-walled carbon nanotubes for the adsorption of heavy metals in aqueous solution. *Applied Surface Science*, *282*, 38–45. doi:10.1016/j.apsusc.2013.05.028

Yang, X., Chen, C., Li, J., Zhao, G., Ren, X., & Wang, X. (2012). Graphene oxide-iron oxide and reduced graphene oxide-iron oxide hybrid materials for the removal of organic and inorganic pollutants. *RSC Advances*, *2*(23), 8821–8826. doi:10.1039/c2ra20885g

Yan, H., Li, H., Yang, H., Li, A., & Cheng, R. (2013). Removal of various cationic dyesfrom aqueous solutions using a kind of fully biodegradable magnetic compositemicrosphere. *Chemical Engineering Journal*, *223*, 402–411. doi:10.1016/j.cej.2013.02.113

Yantasee, W., Warner, C. L., Sangvanich, T., Addleman, R. S., Carter, T. G., Wiacek, R. J., & Warner, M. G. et al. (2007). Removal of heavymetals from aqueous systems with thiol functionalized superparamagnetic nanoparticles. *Environmental Science & Technology*, *4*(14), 5114–5119. doi:10.1021/es0705238 PMID:17711232

Yan, Y., Sun, H., Zhang, L., Zhang, J., Mu, J., & Kang, S.-Z. (2011). Effect of multiwalled carbon nanotubes on the photocatalytic degradation of methyl orange in aqueous solution under visible light irradiation. *Journal of Dispersion Science and Technology*, *32*(9), 1332–1336. doi:10.1080/02604027.2010.505868

Yao, R., Meng, F., Zhang, L., Ma, D., & Wang, M. (2009). Defluoridation of water using neodymium-modified chitosan. *Journal of Hazardous Materials, 165*(1), 454–460. doi:10.1016/j.jhazmat.2008.10.052 PMID:19046805

Yao, S., Jia, Y., Shi, Z., & Zhao, S. (2010). Photoccatalytic oxidation of arsenite by a composite of titanium dioxide and activated carbon fiber. *Photochemistry and Photobiology, 86*(6), 1215–1221. doi:10.1111/j.1751-1097.2010.00813.x PMID:21039575

Yao, X. P., Fu, Z. J., Zhao, Y. G., Wang, L., Fang, L. Y., & Shen, H. Y. (2012). Use of tetraethylenepentamine-functional Fe_3O_4 magnetic polymers for matrix solid phase dispersion extraction and preconcentration of Cr(VI) in water samples at ultra-trace levels. *Talanta, 97*, 124–130. doi:10.1016/j.talanta.2012.04.006 PMID:22841056

Yao, Y., Bing, H., Feifei, X., & Xiaofeng, C. (2011). Equilibrium and kinetic studies of methyl orange adsorption on multiwalled carbon nanotubes. *Chemical Engineering Journal, 170*(1), 82–89. doi:10.1016/j.cej.2011.03.031

Yao, Y., Miao, S., Liu, S., Ma, L. P., Sun, H., & Wang, S. (2012). Synthesis, characterization,and adsorption properties of magnetic Fe_3O_4@graphene nanocomposite. *Chemical Engineering Journal, 184*, 326–332. doi:10.1016/j.cej.2011.12.017

Yao, Y., Yang, Z., Sun, H., & Wang, S. (2012). Hydrothermal synthesis of Co_3O_4–graphene for heterogeneous activation of peroxymonosulfate for decomposition of phenol. *Industrial & Engineering Chemistry Research, 51*(46), 14958–14965. doi:10.1021/ie301642g

Yen, M. T., Yang, J. H., & Mau, J. L. (2008). Antioxidant properties of chitosan from crab shells. *Carbohydrate Polymers, 74*(4), 840–844. doi:10.1016/j.carbpol.2008.05.003

Yen, T.-H., Lin-Tan, D.-T., & Lin, J.-L. (2011). Food safety involving ingestion of foods and beverages prepared with phthalate-plasticizercontaining clouding agents. *Journal of the Formosan Medical Association, 110*(11), 671–684. doi:10.1016/j.jfma.2011.09.002 PMID:22118310

Yoon, J.-W., Baek, M.-H., Hong, J.-S., Lee, C.-Y., & Suh, J.-K. (2012). Photocatalytic degradation of azo dye using TiO2 supported on spherical activated carbon. *Korean Journal of Chemical Engineering, 29*(12), 1722–1729. doi:10.1007/s11814-012-0076-2

Yoon, Y., Park, W. K., Hwang, T.-M., Yoon, D. H., Yang, W. S., & Kang, J.-W. (2016). Comparative evaluation of magnetite–graphene oxide and magnetite-reduced graphene oxide composite for As(III) and As(V) removal. *Journal of Hazardous Materials, 304*, 196–204. doi:10.1016/j.jhazmat.2015.10.053 PMID:26551223

Yoshizuka, K., Lou, Z., & Inoue, K. (2000). Silver-complexed chitosan microparticles for pesticide removal. *Reactive & Functional Polymers, 44*(1), 47–54. doi:10.1016/S1381-5148(99)00076-0

Younes, I., & Rinaudo, M. (2015). Chitin and chitosan preparation from marine sources. Structure, properties and applications. *Marine Drugs, 13*(3), 1133–1174. doi:10.3390/md13031133 PMID:25738328

Yousef, J. M., & Danial, E. N. (2012). In vitro antibacterial activity and minimum inhibitory concentration of zinc oxide and nano-particle zinc oxide against pathogenic strains. *Journal of Health Science*, *2*(4), 38–42.

Yuan, P., Fan, M., Yang, D., He, H., Liu, D., Yuan, A., & Chen, T. et al. (2009). Montmorillonite-supported magnetite nanoparticles for the removal of hexavalent chromium [Cr (VI)] from aqueous solutions. *Journal of Hazardous Materials*, *166*(2), 821–829. doi:10.1016/j.jhazmat.2008.11.083 PMID:19135796

Yu, J. X., Wang, L. Y., Chi, R. A., Zhang, Y. F., Xu, Z. G., & Guo, J. (2013). Competitive adsorption of Pb^{2+} and Cd^{2+} on magnetic modified sugarcane bagasse prepared by two simple steps. *Applied Surface Science*, *268*, 163–170. doi:10.1016/j.apsusc.2012.12.047

Yu, J.-G., Zhao, X.-H., Yang, H., Chen, X.-H., Yang, Q., Yu, L.-Y., & Chen, X.-Q. et al. (2014). Aqueous adsorption and removal of organic contaminants by carbon nanotubes. *The Science of the Total Environment*, *482–483*, 241–251. doi:10.1016/j.scitotenv.2014.02.129 PMID:24657369

Yu, K., Ho, J., McCandlish, E., Buckley, B., Patel, R., Li, Z., & Shapley, N. C. (2013). Copper ion adsorption by chitosan nanoparticles and alginate microparticles for water purification applications. *Colloids and Surfaces. A, Physicochemical and Engineering Aspects*, *425*, 31–41. doi:10.1016/j.colsurfa.2012.12.043

Yu, L., Peng, X., Ni, F., Li, J., Wang, D., & Luan, Z. (2013). Arsenite removal from aqueous solutions by γ-Fe_2O_3–TiO_2 magnetic nanoparticles through simultaneous photocatalytic oxidation and adsorption. *Journal of Hazardous Materials*, *246*, 10–17. doi:10.1016/j.jhazmat.2012.12.007 PMID:23276789

Yu, S., Heather, M. B., Li, G., Zhou, X.-D., Amonette, J. E., John, L. F., & Conrod, Z. et al. (2011). Accelerated cellulose depolymerization catalyzed by paired metal chlorides in ionic liquid solvent. *App Catal Gen.*, *391*(1-2), 436–442. doi:10.1016/j.apcata.2010.09.021

Yu, X., Qin, A., Liao, L., Du, R., Tian, N., Huang, S., & Wei, C. (2015). Removal of organic dyes by nanostructure ZnO-bamboo charcoal composites with photocatalysis function. *Advances in Materials Science and Engineering*.

Zabaniotou, A. A., Antoniou, N., & Stavropoulos, G. (2014). Novel sorbent materials for environmental remediation via depolymerization of used tyres. *Desalination and Water Treatment*, 1–10.

Zainal, Z., Hui, L. K., Hussein, M. Z., Abdullah, A. H., & Hamadneh, I. M. K. R. (2009). Characterization of TiO_2–Chitosan/Glass photocatalyst for the removal of a monoazo dye via photodegradation–adsorption process. *Journal of Hazardous Materials*, *164*(1), 138–145. doi:10.1016/j.jhazmat.2008.07.154 PMID:18809254

Zapata, D. A., Huang, Y., Gonzalez-Borja, M. A., & Resosca, D. E. (2013). Silyted hydrophobic zeolites with enchanced tolerance to hot liquid water. *Journal of Catalysis*, *308*, 62–97. doi:10.1016/j.jcat.2013.05.024

Zargoosh, K., Abedini, H., Abdolmaleki, A., & Molavian, M. R. (2013). Effective removal of heavy metal ions from industrial wastes using Thiosalicylhydrazide-Modified Magnetic Nanoparticles. *Industrial & Engineering Chemistry Research*, *52*(42), 14944–14954. doi:10.1021/ie401971w

Zarpelon, F., Galiotto, D., Aguzzoli, C., Carli, L. N., Figueroa, C. A., Jacob, I., & Giovanela, M. et al. (2016). Removal of coliform bacteria from industrial wastewaters using polyelectrolytes/silver nanoparticles self-assembled thin films. *J. Environ. Chem. Eng.*, *4*(1), 137–146. doi:10.1016/j.jece.2015.11.013

Zawadzki, M. (2008). Preparation and characterization of ceria nanoparticles by microwave-assisted solvothermal process. *Journal of Alloys and Compounds*, *454*(1-2), 347–351. doi:10.1016/j.jallcom.2006.12.078

Zeng, D., Wu, J., & Kennedy, J. F. (2008). Application of a chitosan flocculant to water treatment. *Carbohydrate Polymers*, *71*(1), 135–139. doi:10.1016/j.carbpol.2007.07.039

Zhang, H. (2014). *Biosorption of heavy metals from aqueous solutions using keratin biomaterials* (Doctoral Thesis). Departament de Quinica, Facultat de Ciencies, Universitat Autonoma de Barcelona, Barcelona, Spain.

Zhang, X., Zhang, P., Wu, Z., Zhang, L., Zeng, G. & Zhou, C. (2013). Adsorption of methylene blue onto humic acid-coated Fe_3O_4 nanoparticles. *Colloids Surf. A Physicochem. Eng. Aspects*, 85–90.

Zhang, D., Wei, S., Kaila, C., Su, X., Wu, J., Karki, A. B., & Guo, Z. et al. (2010). Carbon-stabilized iron nanoparticles for environmental remediation. *Nanoscale*, *2*(6), 917–919. doi:10.1039/c0nr00065e PMID:20648288

Zhang, D., Wen, M., Zhang, S., Liu, P., Zhu, W., Li, G., & Li, H. (2014). Au nanoparticles enchanced rutile TiO2 nanorod bundles with high visible light photocatalytic performance for NO oxidation. *Applied Catalysis B: Environmental*, *147*, 610–616. doi:10.1016/j.apcatb.2013.09.042

Zhang, G., Qu, J., Liu, H., Cooper, A. T., & Wu, R. (2007). $CuFe_2O_4$/activated carbon composite: A novel magnetic adsorbent for the removal of acid orange II and catalytic regeneration. *Chemosphere*, *68*(6), 1058–1066. doi:10.1016/j.chemosphere.2007.01.081 PMID:17397903

Zhang, H. M., Liu, J. B., Wang, H., Zhang, W. X., & Yan, H. (2008). Rapid microwave-assisted synthesis of phase controlled BiVO4 nanocrystals and research on photocatalytic properties under visible light irradiation. *Journal of Nanoparticle Research*, *10*(5), 767–774. doi:10.1007/s11051-007-9310-y

Zhang, J., Zhai, S., Li, S., Xiao, Z., Song, Y., An, Q., & Tian, G. (2013). Pb(II) removal of Fe_3O_4@SiO_2–NH_2 core–shell nanomaterials prepared via a controllable sol–gel process. *Chemical Engineering Journal*, *215–216*, 461–471. doi:10.1016/j.cej.2012.11.043

Zhang, J., Zhou, Q., & Ou, L. (2011). Kinetic, isotherm, and thermodynamic studies of the adsorption of methyl orange from aqueous solution by chitosan/alumina composite. *Journal of Chemical & Engineering Data*, *57*(2), 412–419. doi:10.1021/je2009945

Zhang, K., Dwivedi, V., Chi, C. Y., & Wu, J. S. (2010). Graphene oxide/ferric hydroxide composites for efficient arsenate removal from drinking water. *Journal of Hazardous Materials*, *182*(1-3), 162–168. doi:10.1016/j.jhazmat.2010.06.010 PMID:20580161

Zhang, L., Cao, X.-F., Chen, X.-T., & Xue, Z.-L. (2011). BiOBr hierarchical microspheres: Microwave-assisted solvothermal synthesis, strong adsorption and excellent photocatalytic properties. *Journal of Colloid and Interface Science*, *354*(2), 630–636. doi:10.1016/j.jcis.2010.11.042 PMID:21145062

Zhang, L., Jiang, Y., Ding, Y., Povey, M., & York, D. (2007). Investigation into the antibacterial behaviour of suspensions of ZnO nanoparticles (ZnO nanofluids). *Journal of Nanoparticle Research*, *9*(3), 479–489. doi:10.1007/s11051-006-9150-1

Zhang, L., Li, T., Li, B., Li, J., & Wang, E. (2010). Carbon nanotube–DNA hybrid fluorescent sensor for sensitive and selective detection of mercury (II) ion. *Chemical Communications (Cambridge)*, *46*(9), 1476–1478. doi:10.1039/b921191h PMID:20162153

Zhang, Q., & Jiao, T. (2013). *Polymer-supported organic-inorganic nanomaterials: fabrication, characterization and environmental application. In Materials and processes for energy: communicating current research and technological developments* (pp. 903–912). Formatex Research Center.

Zhang, W. X., & Elliot, D. W. (2006). Applications of iron nanoparticles for groundwater remediation. *Remediation Journal*, *16*(2), 7–21. doi:10.1002/rem.20078

Zhang, X., & Bai, R. (2003). Mechanisms and kinetics of humic acid adsorption onto chitosan-coated granules. *Journal of Colloid and Interface Science*, *264*(1), 30–38. doi:10.1016/S0021-9797(03)00393-X PMID:12885516

Zhang, X., Li, W., Blatchley, E. R. III, Wang, X., & Ren, P. (2015). UV/chlorine process for ammonia removal and disinfection by-product reduction: Comparison with chlorination. *Water Research*, *68*, 804–811. doi:10.1016/j.watres.2014.10.044 PMID:25466638

Zhang, X.-Y., Li, H.-P., Cui, X.-L., & Lin, Y. (2010). Graphene/TiO2 nanocomposites synthesis, characterization and application in hydrogen evolution from water photocatalytic splitting. *Journal of Materials Chemistry*, *20*(14), 2801–2806. doi:10.1039/b917240h

Zhang, X., Zhao, X., & Su, H. (2011). Degradation characteristic of TiO_2-chitosan adsorbent on Rhodamine B and purification of industrial wastewater. *Korean Journal of Chemical Engineering*, *28*(5), 1241–1246. doi:10.1007/s11814-010-0501-3

Zhang, Y., & Zhou, J. L. (2005). Removal of estrone and 17β-estradiol from water by adsorption. *Water Research*, *39*(16), 3991–4003. doi:10.1016/j.watres.2005.07.019 PMID:16126247

Zhang, Z., & Kong, J. (2011). Novel magnetic Fe_3O_4@C nanoparticles as adsorbentsfor removal of organic dyes from aqueous solution. *Journal of Hazardous Materials*, *193*, 325–329. doi:10.1016/j.jhazmat.2011.07.033 PMID:21813238

Zhang, Z., Zongboa, K., & Zhao, K. (2009). Solid acid and microave-assisted hydrolysis of cellulose in ionic liquid. *Carbohydrate Research*, *344*(15), 2069–2072. doi:10.1016/j.carres.2009.07.011 PMID:19703684

Zhan, H., Jiang, Y., & Ma, Q. (2014). Determination of adsorption characteristics of metal oxide nanomaterials: Application as adsorbents. *Analytical Letters*, *47*(5), 871–884. doi:10.1080/00032719.2013.850090

Zhan, S., Yang, Y., Shen, Z., Shan, J., Li, Y., Yang, S., & Zhu, D. (2014). Efficient removal of pathogenic bacteria and viruses by multifunctional amine-modified magnetic nanoparticles. *Journal of Hazardous Materials*, *274*, 115–123. doi:10.1016/j.jhazmat.2014.03.067 PMID:24769848

Zhao, Y., Zhu, J., Hong, J., Bian, N., & Chen, H. (2004). Microwave-Induced Polyol-Process Synthesis of Copper and Copper Oxide Nanocrystals with Controllable Morphology. *Eur. J. Inorg. Chem.*, 4072–4080.

Zhao, G., Ren, X., Gao, X., Tan, X., Li, J., Chen, C., & Wang, X. et al. (2011). Removal of Pb(II) ions from aqueous solutions on few-layered graphene oxide nanosheets. *Dalton Transactions (Cambridge, England)*, *40*(41), 10945–10952. doi:10.1039/c1dt11005e PMID:21918761

Zhao, X., Lv, L., Pan, B., Zhang, W., Zhang, S., & Zhang, Q. (2011). Polymer-supported nanocomposites for environmental application: A review. *Chemical Engineering Journal*, *170*(2), 381–394. doi:10.1016/j.cej.2011.02.071

Zheng, S. K., Yang, Z. F., Jo, D. H., Park, Y. H., & Ho, Y. S. (2005). Comment on Removal of chlorophenols from groundwater by chitosan sorption. *Water Research*, *39*(1), 264–268. doi:10.1016/j.watres.2004.07.018 PMID:15607185

Zhong, H., Shaogui, Y., Yongming, J., & Cheng, S. (2009). Enhancement of visible-light-induced photodegradation over hierarchical porous TiO2 by nonmetal doping and water-mediated dye sensitization. *Journal of Environmental Sciences (China)*, *21*, 268–272. PMID:19402433

Zhong, J. S., Wang, Q. Y., Zhou, J., Chen, D. Q., & Ji, Z. G. (2016). Highly efficient photoelectrocatalytic removal of RhB and Cr(VI) by Cu nanoparticles sensitized TiO_2 nanotube arrays. *Applied Surface Science*, *367*, 342–346. doi:10.1016/j.apsusc.2016.01.189

Zhou, C., Gao, X. L., Li, S. S., & Gao, C. J. (2013). Fabrication and characterization of novel composite nanofiltration membranes based on zwitterionic O-carboxymethyl chitosan. *Desalination*, *317*, 67–76. doi:10.1016/j.desal.2013.02.026

Zhou, G., Li, W., Wang, Z., Wang, X., Li, S., & Zhang, D. (2015). Electrosorption for organic pollutants removal and desalination by graphite and activated carbon fiber composite electrodes. *International Journal of Environmental Science and Technology*, *12*(12), 3735–3744. doi:10.1007/s13762-015-0811-4

Zhou, H., & Smith, D. W. (2002). Advanced technologies in water and wastewater treatment. *Journal of Environmental Engineering and Science*, *1*(4), 247–264. doi:10.1139/s02-020

Zhou, L., Jin, J., Liu, Z., Liang, X., & Shang, C. (2011). Adsorption of acid dyes from aqueous solutions by the ethylenediamine-modified magnetic chitosan nanoparticles. *Journal of Hazardous Materials*, *185*(2-3), 1045–1052. doi:10.1016/j.jhazmat.2010.10.012 PMID:21035264

Zhu, H., Jiang, R., Fu, Y., Guan, Y., Yao, J., Xiao, L., & Zeng, G. (2012). Effective photocatalytic decolorization of methyl orange utilizing TiO_2/ZnO/chitosan nanocomposite films under simulated solar irradiation. *Desalination*, *286*, 41–48. doi:10.1016/j.desal.2011.10.036

Zhu, J., Wei, S., Gu, H., Rapole, S. B., Wang, Q., Luo, Z., & Guo, Z. et al. (2011). One-pot synthesis of magnetic graphene nanocomposites decorated with core@ double-shell nanoparticles for fast chromium removal. *Environmental Science & Technology*, *46*(2), 977–985. doi:10.1021/es2014133 PMID:22126606

Zia, K. M., Bhatti, H. N., & Bhatti, I. A. (2007). Methods for polyurethane and polyurethanev-composites, recycling and recovery: A review. *Reactive & Functional Polymers*, *67*(8), 675–692. doi:10.1016/j.reactfunctpolym.2007.05.004

Zubieta, C. E., Messina, P. V., Luengo, C., Dennehy, M., Pieroni, O., & Schulz, P. C. (2008). Reactive dyes remotion by porous TiO_2-chitosan materials. *Journal of Hazardous Materials*, *152*(2), 765–777. doi:10.1016/j.jhazmat.2007.07.043 PMID:17764835

Zuo, X., Peng, C., Huang, Q., Song, S., Wang, L., Li, D., & Fan, C. (2009). Design of a carbon nanotube/magnetic nanoparticle-based peroxidase-like nanocomplex and Its application for highly efficient catalytic oxidation of phenols. *Nano Res*, *2*(8), 617–623. doi:10.1007/s12274-009-9062-3

Zu, Y., Yang, P., Wang, J., Liu, X., Ren, J., Lu, G., & Wang, Y. (2014). Efficient production of the liquid fuel 2,5-dimethylfuran from 5-hydroxymethylfurfural over Ru/CO3O4 catalyst. *Applied Catalysis B: Environmental*, *146*, 244–248. doi:10.1016/j.apcatb.2013.04.026

About the Contributors

Tawfik Abdo Saleh is a faculty member in the Chemistry Department at King Fahd University of Petroleum and Minerals. He has published many books and several papers in international refereed journals and conference proceedings, authored and edited several books and book chapters, and he is a member of several local and international scientific societies. He has supervised many graduate students. His expertise includes the synthesis, design, and preparation of materials, nanomaterials, nanocomposites, and hybrid materials and their applications. His expertise also includes materials characterization using various techniques.

***

Salawu Omobayo Adio had his bachelor's degree from the University of Maiduguri, Nigeria. He was a volunteer and later became the Program Coordinator at the Centre for Water and Environment Development, Nigeria. He had his masters from the department of earth sciences, King Fahd University of Petroleum and Minerals in 2016. His research interests include but not limited to colorimetry, soil and water engineering, wastewater treatment using advanced oxidation, membrane technology and nanotechnology/bionanotechnology. He is a member of different professional organizations in Nigeria and abroad.

Samira Bagheri is a senior lecturer in the Nanotechnology & Catalysis Research Centre, University of Malaya. She received her PhD (2013) from Nanotechnology & Catalysis Research Centre, University of Malaya. Samira's main research interests are in the areas of carbon nanomaterials, such as carbon nanotubes, graphene oxide and graphene nanosheets, metal oxide nanocomposites advance smart nanohybrids, especially apply in electrochemical sensors, supercapacitors, fuel cells and environmental pollution management.

Gaddafi I. Danmaliki is currently a Research Assistant in Biochemistry Department at the University of Alberta, Canada. He has obtained his Masters degree in Environmental Science from King Fahd University of Petroleum and Minerals, Saudi Arabia (2013 - 2015). His research focuses on the application of nanomaterials derived from waste materials such as waste rubber tires to selectively isolate refractory sulfur compounds from fuels.

Anupreet Kaur worked as Assistant Professor in Basic and Applied Sciences Department, Punjabi University-Patiala. Her area of interest is synthesis of nanomaterials and their use for separation and preconcentration of various environmental pollutants.

Chispin Kowenje is a Senior Research Scientist/Senior Lecturer (Physical/ Materials Chemistry), Head of Chemistry Department Maseno University, Kenya.

Shafquat Majeed is currently working as an assistant professor in the Department of Nanotechnology, University of Kashmir (J & K, INDIA). He received his PhD from Indian Institute of Science (Bangalore, India). He worked on microwave-assisted synthesis of rare-earth based nanomaterials and their optical, magnetic and biomedical properties. Majeed's main research interests are in the areas of nanomaterial synthesis (Solvothermal, Low Temperature, Sonochemical and Microwave-Assisted), thin film deposition through solution-based routes, their characterization/properties and application in optical, magnetic and biomedical fields.

Arunima Nayak is currently working as a Marie Curie experienced researcher at *Innoviacio i Recerca Industrial I Sostenible*, Spain and is simultaneously attached to Graphic Era University, Dehradun as Assistant Professor. She received her PhD from Indian Institute of Technology, Roorkee in the year 2012 and her research area is development of advanced low cost, green materials and techniques for wastewater treatment. Her research work has resulted in 30 publications in high impact factor journals of international repute.

Prakash Parthasarathy is an independent researcher from India. He completed his Ph.D. degree in Chemical Engineering (2014) from National Institute of Technology, Tiruchirappalli (NIT-T), India. He worked as a postdoctoral research fellow at Yonsei University, South Korea (2015-2016). His area of interest is pyrolysis and gasification of biomass, energy recovery from waste tyres and upgradation of low grade coal.

Kamlesh K. Shrivas obtained his Master degree in Chemistry (2000) and Ph.D. degree in Chemistry (2004) from Pt. Ravishankar Shukla University, India. He worked as quality control officer, Gharda Chemical Ltd., India (2005-2006). He worked as a post-doctoral fellow at Tamkang University, Taiwan and National Sun-yat Sen University, Taiwan (2006-2008), at Food Drug and Administration, USA (2008-2009), at Hamamatsu University School of Medicine, JAPAN (2009-2011). Since 2009, he has been an Assistant Professor at the Guru Ghasidas University, India. His current interests include the analytical development of chemical sensors for the detections of metal ions, drugs and biomolecules. In addition, imaging and identification of biomolecules (peptides, proteins, lipids and drugs) in tissue sample using mass spectrometry. He has published 44 research papers in highly impact journals and 3 book chapters.

Index

T

W

Z

www.ingramcontent.com/pod-product-compliance
Lightning Source LLC
LaVergne TN
LVHW080606300926
849162LV00005B/1060